Pork Production Systems

Efficient Use of Swine and Feed Resources

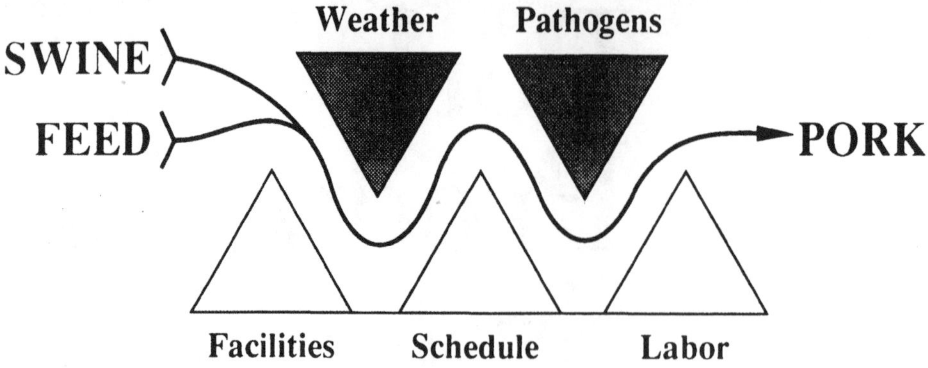

Explanation of Frontispiece

Pork production is accomplished by the synchronization and conversion of swine breeding stock and feed resources minimizing antagonistic effects of weather and pathogens. Producers and researchers have developed systems of facilities and labor coordinated with appropriate scheduling of management and husbandry practices to support this conversion and to counteract the negative forces. The objective is efficient and, thus, profitable utilization of swine, feed, facility, labor, and capital resources to obtain nutritious lean products for the consumer. This book strives to provide the current biological knowledge and principles to support the design and implementation of improved systems of the future.

Pork Production Systems

Efficient Use of Swine and Feed Resources

Wilson G. Pond
United States Department of Agriculture–Agricultural Research Service, Children's Nutrition Research Center, Houston, Texas

Jerome H. Maner
Winrock International, Morrilton, Arkansas

Dewey L. Harris
United States Department of Agriculture–Agricultural Research Service, Roman L. Hruska United States Meat Animal Research Center, Clay Center, Nebraska

An avi Book
Springer Science+Business Media, LLC

An AVI Book

Originally published by Van Nostrand Reinhold in 1991

ISBN 978-0-442-00115-5

16 15 14 13 12 11 10 9 8 7 6 5 4 3 2 1

Library of Congress Cataloging in Publication Data

Pond, Wilson G., 1930-
Pork production systems: efficient use of swine and feed resources / Wilson G. Pond, Jerome H. Maner, Dewey L. Harris.
p. cm.
"An AVI book"—T.p. verso.
Includes bibliographical references and index.
ISBN 978-0-442-00115-5 ISBN 978-1-4615-3932-2 (eBook)
DOI 10.1007/978-1-4615-3932-2
1. Swine. 2. Swine—Feeding and feeds. 3. Pork. 4. Swine--United States. I. Maner, J. H., 1935- . II. Harris, Dewey L., 1933
III. Title.
SF395.P56 1991
636.4'0855—dc20 90-47270
CIP

We dedicate this book to our respective mentors, Dr. J. C. Hillier, Professor Emeritus, Oklahoma State University, Dr. J. K. Loosi, Professor Emeritus, Cornell University, and Dr. L. N. Hazel, Professor Emeritus, Iowa State University, with deep appreciation for their influence and caring guidance. We humbly acknowledge that each of us "drinks from wells dug by others." We further dedicate it to our wives (Marsha Pond and Jeanne Harris) and mother (Gladys Maner) whose loving kindness, perseverance, and enduring reassurance and support have been inspirational through the years.

Contents

Appendices

Preface

Pork continues to occupy an important position as a food source in affluent societies as well as in developing countries with slower economic growth. The growth of the world swine population continues at a faster rate than that of the human population, a reflection of the sustained demand for pork in all parts of the world. The technical basis for commercial production of swine was presented in our two earlier textbooks—*Swine Production in Temperate and Tropical Environments,* by Pond and Maner, 1974, and *Swine Production and Nutrition,* by Pond and Maner, 1984.

In view of rapidly advancing technology and an appreciation for the systems approach in industry and agriculture, this third book has been restructured to provide the student and practitioner with an integrated concept of pork production. We have attempted to blend the fundamental principles from genetics, physiology, nutrition, and biotechnology into the modern concepts of systems analysis and simulation modeling. The objective is to create a teaching approach which emphasizes the integrated synthesis of biological with physical and environmental sciences and economics. This approach is expected to provide an overall pork production systems view that individual producers can adapt to their specific resources, needs, and goals. Our new co-author, Dr. Dewey Harris, has used his expertise and perspective on interacting systems to change the complexion of the book to fulfill this objective. In addition, Dr. Harris has updated and enhanced the genetics chapter, stressing the evolution of practices to the modern technology being utilized by progressive elements of the breeding service industry.

The section on Feedstuffs, which in the 1984 book included more than 350 pages, has been reduced in size and updated. In the interest of brevity and economy, the entire book has been condensed to about 350 pages and the new format is tailored to a systems approach. The information on feedstuff composition and feeding value for swine contained in the 1984 book dictates that it remain as a classic reference document and we anticipate that book's continued use by students, producers, and nutritionists to supplement the more condensed information on feed resources in our present book.

We intend for this book to contribute in a unique way to the continued growth of pork production in improving human nutrition and in stimulating improved use of existing, new, and alternative feed and breeding stock resources in complex and varied plant-animal ecosystems designed by man for increased efficiency of resource utilization. We believe pork production will continue to have an important role through innovative approaches to meeting food and cultural needs of an expanding human population as we approach and enter the 21st century.

We express our sincere appreciation to the U.S. Department of Agriculture for making available office and library facilities. We are deeply grateful to Sherry Hansen and Sherry Kluver for technical assistance with manuscript preparation, to Pat Sheridan for her valuable librarian skills, and to Dr. Jackson Dzakuma for statistical and simulation modeling assistance.

Pork Production Systems

Efficient Use of Swine and Feed Resources

1

The Pig: Past, Present, and Future

DOMESTICATION

The evolution and domestication of the pig has been described in several books (Towne and Wentworth 1950; Mellen 1952; Briggs 1969; Mason 1969), and the student interested in a detailed account of this fascinating story is referred to them. Recorded domestication appears in Biblical accounts as early as 2000 B.C. Despite some ancient cultural and religious taboos forbidding the consumption of pork, the domestication of the pig as a source of human food has persisted, and the continued increases in numbers of swine throughout the world up to the present time provide evidence of the contribution of the pig to human nutrition through the ages.

The pig has been traditionally a scavenger, and in early domestication it was raised as a means of utilizing human food wastes. In many parts of the world the pig still performs this function as a "backyard" inhabitant (Fig. 1.1). Even in the developed countries some swine produced for commercial pork are fed kitchen wastes. However, such enterprises are closely controlled by local and state government health regulations (such as those requiring that raw garbage be cooked) to prevent the spread of diseases to humans and other animals.

The proportion of commercially produced swine fed on garbage (kitchen wastes) in the developed countries is less than 1% of total production. Most commercial swine production in the developed countries is located in areas where cereal grains, maize (corn), and other high-energy feeds are grown in large quantities and are available at low cost. A classic example of the relatively recent development of commercial swine production on a large scale based on the availability of a source of high-energy feedstuffs is that of the "corn belt" of the United States. The corn belt, located in the north-central section of the United States, has soil and climate conditions that are ideal for the production of large amounts of high-yielding corn (*Zea mays*) hereafter referred to as maize. The cultivation of

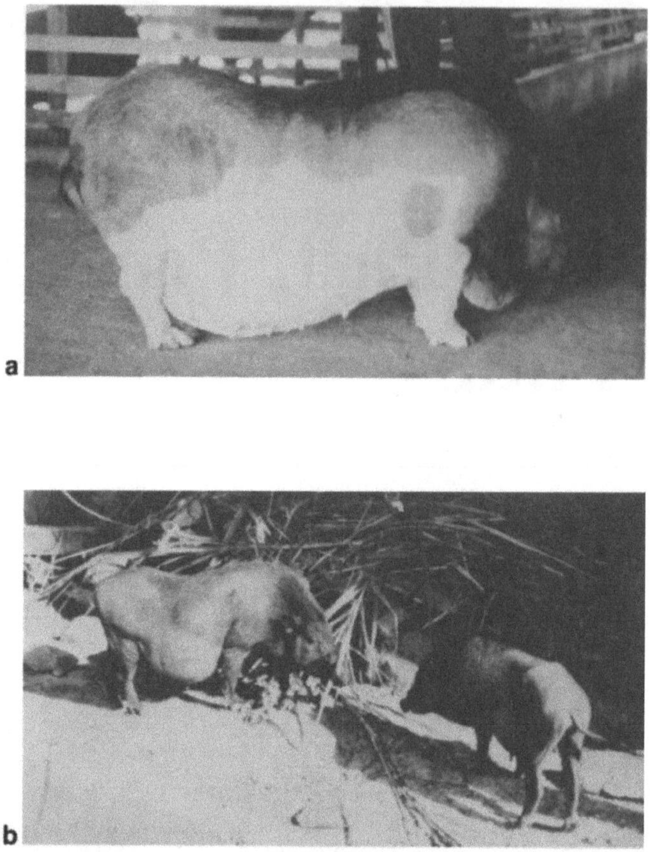

FIGURE 1.1 (a) Pigs of Thailand; note the swayback typical of pigs in Southeast Asia. (b) Pigs in North Sumatra.

hybrid maize and the use of fertilizer have resulted in yields of feed energy per acre far above anything realized before their introduction. The abundance of favorably priced feedstuffs provides an opportunity to convert this energy to pork for sale at a greater profit than could be obtained by selling the maize as a cash crop.

Pork production in the United States continues to be concentrated in the areas of maize and sorghum production, even though generally the population centers are remote from these areas. Pork production around the world tends to be concentrated in grain producing areas and in areas where sufficient grain can be imported. The total land area devoted to grain production and the yield per hectare and the total tonnage of grain production in the developed and developing countries of the world in various periods from 1947 to 1980 are shown in Fig. 1.2: a, b, and c (Barr 1981). Among the industrialized countries, the United States is currently producing about 53% of the grain and will be producing about 59% by 2000. Countries with central planning as well as less-developed countries, with a few exceptions are currently consuming more grain than they produce, and the deficit will be more severe by 2000. Most of the population growth (92%) will be in the less-developed countries (Global 2000 Report 1980). Of the projected 6.35 billion humans alive in

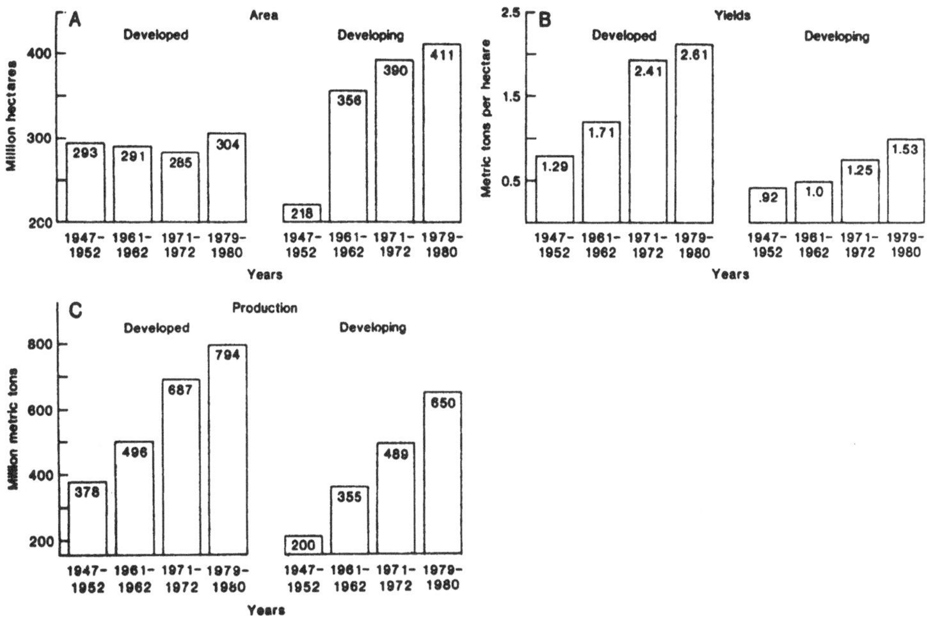

FIGURE 1.2 World grain in developed and developing countries: (a) area under cultivation; (b) yields; and (c) production. From Barr, 1981. Reprinted with permission of the author and the American Association for the Advancement of Science.

2000, 5 billion will live in these less-developed countries. As the human population of the world continues to increase, the proportion of plant products that can be fed to pigs for conversion to meat can be expected to decline, since the pig competes directly with humans for these feeds as a source of carbohydrate (energy) and protein. However, protein is relatively more limited than carbohydrates as a nutrient for humans and, in general, plant products, except some legume seeds, do not contain an adequate quantity or quality of protein to meet human requirements. The pig serves as an efficient synthesizer and storage depot for fat, proteins, vitamins, and minerals which, taken separately, would be very expensive. Furthermore, no one has yet produced, at any price, a synthetic ham or pork chop equal in acceptability to the real thing. Strong dietary preferences for traditional foods exist within individual regions of the world so that there is reluctance to accept esthetically inferior replacements. Since animal products when available at reasonable prices are readily consumed by people throughout the world, this potential for providing adequate nutrients cannot be overlooked.

Growing consumer concern in the early 1980s about the health aspects of meat, including pork, forced a change in emphasis in the pork industry from a production-oriented to a market-oriented mode. Surveys of consumer perceptions of pork suggest that a high proportion believe that pork is high in fat, sodium, cholesterol, and calories. These perceptions persist despite dramatic increases in carcass leanness and decreased lard production of U.S. pigs during the past 10 years. Per capita consumption of pork in the United States during the 10 years from 1978 to 1987 averaged 61.4 lb (USDA, 1988). A pork advertising campaign launched in 1987 by the National Pork Producers Council highlighted by the slogan "the other white meat," may have had a positive impact on consumer attitudes toward pork which may be reflected ultimately in increased pork consumption (Harness, 1988). There has been a dramatic increase in per capita meat consumption in all parts of the world since 1960, especially in those countries whose economic growth has been most rapid. Figure 1.3 (Barr 1981) shows the per capita meat consumption (in kilograms) in 34 countries plotted against the percentage change for each country from 1961 to 1980. World meat consumption grew at an annual rate of nearly 3%, and poultry consumption grew more than 5% during that period (Barr 1981). Countries reporting the most rapid increase in pork consumption during the past decade were Greece, Bulgaria, and Japan, each of which doubled its intake, and Spain, Italy, Taiwan, and Venezuela, each of which recorded gains of 50% or more. This trend toward indirect consumption of grains (marketing grains through livestock) has resulted in increased demand for grains and oilseed meals and other by-product feedstuffs. The degree to which animal protein production can continue to increase in response to human demand will depend on economic as well as esthetic factors.

The amounts of pork needed to supply human demands and the number of animals required to produce it in 1965, 1986, and 2000 (as projected) are shown in Table 1.1. The rapid increase in human population ensures a substantial increase in demand for pork even with static per capita consumption. If per capita consumption of 30 kg, as currently recorded in the United States, were extended to the entire world, the number of pigs required to meet the demand would increase

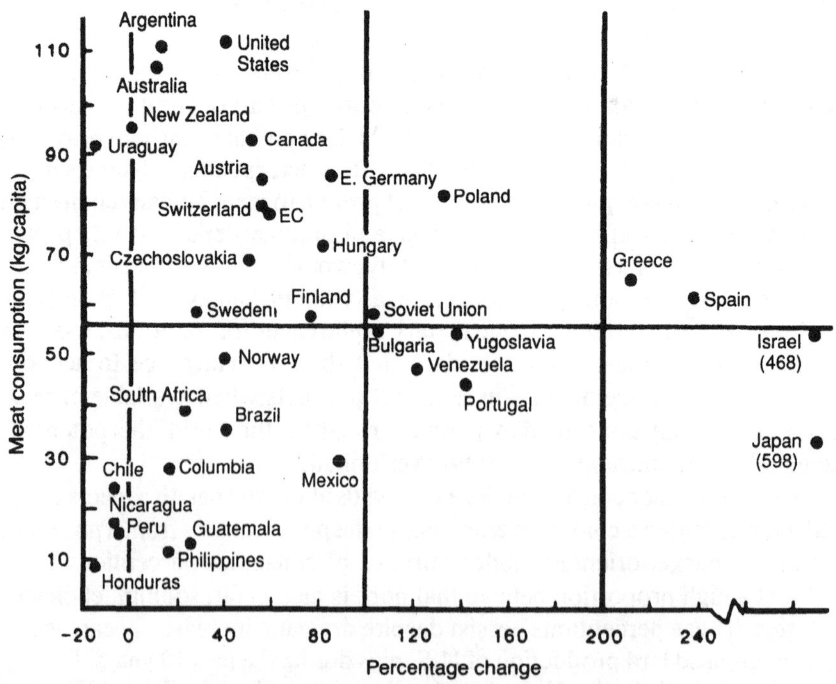

FIGURE 1.3 Per capita meat consumption and percentage change, 1961–1980 (EC = European Economic Community). From Barr 1981. Reprinted with permission of the author and the American Association for the Advancement of Science.

TABLE 1.1 Amounts of Pork Needed to Supply Human Demands in 1965, 1986, and 2000

	1965	*1986*	*2000*[a]	
Human population, in billions	3,281	4,916	6,351	
World per capita consumption of pork, kg	9	9	9	30
Amount of pork needed, in billions of kg	29.5[c]	44.2	57.2	190.5
Amount of pork protein needed, in billions of kg[b]	4.7	7.1	9.1	30.5
Number of animals needed to supply it, in millions[c]	468	822	908	3024

[a] Two amounts are predicted for the year 2000: the first is based on the assumption that the world per capita consumption stays at 9 kg/year; the second is that the total world consumption increases to the present U.S. per capita consumption of 30 kg/year.

[b] Based on the assumption that edible carcass contains 16% protein.

[c] Based on the assumption that the extraction rate is 100 and animals are slaughtered at 90 kg liveweight and have a dressing percentage of 70.

Sources: FAO Production Yearbook 1967, 1986.

dramatically to about 3 billion (about 4.5 times the present number). Such an increase seems highly unlikely, but even if per capita consumption of pork remains at the current level of 9 kg, the total world demand may nearly double by the year 2000. (Fig. 1.4).

The factors presented above do not include many unrealized potential applications. For example, in swine production large quantities of waste bananas, cassava, and other root crops could be used as sources of energy, and legume seeds could be used as sources of protein. This would provide large quantities of protein and other nutrients for human consumption at relatively low cost. In addition to these examples of tropical applications, there are many potential uses of temperate plant materials. Because the pig bred for production in temperate environments adapts well to the tropics, it is not necessary to incorporate genes of indigenous animal populations for maximum productivity, as it is with other food animals.

ALTERNATIVE PROTEIN SOURCES FOR HUMANS

The effort to feed a rapidly expanding human population during the coming decades will be influenced greatly by the ability of science and technology to improve conventional methods and to develop new methods of food production. The chief conventional method of food production until now has been based on agriculture. Except in the developed countries, much of agriculture has been—and still is—at subsistence level with inefficient methods and primitive technology. The sophisticated agricultural technology of the more-developed countries is being applied with ever increasing pressure in the less-developed countries, although it is increasingly clear that such technology transfer is slow and under constraints of local cultures and resources.

Nonetheless, unconventional potential applications of agricultural technology

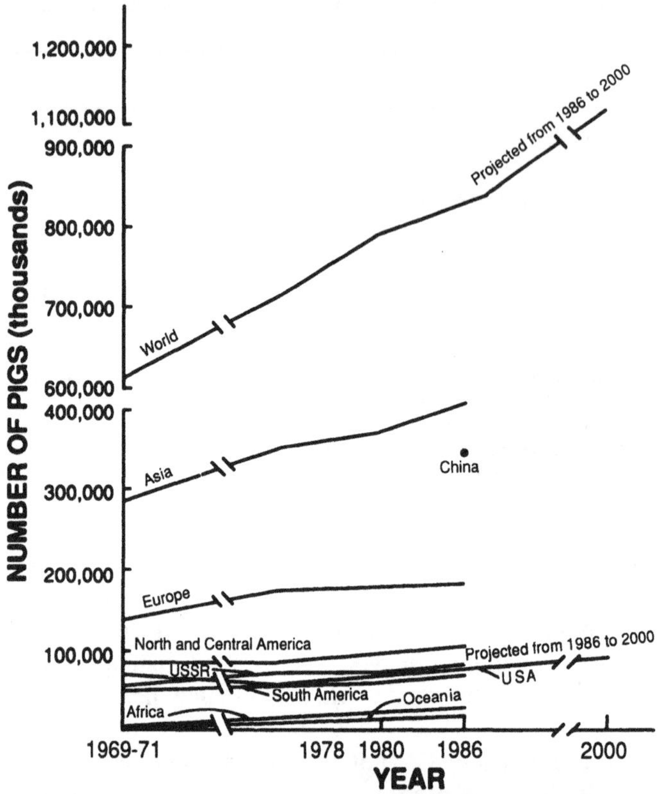

FIGURE 1.4 World and regional pig population in 1969–1971, 1978, 1980, 1986, and projected to 2000, based on assumed constant annual per capita pork consumption of 9 kg worldwide. According to Food and Agriculture Organization (FAO) statistics, there were 783,485,000 pigs in the world in 1982, and 822,443,000 in 1986. In 1986, Asia had 49.1%, Europe (excluding USSR) 21.8%, USSR 9.5%, North and Central America 11.1%, South America 6.5%, Africa 1.3%, and Oceania 0.6% of the total. Total pork production (FAO statistics) in 1982 was 55,263,000 tons, of which Asia produced 37.7%, Europe 33.5%, USSR 9.2%, North and Central America 14.3%, South America 3.2%, Africa 0.8%, and Oceania 0.5%. Corresponding figures for 1986 were 61,209,000 tons total, of which Asia produced 37.8% (81.3% of which was in China), Europe 33.5%, USSR 9.6%, North and Central America 14.4%, Oceania 0.6%, South America 3.2%, and Africa 7%.

to meeting human food needs include many exciting possibilities. Noncompetitive sources of feed and ways in which their usefulness can be extended have been reviewed (Altschul and Wilcke 1981). Such approaches include amino acid fortification of feeds and utilization of animal wastes and of microorganisms in animal feeds. Other examples are the domestication of selected species of wildlife or the control of their ecology in their natural environment; the utilization of exotic plant materials; the employment of different methods to process existing plant materials; and the genetic manipulation of existing plant species by conventional methods of breeding as well as by exciting new recombinant DNA (gene-splicing) techniques.

Man must depend mostly on agricultural technology as a major source of food (either by direct consumption or by feeding to animals) until nonagricultural techniques of food production become advanced to the point of making a major contribution. Innovative new food sources include single-cell protein (yeasts, bacteria, and algae), cultivated fish (in fresh and salt water), and petroleum by-products (which in combination with inorganic nitrogen can be used to produce single-cell protein). Petroleum by-products and other fossil energy sources obviously have restricted usefulness because they are available in limited quantities. Food sources (such as blue-green algae) generated by photosynthesis (solar energy), just as those produced by conventional agriculture except for the improved efficiency, offer greater potential application. Other energy sources (e.g., nuclear) undoubtedly also will make significant contributions to human food production in the future. Most nonagricultural food sources, while technologically sound, will be too expensive to compete for consumer purchasing power for many years, especially in the less-developed countries where foodstuffs of high nutritional value are in short supply. Crude preparations of many of these nonagricultural food sources first may find application as feedstuffs for livestock.

Thus, one can conclude that the war against hunger still has as a major participant conventional agricultural technology, including pork production.

PIG POPULATION DISTRIBUTION IN THE UNITED STATES

The total pig slaughter in the United States in 1987 was 81.4 million, of which 21.2 million (26.2% of total) was produced in the state of Iowa and followed by Illinois, 8.2 million; Minnesota, 7.2 million; Indiana, 6.6 million; Nebraska, 6.4 million; and Missouri, 4.9 million (USDA 1988). These six states, which together produce nearly two-thirds of the total pork in the U.S., are all located in the Midwestern section of the United States, often referred to as the corn belt. Other states ranking in the top 10 in pig production in 1987 were North Carolina, 3.8 million; Ohio, 2.9 million; South Dakota, 2.6 million; and Kansas, 2.3 million. Seventy percent of the pigs marketed in the U.S. in 1987 were produced on farms marketing 1,000 or more pigs annually (Rhodes et al. 1987), compared with only 13% in 1969. This trend for larger swine enterprises is associated with a steady reduction in number of farms reporting swine from 1,846,980 in 1959 to 532,204 in 1969 and 332,760 in 1987 (Christian et al. 1988). In 1987, 20.2% of U.S. swine farms had fewer than 25 sows, while 20.4% had 25 to 50, 29.0% had 51 to 100, and 30.4% had more than 100 sows (Holden et al. 1988). The same trend is under way in other countries with a large swine industry.

HUMAN POPULATION DISTRIBUTION IN THE UNITED STATES

All 50 states produce some pork, but the states with largest human populations, i.e., New York, California, Florida, generally are not large contributors to the total pork supply; an exception is Illinois. This disparity between human and pig population density results in shipment of pork out of production centers to the consumption centers. Demographics predict a continued tilt toward the South and West in the U.S. population in the next decade (Putnam and Van Dress 1984). For example, in 1970, 22.3% of the population lived in the Northeast; a reduction to 17.3% is predicted for the year 2000. Similarly, the number of people living in the North Central states is predicted to decline from 26.5% in 1970 to 22.3% in 2000. These shifts will be associated with a population increase from 32.5% to 37.0% in the South and from 18.7% to 23.4% in the West from 1980 to 2000. These percentage changes take on added significance in view of the projected population increase from 228 million in 1980 to 267.5 million in 2000. These demographic projections suggest an accelerated movement of pork and other livestock products from the point of production to consumers located elsewhere.

LIFE CYCLE AND GENERAL CHARACTERISTICS OF THE PIG

The domestic pig[1] has evolved by natural selection and—more recently—by the application by animal breeders of the principles of population genetics. Compared with other farm mammals, the pig has a short gestation period and is highly prolific. Pregnancy is about 114 days (range 110–118 days); lactation is 6 to 10 weeks, but in commercial production, piglets are typically weaned at 3 to 5 weeks. Males and females reach puberty at 5 to 6 months in European and U.S. breeds (3 to 5 months in some Chinese breeds), but mating is usually delayed 1 or 2 months to permit attainment of greater body size. Average litter size is 7 to 8 for primiparous females, 10 to 12 for older females of the European and U.S. breeds, and 15 to 18 for females of some Chinese breeds. Estrus recurs at about 3-week intervals except during pregnancy and lactation and resumes a few days following the weaning of the litter. Productive sows are often kept in the herd to produce two litters per year for four or five years before sale for meat. Life span is much longer than this in animals allowed to live out their normal life cycle, probably approaching 15 years, although the available data on this question are limited for economic reasons. Details of each phase of the life cycle of swine are addressed in appropriate chapters throughout the book, and specific features of the biology of the pig in relation to its use in biomedicine are described elsewhere (Pond and Houpt 1978; Tumbleson 1986). The time sequence of the life cycle of pigs under most production systems is listed in Table 1.2. There are currently probably more than 100 breeds of domestic pigs in the world, most of them in China, North America, and Europe. In addition, there are another 225 or more groups of pigs not recognized as breeds but each having unique characteristics, appearance, or geographic location (Mason 1969). Representative breeds and types of pigs are illustrated in Fig. 1.5.

CHROMOSOMES

The domestic pig has 38 somatic chromosomes (19 pairs) (Jimenez et al. 1962). As in other mammals, the male determines the sex of the offspring by producing both X and Y chromosomes. The karyotype of the domestic pig is shown in Fig. 1.5.

[1] Phylum, Chordata; class, Mammalia; order, Artiodactyla; family, Suidae; genus, *Suis*; species, domesticus (suis).

TABLE 1.2 The Life Cycle of the Pig

Period	*Duration*
Prenatal period, days	114 ± 4
Suckling period, wks.	3—8[a]
Growing–finishing period (to 105-kg body wt), days	100—160
Age at 105 kg, days	135—205[b]
Age at puberty	150—200[c]
Optimum age at first service, months	7—8[d]
Reproductive longevity, years	4—8
Longevity, years	12—15[e]

[a] Pigs can be weaned to cow milk or synthetic diets at birth but it is more common in commercial production to wean at three weeks or older.

[b] Boars and male castrates grow slightly faster than females, so can be expected to reach 105 kg body wt. several days earlier.

[c] Chinese breeds may reach puberty at less than 100 days.

[d] Since puberty is reached before growth stops, better reproductive performance is obtained (large litters in females and more sperm produced in males) if reproductive service is not initiated before 7 months of age.

[e] Most female and male breeding stock are slaughtered when their reproductive efficiency begins to decline, so that the longevity figure is based on a very small sample of the total swine population.

A high proportion (up to 10%) of blastocysts from normal pigs have grossly detectable chromosome defects. The proportion of these defects associated with the high early embryonic mortality in pigs is not known. However, since they have not been reported in living adult pigs, many are probably lethal.

BODY SIZE

Much genetic variation exists among and within breeds in mature size and characteristics. The average mature weight of conventional pigs raised for meat is approximately 200–300 kg, whereas the breeds of miniature pigs developed for use in biomedical research have a mature weight of less than 100 kg. This wide variation makes it possible to breed selectively for small or large body size and is the basis for the development of miniature pigs.

BEHAVIOR TRAITS

The behavior of swine has been described in detail (Pond and Houpt 1978; Houpt and Wolski, 1982; Hafez et al. 1962). The general behavior patterns described here are pertinent to commercial swine production. Pigs are notoriously curious. This trait is probably reflected in the tendency of the suckling pig to sniff and nibble at clean earth or feed materials to which it has access. Older pigs given access to pasture will spend hours rooting through the soil and will inevitably locate an open gate or a flaw in the fence. Pigs readily learn to open the lid of a self-feeder or push a valve for water. Even newborn pigs, whose suckling instinct is highly developed, can be taught readily to drink from a bowl when weaned from their dam. Their urge to suckle makes it highly desirable to cage pigs individually when weaned at a few

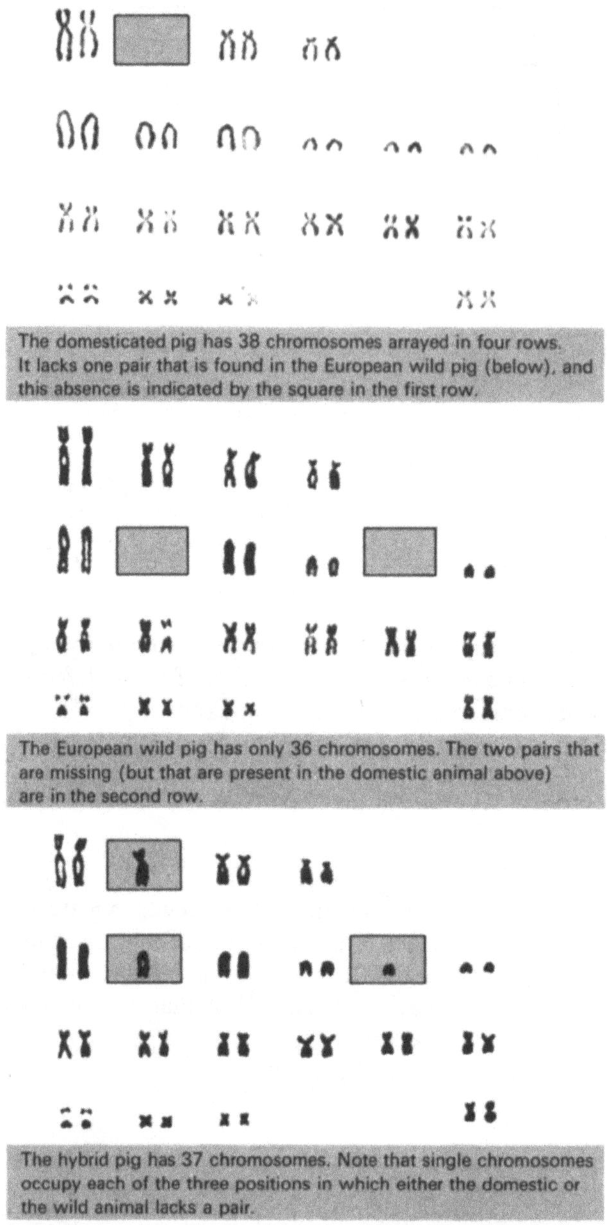

FIGURE 1.5 Chromosomes of the domestic pig, European wild pig, and hybrid pig. Courtesy of AEC Agricultural Research Laboratory, University of Tennessee, Knoxville.

days of age, since they constantly spend effort sucking the ears and navels of other pigs if kept together. Individual pigs within litters suckling the sow establish teat position within the first week of age so that possession is jealously guarded and attempted takeover of a teat by a littermate is met with resistance and a struggle for maintenance of position.

The pig is a "nibbler." That is, if feed is offered *ad libitum* from a self-feeder the total daily intake will be spread over many small meals throughout a 24-hr period. If given only one or two meals per day to satisfy appetite, daily food intake is reduced in young pigs, but in pigs approaching market weight it is maintained at a level equal to that achieved with continuous *ad libitum* feeding.

Pigs tend to be good "housekeepers" when given the chance. If the size of the nesting area is restricted they usually use a corner of the pen apart from the nesting area to defecate and urinate. Since the pig cannot efficiently dissipate body heat by sweating, high ambient temperatures encourage rooting in cool, moist earth or wallowing in mud, if available, in an effort to keep cool.

During gestation sows tend to become more docile, especially as they approach term. About one day before farrowing, the sow prepares a nest by rooting and pawing straw or other bedding into a small area of the pen. Colostrum usually can be milked from the teats at the time of nest-building. Nest-building is associated with distress and irritability and is followed, shortly before farrowing, by a period of up to several hours during which the sow lies quietly before labor begins. Labor includes restlessness, repeated rising and lying down, and straining. When farrowing begins, the sow usually remains lying down until the last pig is born, sometimes a period of several hours. Milk letdown is continuous during farrowing and the first-born pigs find their way to a teat before their littermates are born. The sow is usually indifferent to their presence. Occasionally sows are hostile to their pigs during farrowing and may bite them or throw them violently across the pen, causing serious injury or death. When this happens, pigs should be protected from the sow until farrowing is completed and then allowed free access to the sow only with caution.

A type of social dominance order, such as in the establishment of teat position, is evident very early in the life of the pig. This same social dominance order prevails when strange pigs are penned together after weaning. Invariably, there is fighting, manifested mainly by biting and head confrontation, which ultimately ends in a social hierarchy. The same pattern exists among sows kept together and fed with limited feeder space. Strange boars fight fiercely, often resulting in serious injury or death, and should not be penned together unless introduced as immature animals.

A common problem among growing pigs kept in confinement is tail biting. The cause of this peculiar type of cannibalism is not known, but hanging a rubber tire or similar object in the pen sometimes seems to alleviate the problem partially. Usually one or two pigs in a group are responsible for starting the tail biting and the habit seems to spread to others. The infections and occasional deaths that result have led most commercial producers to the practice of removing the tail from each pig at birth to eliminate the problem.

GROSS ANATOMY

The anatomy of the pig is thoroughly covered by Sisson and Grossman (1953), so that the description of anatomical details need not be included here. Gross anatomy of the pig fetus is described in several dissection guides for students (Walker 1964; Holstad 1956; Stuart and Pinder 1948; Gilbert 1966; Harrison 1971). Some details of structure and functions of some of the liver and digestive tract of the pig have been provided by Moran (1982).

There is considerable variation among pigs in the number of ribs and thoracic and lumbar vertebrae. Results of extensive X-ray studies of live animals and examination of carcasses by several workers (Shaw 1930; Berge 1948; Freeman 1939) show that lumbar and thoracic vertebrae are variable in number. With an increase in the number of ribs the number of lumbar vertebrae decreases. The heritability of pre-sacral vertebrae number is high and has been estimated at 0.74 (Shaw 1930). (For a definition and discussion of heritability, see Chapter 6.) There are breed but not sex differences in average vertebrae number, additional ribs being generally associated with breeds of larger size. Selection for carcass length (distance from first cervical vertebrae to aitch bone or from first rib to aitch bone) has probably resulted in an increase in the average number of vertebrae in selected populations.

A problem of practical importance in some strains of pigs is the tendency toward unequal development in the size of the members of digit pairs (toes) on the front and/or hind hoofs. Usually the inside toe is shorter than the outside toe so that the two do not make equal contact with the ground, and body weight is unequally distributed, sometimes resulting in lameness and soreness. When the pasterns (second and third phalanges) are long the defect is more noticeable than when the pasterns are short.

PLASTICITY OF THE PIG

Swine breeders have long recognized the fact that to meet changing consumer demands, changes in the general shape and type of pig can be made by selection based on visual appraisal. This is exemplified by the cyclic changes that have occurred in the type of pig produced in the United States during the past half-century. When lard was high priced, selection pressure was exerted for short, chuffy pigs yielding a high percentage of fat. From this extreme the pendulum swung to the long, narrow, slowly maturing hog. In World War II, the demand for lard was again increased and swine breeders responded by selecting fat pigs to meet this demand. With the present demand for lard greatly reduced in the United States by competition from vegetable oils and synthetic detergents, the goal is for pork with a minimum amount of fat. In some countries lard is highly valued and pigs are selected accordingly. Rapid changes can be made in the kind of pig produced as a result of the high degree of heritability of body type (Hetzer et al. 1944), the relatively short generation interval of pigs, and their high prolificacy.

For example, the selection of swine for small size can be accomplished in a relatively short period of time as demonstrated by Dettmers et al. (1965) at the University of Minnesota in cooperation with Hormel Institute, Austin, Minnesota. The purpose of this selection program was to develop a small pig for use in biomedical research. Over the course of 11 generations, the average 140-day weight was reduced from 39.2 to 17.5 kg. Birth weight, but not number per litter, was reduced as well, although the only criterion of selection was body weight at 140 days of age. Carcass traits can also be altered by selection as clearly shown by Clausen (1970) based on the results of the Danish progeny-testing program. Similar responses to selection for back fat have been reported in the United States by Hetzer and Harvey (1967) and Berruecos et al. (1970). Extreme differences in back-fat thickness between individuals from populations selected for low or high fatness are evident not only from carcass measurements (Fig. 1.6) but from general appearance (Fig. 1.7).

HAIR AND SKIN COLOR

The hair color and skin pigmentation of pigs vary greatly. Skin pigmentation generally is associated with hair color; that is, white-haired pigs usually have very little skin pigment. However, it is not uncommon for white pigs, even within recognized breeds, to have one or more small pigmented spots on the skin even

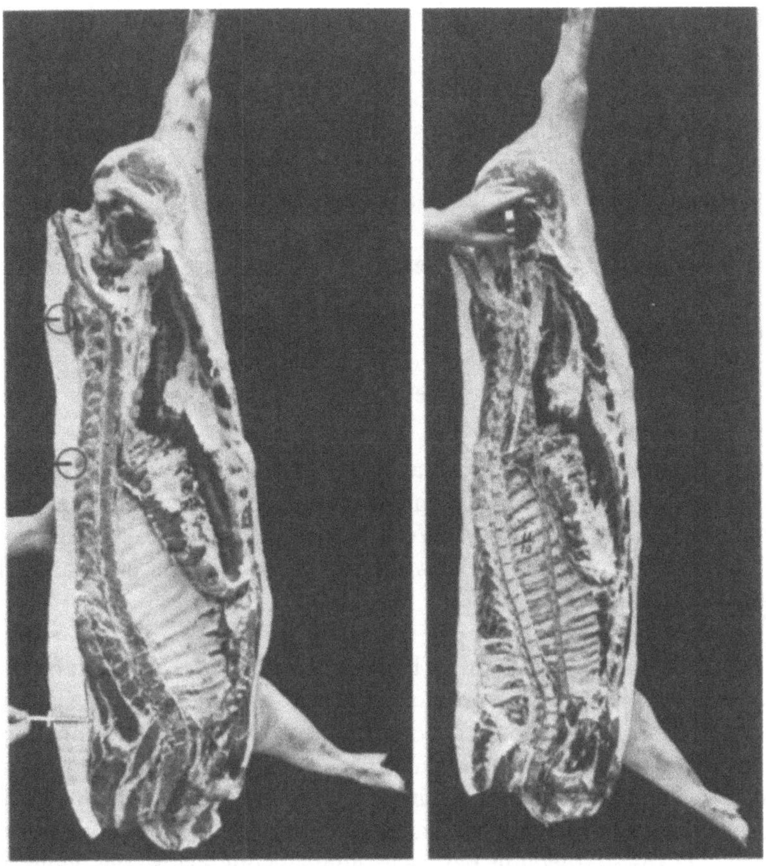

FIGURE 1.6 Carcasses of fat and lean pigs, showing points of reference for carcass measurements.

FIGURE 1.7 Representative Duroc × Yorkshire crossbred gilts from populations of Durocs and Yorkshires selected for about 20 generations for low or high backfat thickness. Note the lean, long appearance of the gilt on the left (low backfat) compared with the fat, short appearance of the gilt on the right (high backfat). Photo courtesy of U.S. Department of Agriculture.

though the hair covering these spots is white. The inheritance of coat color in pigs was described in the first of a series of papers by Hetzer (1945) dealing with results of various breed crosses. The following categories of coat color were given:

1. Tortoise-shell patterns: (a) complete extension of black (self black) as in unbelted Hampshires, (b) partial extension of black (black spotted) as in Poland China and Berkshire, and (c) complete restriction black (self red) as in Durocs and Tamworths. Differences between these patterns were described as being due primarily to a series of triple alleles. Self black and black spotted are apparently both dominant to red.
2. White patterns: (a) self white of the U.S., British, German, and Scandinavian breeds; (b) slaty or dirty white on dark skin found in the Mangalitza breed; (c) white belt of the Hampshire, Wessex Saddleback, Essex, and Hanovarian; and (d) white responsible for the white feet, nose, and tail of Poland China or Berkshire breeds.

Self white is dominant over most black and red breeds, although blue-gray or blue-roan individuals have been produced in some crosses. Also, the slaty white of the Mangalitza has been reported dominant over the self white of Norwegian Landrace. The width of the white shoulder belt of Wessex Saddleback (Donald 1951) and presumably of the Hampshire is apparently associated with three genotypes: homozygous belted pigs have a medium-to-wide belt; heterozygous belted pigs have a narrow-to-medium belt; and homozygous black pigs have no belt. Persistence of occurrence of solid black pigs apparently is due to the breeder's preference for a narrow or medium belt in selection.

The wide range of color patterns (white, black, red, gray, brown, and all combinations) among crossbred pigs on commercial farms in the United States is ample evidence for the complex nature of coat color inheritance in pigs.

The wild European hog has characteristic longitudinal stripes which are more conspicuous at birth than later. Occasionally the texture of the hair of newborn pigs of some domestic pigs also suggests the presence of this characteristic hair pattern.

A peculiar type of curly (woolly) hair occurs in Conastras (Brazilian) pigs, not associated with a particular color or sex, but apparently under the influence of a single completely dominant gene. The hair shaft of "woolly" hair is oval or flat in cross-section as compared to the round shaft of normal, straight hair.

Pigs in which hair very sparsely covers the body are represented by the Extremadura breed of Spain and Tatu breed of Brazil. The trait is apparently contolled by an incompletely dominant gene for hair, so that the heterozygous individual is midway between the two homozygous extremes in the amoung of hair covering. Mating two hypotrichotic pigs results in 100% hypotrichotic offspring.

FUTURE PROSPECTS FOR SWINE AND PRODUCTS

Recent estimates (Global 2000 Report 1980) put the world human population by A.D. 2000 at approximately 6.35 billion (approximately 4.9 billion in 1986) (FAO, 1986). The amounts of calories and protein needed to feed people properly in the year 2000 based on published nutritive requirements appear in Table 1.3, adapted from a report of the President's Science Advisory Committee (The World Food

TABLE 1.3 Amounts of Energy and Protein Needed to Feed the Human Population of the World Adequately in 1970, 1985, and 2000

Requirements	*Year*		
	1970	*1985*	*2000*[a]
Protein needs per person[b] per day, grams			
FAO Reference Protein	35.4	35.1	35.1
Protein with NPU of 62[c]	57.1	56.6	56.6
Total world protein needs per day, metric tons			
FAO Reference Protein	127,922	168,139	211,325
Protein with NPU of 62	205,600	272,000	341,863
Energy needs per day, kilocalories			
Per person	2,354	2,393	2,393
Total world needs	7,787,000,000	11,118,000,000	13,974,000,000 [d]

[a] Values for 2000 are computed on the basis of projections for human population of The World Food Problem 1967. The low estimate of 5.965 billion was chosen for the computations.

[b] Assumptions: Adult male weighs 65 kg; adult female weighs 55 kg; children weigh according to FAO standards.

[c] NPU is net protein utilization based on egg protein = 100. The different values listed for 1970 and 1985 represent changes in median age of the population.

[d] The appropriate figure would be 15,195,550,000 if the population were 6.35 billion as projected by Global 2000 Report (1980).

Source: Values for 1970 and 1985 taken from The World Food Problem 1967.

Problem 1967). Figure 1.4 shows the changes in pig numbers by continents and in time since 1969–1971 (FAO Production Yearbook 1980, 1986) with projections to 2000 for the world as a whole and for the United States. The required pork production could be achieved by increased land area devoted to these feed crops or by increased yield per unit land area or by both. Undoubtedly, both factors will be involved in increased future production, along with changes in proportions of what is produced and the introduction of other sources of feed.

The efficiency of utilization of feed for pork production depends on the provision of a balanced diet and on husbandry conditions and genetic selection conducive to high efficiency. The pig is by far the most efficient among the farm animals in conversion of feed energy to body energy and ranks only behind poultry and dairy cattle in efficiency of conversion of feed energy to protein. These comparisons are shown in Table 1.4 (Holmes 1970).

A useful measure of efficiency of swine production for a country or region is the *extraction rate* (the number of pigs slaughtered per year/the number of pigs on hand at a particular time). The developed countries of northern Europe and North America (Belgium, Denmark, United Kingdom, United States), for example, have extraction rates greater than 150 whereas some countries (i.e., Mexico, Brazil, Venezuela, Viet Nam) have extraction rates under 50. The low values reported in many developing countries reflect a need to improve efficiency of production, for as efficiency improves unit cost of production declines. The low efficiency is probably due to a variety of factors, but it seems clear that significant improvements can be made in most cases by appropriate adjustments in feeding, breeding, and management.

The pig population of 822,433,000 in 1986 was distributed 40.8% in developed countries and 59.2% in developing countries; pig meat production of 61,209,000 metric tons in 1986 was distributed 58.2% in developed countries and 41.8% in developing countries (FAO Yearbook, 1986). This discrepancy between pig numbers and pig meat produced reflects the higher efficiency of pork production in developed countries.

The top 20 countries in the world in 1986 for pig numbers, pig slaughter, and pork production are listed in Table 1.5. Discrepancies in ranking for pig numbers (census figures) and pigs slaughtered reflect differences in efficiency of production

TABLE 1.4 Efficiency of Energy and Protein Utilization and the Energy Cost of Protein for Beef, Lamb, Poultry, Eggs, and Milk

Product	*Edible protein (g/100 g consumed)*	*Edible energy (kcal/100 kcal ME consumed)*[a]	*Edible protein (g/100 kcal ME consumed)*[a]
Beef	6.0	7.0	2.6
Lamb	3.0	3.0	1.3
Pork	12.0	23.0	6.0
Poultry	20.0	13.0	11.0
Eggs	16.0	15.0	11.0
Milk	23.0	21.0	10.0

[a] ME = metabolizable energy.

Source: Robinson and Vohra (1976), adapted from Holmes (1970).

TABLE 1.5 Top 20 Countries in Pig Numbers, Pig Slaughterings, and Pork Production, 1986

Pig numbers (thousands)		*Pig slaughterings (thousands)*		*Pork production (thousand tons)*	
China	338,074	China	267,726	China	18,810
USSR	77,772	USA	80,000	USA	6,379
USA	52,313	USSR	72,960	USSR	5,910
Brazil	33,000	West Germany	39,421	West Germany	3,334
West Germany	24,282	Japan	20,995	France	1,835
Poland	18,949	France	20,860	Poland	1,730
Mexico	18,631	Poland	18,500	Japan	1,550
Romania	15,766	Mexico	18,199	Netherlands	1,430
East Germany	12,946	Netherlands	17,800	Mexico	1,327
Netherlands	12,908	Denmark	16,099	East Germany	1,250
Japan	11,061	Spain	15,985	Spain	1,167
France	10,956	Brazil	15,700	Italy	1,160
Canada	10,721	United Kingdom	15,642	Brazil	1,100
Spain	10,367	East Germany	14,615	United Kingdom	993
Denmark	9,321	Yugoslavia	14,200	Romania	981
Italy	9,169	Canada	14,050	Hungary	960
India	8,700	Romania	12,577	Canada	875
Hungary	8,280	Italy	10,880	Czechoslovakia	833
United Kingdom	7,930	Philippines	8,800	Denmark	833
Yugoslavia	7,821	Belgium/Luxembourg	8,489	Yugoslavia	791

Source: FAO Production Yearbook (1986).

(number of pigs slaughtered per breeding female and age at which market weight is reached). The United States ranked second, behind China, in total pork production and pigs slaughtered in 1986, but was third, behind the USSR, in pig numbers. The developed countries dominate the world in pork production; but, as pig numbers increase in the developing countries, changes can be expected in rankings for pork production. Total pork production of 53.248 million tons from the top 20 countries in 1986 (FAO Yearbook, 1986) may be increased considerably in the coming decades, primarily by increased pork production in countries with rapidly expanding swine industries.

The pig will continue to contribute a significant proportion of energy and protein in the diets of developed and developing countries well into the twenty-first century (Pond 1983; Fredeen and Harmon 1983). In addition, numerous nonfood products are significant contributors to society, including insulin, gastrointestinal peptides, heparin, and other pharmaceutical and diagnostic products; skin for leather goods, hair for brushes, heart valves for replacement of damaged valves in human hearts, and many others (Banis 1987). In addition to tangible products of the pig as described in this chapter, one cannot ignore the colorful history and unique characteristics of the pig, playfully yet eloquently described by Bustad and Horstman (1986) who reached the following evaluation:

> As a result of many decades of experience, observations and extensive reading, we have reached the conclusion that the pig is gregarious, perceptive, quick to learn,

sensitive, very responsive to gentle care and direction, playful and courageous, and neat and clean if given a chance. They are also one of the best experimental models for people—anatomically, physiologically, psychologically and socially. This remarkable animal deserves our respect, our enlightened stewardship and informed gentleness, as well as our gratitude for its continuing remarkable contributions to our health and well-being.

REFERENCES

Altschul, A. M., and Wilcke, H. L. 1981. New Protein Foods, vol. 4: Animal Protein Supplies, part B. Academic Press, New York.

Banis, R. J. 1987. Pharmaceutical and Diagnostic By-Products. *In* The Science of Meat and Meat Products (ed. J. F. Price and B. S. Schweigert), 3rd ed., pp. 565–578. Food and Nutrition Press, Westport, Conn.

Barr, T. N. 1981. Science *214,* 1087.

Berge, S. 1948. J. Anim. Sci. *7,* 233.

Berruecos, J. M., Dillard, E. U., and Robison, O. W. 1970. J. Anim. Sci. *30,* 844.

Briggs, H. M. 1969. Modern Breeds of Livestock, 3rd ed. Macmillan, New York; Collier-Macmillan, London.

Bustad, L. K., and Horstman, G. 1986. Pigs: From B. C. to 2000 A.D.: From Outhouse to Penthouse. *In* Swine in Biomedical Research (ed. M. E. Tumbleson), pp. 3–15. Plenum Press, New York.

Christian, L., Holden, P., Futrell, G., and Kliebenstein, J. 1988. Iowa Pork Industry: Competitive Situation and Prospects. Iowa State University, STFE, Ames, pp. 35–42.

Clausen, H. 1970. Sammenlignende forsog med. svin. *379,* 74–75. Nat. Agric. Research Laboratory, Copenhagen, Denmark.

Dettmers, A. E., Remple, W. E., and Comstock, R. E. 1965. J. Anim. Sci. *24,* 216.

Donald, H. P. 1951. J. Agric. Sci. *41,* 214.

FAO Production Yearbook. 1961. Food and Agric. Organization, Rome, Italy.

FAO Production Yearbook. 1967. Food and Agric. Organization, Rome, Italy.

FAO Production Yearbook. 1982. Food and Agric. Organization, Rome, Italy.

FAO Production Yearbook. 1986. Food and Agric. Organization, Rome, Italy.

Fredeen, H., and Harmon, B. G. 1983. J. Anim. Sci. *57* (Suppl. 2), 100.

Freeman, V. A. 1939. J. Hered. *30,* 61.

Gilbert, S. G. 1966. Pictorial Anatomy of the Fetal Pig, 2nd ed. University of Washington Press, Seattle.

Global 2000 Report to the President. 1980. Entering the Twenty-First Century, vol. I. Government Printing Office, Washington, D.C.

Hafez, E. S. E., Sumptron, L. H., and Jakway, J. S. 1962. *In* The Behavior of Domestic Animals (ed. E. S. E. Hafez), p. 334. Williams and Wilkins, Baltimore.

Harness, C. R. 1988. The Other White Meat. Marketing U.S. Agriculture. 1988 Yearbook of Agriculture, U.S. Government Printing Office, Washington, D.C.

Harrison, B. M. 1971. Embryology of the Chick and Pig. William C. Brown, Dubuque, Iowa.

Hetzer, H. O. 1945. J. Hered. *36,* 121; *26,* 187; *36,* 255; *36,* 302; *38,* 120; *39,* 122.

Hetzer, H. O., Dickerson, G. E., and Zeller, J. H. 1944. J. Anim. Sci. *3,* 390.

Hetzer, H. O., and Harvey, W. R. 1967. J. Anim. Sci. *26,* 1244.

Holden, P., Christian, L., and Meyer, V. 1988. Iowa Pork Industry Competitive Situation and Prospects. Iowa State University, STFE, Ames, pp. 43–48.

Holmes, W. 1970. Proc. Nutr. Soc. *29,* 237.

Holstad, G. E. 1956. The Fetal Pig. An Introduction to the Anatomy of the Fetal Pig. Burgess, Minn.

Houpt, K. A., and Wolski, T. R. 1982. Domestic Animal Behavior, Iowa State University Press, Ames, pp. 1–356.

Jimenez, M., Lopez-Saez, J. F., and Monge, E. G. 1962. J. Hered. *53,* 281.

Legault, C. 1983. 34th Annual Meeting of European Assoc. Anim. Prod., Madrid.

Mason, I. L. 1969. A World Dictionary of Livestock Breeds, Types, and Varieties. Tech. Communication No. 8, Commonwealth Bureau of Animal Breeding and Genetics, Edinburgh. Commonwealth Agric. Bureau, Farnham Royal, Bucks., England.

Mellen, I. M. 1952. The Natural History of the Pig. Exposition Press, New York.

Moran, E. T., Jr. 1982. Comparative Nutrition of Fowl and Swine. The Gastrointestinal Systems. Office of Educational Practice, University of Guelph, Guelph, Ontario, Canada, pp. 1–253.

Pond, W. G. 1983. Sci. Am. *248,* 96.

Pond, W. G. and Houpt, K. A. 1978. The Biology of the Pig, pp. 1–371. Cornell University Press, Ithaca, N.Y.

Putnam, J. J., and van Dress, M. G. 1984. National Food Review (NFR-26). USDA, Econ. Res. Service, pp. 15–20.

Rhodes, V. J., Flottman, D., and Procter, M. H. 1987. Basic data on U.S. mid/large size hog operations, 1986–87, AEWP, 1987–17, Agric. Econ. Dept., Univ. Missouri, Columbia, Mo.

Robinson, D. W., and Vohra, P. 1976. Feedstuffs, February 23, p. 20.

Shaw, A. M. 1930. Sci. Agr. *10,* 690.

Sisson, S., and Grossman, J. D. 1953. The Anatomy of the Domestic Animals, 4th edition. W. G. Saunders, Philadelphia.

Stuart, R. R., and Pinder, D. E. 1948. The Anatomy of the Fetal Pig. Denoyer-Geppert, Chicago.

The World Food Problem. 1967. A Report of the President's Science Advisory Committee, Vol. II, Report of the Panel on the World Food Supply. Superintendent of Documents, U.S. Government Printing Office, Washington, D.C.

Towne, C. W., and Wentworth, E. N. 1950. Pigs from Cave to Cornbelt. University of Oklahoma Press, Norman.

Tumbleson, M. E. 1986. Swine in Biomedical Research. Vols. I, II, III. Plenum Press, New York, pp. 1–1988.

United Nations Department of Social Affairs, Population Division. 1966. Population Studies No. 41, Washington, D.C.

USDA. 1988. Agricultural Statistics. U.S. Government Printing Office, Washington, D.C.

Walker, W. F., Jr. 1964. Dissection of the Fetal Pig. Laboratory Studies in Biology. W. H. Freeman, San Francisco.

2

Pork as Human Food

Pork, along with other animal products, provides protein of higher nutritive value than that present in most plant proteins and also provides other nutrients not found in adequate amounts in plants. Pork supplies nearly one-half of the meat production from domestic animals in the world and about 40% of the meat (excluding poultry) in the United States. The long-term acceptance of pork as a major food source has been a result of its high nutritive value as well as the variety of processing and cooking methods available for its inclusion in many cultures. Those associated with the swine industry must not lose sight of the fact that the consumer ultimately determines the amount and kind of pork produced. The purpose of this chapter is to characterize pork as to its nutrient composition and other properties that affect its acceptability to humans.

NUTRIENT COMPOSITION OF PORK

Pork is an excellent source of high-quality protein, B vitamins, and trace elements. Its chemical composition has been described by Leverton and Odell (1958), Watt and Merrill (1963), Kiernat et al. (1964), and more recently by Moss et al. (1983) and the USDA (1983), from analyses of carcasses from contemporary swine of considerably leaner composition than those used in earlier studies. The dramatic changes that have occurred in the fat content of pork since the 1960s are illustrated in Fig. 2.1. Rice (1971) reviewed much of the earlier work on the nutritional value of pork and other meat products. Table 2.1 contains average percentages of water, protein, and fat in raw fresh retail cuts of pork, and the average calorie content of each cut from lean (U.S. Grade 1), moderate (U.S. Grade 2), and fat (U.S. Grade 3) pigs. It is clear that the overall gross chemical composition of these cuts is similar. Cooking reduces the water content so that percentages of protein, fat, and ash are increased appreciably and calorie concentration is increased. A given weight of roasted lean pork may contain less than 80% of the water and more than 130% of

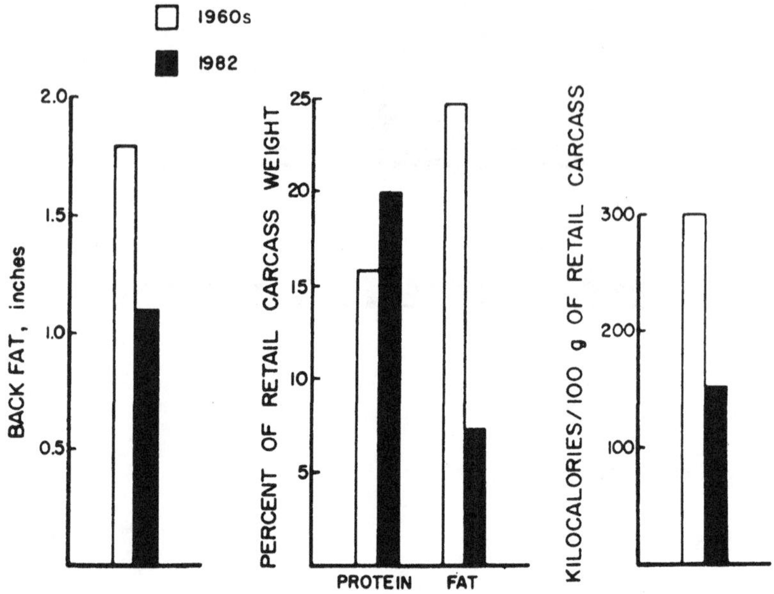

FIGURE 2.1 Changes in fat content of pork since the 1960s as illustrated by back fat thickness, percentage of protein and fat in the retail carcass, and kilocalories of energy per gram of retail carcass.

the protein of raw pork. This is illustrated in Table 2.2 in the columns headed "Fresh Loin" and "Roasted Loin." It is also noteworthy that the water, protein, mineral, vitamin, and fatty acid content of pork taken from different parts of the carcass are similar (for example, note the fresh leg [ham] and loin data in Table 2.2).

The type of cooking will, of course, have an influence on the nature of the

TABLE 2.1 Water, Protein, Fat and Calorie Content of Raw and Cooked Fresh Retail Cuts of Pork

Retail cut	*USDA grade*	*Back fat (cm)*	*Protein (%)*	*Fat (%)*	*Water (%)*	*kcal/100g*
Picnic	1	2.78	19.9	7.2	73.6	150
	2	4.01	19.4	7.4	73.3	150
	3	4.51	19.2 (15.8)[a]	6.8 (24.7)	74.0 (58.9)	144 (290)
Ham	1	2.78	20.0	5.5	74.0	140
	2	4.01	19.4	5.6	73.9	135
	3	4.51	20.5 (15.9)	6.2 (26.6)	73.6 (56.5)	144 (308)
Loin	1	3.00	20.7	6.9	71.5	151
	2	4.09	20.6	8.1	70.7	161
	3	4.74	20.2 (17.1)	9.8 (24.9)	69.2 (57.2)	175 (298)
Spare rib	1	2.49	17.8	26.2	57.7	312
	3	4.23	17.0 (14.5)	25.2 (33.2)	58.1 (51.8)	300 (361)

[a] Values in parentheses are from Kiernat et al. (1964) based on data from pork produced in the 1950s and early 1960s. Note the dramatic decrease in fat content of picnic, ham, and loin and the increase in protein content during the 20-year period. These dramatic changes have resulted from improved feeding and genetic selection for leanness.

Source: USDA (1983)

TABLE 2.2 Water, Energy, Fat, Mineral, Vitamin and Fatty Acid and Cholesterol Content of Pork Cuts

	Carcass, Fresh (lean & fat unseparated)		*Leg (Ham), Fresh Separable lean*		*Loin, Fresh Separable lean*		*Loin, Roasted Separable lean*	
	100 g portion	*1 oz (28.35 g)*	*100 g portion*	*1 oz (28.35 g)*	*100 g portion*	*1 chop (91 g)*	*100 g portion*	*1 chop (91 g)*
Water, g	47.9	13.6	72.9	20.7	70.6	64.2	55.5	36.6
Energy, kcal	398	113	136	39	156	142	257	169
Protein, g	13.4	3.8	20.5	5.8	20.7	18.8	27.8	18.4
Fat, g	37.8	10.7	5.4	1.5	7.5	6.9	15.3	10.1
Minerals								
Calcium, mg	5	2	6	2	7	7	7	5
Iron, mg	0.7	0.2	1.0	0.3	0.85	0.78	0.93	0.61
Magnesium, mg	13	4	25	7	23	21	29	19
Phosphorus, mg	152	43	229	65	230	209	279	184
Potassium, mg	244	69	369	105	353	322	418	276
Sodium, mg	44	13	55	16	64	58	75	49
Zinc, mg	1.6	0.5	2.3	0.6	2.1	2.0	2.9	1.9
Copper, mg	0.06	0.02	0.08	0.02	0.08	0.07	0.10	0.07
Manganese, mg	0.01	0.002	0.03	0.01	0.01	0.01	0.01	0.01
Vitamins								
Ascorbic acid (vit. C), mg	0.5	0.1	0.9	0.3	0.9	0.8	0.3	0.2
Thiamin (B_1), mg	0.57	0.16	0.88	0.25	1.00	0.92	0.97	0.64
Riboflavin (B_2), mg	0.21	0.06	0.23	0.07	0.30	0.27	0.42	0.28
Niacin, mg	3.9	1.1	5.3	1.5	5.4	4.9	5.95	3.93
Pantothenic acid, mg	0.9	0.25	0.81	0.23	0.81	0.74	0.91	0.60
Pyridoxine (B_6), mg	0.3	0.08	0.50	0.14	0.53	0.48	0.46	0.30
Folacin, mcg	4	1	9	3	5	4	5	4
Vitamin B_{12}, mcg	0.7	0.2	0.71	0.20	0.86	0.79	1.08	0.71
Vitamin A, IU	8	2	6	2	7	6	8	5
Fatty acids								
Saturated, g	13.8	3.9	1.9	0.5	2.6	2.4	5.3	3.5
Monounsaturated, g	17.6	5.0	2.5	0.7	3.4	3.1	6.9	4.5
Polyunsaturated, g	4.2	1.1	0.6	0.2	0.8	0.7	1.9	1.2
Cholesterol, mg	74	21	68	19	60	55	95	63

Source: USDA Handbook No. 8–10 (1983).

weight loss. Any cooking method, such as frying or charcoal broiling, that results in dripping losses of fat would result in a reduction in amount of fat as well as of water in the final product. Table 2.3 contains data on percentages of connective tissue, lean, bone, and intramuscular and subcutaneous fat in retail cuts of U.S. Grade 1 pork carcasses (USDA 1983). The protein content of 20–30% on a fresh basis is higher than nearly all plant materials and similar to most foods of animal origin. The National Research Council Food and Nutrition Board (1989) recommends a daily protein intake of 56 g for young adult men (70 kg, 177-cm tall) and 46 g for young adult women (55 kg, 163-cm tall). Therefore, the entire daily protein requirement is met by a 150–280-g serving of lean pork. Organs such as brain, heart, and liver are also good sources of protein, vitamins, and minerals (Table 2.4).

The amino acid composition of the protein is of importance, as is total amount of protein. Most animal products, including pork, provide a better balance of essential amino acids than most plant products provide. This is illustrated in Table 2.5 which lists the essential amino acid composition of pork and casein, which is the major protein in milk (used to represent animal proteins), and that of maize and peanut meal (used to represent plant proteins). Both pork protein and casein have adequate amino acid balance to support good growth, although casein is slightly deficient in sulfur-containing amino acids (methionine and cystine). Most plant proteins are deficient in lysine, tryptophan, threonine, or methionine-cystine, or a combination of two or more of them. A comparison of the amounts of lysine and methionine-cystine in peanut meal and of lysine and tryptophan in maize with the amounts of these amino acids in pork protein and casein (Table 2.5) illustrates this characteristic of plant proteins. Thus, a diet consisting entirely of plant foodstuffs is almost certain to be deficient in at least one of the essential amino acids for man. Pork in the diet, even if only a small proportion, can measurably improve the protein quality of the diet.

Pork is also an excellent source of some of the mineral elements and a poor source of others. For example, it is high in phosphorus, but almost devoid of calcium; it is high in potassium, but rather low in sodium; it is a good source of iron, zinc, manganese, and magnesium. Thus, if pork were the only source of food for

TABLE 2.3 Percentages of Lean, Bone, Connective Tissue, and Intramuscular and Subcutaneous Fat in Retail Cuts of U.S. Grade 1 Pork Carcasses

Retail cut	*Lean*	*Connective tissue*	*Bone*	*Fat*		
				Intramuscular	*Subcutaneous*	*Total*
Picnic	55.6	8.9	16.8	9.1	10.4	19.5
Blade loin roast	53.6	4.9	20.7	10.6	10.2	20.8
Ham rump	69.5	4.9	12.8	4.6	8.3	12.9
Ham shank	62.5	7.7	11.6	5.0	13.1	18.1
Rib chop	57.9	3.7	18.5	6.8	14.8	21.6
Shoulder blade steak	68.0	5.5	9.2	10.8	9.0	19.8
Sirloin roast	57.8	5.9	16.6	4.0	15.7	19.7
Spare rib	58.4	0.7	40.4	7.6	0.03	7.6

Source: USDA (1983)

TABLE 2.4 Water, Protein, Mineral, Vitamin and Lipid Content of Pork Brain, Heart, Liver and Kidney

	Brain		*Heart*		*Liver*		*Kidney*	
	100 g	*1 oz (28.35 g)*	*100 g*	*1 oz (28.35 g)*	*100 g*	*1 oz (28.35 g)*	*100 g*	*1 oz (28.35 g)*
Water, g	78.4	22.2	76.2	21.6	71.1	20.1	80.1	22.7
Lipid, g	9.2	2.6	4.4	1.2	3.7	1.0	3.3	0.9
Protein, g	10.3	2.9	17.3	4.9	21.4	6.1	16.5	4.7
Minerals								
Calcium, mg	10	3	5	2	9	2	9	3
Iron, mg	1.6	0.5	4.7	1.3	23.3	6.6	4.9	1.4
Magnesium, mg	14	4	19	5	18	5	17	5
Phosphorus, mg	282	80	169	48	288	82	204	58
Potassium, mg	258	73	294	83	273	71	229	65
Sodium, mg	120	34	56	16	87	25	121	34
Zinc, mg	1.3	0.4	2.8	0.8	5.8	0.62	2.8	0.78
Copper, mg	0.24	0.07	0.41	0.12	0.68	0.19	0.62	0.18
Manganese, mg	0.09	0.03	0.06	0.02	0.34	0.10	0.12	0.04
Vitamins								
Ascorbic acid, mg	13.5	3.8	5.3	1.5	25.3	7.2	13.3	3.8
Thiamin (B_1), mg	0.15	0.04	0.61	0.17	0.28	0.08	0.34	0.10
Riboflavin (B_2), mg	0.28	0.08	1.19	0.34	3.01	0.85	1.70	0.48
Niacin, mg	4.28	1.21	6.77	1.92	15.3	4.3	8.21	2.33
Pantothenic acid, mg	2.80	0.79	2.52	0.71	6.65	1.89	3.13	0.89
Pyridoxine (B_6), mg	0.19	0.05	0.39	0.11	0.69	0.20	0.44	0.13
Folacin, mcg			4	1	212	60	42	12
Vitamin B_{12}, mcg	2.19	0.62	3.79	1.07	26	7.37	8.5	2.4
Vitamin A, IU	0	0	25	7	21,650	6,138	198	56
Cholesterol, mg	2,195	622	131	37	301	85	319	90

Source: USDA Handbook No. 8–10 (1983).

TABLE 2.5 Amino Acid Composition of Fresh Pork, Casein (a Good-quality Protein), Peanut Meal (a Poor-quality Protein), and Maize (a Poor-quality Protein)

Amino acid	*Percent of protein*			
	Pork	*Casein*	*Peanut meal*	*Maize*
Arginine	6.4	3.9	12.4	4.6
Histidine	3.2	3.2	2.4	2.9
Isoleucine	4.9	5.1	3.8	5.0
Leucine	7.5	7.6	6.8	16.1
Lysine	7.8	8.2	3.7	2.8
Methionine–cystine	3.8	3.1	2.8	4.9
Phenylalanine	4.1	4.1	6.3	5.4
Threonine	5.1	4.1	3.1	3.5
Tryptophan	1.4	1.0	1.0	0.6
Valine	5.0	6.0	4.2	7.2

Source: Schweigert and Payne 1956.

animals and man, a deficiency of calcium would most certainly occur, but the supply of many of the elements would be sufficient.

Fortunately, the biological availability of minerals in pork and in the skeleton (bone) is very high and, unlike the case with some plant sources of minerals, there is no problem with naturally occurring chelating agents which interfere with the utilization of the minerals present. Some organs are especially good sources of particular mineral elements. For example, pork liver is an excellent source of iron. The amounts of trace elements such as iron, copper, zinc, and manganese deposited in the liver of the pig are related to the diet content of each mineral. That is, pigs fed diets high in copper tend to accumulate copper in the liver. Gipp (1971) found levels of copper as high as 1400 ppm (0.14%) in the liver of pigs fed a diet containing 250 ppm copper. Unfortunately, toxic elements such as cadmium, lead, and mercury also tend to accumulate in the liver and kidneys of pigs ingesting them, so that a human health hazard can be created by consumption of organs from such animals. However, a steady diet of liver or kidneys or other organ meats is not likely, so that a single meal of organs from a pig exposed to high levels of a toxic element is of less potential danger. Average concentrations of several required elements present in various organs of pigs are shown in Table 2.4.

The water-soluble vitamin content of pork is high compared to most food products. Also, the organs commonly used for food, especially the liver, are unusually good sources of the vitamins. Average concentration of vitamins in raw and cooked fresh ham and loin and in cured ham are given in Table 2.6, and concentrations in various organs are in Table 2.4. Pork is especially high in thiamin and, as shown in Table 2.6, a 100-g serving of pork loin provides 70% of the daily requirement for adult humans. The same 100-g serving supplies substantial proportions of the daily requirements of other vitamins as well. Liver is an excellent source of vitamin A. Vitamin B_{12} is a component of animal and microbial products only, so unless a meat, dairy, or poultry product is eaten, a vitamin B_{12} deficiency may result. For comparison, the concentrations of three important water-soluble vitamins, thiamin, riboflavin, and niacin, in a variety of plant and animal products is listed in Table 2.7 along with the amount of each vitamin supplied by each

TABLE 2.6 Water-soluble Vitamin Content (mg/100g) of Raw and Cooked Pork [a, b] and Percentage of Daily Adult Human Requirement

Vitamin	Raw pork				Cooked pork			% of daily adult requirement from 100 g of loin[c]
	Fresh ham	Cured ham	Fresh loin	Fresh composite[b]	Fresh ham	Cured ham	Fresh loin	
Thiamin	0.77	0.72	0.98	—	0.51	0.47	1.13	70
Riboflavin	0.19	0.19	0.24	0.32	0.23	0.18	0.33	14
Niacin	4.1	4.1	5.2	—	4.6	3.6	6.8	29
Pantothenic acid	0.72	0.64	2.0	—	—	—	—	—
Vitamin B_6	0.42	0.39	0.50	0.52	—	—	—	25
Vitamin B_{12}	0.0009	0.001	—	0.0009	—	—	—	20
Folic acid	0.009	0.011	0.007	—	—	—	—	2
Biotin	0.0053	0.005	0.0055	—	—	—	—	—

[a] Leverton and Odell 1958; Watt and Merrill 1963; Kiernat et al. 1964.
[b] Moss et al. 1983.
[c] National Research Council 1980.

product as a percentage of that supplied by an equal amount of lean pork. Pork is by far the best source of thiamin among both plant and animal products with the exception of peanuts, and of yeast, which is notable as a B-vitamin source. Beef and lamb contain only 9% and 17% of the amount of thiamin provided by pork. Only some of the seed legumes, including red beans, cowpeas and soybeans, and the cereal grain, wheat, contain half or more of the amount of thiamin supplied by pork. Pork is comparable to milk, beef, and lamb as a source of riboflavin and niacin and is considerably higher than almost all plant products in these two vitamins. Furthermore, the niacin of cereal grains and maize, although approaching the level of that found in pork, is almost completely unavailable to the animal as it is chemically bound in a form not easily broken down in digestion. Thus, the value of pork as a source of niacin is even greater than tabular values of food composition would indicate.

As pork contains a relatively high proportion of fat compared to most plant products, it is important to be aware of the composition of the fat. The controversial question of the relationship between dietary fat and heart disease has not been answered, but it is often assumed that animal fats may enhance the development of atherosclerosis. A large long-term study completed among humans in Framingham, Massachusetts, summarized by Kannel and Gordon (1971) has provided some evidence of diet as a contributing factor.

Other more recent studies suggest a probable relationship between high levels of dietary fat and heart disease in some population groups. In response to such a possible relationship, U.S. pork producers have reduced the average lard produced from about 14% of the carcass in 1960 to less than 5% in 1983 (Fredeen and Harmon 1983). Pork carcass composition can be estimated accurately by measuring backfat thickness and cross-sectional area of the loin muscle in the live animal or carcass (Fahey et al. 1977; Edwards et al. 1981).

It seems appropriate to characterize the fat composition of pork and to clear up

TABLE 2.7 Water-soluble Vitamin Content of Pork Compared to That of Some Other Common Foods

Food	*Thiamin*		*Riboflavin*		*Niacin*	
	mg/100 g	*% of that supplied by lean pork*	*mg/100 g*	*% of that supplied by lean pork*	*mg/100 g*	*% of that supplied by lean pork*
Pork						
Total edible	0.50	57	0.12	57	2.7	59
Separable lean	0.87	—	0.21	—	4.6	—
Separable fat	0.17	20	0.04	19	0.9	20
Apples	0.03	3	0.02	10	0.1	2
Bananas	0.05	6	0.06	29	0.7	15
Beans, red	0.51	59	0.20	95	2.3	50
Beans, snap (yellow or snap)	0.08	9	0.11	52	0.5	11
Beef	0.08	9	0.17	81	4.7	102
Cabbage	0.05	6	0.05	24	0.3	7
Cheese, cheddar	0.03	3	0.46	219	0.1	2
Cheese, cottage	0.03	3	0.25	119	0.1	2
Chicken, light meat without skin	0.05	6	0.09	43	10.7	233
Chicken, dark meat without skin	0.08	9	0.20	95	5.2	113
Clams	0.10	11	0.18	86	1.3	28
Coconut	0.05	6	0.02	10	0.5	11
Corn, yellow	0.15	17	0.12	57	1.7	37
Cowpeas	0.43	49	0.13	62	1.6	35
Eggs, whole	0.11	13	0.30	143	0.1	2
Grapefruit	0.04	5	0.02	10	0.2	4
Halibut	0.07	8	0.07	33	8.3	180
Herring	0.02	2	0.15	71	3.6	78
Ice cream (12% fat)	0.04	5	0.19	90	0.1	2
Lamb	0.15	17	0.20	95	4.8	104
Liver, beef	0.25	29	3.26	1,552	13.6	296
Liver, chicken	0.19	22	2.49	1,186	10.8	235
Liver, lamb	0.40	46	3.28	1,562	16.9	367
Liver, pork	0.30	34	3.03	1,443	16.4	357
Milk, whole fluid	0.03	3	0.17	81	0.1	2
Milk, skim fluid	0.04	5	0.18	86	0.1	2
Oranges	0.10	11	0.04	19	0.4	9
Peanuts	0.99	114	0.13	62	15.8	343
Potatoes	0.10	11	0.04	19	1.5	33
Rice, brown	0.34	39	0.05	24	4.7	102
Rice, white	0.07	8	0.03	14	1.6	35
Soybeans	0.44	51	0.16	76	1.4	30
Sugar, beet or cane	0.01	1	0.03	14	0.2	4
Sweet potatoes	0.10	10	0.06	29	0.6	13
Wheat	0.55	63	0.12	57	4.3	93
Yams	0.10	11	0.04	19	0.5	11
Yeast, dried or bakers'	2.33	267	5.41	2,576	36.7	798

Source: Watt and Merrill 1963.

some often-stated but unfounded claims regarding pork. Among those claims are these two:

1. Pork is high in saturated fatty acids and cholesterol and low in unsaturated (polyunsaturated) fatty acids. *Answer:* False. Pork is higher in polyunsaturated fatty acids than beef, lamb, or milk and is lower in cholesterol than butter, cheddar cheese, eggs, organ meats, and many seafoods (Table 2.8)
2. Pork fat is difficult to digest. *Answer:* False. Digestibility experiments with animals and humans have failed to show differences between pork fat (lard) and other animal and plant fats in apparent digestibility.

Average fatty acid composition of typical pork fat taken from various body sites is shown in Table 2.9. The pattern of fatty acid distribution is similar among sites (loin, belly, ham, and shoulder) and between subcutaneous, intermuscular, and intramuscular fat within each site. Linoleic acid makes up from 8% or 9% to

TABLE 2.8 Cholesterol Content of Pork Compared to That of Some Other Common Foods

Food	*Cholesterol content per 100 g of edible portion (mg)*
Pork	61[a]
Beef	70
Brain	>2000
Butter	250
Caviar	>300
Cheese, cheddar	100
Cheese, cottage or creamed	15
Chicken	60
Corn (maize)	0
Eggs, whole	550
Egg, white	0
Egg, yolk	1500
Fish	70
Heart	150
Lamb	70
Margarine, all-vegetable	0
Lard	95
Liver	300
Milk, whole	11
Milk, skim	3
Orange	0
Oyster	>200
Rice	0
Shrimp	125
Veal	90

[a] Moss et al. 1983.

Source: Watt and Merrill 1963 (animal products only).

TABLE 2.9 Fatty Acid Composition of Pork Fat

Tissue	Fatty acids[a] (% of total contained in tissue)								
	Myristic ($C_{14:0}$)	*Palmitic* ($C_{16:0}$)	*Palmitoleic* ($C_{16:1}$)	*Stearic* ($C_{18:0}$)	*Oleic* ($C_{18:1}$)	*Linoleic* ($C_{18:2}$)	*Linolenic* ($C_{18:3}$)	*Arachidonic* ($C_{20:4}$)	*Miscellaneous*
Loin									
Subcutaneous	1.3	22.5	2.8	14.1	42.5	13.5	0.9	0.3	2.1
Intermuscular	2.0	22.5	2.2	14.2	42.9	11.9	0.9	0.4	2.0
Intramuscular	1.8	21.3	4.2	13.4	32.8	13.4	0.3	3.2	4.6
Belly									
Subcutaneous	1.1	23.3	3.5	8.7	50.2	10.8	0.8	0.2	1.4
Intermuscular	1.6	25.7	1.7	11.5	40.5	10.3	0.7	0.4	7.6
Intramuscular	1.3	22.7	4.3	10.9	46.5	9.4	0.7	0.5	3.7
Ham									
Subcutaneous	2.2	22.5	3.5	9.4	47.8	11.6	0.8	0.3	1.9
Intermuscular	2.2	21.3	3.1	12.6	36.9	13.3	1.1	0.7	8.8
Intramuscular	2.1	21.2	4.3	11.5	38.8	13.7	0.6	3.8	4.0
Shoulder									
Subcutaneous	1.7	21.9	3.4	11.5	45.8	12.8	1.0	0.4	1.5
Intermuscular	1.6	24.3	2.5	14.9	43.5	11.5	0.8	0.3	0.6
Intramuscular	2.1	22.9	3.8	13.4	43.5	10.0	0.7	1.2	2.4

[a] Number before colon refers to the length of the carbon chain; number after colon refers to the number of double bonds; for example, $C_{14:0}$ is a fatty acid with 14 carbons and no double bonds.

Source: Austin et al. 1964; Rice 1971.

more than 13% of total fatty acids in a typical pork fat sample. Linolenic and arachidonic acids, the other two predominant polyunsaturated fatty acids, contribute an additional 1% to 4 or 5% of total fatty acids in the fat from a typical pig. Thus, pork fat can be considered a good source of polyunsaturated fatty acids. Fatty acid composition of backfat is affected by sex (boars have a higher content of linoleic acid than gilts), sire, and diet (Koch et al. 1967). As shown in Table 2.10, the fatty acid composition of pork fat is affected to a great extent by diet. The fatty acid pattern of the carcass fat tends to resemble that of the dietary fat. For example, it is possible to increase the percentage of linoleic acid in the depot fat substantially by feeding a diet high in linoleic acid. Feeding a diet containing 3% safflower oil, which contains more than 50% linoleic acid, results in a carcass containing 17% or more of linoleic acid in the depot fat. This flexibility in the fatty acid composition of pork allows the production of pork to meet consumer demands for polyunsaturated fat. Although there is still some stigma attached to "soft pork" by meat packers and others, taste panel studies of pork produced from pigs that had been fed diets of whole maize and cooked soybeans, which resulted in a linoleic acid content of 14.1% in the backfat, failed to show any objectionable flavor or any decline in acceptability of the pork. There are potential problems with storage and rancidity of "polyunsaturated pork," but proper refrigeration and handling should avoid such problems.

The proportions of ash, protein, water, and fat in the pig carcass are controlled by genetics as well as by environmental factors. Suffice it to say, here, that the proportions of fat and water in the edible carcass vary inversely with each other, so that as fattening of the animal progresses, water content declines. Percentages of

TABLE 2.10 Fatty Acid Composition (%) of Pork Back Fat from Pigs Fed Various Diets

Fatty acid	Composition in typical pork fat[a]	*Composition in fat from pigs fed various diets* *Diet description*			
		3% coconut fat[b]	*3% safflower oil*[b]	*Maize-soybean meal*[b]	*Maize-whole soybeans*[c]
Capric ($C_{10:0}$)	—[d]	0.18	0.20	0.11	—
Lauric ($C_{12:0}$)	—	0.63	0.12	0.12	—
Myristic ($C_{14:0}$)	—	3.85	2.00	1.76	1.2
Palmitic ($C_{16:0}$)	28	20.00	15.60	19.31	28.7
Palmitoleic ($C_{16:1}$)	3	16.47	13.01	12.68	1.5
Stearic ($C_{18:0}$)	13	55.93	50.58	55.76	9.4
Oleic ($C_{18:1}$)	46				
Linoleic ($C_{18:2}$)	10	0.93	17.15	8.32	14.1
Linolenic ($C_{18:3}$)	0.7	0.09	0.09	0.09	—
Eicosamonoenoic ($C_{20:1}$)	—	0.76	0.56	1.08	—
Eicosatrienoic ($C_{20:3}$)	—	0.11	0.46	0.49	—
Arachidonic ($C_{20:4}$)	2.0	0.07	0.19	0.23	—
Total polyunsaturated	—	1.96	1.30	10.21	—

[a] Dugan 1957.
[b] Babatunde et al. 1968.
[c] Wahlstrom et al. 1971.
[d] Value not available.

protein and ash remain rather constant, but tend to decline slightly in fatter animals.

The proportion of the live animal that is available as edible lean meat is often referred to as "percent lean cuts." These lean cuts are the shoulder (Boston butt plus picnic), loin, and ham. Other important cuts from the carcass are the jowl, clear plate, spareribs, belly, backfat, feet, and organs such as brain, liver, and kidney. The wholesale cuts—showing the location of each component on the carcass—are indicated in Figs. 2.2 and 2.3. The "lean cuts" constitute 40% or more of the weight of a U.S. No. 1 hog but 75% or more of the value under U.S. marketing conditions. The fat trim of a U.S. No. 1 hog is 17% of the weight and 5% of the value, but as fatness increases the fat trim makes up 21% of the weight and 7% of the value of a U.S. No. 3 hog. A description and the specifications for official U.S. market grades of hogs are given in Appendix 3. The relative weights and value of wholesale cuts of pork from a U.S. No. 1 pig are shown in Table 2.11. In fatter pigs (U.S. Nos. 2 and 3) the weights of the trimmed ham and liver as a percentage of liveweight tend to decline while the fat trim and leaf fat increase. The lean cuts therefore represent a smaller proportion of the total weight and total value of the carcass, as the carcass increases in fatness. Most commercial swine slaughtering and processing businesses have developed their own live and carcass grading and pricing systems, based on principles of leanness and indices of body composition addressed by the official USDA grading standards. In recent years, a pork value chart, developed from the results of a study done jointly by the National Pork Producers Council and the Agricultural Research Service of the U.S. Department of Agriculture, has been used commonly to establish value of carcasses of different composition and weight. The chart (Table 2.12) is based on the relationship between one backfat measurement (taken at last rib), carcass weight, and a subjective muscling score. A carcass weighing 167 to 173 lb (75.8 to 78.5 kg) with a backfat of 0.6 in (1.5 cm) would be considered top value; value declines in carcasses lighter or heavier at the same backfat and in carcasses of the same weight but progressively more backfat. This procedure reflects the differences in yield of the various wholesale cuts weighted by their average prices. Such a procedure shows the difference in value of various cuts and organs in relation to size of wholesale cuts (Table 2.13). Future pricing systems may reflect the difference in lean content of the individual cuts.

PROPERTIES AFFECTING THE ACCEPTABILITY OF PORK TO HUMANS

Pork Quality Factors

In addition to the high nutritive value of pork as discussed above, there are some properties of pork that make it inherently a sought-after food when the consumer can afford to purchase it. These properties are shared with other animal products (beef, lamb, poultry), and the amount of pork consumed in relation to alternative meat sources is dependent on esthetic, economic, religious, and cultural factors. The esthetic aspects will be considered here. Those working in meat technology

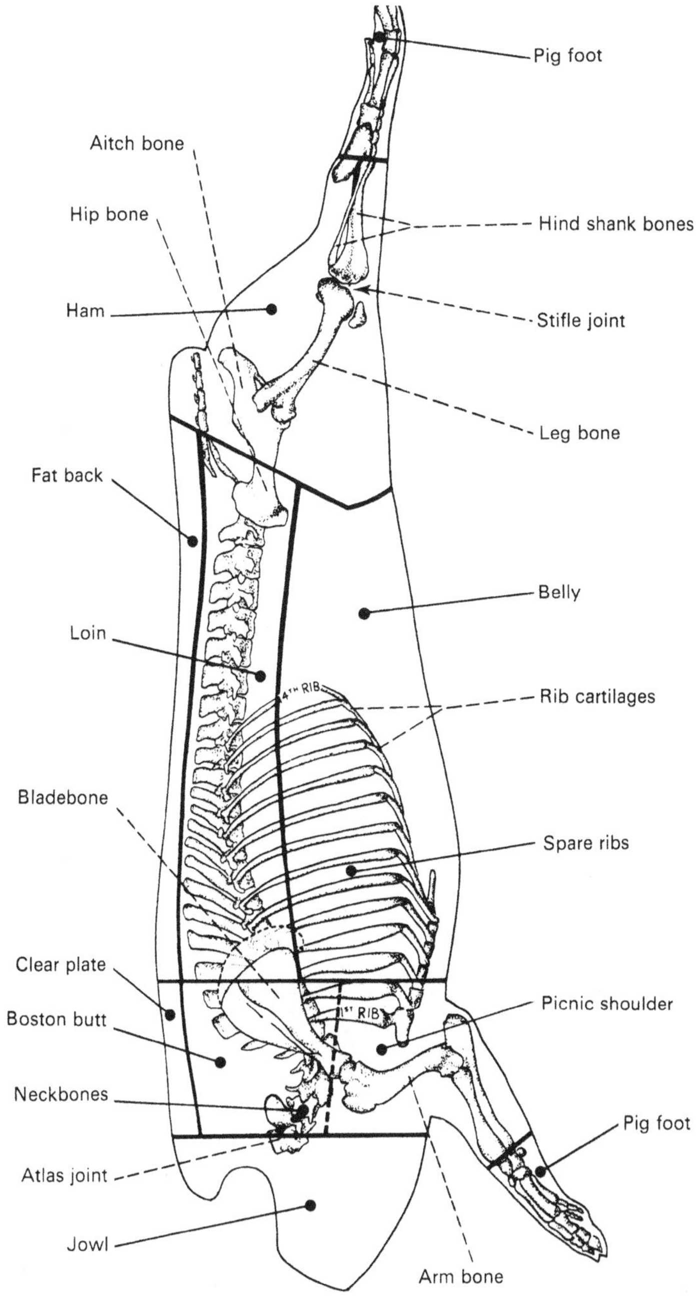

FIGURE 2.2 Carcass diagram showing location of wholesale cuts. Courtesy of J. R. Stouffer, Cornell University.

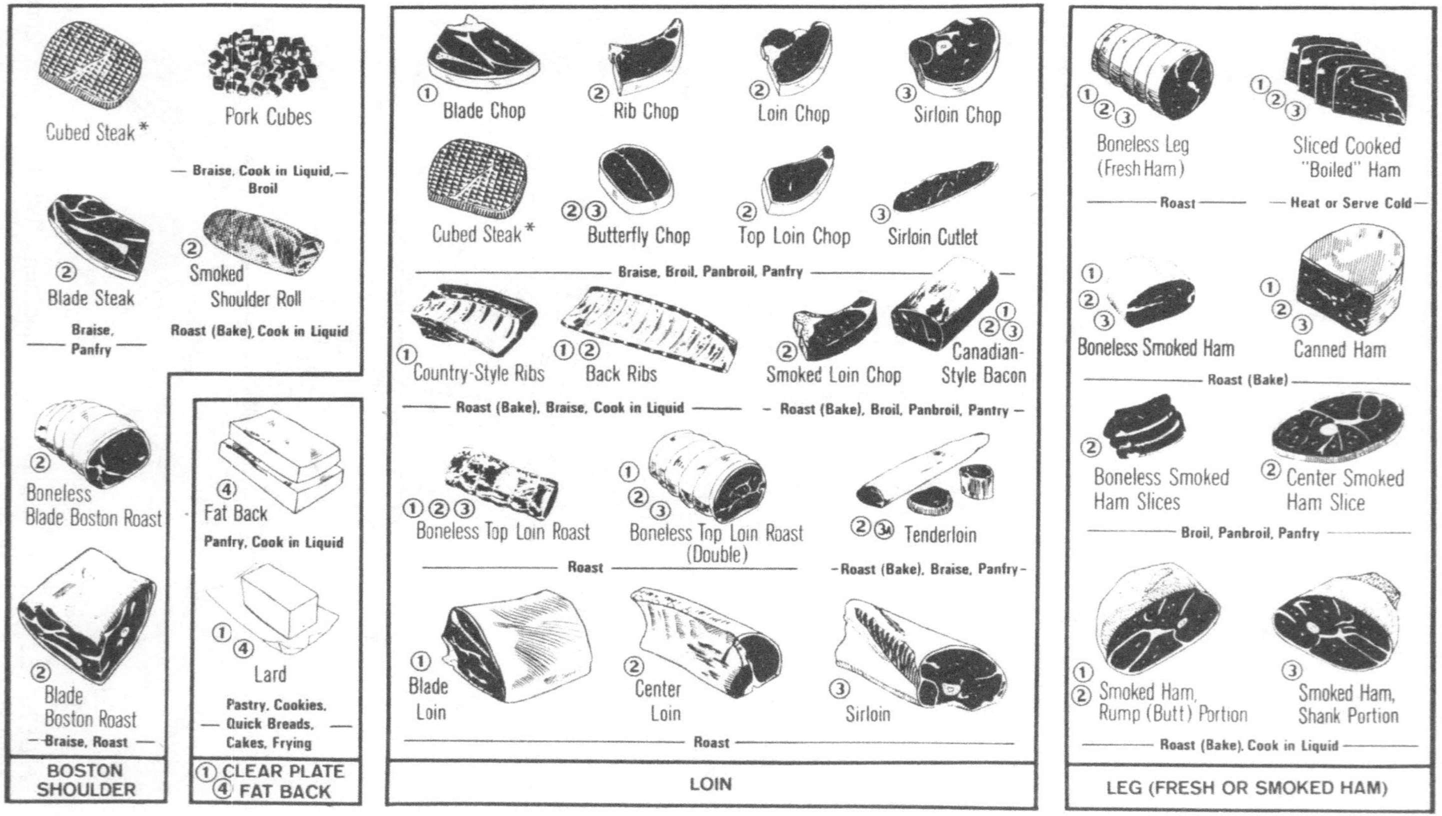
Cubed Steak*
Pork Cubes
— Braise, Cook in Liquid, — Broil
② Blade Steak
Braise, Panfry
② Smoked Shoulder Roll
Roast (Bake), Cook in Liquid
② Boneless Blade Boston Roast
② Blade Boston Roast
— Braise, Roast —
BOSTON SHOULDER
④ Fat Back
Panfry, Cook in Liquid
①④ Lard
Pastry, Cookies, Quick Breads, Cakes, Frying
① CLEAR PLATE
④ FAT BACK
① Blade Chop
② Rib Chop
② Loin Chop
③ Sirloin Chop
Cubed Steak*
②③ Butterfly Chop
② Top Loin Chop
③ Sirloin Cutlet
Braise, Broil, Panbroil, Panfry
① Country-Style Ribs
①② Back Ribs
② Smoked Loin Chop
①②③ Canadian-Style Bacon
Roast (Bake), Braise, Cook in Liquid
Roast (Bake), Broil, Panbroil, Pantry
①②③ Boneless Top Loin Roast
①②③ Boneless Top Loin Roast (Double)
②③④ Tenderloin
Roast
-Roast (Bake), Braise, Panfry-
① Blade Loin
② Center Loin
③ Sirloin
Roast
LOIN
①②③ Boneless Leg (Fresh Ham)
①②③ Sliced Cooked "Boiled" Ham
Roast
Heat or Serve Cold
①②③ Boneless Smoked Ham
①②③ Canned Ham
Roast (Bake)
② Boneless Smoked Ham Slices
② Center Smoked Ham Slice
Broil, Panbroil, Panfry
①② Smoked Ham, Rump (Butt) Portion
③ Smoked Ham, Shank Portion
Roast (Bake), Cook in Liquid
LEG (FRESH OR SMOKED HAM)

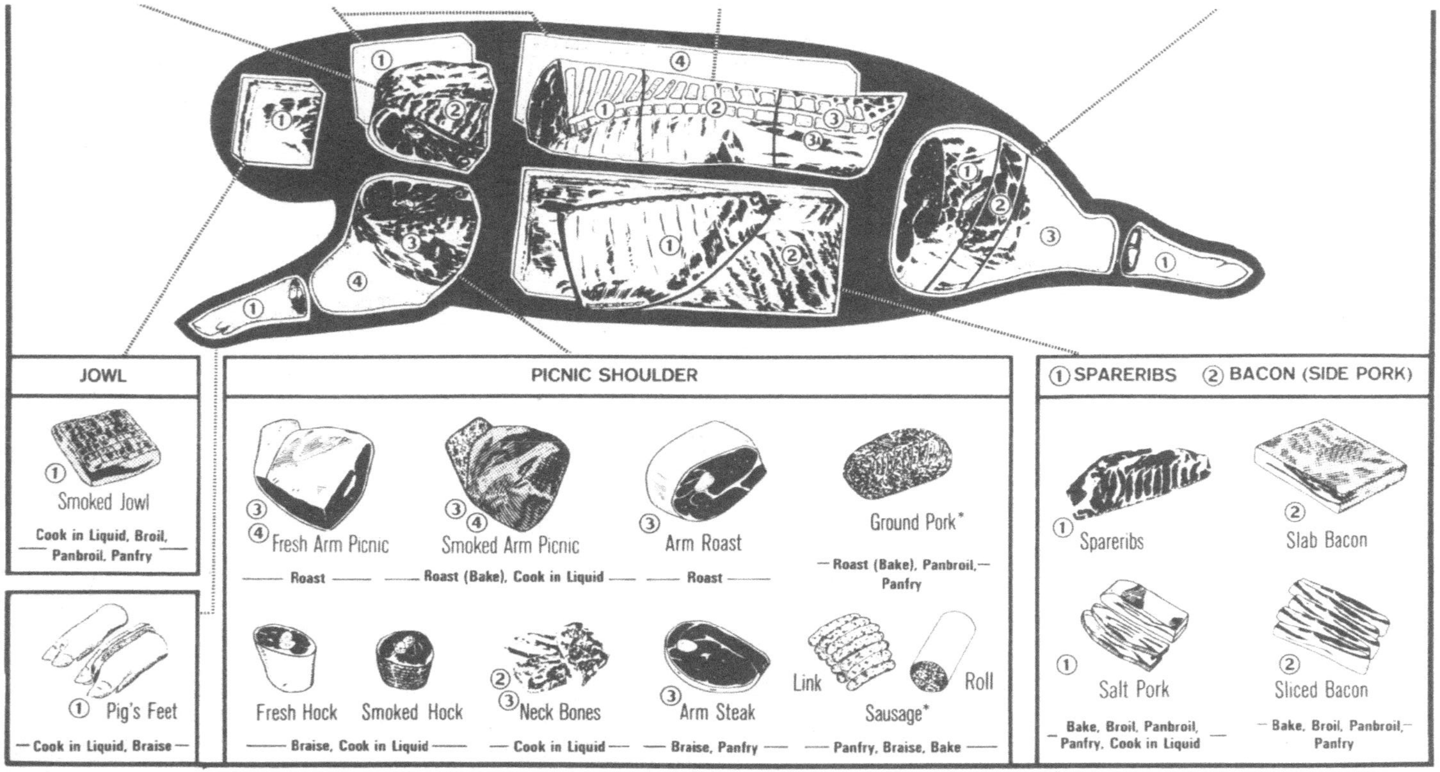

*May be made from Boston Shoulder, Picnic Shoulder, Loin or Leg.

FIGURE 2.3 Wholesale cuts of pork and their uses. Courtesy of National Livestock and Meat Board, Chicago, Illinois.

TABLE 2.11 Weight and Value of Wholesale Pork Cuts from U.S. No. 1 Slaughter Hogs

Cut	*Weight (% of liveweight value)*	*Value (% of total carcass value)*
Ham, skinned	14.7	30.3
Loin	11.8	24.9
Boston butt plus picnic[a]	12.8	18.0
Belly (bacon)	10.0	14.1
Lean trim	1.8	1.8
Fat trim plus leaf fat	16.6	5.0
Spareribs	1.9	3.5
Jowls	2.0	1.1
Neckbones	1.3	0.6
Feet	1.9	0.6
Lean cuts[b]	89.3	73.2

[a] New York shoulder is Boston butt and picnic left unseparated.
[b] Ham, loin, and Boston butt and picnic combined.
Source: Pearson 1966.

research generally divide palatability factors into tenderness, juiciness, color, aroma, and flavor. Pork is less variable in these criteria than are beef and lamb so that differences in age, breed, and environment have a relatively small effect on the quality of pork. Desirable quality pork has a firm, dry surface with a pinkish red color (Briskey and Kauffman 1971). Color, firmness, and water-holding capacity (exudation) are included together as a general appraisal of pork in quality ranging from pale, soft, exudative (PSE) pork to dark, firm, dry pork (DFD). Between these extremes is the most desirable. The amount of fat within the muscle is rather consistent and no differences have been observed in protein, fat and water content between PSE and normal muscle (Briskey et al. 1959; Briskey and Kauffman 1971.) Although PSE muscle shows very little visible intramuscular fat, ether extraction of the muscle fails to reveal differences in fat content of PSE versus normal or DFD muscle, indicating that the fat in PSE muscle is somehow masked.

Pale, Soft, Exudative (PSE) Pork

The pale color and excessive exudation of PSE pork creates merchandising problems in the retail store. The development of PSE pork is related to a more rapid than normal depletion of muscle glycogen postmortem, which results in a reduced pH of the muscle related to lactic acid production. Muscle pH is closely related to its water-holding capacity (Herring et al. 1971.) This condition is apparently partly under genetic control and seems to be associated with stress resistance in live animals. Stress-susceptible pigs are prone to sudden death when exposed to excitement in a warm environment. The muscles of these animals show the typical PSE syndrome and the adrenal hormone secretions are reduced compared to those of normal animals. Thus, we see that physiological response to stress in pigs can have an influence on the acceptability of pork by humans. Much remains unknown concerning all of the factors involved in determining pork quality, the relative

TABLE 2.12 Lean Guide to Pork Value Chart[a]

		Last Rib Fat Thickness (in.)								
Live Wt, lb	*Carcass Wt, lb*	*0.6*	*0.7*	*0.8*	*0.9*	*1.0*	*1.1*	*1.2*	*1.3*	*1.4*
190–199	139–145	88.7	88.1	87.8	86.8	86.6	85.8	84.1	83.8	83.6
200–208	146–152	97.5	97.0	96.6	95.6	95.4	94.6	92.8	92.5	92.3
209–218	153–159	100.1	99.6	99.2	98.2	98.0	97.2	95.5	95.2	95.1
219–227	160–166	102.5	101.9	101.6	100.6	100.4	99.5	97.9	97.5	97.4
228–237	167–173	102.9	102.3	102.0	101.0	100.8	100.0	98.3	98.1	97.8
238–247	174–180	102.3	101.8	101.4	100.4	100.2	99.4	97.7	97.5	97.3
248–256	181–187	102.3	101.8	101.4	100.4	100.2	99.4	97.7	97.5	97.2
257–266	188–194	101.7	101.1	100.7	99.8	99.6	98.7	97.0	96.7	96.6
267–275	195–201	101.3	100.7	100.3	99.4	99.2	98.3	96.7	96.5	96.2

[a] Muscle and fat quality assumed to be acceptable. Adjusted values up by 1.0% for thick muscling, and down by 1.0% for thin muscling.

Source: Guidelines for Uniform Swine Improvement Programs, National Swine Improvement Federation (1987).

TABLE 2.13 Mean Prices (¢/lb) with Ranges and Coefficients of Variation for 1987, 1988, 1989 from *NATIONAL PROVISIONER*

Cuts and Organs	*No. Quotes*	*Mean*	*Min.*	*Max.*	*C.V.*
Skinned Hams, lb					
14–17	733	75.7	59	118	15.2
17–20	757	73.0	54	118	15.6
20–26	757	70.2	48.25	98	15.1
26–32	756	66.0	49.5	91	14.3
Greater than 32	757	63.6	46.25	90	14.9
Loins, lb					
Less than 14	758	102.3	72	140	13.7
14–18	759	101.9	71	140	13.8
18–22	752	100.6	70	135	14.0
Greater than 22	748	79.3	50	108	15.1
Picnics, lb					
4–8	754	47.6	35	64	13.5
Greater than 8	758	46.3	34.5	63	14.9
Boston Butts, lb					
4–8	758	71.3	52	114	17.8
Greater than 8	747	66.2	49.5	97	15.1
Skinned Shlders, lb	759	61.4	47.5	87	13.6
Bellies, lb					
10–12	754	42.2	20	87	36.1
12–14	757	46.1	22.5	91	34.1
14–16	757	46.2	20.5	91	33.9
16–18	758	43.7	20.75	88.5	34.4
18–20	756	39.5	19.5	80	36.3
20–25	754	35.0	16	73	38.4
50% lean trim	760	30.3	14	57	34.2
80% lean trim	760	67.8	44	100	16.4
Spareribs, lb					
Less than 3	751	112.5	82	158	16.9
3–5	741	94.8	69.5	143	18.6
Greater than 5	753	60.7	38	92	22.5
Hearts, lb	759	29.3	18	47	27.3
Kidneys, lb	759	11.9	9.5	15	12.4
Livers (Dom), lb	756	16.4	12	24	17.5
Stomachs (Sc 30/35), lb	756	21.9	14.5	35	23.0
Chitterlings, lb	751	31.6	22	42	14.7
Melts, lb	759	12.2	10	13	9.0
Neckbones, lb	759	13.7	9.5	22	27.7
Skins, lb	490	8.8	7	10	12.2
Lard, lb	156	15.1	12.5	18.5	8.3

importance of each factor, and the degree to which pork acceptability can be controlled through husbandry practices, marketing and slaughter procedures, and meat processing and technology.

Boar Odor

Meat from mature boars has an objectionable odor and flavor when cooked. This odor is produced by at least one fat soluble substance, 5-α-androst-16ene-3-one, characterized by Craig et al. (1962) and later identified by Patterson (1968). In taste-panel studies designed to detect "boar odor" in pork, the odor is detected in most boars weighing more than 90 kg (6 months of age) and occasionally in gilts, sows, and barrows. Since boars grow faster, require less feed per unit of body weight gain, and produce leaner carcasses than barrows, it would be a distinct advantage to the swine producer and consumer alike if a means of eliminating or reducing the boar odor problem without castration could be developed. This is being studied by feeding or implantation of hormones in the boar at various ages and under a variety of conditions. Until suitable methods of circumventing the problem of boar odor are developed, the current management practice of castrating male pigs at an early age is likely to continue in the United States; boar castration is less widely practiced in several other countries. Meat from mature boars sold for slaughter after outliving their usefulness as breeding animals is perfectly safe and nutritious and is normally sold as a component of sausage. In this way its presence goes unnoticed and the nutritional value of the meat is salvaged.

Trichinosis

Perhaps no other factor has been as detrimental in the past to the consumption of pork as the threat of infection with the parasite, *Trichinella spiralis,* from eating infected pork which has been improperly cooked. Due to strict garbage cooking laws and better education of the public about trichinosis, the incidence of this infection in humans has been drastically reduced. In the United States, the annual number of reported human cases has declined steadily in recent years so that fewer than 100 cases per year are now being reported. Suffice it to say, here, that the current threat to human health is not serious when one considers that only a small fraction of all swine is infected with trichinella and that heating pork to an internal temperature of 77°C or freezing for 20 days at −15°C destroys the parasite (Price and Schweigert 1971). Also, there is now a drug, thiabendazole, available for successful treatment of humans infected with trichinella. Thiabendazole can also be used to treat infected animals. Exposure of pork carcasses to high energy radiation has been shown to destroy trichinella without adversely affecting the nutritive value or palatability of the pork. This technology is currently available for use by the meat processing industry and may become an additional tool for producing trichinella-free pork.

Thus, we see that the threat of trichinosis, while real, remains grossly exaggerated as a stigma attached to consumption of pork. The newly developed procedure for identifying infected pigs in the slaughterhouse will undoubtedly do much toward eliminating this parasite as a deterrent to pork consumption.

REFERENCES

Austin, F. L., Ng, C. S., Ostrander, J. A., and Slover, H. R. 1964. Proc. 16th Am. Meat Inst. Res. Conf., Am. Meat Inst. Found. Circ. no. *76*, Chicago.

Babatunde, G. M., Pond, W. G., Walker, E. F., Jr., and Chapman, P. 1968. J. Anim. Sci. *27*, 1290.

Briskey, E. J., Bray, R. W., Hoekstra, W. G., Phillips, P. H. and Grummer, R. H. 1959. J. Anim. Sci. *18*, 146.

Briskey, E. J., and Kauffman, R. G. 1971. *In* The Science of Meat and Meat Products, 2nd ed. (J. F. Price and B. S. Schweigert ed.). W. H. Freeman, San Francisco.

Craig, H. B., Pearson, A. M., and Webb, N. B. 1962. J. Food Sci. *27*, 29.

Dugan, L. R., Jr. 1957. Am. Meat Inst. Found. Circ. no. *36*, Chicago.

Fahey, T. J., Shaefer, D. M., Kauffman, R. G., Epley, R. J., Goris, P. F., Romans, J. R., Smith, G. C., and Topel, D. G. 1977. J. Anim. Sci. *44*, 8.

Edwards, R. C., Smith, G. C., Cion, H. R., and Carpenter, Z. L. 1981. J. Anim. Sci. *52*, 703.

Fredeen, H. T., and Harmon, B. G. 1983. J. Anim. Sci. *57* (Suppl. 2), 100.

Gipp, W. F. 1971. The effect of supplemental copper on growth, hematology and iron utilization in swine. Ph. D. Thesis, Cornell University, Ithaca, N. Y.

Herring, H. K., Haggard, J. H., and Hanson, L. J. 1971. J. Anim. Sci. *33*, 578.

Kannel, W. B., and Gordon, T. 1971. Section 24: The Framingham diet study; diet and regulation of cholesterol. U.S. Dep. of Health, Education and Welfare, Public Health Service, Natl. Inst. of Health Publication. Bethesda, Md.

Kiernat, B. H., Johnson, J. A., and Siedler, A. J. 1964. Am. Meat Inst. Found. Bull. no. *57*, Chicago.

Koch, D. E., Parr, A. F., and Merkel, R. A. 1967. J. Food Sci. *33*, 1.

Leverton, R. M., and Odell, G. V. 1958. Okla. Agric. Exp. Stn. Misc. Publ. *MP-49*, 90 pp.

Moss, M., Ono, K., Cross, R., Slover, H., Berry, B., Lanza, E., Thompson, R., Wolf, W., Vanderslice, J., Johnson, H., Holden, J. M., and Stewart, K. 1983. Nutrient composition of fresh retail pork. J. Food Sci. *48*, 1767.

National Research Council. 1968. Recommended Dietary Allowances, 7th ed. National Academy of Sciences. publ. *1694*. Washington, D. C.

National Research Council. 1989. Recommended Dietary Allowances, 10th ed. Natl. Acad. Sci., Washington, D. C.

Patterson, R. L. S. 1968. J. Sci. Food Agric. *19*, 31.

Pearson, A. M. 1966. Meats. *In* Introduction to Livestock Production (ed. H. H. Cole), 2nd ed. W. H. Freeman, San Francisco.

Price, J. F., and Schweigert, B. S., editors. 1971. The Science of Meat and Meat Products. 2nd ed. W. H. Freeman, San Francisco.

Rice, E. E., 1971. *In* The Science of Meat and Meat Products (ed. J. F. Price and B. S. Schweigert), 2nd ed. W. H. Freeman, San Francisco.

Schweigert, B. S., and Payne, B. J. 1956. Am. Meat Inst. Found. Bull. No. *30,* Chicago.

USDA. 1983. Composition of Foods, Pork Products: Raw, Processed, Prepared. Agricultural Handbook Nos. 8–10, Consumer Nutrition Division, Human Nutrition Information Service, U.S. Government Printing Office, Washington, D.C. 20402.

Watt, B. K., and Merrill, A. L. 1963. Composition of Foods: Raw, Processed, Prepared. Agric. Handbook No. 8. Consumer and Food Economics Research Division, Agric. Res. Serv., USDA, Washington, D.C.

3

Production Systems: Perspective and Components

Some decades ago, "raising pigs" was primarily a subsidiary farm activity that took place in mud lots (Fig. 3.1), with minimal shelter, to utilize garbage and leftover grains to produce meat for the farm family. Pasture production systems (Fig. 3.2) predominated through the 1950s, with more intensive systems starting in the 1960s. From these origins, "pork production" has evolved to be a primary enterprise on many specialized farms to convert processed feedstuffs and breeding stock resources into marketable pork products. This modern production often occurs in insulated, ventilated facilities with automated feed delivery and manure disposal. This requires labor competent to care for the stock and operate the equipment which is supervised by managers (who may be the owners) with considerable expertise. This trend is likely to continue.

In this chapter, we will identify the major components of pork production systems, including the associated costs and income. This will be done with a view toward Chapters 4 through 9 where the biological principles pertinent to these components are elaborated. Although pork production has unique features, the recognition of the conversion of feedstuffs and breeding stock resources into pork products as being a production system analogous to the manufacture of products in other industries, leads to a "systems approach" for evaluating the merits of alternatives in the production process, as is done in many modern industries. Recognition of how the major components interact with each other and the implications of these interactions upon the pork production system will be addressed in Chapter 10.

FIGURE 3.1 Typical pig production of some decades ago with mud lots, portable shelters, plentiful space, minimal protection from weather and low investment costs.

THE SYSTEMS PERSPECTIVE

To provide a broad definition for analysis of a system, Spedding (1979) suggested: "A system is a group of interacting components, operating together for a common purpose capable of reacting as a whole to external stimuli: it is unaffected directly by its own outputs and has a specified boundary based on the inclusion of all significant feedbacks." To start development of the concept of a pork production

FIGURE 3.2 Typical pasture production system with minimal shelter and low investment costs. Although such systems still occur, especially when weather conditions are not extreme, they are less frequent than in the past.

enterprise as a system, we first need to examine the purpose of pork production systems. Like most production systems (manufacturing systems), the purpose of pork production is to efficiently convert raw materials into finished products. The products of pork production are the wholesale and retail cuts of pork that are offered for sale. The major input components (raw materials) into this system are breeding stock, feedstuffs, facilities, labor, and management, including financial resources. Efficiency of the system reflects the magnitude and quality of output relative to the combined magnitude of all of the inputs. The primary purpose of all production units in the view of the owner or manager is to realize a reasonable profit for the enterprise. Profit considerations lead to expressing efficiency in economic terms. Economic efficiency is achieved when the value (prices received) for the output products reasonably exceeds the total costs (prices paid) for the inputs. The fact that prices paid and received fluctuate over time in response to external influences contributes to the complexity of designing and managing a pork production system. This is true of nearly all production or manufacturing systems.

The precise definition presented in the previous paragraph does not lead to a single precise optimum pork production system. This imprecision results from the following anticipated difficulties of designing a single optimum system:

1. Different starting resources of facilities, capital and labor
2. Prices and values change over time and from one production region to another
3. Climatic and other environmental conditions differ between regions and change over the seasons of the year
4. Considerable interrelationships, interdependencies, interactions and feedbacks occur between system components
5. Knowledge is often incomplete for present conditions, future changes, and the details of complex biological processes

Differences in managerial skills also exist. Thus, this chapter (and the remainder of this book) will not present a precise definition of one ideal production system. The book will describe current practical systems and current trends towards improved systems, elaborate upon the principles, and review the understanding of the biological processes in the growing and reproducing pig that impact upon system performance. This will be done with a view to providing managers and potential managers (students) the basic knowledge for developing efficient pork production systems for their specific resources and situations. The systems perspective presented here will emphasize the coordination and synchronization among systems components that need to be accomplished to achieve full economic efficiency. Even when a near optimum system is planned, the intelligent manager will frequently have to adjust specifics of the system in response to changing prices, unusual circumstances such as disease, availability of new products to enhance efficiency of production, etc. The newly emerging tool of systems analysis through computerized simulation modeling will be described in Chapter 10 because this tool is expected to have a greater role in designing and managing pork production systems in the future.

SYSTEM INPUTS AND EXPENSES

The "specified boundary" of Spedding's (1979) definition of a system will usually be considered as the boundary of a specific farm enterprise producing market swine and will include the biological functioning of both parent breeding stock and their progeny that are grown and finished to be marketed. Sometimes pork production is more specialized, with some producers managing breeding stock herds to produce feeder pigs for marketing to separate producers growing and finishing those feeder pigs for slaughter. In many enterprises, the production of the grain ingredients of the feeds (and, maybe, even the protein sources) is an integral part of the same farm enterprise. For full attention to the yield and quality of the different pork products, the processors' who convert the live animal into pork products might be considered within the "boundary" of the system. Alternately, the influence of those characteristics of the market animals pertinent to the processor's economic efficiency needs to be considered so as to reflect the full impact of the body composition of growth. Ideally, the processors' payment for market animals should adequately compensate the producer for the value of production outputs as inputs to the processing sub-system.

The inputs (external stimuli of Spedding's [1979] definition) to the pork production systems and the expenses associated with those will be discussed in five groupings:

1. Controllable inputs
2. Extraneous influences
3. Environmental modification
4. Animal performance
5. Facility utilization

CONTROLLABLE INPUTS

The characteristics of the input components of the production system will affect the performance of the system. Thus, these characteristics need to be chosen so as to control and modify the performance of the system to the producer's advantage. The major controllable inputs are the breeding system, the breeding stock, the feedstuffs, and the formulated diets.

Breeding Stock and System

The genetic characteristics of the breeding stock may be described as providing both potential and limits upon the performance of the system. The several breeds of swine available to pork producers differ in their genetically controlled performance characteristics, with most breeds being superior in some traits but possibly deficient in others. Most production is currently done with crossbred swine because the performance of crossbreds will exceed purebreds in most traits, especially those associated with reproduction. These crossbred production animals will result from matings of two or more pure breeds. Much of the breed differences in component traits will be manifested in differences in crosses derived from those breeds, along with the heterotic effects of crossbreeding.

However, breeds of swine are never uniform in performance or uniform in the genetic attributes that control that performance. Most breeds have been developed to be uniform for specific color patterns and ear shapes. Uniformity seldom goes beyond those precisely controlled genetic characteristics. Considerable genetic variation within all breeds remains for most characteristics contributing to economic efficiency of the production system. For the breeders, the remaining genetic variation is desirable raw material as they strive to further improve their herds as components of their breed.

The tools available to the modern breeder as he strives to increase the genetic capabilities of his breeding stock will be elaborated in Chapter 6. For now, we will stress the importance for the producer to secure improved breeding stock as a fundamental input into his production system. The producer needs to choose breeds, crosses, breeding herds, and specific animals that will contribute positively to improving the performance of the production system. Better breeding stock is likely to be more expensive; but, if the manager is wise in choices, the value of system outputs and the efficiency of producing them will more than reimburse for the added expenses of superior stock.

Since most production systems involve crossbreeding, the breeding stock necessarily involves two separate, but interacting, components. These are: (1) the parent boars, and (2) the parent gilts and sows. Some breeding systems allow the producer to select replacement gilts from within his herd while other breeding systems usually require the purchase of replacement gilts from breeders. In either approach, the parent females usually will also be crossbred because of the advantages of crossbreds over purebreds for reproduction. Characteristics of primary importance in parent females are those concerned with efficient reproduction and lactation. Of secondary importance in parent females, but still of major concern, are the characteristics which influence the efficiency of the progeny during growing and finishing, such as greater lean growth rate, limited tendency to fatten, and greater efficiency of nutrient utilization. These latter, genetically controlled, characteristics become of greater concern in breeding boars for use as sires of the growing-finishing animals. Other traits of concern are those concerned with reproduction in the male, which may or may not be crossbred. Careful analysis of the contributions of combinations of these traits to the economic efficiency of the producer's specific system will determine his ability to pay for demonstrated genetic superiorities. These fundamental characteristics will be further elaborated on in the later section on animal performance.

Feedstuffs and Diets

The primary consideration in the conversion of raw materials into salable products in livestock production is the conversion of plant materials into animal products (Figs. 3.3 and 3.4). The products of concern are not just the live animals out the chutes of Figure 3.4 but the carcasses of Figure 1.6, the wholesale cuts of Figure 2.2, and the retail cuts of Figure 2.3. The protein and energy contents of pork are

FIGURE 3.3 Feed is the most costly input into modern pork production systems; formulation for efficient conversion is mandatory. Photo: Sands Livestock Systems, Inc., Columbus, Nebraska.

FIGURE 3.4 Primary product output for pork production systems is market hogs shipped and sold to processor for conversion into wholesale and retail cuts of pork. Photo: Sands Livestock Systems, Inc., Columbus, Nebraska.

the primary constituents of its nutritive value to humans. Other important constituents are trace mineral elements and vitamins (see Chapter 2). The marketing of pork products has evolved to the highest prices usually being paid for those cuts with greatest lean pork content. When analyzed chemically, lean pork (as well as other lean meats) is composed primarily of protein (which, in turn, is composed of amino acids) plus water. Other components of the slaughter animal are bone, cartilage, visceral organs, skin, and fat, all of low value, and a broad array of substances used widely in medicinal and pharmaceutical products. At one time, pork fat, in the form of lard, held a primary role as a pork product. However, lard for most uses has been replaced by fats and oils of plant origin. Now, the value of lard derived from the fat trim from the carcass is one-third to one-seventh of the value per unit weight of the lean cuts. The lean cuts contain some fat, but excessive amounts make the cuts less desirable to the consumer. The demands of the consumer have changed pork fat from being a primary product to being a low-valued by-product. When the feed costs for producing it are considered, pork fat becomes a source of inefficiency in the production process. Some body processes, especially at the cellular level require lipids, the primary component of fat. Stored body fat is sometimes physiologically desirable, as it provides energy reserves for situations when feed sources of energy may be deficient, such as in the lactating sow.

Since the pig is a monogastric (simple-stomach) animal, many of the amino acids comprising pork protein must be obtained from the diet or the animal cannot grow. These essential amino acids must be in the daily ration in proportions related to their proportions in pork protein. **Diets**—as used in this book—refer to formulated feed mixtures; **rations** refer to amount of diet provided daily (daily rations). Thus, swine feeds must include better balanced protein sources than the ruminant (complex-stomach) animals, such as cattle and sheep. The feed ingredients (such as soybean meal) that include large percentages of well-balanced protein are likely to be more expensive than the ingredients (such as corn) that may be high in energy but low in protein or low in protein quality (amino acid balance). The nonessential amino acids may be synthesized from other dietary amino acids and nitrogen-containing precursors.

Feed proteins can be used as energy sources for the pig's body processes requiring energy. But, because of the usual higher prices for feed ingredients high in protein content, proteins will usually not be cost-effective in meeting energy needs. Other sources of energy should be included in the diet. The nutritional requirements and nutritional physiology associated with modern pork production are discussed in Chapter 8. The feed resources available for use in formulating balanced diets for all phases of the life cycle of the pig are described in Chapter 9. For the current chapter giving the systems view, the diets are seen as the primary cost-incurring input, but they also provide major controls on the system. Control occurs both by varying the composition of the diet or by limiting the amount of feed consumed (ration) to less than that dictated by voluntary intake (appetite). Within the limits set by the genetic influences on performance, the diets need to be formulated to support near-maximum growth of lean tissue (protein) and associated bone growth (minerals). During the reproduction phase, potential reproduction and lactation needs should also be nutritionally supported. However, nutrition substantially in excess of needs to support lean growth, reproduction, and lactation will not be cost-effective because of usual ingredient prices and because excess nutrition (both protein and nonprotein energy) is likely to be used for wasteful fat production. The conversion of expensive feed ingredients into low-valued fat has become an undesirable part of pork production systems. Procedures to reduce this occurrence are being researched and striven for by the industry.

In summary, the content of ration inputs must be synchronized with the performance needs and potentials under genetic control, so as to maximize lean growth, reproduction, and lactation; but, at the same time, to minimize costs by limiting the expensive ingredients in excess of those needed to support desirable potential performance. Since both excess energy and excess protein tend to be deposited as low-valued fat, alternative controls on feed intake in excess of needs for potential lean growth and reproduction will increasingly be used by the pork producer as he develops his production system. This needs to be done in coordination with breeding stock potentials for high lean growth rate and high reproduction capabilities.

EXTRANEOUS INFLUENCES

In this section we will focus attention on two sets of extraneous factors that also influence the efficiency of pork production systems. These are:

1. Climate and weather
2. Disease-producing pathogens and parasites

These are less controlled "external stimuli" of Spedding's definition of a system and are usually negative in their effects.

Climatic Influences

As will be discussed further in Chapter 5, temperature deviations in either direction from a desirable range have adverse influences upon swine. Extremely cold temperatures, especially for young pigs, can result in hypothermia and death. Less extreme cold temperatures lead to increased metabolism and activity in order for the animal to maintain body temperature. Nutritional support for this has to come from the energy content of the feed intake. To a partial extent, pigs can increase consumption to compensate for these needs. But, the increase in feed used, just to keep warm, represents a cost to the production system for which there is no return.

As the pig grows to larger size, with an external layer of fat serving as insulation, dissipation of excess heat from digestion and other metabolic processes can become a problem with high ambient temperatures. The animal will reduce feed consumption and metabolism to compensate and can increase respiration to a degree to help dissipate the heat load. However, these responses can lead to reduced growth, fertility, implantation, and lactation—depending on the stage of life-cycle for the animal. Besides direct heat loss to surroundings, dissipation of heat comes through evaporation of water, either in the lungs or on the skin, if wet. Thus, a combination of high ambient temperatures with high humidity, reducing the possible evaporative cooling, can have adverse consequences, including death.

Extremes for temperature, in either direction, can thus reduce performance of pigs in various stages to something less than would be possible for the particular genotype on a specified ration. In fact, planning for a production system should include consideration for reduced food intake and use of feed energy to maintain body temperature expected for the location and season of the year as an alternative control on the system when full environmental control is not practical. Additional adjustments may be necessary when weather conditions depart from what is expected.

Disease-producing Pathogens and Parasites

It is beyond the scope of this book to review all the diseases to which swine are subject. The list would include diseases resulting from bacteria, viruses, coccidia, mycoplasma, fungi and parasites. Disease-causing organisms are collectively called pathogens. The basic definition of diseases states that they impair the performance of one or more vital body functions. Thus, with the presence of any pathogen, the possible manifestation of the disease condition will reduce the performance of, at least, some pigs in the production system to a level less than would otherwise be expected for the specified genotype and feeding program. The effects of the disease may be increased mortality, reduced appetite, reduced efficiency of nutrient utilization, reduced growth rate, reduced fertility, reduced embryonic and fetal survival, or reduced lactation. Any of these consequences result in reduced efficiency of the production system. The veterinary profession and the pharmaceutical industry have developed and continue to develop a vast array of vaccines, pharmaceuticals, disinfectants, sanitation programs, and treatments to prevent, control, diagnose, treat, or reduce the impact of most of these conditions. Thus, a necessary part of a pork production system is the combination of eradication, hygiene, vaccination, treatment, or acceptance for each of the pathogens of risk. Although the effects of all diseases are negative and infection by most pathogens is highly risky, most methods of prevention or treatment involve costs to the production system. There are often several alternative avenues to control among which the producer must choose.

A thorough review of swine diseases is in Leman et al. (1986). Fact sheets about several major diseases are included in the Pork Industry Handbook (see Appendix 5a) which can be obtained from state extension offices.

ENVIRONMENTAL MODIFICATION

Although the weather and pathogens of the previous section are not fully controllable, their impact may be partially modified. The evolution in recent decades of pork production from pasture systems to increasing use of facilities for environmental control has occurred for two reasons. They are:

1. To reduce the necessary labor and land input to the production enterprise
2. To reduce the detrimental effects of fluctuating temperature and other weather

conditions at that location upon the feed conversion performance of the swine and allow year-round production

Labor savings largely come from more automated feed delivery, but manure disposal might become more costly. The cost of equipment for automating feed delivery and manure disposal must be less than the labor savings to be cost effective. The housing with insulation, ventilation and, possibly, supplemental heat and cooling for temperature control must pay for itself through lower feed costs, higher growth rate, greater reproduction or lower mortality to be cost effective. Accompanying the trend to more expensive housing and equipment has been a trend to reducing the space and time requirements per animal so that the facility costs can be prorated over more animals during the expected life span of the facilities. Some disease conditions may be lessened with such changes; for example, the parasites of the mud lot or cold-weather-induced respiratory problems. But placing more pigs in close proximity to each other, as is done in most modern facilities, can increase the incidence of contagion between animals for many pathogens. Reducing the time requirement for a growing-finishing animal to reach market weight through increased growth rate can reduce costs per animal.

Temperature and humidity control are important functions of the buildings and associated equipment. With optimum temperature ranges for each age group of animals, the requirements shift from increasing ambient temperatures relative to outside temperatures during cold weather to dissipation of excess body heat during hot weather. For older animals, raising temperature can usually be accomplished by retaining dissipated body heat within the facility. Well-insulated facilities can usually accomplish this with minimum air movement. For younger animals, some supplemental heat is usually needed because they produce less body heat and are quite vulnerable to low temperatures. During hot weather, the emphasis shifts to allowing or creating enough air movement to exhaust the excess body heat and enough shading and insulation to minimize the radiant heat from sunlight, both of which increase the temperature load. Evaporation of moisture can lower temperatures. The pigs do this in their lungs. Their body heat can also evaporate water from the skin, but this has to be provided by sprinklers or drippers, since pigs do not have the capacity to sweat as do some other animals. Air intake through evaporative pads can reduce the temperature of incoming air. The resulting incoming air has higher humidity which might reduce the lung and skin evaporation. A liquid manure handling system might add considerable humidity to the ambient environment.

Situations of dust, odors, and noxious gases, which occur in most swine operations, can increase to become serious problems with controlled environment facilities. Besides being offensive and, possibly, a health hazard to caretakers, these conditions can have detrimental effects upon animal performance. Reduced air movement, especially that needed for maintaining internal temperatures, can also retain dust and odors within the building. Thus, winter conditions lead to compromises between maintaining sufficient air movement to exhaust these undesirable products from the presence of the animals and restricting excessive air intake to allow temperatures to stay above the lower end of the optimum range.

Since most of the dust comes from the feed, the form of the diet and the feed-handling system impact that problem. The grinding of feed leads to more efficient digestion of nutrients, but creates the possibility of dust. Pelleted feed can reduce the dust problem, but the process can be expensive. Also, oils or fats in the ration reduce dust. Because the manure is the source of most of the odors and gases, a good manure-handling system can reduce that potential problem. A common manure-handling procedure in modern controlled environment houses involves fully slotted floors. This allows manure to drop through the slots to liquid or slush pits below with the liquid manure pumped frequently from the pits to external lagoons (Fig. 3.5). Alternatively, aerobic or anaerobic techniques can be used to digest the manure. Aerobic procedures are usually more effective in reducing odors, but also more costly. Noxious odors from swine operations can be offensive to neighbors and need to be controlled. Ground water contamination is another concern.

Of course, manure has potential value. This value primarily comes from its use as a fertilizer, since it contains considerable nitrogen and organic matter and some minerals. There are some attempts at making practical the techniques for using swine manure as a biological input into methane production units. Currently, numerous alternative schemes are in use for disposing of manure, with considerable variation in the advantages and disadvantages of odor control, salvage value, and costs.

Modern facilities have evolved to include separate facilities for the different stages in the life cycle. They might include six different types of buildings corresponding to six stages and functions in a production complex. These are breeding, gestation, farrowing, nursery, growing, and finishing.

This separation of different stages of the life-cycle into separate buildings facilitates differing temperature and ventilation controls, differing arrangement of pens and other equipment, different feeding programs, and specialization of labor duties. Figure 3.6 shows an aerial view of facilities for these different stages in

FIGURE 3.5 Aerated manure lagoon for aerobic digestion; note windowless building in background with exhaust fans on side and evaporative cooling intakes on roof. Photo: Sands Livestock Systems, Inc., Columbus, Nebraska.

modern production farms. Figure 3.7 shows a schematic example of a similar set of facilities. Not all facilities in current use are as sophisticated as in these examples, but the trend is towards this kind of facility as new operations are constructed. At least one large modern facility has a developing unit between growing and finishing. There is variation in the duration of time animals remain in each of these facilities, but a general pattern will be described here and the alternatives of scheduling will be elaborated in Chapter 10. The breeding herd (parent stock) will be located in breeding, gestation, and farrowing buildings with the boars remaining continuously in breeding. Gilts, after development, and sows, after weaning their previous litter, are introduced to the breeding facility (Fig. 3.8) for mating to boars. These gilts or sows remain in this facility until it is reasonably certain that they are pregnant for the next litter. After either a positive pregnancy diagnosis or, at least, the lapse of a reasonable period since last mating without an observed estrus, they are moved to a gestation facility in the same building or in a separate building for most of the remaining gestation period. A few days before expected farrowing, the pregnant sows will be moved to the farrowing facility until the litter is farrowed and suckled until weaning (Fig. 3.9). Usually, creep feed is available to the pigs after they are around two to three weeks old. The specified age or the average weight of pigs in the litter at weaning differs between operations. The longer the pigs suckle the sow, the larger they become and the more easily they adjust to the transition of weaning from nursing to dry feed. However, the farrowing facility is likely to be the most expensive facility, with expensive farrowing stalls and supplemental heat to provide a good environment for the vulnerable young pigs. Older ages of weaning

FIGURE 3.6 Aerial view of confinement facilities for a modern production system. Primary functions of the buildings from left to right are: Breeding-gestation, farrowing, nursery, growing, and finishing, with manure lagoon in background and office and work area attached to breeding-gestation facility. Photo: Sands Livestock Systems, Inc., Columbus, Nebraska.

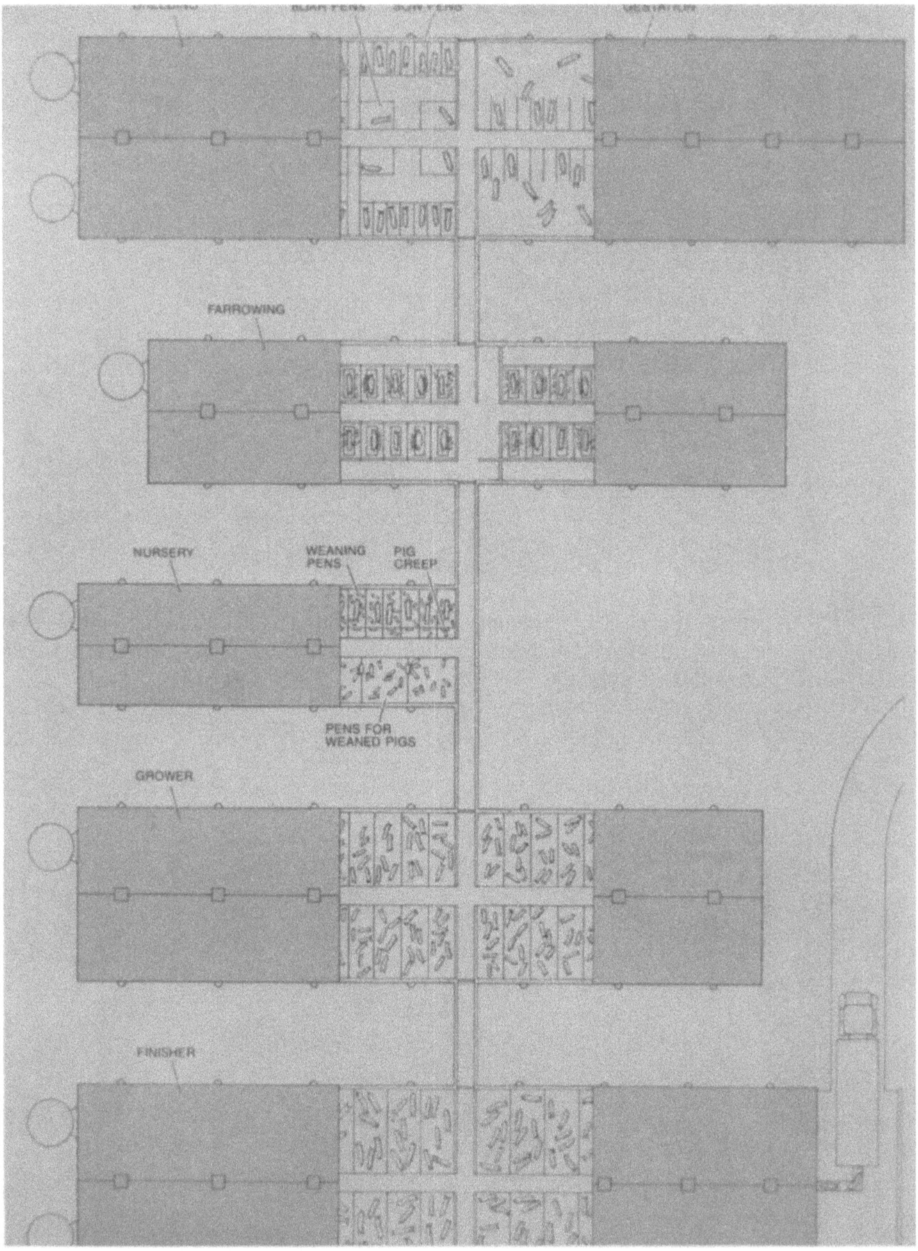

FIGURE 3.7 The total-confinement unit depicted keeps pigs indoors from birth until they are sent to market. The perspective is straight down on a five-building complex, with sections of the roofs removed to show the activity inside. In the farrowing unit, pigs are born and suckled while the sow remains in a metal crate that prevents her from rolling onto her piglets. In the nursery, the grower, and the finisher, the pigs are fed carefully formulated diets designed to bring them to the slaughter weight of 220 lbs or more at the age of 5 or 6 months. Many farms that raise pigs in this way have the grower and finisher units in one building. Figure reproduced from Pond, 1983 (*Scientific American*) with permission of the author and the publisher.

FIGURE 3.8 Breeding area of confinement facility with gilts and sows in individual stalls and boars in close proximity for stimulation and detection of estrus. Photo: Sands Livestock Systems, Inc., Columbus, Nebraska.

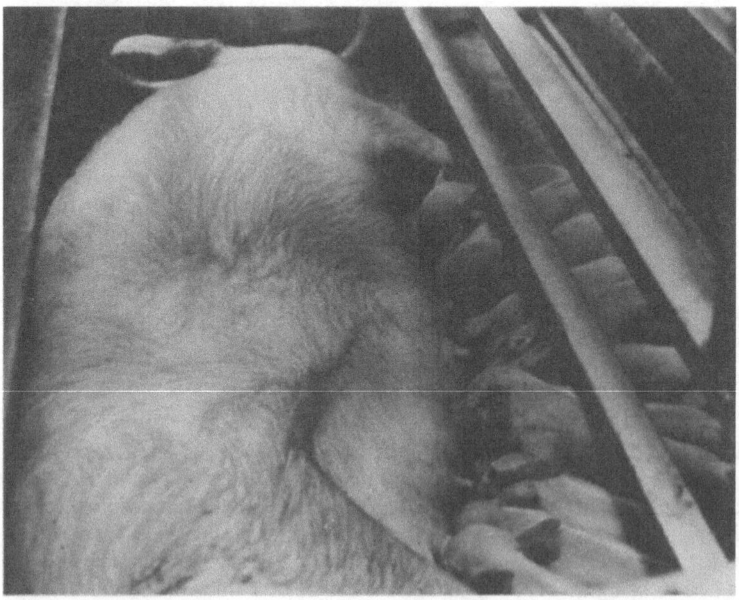

FIGURE 3.9 Farrowing crate with sow and litter in farrowing building; note guard rail to protect pigs but allow nursing of pigs by sow. Photo: Sands Livestock Systems, Inc., Columbus, Nebraska.

reduce the number of litters that can be produced per sow and per stall over a fixed period of time, increasing costs per litter and per pig. Thus, recent trends have been to earlier weaning with greater efforts at speeding the adjustment to prepared diets.

If the breeding system involves saving replacement gilts from the market animals, an area needs to be available for development of these gilts from market age or weight until they are sexually mature and ready to breed for the first time. This might be in the finishing facility, in the breeding facility, or in a separate facility.

Culling of sows from these facilities might be of four types:

1. Culling from breeding if they fail to mate or conceive in a reasonable time
2. Culling from the farrowing facility for lactation failure or after the planned number of parities
3. Culling for illness, as seems appropriate, at any point in the cycle
4. Culling for structural or udder unsoundness

The progeny are moved sequentially as they grow from farrowing to nursery to growing to finishing facilities to market. In the nursery (Fig. 3.10), the pigs complete the transition from nursing to prepared feeds. Because of the sensitivity of these young animals to temperature extremes, especially cold, these facilities usually will have more capacity to provide warm and draft-free conditions. Pigs might stay in such a facility from 7 kg (15 lb) to around 23 kg (50 lb), but this may differ. The growing facility (Fig. 3.11) would likely have less temperature control than in the nursery. They might stay in the facility up until around 55 kg (120 lb). The finishing facility (Fig. 3.12), where they stay until ready to market, usually will involve less temperature control—with the protection from cold conditions coming

FIGURE 3.10 Compartment in nursery building with supplemental heat and exhaust for ventilation. Photo: Sands Livestock Systems, Inc., Columbus, Nebraska.

FIGURE 3.11 Pen in growing building with growing pigs; note slotted floors for manure disposal. Photo: Sands Livestock Systems, Inc., Columbus, Nebraska.

by retaining much of the dissipated body heat. In fact, dissipation of body heat from the older, more metabolically active animals during warmer weather may be a greater concern, requiring greater ventilation capacity.

Separate growing and finishing facilities is optional. Labor requirement for moving is less with a combined facility, but floor space utilization and body heat

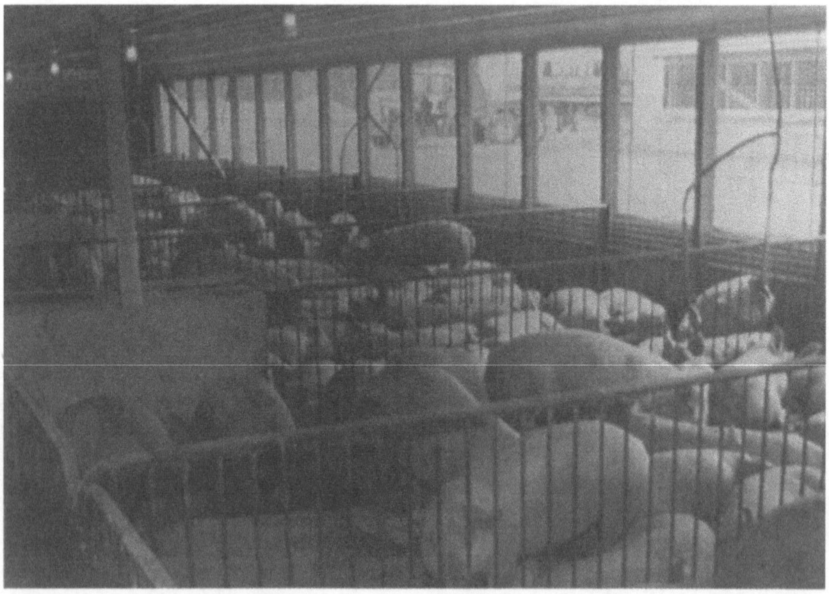

FIGURE 3.12 Finishing building for animals nearing market weight; note open sides for natural summer ventilation. Photo: Sands Livestock Systems, Inc., Columbus, Nebraska.

retention may be less if young, growing animals are allowed all the floor space they will need as they approach market weight.

Often only one diet is available for each building. Thus, the subdivision into several facilities gives some flexibility of nutrition by having diet changes coincide with facility changes. Alternative diets within a building would give even greater flexibility to meeting nutritional needs, but would complicate the feed delivery operation. Sometimes the breeding and gestation functions are combined in one building and, as mentioned above, sometimes the growing and finishing functions are also combined. To facilitate the flow of animals through these facilities over time, the size of these buildings and the pen arrangements within them have to be matched with the expected number of animals and the duration of time in those buildings. To facilitate sanitation measures for disease control, scheduling should include time periods when the buildings are empty for thorough cleaning and disinfection.

Considerably more thorough reviews of the current knowledge of environmental physiology in food animals, including swine, and the management of the animal's environment can be found in Curtis (1981) and Woods (1982). Specific housing recommendations are included in the Pork Industry Handbook (Appendix 5a).

ANIMAL PERFORMANCE

Before focusing on the characteristics of animal performance that contribute to the economic efficiency of the production system, we shall review the general categories of costs associated with the separate inputs to the system. These are:

1. Breeding stock with costs in proportion to the number and expected performance of the breeding stock or their progeny
2. Feedstuffs with costs associated with the amount and nutritional content of the prepared rations and, thus, their influence upon performance
3. Buildings with costs associated with the capacity and the capabilities for modifying the environmental influences, especially the detrimental extremes of temperature and the risk of disease
4. Equipment with costs associated with the capabilities for moving feed, manure, etc., whose costs are expected to relate to the labor saved by this equipment
5. Labor with costs associated with the number and time of employees, but with wages per unit time in proportion to skill level
6. Capital with costs associated with the interest rate for an investment (such as for facilities) for the expected life span of that investment
7. Veterinary costs for disease prevention, control, and/or treatment

A large part of the system efficiency lies in the efficiency of utilizing each of breeding stock, feed, labor, equipment, and facilities. Note that the costs of buildings, equipment, labor, and capital can be prorated for the capacity for number of animals and for the duration of time the facilities will be used. Thus, on a per-animal-capacity basis, these costs are largely functions of time. Thus, costs per animal produced for labor, facilities, and capital are reduced by an increase in "through-put" of animals.

Breeding Herd

The costs for the breeding herd per litter produced will largely be determined by breeding stock costs, feed consumed, and time spent to produce each litter. These costs per litter must be prorated over the number of pigs weaned when determining the costs per animal produced. The costs of producing or purchasing a ready-to-breed gilt are partially compensated by the salvage value when she is sold as a cull sow leaving the breeding herd. The difference between salvage value and cost is prorated over the number of litters produced before being culled. A shorter period of time from farrowing to weaning reduces costs to produce an average weaned pig. Also reducing the time to first conception or the time to rebreeding and conception after weaning reduces feed, labor and facilities costs for the average litter. The objective lies primarily in reducing costs per pig weaned and not just costs per litter, but the value of these weaned pigs are influenced in part by their weaned size and potential for survival and further growth.

Thus, we see several performance characteristics of female parent stock with increased performance potential of each gilt/sow contributing to efficiency of the production system. These are all genetically influenced, but can be influenced by other factors. These are:

1. Early sexual maturity
2. Early capacity to conceive
3. Larger litter size at birth
4. Greater lactation potential and other aspects of pig survival and growth
5. Short interval from weaning to rebreeding and conception
6. Longevity of efficient reproduction and lactation

Larger litter size at birth can be further subdivided into the more fundamental characteristics of:

1. Greater ovulation rate
2. Greater ovum fertilization
3. Greater ovum survival and implantation
4. Greater embryo survival and growth
5. Greater fetal survival and development

These are primarily characteristics of the parent female, but fetal survival and development may also be influenced by the genotype of the fetus. Growth and survival of embryos and fetuses are greatly determined by a complex of factors in the interaction between the sow's uterus and the embryos and fetuses contained in it. Pig survival and development after birth are influenced by both the genetics of the dam and the genetics of the pig. This, in turn, is partially determined by the genetic characteristics of the boar. Other characteristics of the boar which affect system efficiency are:

1. Early sexual maturity
2. Greater libido and aggressiveness
3. Greater sperm volume, activity, etc., affecting fertility of each mating
4. Greater capacity to service females including recovery time between matings

Boar costs per pig produced are greatly influenced by the number of gilts or sows the boar successfully services.

Post-weaning Progeny

Characteristics for the post-weaning progeny are largely determined by the genetic characteristics of both the parents, including heterosis relative to the parents' performance, but with genetic variation among litter mates with the same parents. The prorated costs per animal for the buildings (nursery, growing, and finishing), equipment, labor, and associated capital investment all are reduced by the number of animals utilizing those facilities. The total number of animals that can use a facility in a specified period of time is determined by:

1. The maximum capacity of the building
2. The duration of time an average animal occupies space in the facility
3. The average fraction occupancy of the building over time, including the time a building may be empty between cycles for cleanup and sanitation or because of scheduling

Since movement between facilities often occurs at specified sizes for the animals, faster growth rates can reduce the duration of time spent in the post-weaning facilities and, thus, reduce the prorated costs for the post-weaning period for these animals. But, since increase in fat content of the carcass is of low value and will usually be negative in net value, the desired trait is not just rapid live-weight gains but rapid lean-muscle growth. Rapid lean-muscle growth will be associated with necessary supporting visceral and bone growth, but does not necessarily have to be associated with fat deposition. Excess fat deposition occurs when excess appetite (voluntary feed intake) leads to nutrient intake, digestion, and metabolism in excess of needs for body maintenance and growth of lean muscle tissue and associated bone and viscera.

Thus, the primary traits of concern for the post-weaning growing and finishing period are:

1. Rate of lean muscle and associated tissue growth
2. Appetite (voluntary feed intake) leading to tendency to fatten

Fundamental to lean tissue growth is the rate of protein accretion. The protein content of tissues is metabolically active, with degradation and synthesis continuously occurring. Accretion rate is the difference between the rates of synthesis and degradation. Protein synthesis is controlled by the DNA (deoxyribonucleic acid) genetic material in the chromosomes of the nuclei of the cells. Lean tissue growth occurs by a combination of division of cell nuclei (hyperplasia) and increase in protein and associated water and minerals per nucleus (hypertrophy). Increase in protein accretion must be supported by intake of the appropriate amino acids and metabolizable energy in proper balance. Thus, the genetically controlled potential for protein growth may be limited by inadequate feed intake. The upper limit for intake is determined by appetite, with excess intake leading to excess fatness. Enough appetite to give consumption of nutrients necessary to support rapid lean growth is desirable, but appetite in excess of that can be detrimental to efficient performance unless intake is controlled by other means. Both the potential rate of protein accretion and relative appetite seem largely under genetic control. A trait combining these fundamentals could be termed lean feed efficiency, the ratio of growth in lean tissue to feed intake.

Closely associated with the rate of protein or lean tissue growth is the expected mature protein or lean mass of the animal. Faster growing animals often have larger mature sizes. It is difficult to define a constant mature size or weight of an animal due to variability in fatness due to past history. Protein mass with associated water content and bone mass seems a stable measure of the upper limit for tissue growth. The principles involved in this perspective will be reviewed and extended in Chapters 5 and 8.

The characteristics observed for individual animals in the process of monitoring the production system are usually not as precise and fundamental as relative appetite and rate of protein accretion. Visual observations primarily involve size and indications of relative fatness and muscling. More precise measures would be actual weights or backfat measures of the live animal at specific ages or of the carcass after slaughter. But the reader should realize that even these measures are relatively crude indicators of the fundamental physiological processes underlying these outward manifestations. The suggested traits of protein accretion rate, appetite, and mature protein mass are even oversimplifications of the physiological system for a growing-finishing animal. But, these fundamental traits seem to include a large amount of pertinent fundamental biology for production system considerations.

Since feed costs are the major costs of the production system, especially for the post-weaning period, the efficiency of converting feed to weight gain is an attribute that has received considerable attention both in research and in the monitoring of production systems. This is usually crudely expressed as total gain divided by total feed. There are many underlying contributing factors to this simple ratio. One of the more important is the composition of the gain in terms of fatness. But even if attention is focused on efficiency of lean gain (excluding fat), the composition of the ration, the stage of development between specified ages and weights, and the environmental conditions all can influence this ratio. Adequate intake to support lean growth potential at a young age is of concern. Because both the protein and energy requirements for maintaining body tissues increase as the amount of lean body tissue increases, lean growth early in life will be most efficient.

Efficiency will decline and go to zero as the animal reaches a mature size. Genetic differences in the ability to utilize specific nutrients seem to be slight. The differences in feed efficiency result primarily from the relative utilization of protein and energy consumption for maintenance, lean growth, and fat deposition. A composite trait including the feed for the breeding herd is herd feed efficiency (equals weight of feed consumed divided by weight of pigs sold). This function also includes the impact of reproductive efficiency.

Post-weaning survival also becomes a characteristic of importance to the system. Embryonic, fetal, and pre-weaning survival was stressed in the previous section. When the production herd is healthy, survival after weaning is usually high except for some loss of weak pigs unable to make the transition to consuming prepared feeds. However, disease incidence can cause reductions in survival at most stages of the life cycle. These can be drastic because for a dead animal there is no return for the costs incurred up to that point. Treatment of disease conditions can result in large costs to the system. Sometimes, severe disease conditions can lead to the necessity of depopulating individual facilities with a period of reduced productivity of the enterprise while drastic sanitation measures are imposed. Depending on the nature of disease, repopulation of the whole herd with new breeding stock may be necessary with additional capital outlay. The risk of disease and repopulation has such severe consequences that the trend in the industry is toward stringent measures for disease prevention.

FACILITY UTILIZATION

In the past, whether or not an inexpensive pasture was occupied was a small consideration leading to considerable in-and-out swine production, depending upon the producer's assessment of likely profit from feed availability and predictions of relative feed prices and market hog prices. Hogs, other than for family use, were an alternate way of marketing grain. With the trend to more elaborate, and thus costly, facilities, consideration of facility utilization has become more important. Many of the costs (depreciation, taxes, electricity, interest on investment, and, maybe, labor) continue whether the facility is full, half-full, or empty. When empty, there will be no return to compensate for these costs. Thus, facility utilization or "through-put" becomes a greater and greater concern of producers.

As discussed in the previous section, rapid growth or rapid reproduction of litters can contribute to the efficiency of facility utilization. But without appropriate adjustments in scheduling, more rapid growth or reproduction might only result in facilities being empty for longer periods of time. Therefore, to achieve efficient facility utilization, numbers of animals and the movement of animals through the complex of facilities need to be scheduled in coordination with performance so that, even with rapid turnover, all the facilities are filled to capacity much of the time.

Greatest facility utilization might be achieved if buildings were continuously used and each pen refilled shortly after the previous group of pigs was moved out; thus, the building would never be empty. However, such a plan would be at risk for contagion by pathogens among diverse ages of pigs having pen-to-pen contact and due to limited opportunity for cleanup and sanitation practices. For reduction of mortality, most facilities, especially these involving young pigs more vulnerable to infection from many pathogens, a cleanup and sanitation (possibly including fumigation) period during which the building is empty between cycles is highly desirable, along with separation by physical distance from other facilities.

For each cycle of pigs in each building, there are usually five periods of time. They are:

1. Building is empty as needed for cleanup and sanitation.
2. Building is cleaned but awaiting a new group of pigs.
3. New pigs are being brought into the building.
4. Building is near capacity.
5. Older pigs are being removed from the building.

The manager needs to schedule and coordinate movement between facilities to minimize the time periods of 2, 3, and 5. Some control over the flow of pigs through facilities can be achieved by mating groups of sows in short time-periods so as to fill each unit of the farrowing facilities (building or room), and then have the nursery, growing, and finishing units (buildings or rooms) matched with the expected number of animals from each farrowing group, and keep the period of time in which animals are being moved from one facility to another at a minimum.

PRODUCT(S) OUTPUT FROM SYSTEM

The output from the production system determines the income to offset the costs plus provide the profit for the enterprise. The income is determined by the following three aspects:

1. Number of individual products (animals) marketed
2. Weight of each product
3. Value (price received) per unit of weight for products

In the past, market hogs were usually sold on the basis of a price per pound of live weight. But differences in visual appearance to the buyer and his offered price per unit weight usually did not adequately reimburse the producer for the differences in value to the processor, who received his income from the yield of wholesale or retail cuts. In Chapter 2, the wholesale cuts and some of the retail products were identified. Table 2.13 gives a summary of average market prices per pound for products sold as boxed pork for a three-year period from 1987 through 1989. After slaughtering and processing the animal into these cuts, the processor's income comes from the relative amount of high-value cuts to the low-value cuts such as lard, skin, bones, and internal organs. In general, the cuts with high lean content—ham, loin, and shoulder (picnic, plus butt)—are the most valuable parts of the animal. Exceptions are the bony spareribs and fat bacon. Up to now, recording the actual lean-cut yield of each producer's animals has seemed impractical (but is technologically possible) as a basis for payment-for-value in today's high speed and high volume processing plants.

In recent years in the United States, following earlier trends in other countries, an increasing volume of market swine are sold to the processor with payment based on carcass weight and backfat thickness, with the base price agreed before slaughter (Table 2.12). Payment is discounted below the base price for backfat depth measures greater than 2.54 cm (1 in) when carcass weights are from 76 to 78 kg (167 to 173 lb). Premiums are paid if the backfat measure is less than 2.54 cm in this carcass weight range. These desired carcass weights correspond to 104 to 107 kg (228 to 237 lb) live animal weights. It is indicative of a major trend that thirty years ago the desired slaughter weight for live animals was 93 kg (205 lb). However, the base price level can be obtained for even heavier animals if the backfat measure is sufficiently low.

This manner of payment is based upon the prediction of wholesale cut yield from animals of designated carcass weight and backfat. In this form, payment is on the number and weight of carcasses with the quality considerations reflecting fatness and uniformity of size governing price per unit weight of the carcass. This manner of payment provides greater incentive for the producer to control fatness and to deliver uniform size groupings of animals to the processor.

In summary, the outputs from the pork production system should be controlled through the inputs into the system such that the value (income) of the outputs exceeds the costs of the inputs by a reasonable profit margin. In the ensuing chapters, the knowledge of the biological aspects of swine will be reviewed to provide a basis for understanding the possible controls on the system. In Chapter 10, consideration will be given to the opportunities for a knowledgable manager to integrate this knowledge into sound procedures for a profitable pork production system.

Pork Industry Handbook

The Pork Industry Handbook (PIH) has been developed to provide recommendations to pork producers and includes concise presentations on specific topics prepared by extension swine specialists from the key pork producing states. The sections in this loose-leaf notebook publication correspond to the major aspects of pork production systems. These sections are: Production Systems, Breeding and Genetics, Reproduction, Nutrition, Management, Housing, Waste Management, Herd Health, Marketing, and Pork and Pork Quality. A detailed list of these fact sheets is presented in Appendix 5a. It is the intent of extension workers to update the PIH publications as recommendations change.

Space limitations and the desire to emphasize biological principles and to emphasize the interrelationships between system components has reduced the presentation in this book of some specific aspects and current recommendations for production, husbandry, and management practices. The authors of this book encourage the reader to pursue the specifics of points in which they are especially interested in the appropriate PIH publication. It has been our intent to provide biological principles and the interrelationships in a manner that will be complementary rather than competitive with PIH.

A similar series specifically oriented to swine genetics has been developed by university personnel and breeders of the National Swine Improvement Federation. These publications will extend the view presented in Chapter 6 and are listed in Appendix 5b.

Both of these series are available from state extension offices.

REFERENCES

Cooperative Extension Service. 1976. Pork Industry Handbook.

Curtis, S. E. 1981. Environmental Management. Animal Environment Services, Box 507, Mahomet, Il.

Leman, A. D., Straw, B., Glock, R. D., Mengeling, W. L., Penny, R. H. C. and Scholl, E, eds. 1986. Diseases of Swine, 6th ed. Iowa State University Press, Ames. 930 pgs.

National Swine Improvement Federation. 1987. Swine Genetics Fact Sheets.

Pond, W. G. 1983. Scientific American 248(3), 96–103.

Spedding, C. R. W. 1979. An Introduction to Agricultural Systems. Appl. Sci. Publ., London.

Woods, W. R. ed. 1982. Proc. Symp. Management of Food Producing Animals. Purdue Univ., West Lafayette, Ind.

4

Prenatal Development

Prenatal development is a critical part of the life cycle of the pig, because the events of this period have an influence on subsequent growth and development and on the productive life of the pig. Complicated and fascinating, yet still incompletely understood, cell multiplication and differentiation occur early in prenatal life to form tissues and organs of specialized structure and function.

This discussion includes only the events of embryonic and fetal development considered to be of special developmental significance in swine production. For detailed embryology of the pig, the student should consult Patten (1948) or Marrable (1971). Some of these events are assembled chronologically in Table 4.1 (Patten 1948; Altman and Dittmer 1968). Many details of embryonic and fetal development beyond the scope of our discussion were reviewed in a symposium (Book and Bustad 1974; Brinster 1974; Ulberg and Rampacek 1974), a conference on swine in biomedical research (Tumbleson 1986), and in two books (Dawes 1968; Hafez 1975). Walker (1964) described the dissection of the fetal pig.

FERTILIZATION

Fertilization (the union of the male and female pronuclei) includes the penetration of the surface of the ovum (zona pellucida) by several sperm, but normally entry by only one. This results in a decrease in permeability and an increase in thickness of the ovum surface (Green and Winters 1946; Goldstein 1926; Heuser 1927), so that only one sperm enters the ovum. Cell division begins soon following fertilization. The nutrient reserve of the fertilized ovum is extremely small, so that almost from the start, development depends on drawing nutrients from the surrounding fluid. Fertilization normally occurs in the oviduct a few hours after mating. The fertilized ova pass down the oviduct and into the uterus by the third day after mating. Cell specialization and rearrangement begin by the sixth day.

TABLE 4.1 Chronological Events in Prenatal Development

Event	*Days*
Passes into uterus (4 cells)	2
Morula (16 cells)	3–4
Blastocyst	4.5
Differentiation of germ layers	7–8
Completed primitive streak, attachment to endometrium	11–12
Primitive gut	11–12
First somite differentiation	14–20 (anterior first)
Auditory pit	15–17
Allantoic arteries and veins; umbilical veins	15–17
Arteriosus, truncus arteriosus, dorsal aorta	15–17
Fusion of chorioamniotic fluids	16
Heart beat recognizable	16
Pancreas	16.5–17.5
Lung buds	16.5–17.5
Esophagus, stomach, intestine	16.5–18
Allantois prominent	16–17
Thyroid primordia	17
Amnion complete	17–18
Limb buds recognizable	17–19
Implantation	18
Optic cup and lens	19
Adrenal cortex	20
Olfactory pit	20
Umbilical veins routed through liver	20
Bronchi branching	20–21
Spleen	20–21
Bulbous arteriosus divides into aortic and pulmonary trunks	20–21
Thymus	21–22
Nostrils and eyes differentiated	21–28
Cecum	22
Cross striations in skeletal muscle	22
Parathyroids primordia	22
Allantois replaces all of exocoelom	25–28
Dental lamina	28
Digits differentiated	28
Epiphysis	28
Glomeruli and convoluted tubules in metanephros	28
Hair follicles recognizable	28
Mammary gland primordia	28
Testes differentiated	28
Ovaries differentiated	28
Activity of blood cell forming centers	30
Heart myoblasts become spindle shaped	30
Elongation of smooth muscle nuclei in longitudinal muscle of esophagus and stomach	32.5
Mandible ossifies	32.5
Adrenal medulla	34.5

TABLE 4.1 (*Continued*)

Event	Days
Coils of large intestine	34.5
Humerus, femur and ribs ossify	34.5
Nasal chamber, nasal septum complete	34.5
Cerebellum	34–35
Skeletal muscle bundles	35
Vertebrae ossify	36
Liver glycogen storage	40
Secretion of epinephrine	40
Scrotum in male, clitoris in female	44
Serum protein synthesized in appreciable amount	50
Secretion of pituitary hormones	50
Secretion of thyroxin	50
Seminal vesicles, prostate, Cowper's glands in male; oviducts, uteri and vagina in female	51
Spinoid processes of vertebrae ossify	63
Fibrinogen synthesis	90
Testes enter inguinal canal in male	95
Birth	114

Source: Patten 1948; Altman and Dittmer 1968.

IMPLANTATION

Eleven-day embryos begin to show signs of attachment to the endometrium. The uterine lining shows a slight indentation at the site of attachment. However, true implantation of the blastocoele to form the placenta does not occur until about day 18 (embryo length of 5–10 mm). The three germ layers (the ectoderm, mesoderm, and entoderm) are by now clearly formed and cell specialization continues. From the ectoderm arise the outer epithelium of the body, the mammary, sebaceous, and sweat glands, the hair and hoofs, intestinal epithelium, teeth enamel, and all of the nervous system. From the entoderm arise the digestive tract (except epithelial covering), pancreas, liver, thyroid, pharynx, trachea, lungs, middle ear, and parathyroids. From the dorsal mesoderm arise the skeleton, skeletal muscle, and connective tissue; from the lateral mesoderm the blood vessels, heart, smooth muscle, connective tissue of blood vessels, blood cells, cortex of the adrenal gland, visceral pleura and peritoneum, and the pericardium; and from the intermediate mesoderm the primary reproductive organs and accessory glands and the kidney tubules and ureters.

By the 20th day the embryo has reached a crown–rump length of about 10 mm and can be successfully used for gross laboratory study. Fetuses at this and subsequent stages of development are diagrammed and photographed in Figs. 4.1 and 4.2.

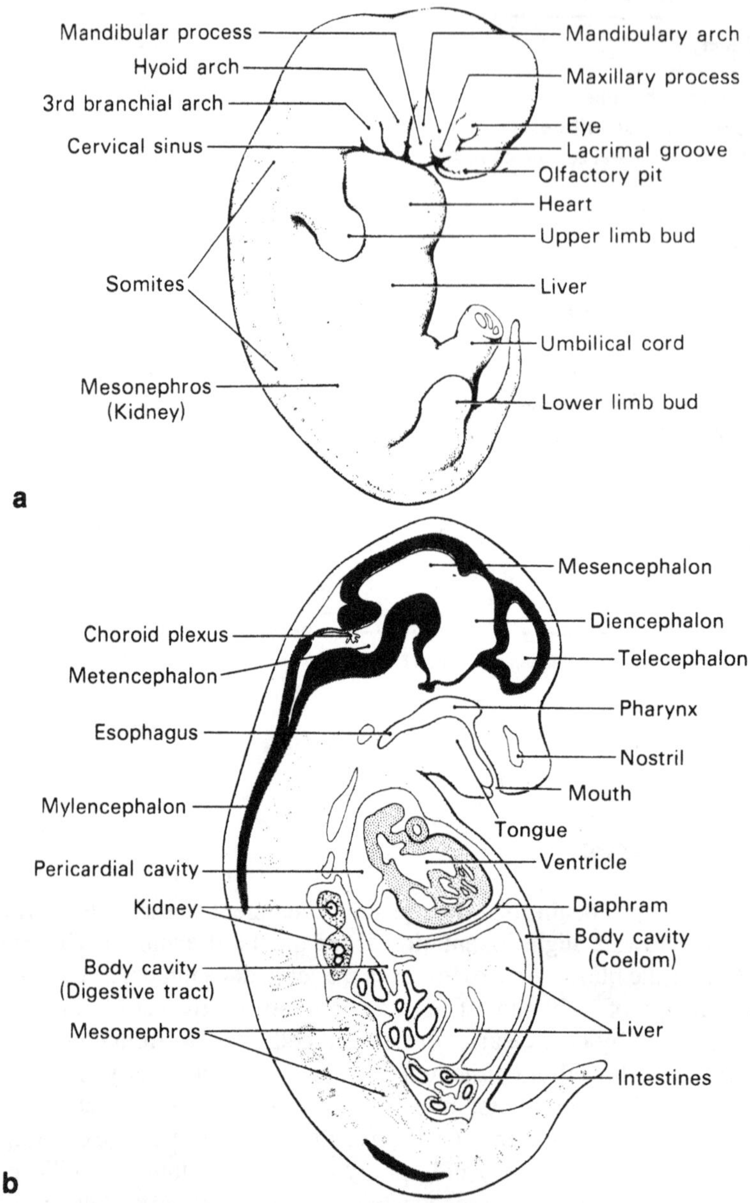

FIGURE 4.1 Stages of development of pig embryo; (a) 10–12 mm embryo (12–15 days); (b) 35 mm embryo (20–25 days). From Harrison 1971.

DIFFERENTIATION OF ORGANS

Many of the major organs can be seen by day 20. These include the heart, pharynx, trachea, esophagus, stomach, lung, liver, eye socket, nasal pit, and anterior and posterior appendage buds. The elongation of the intestines has just begun at this stage. The digestive tube protrudes into the belly stalk and communicates with the

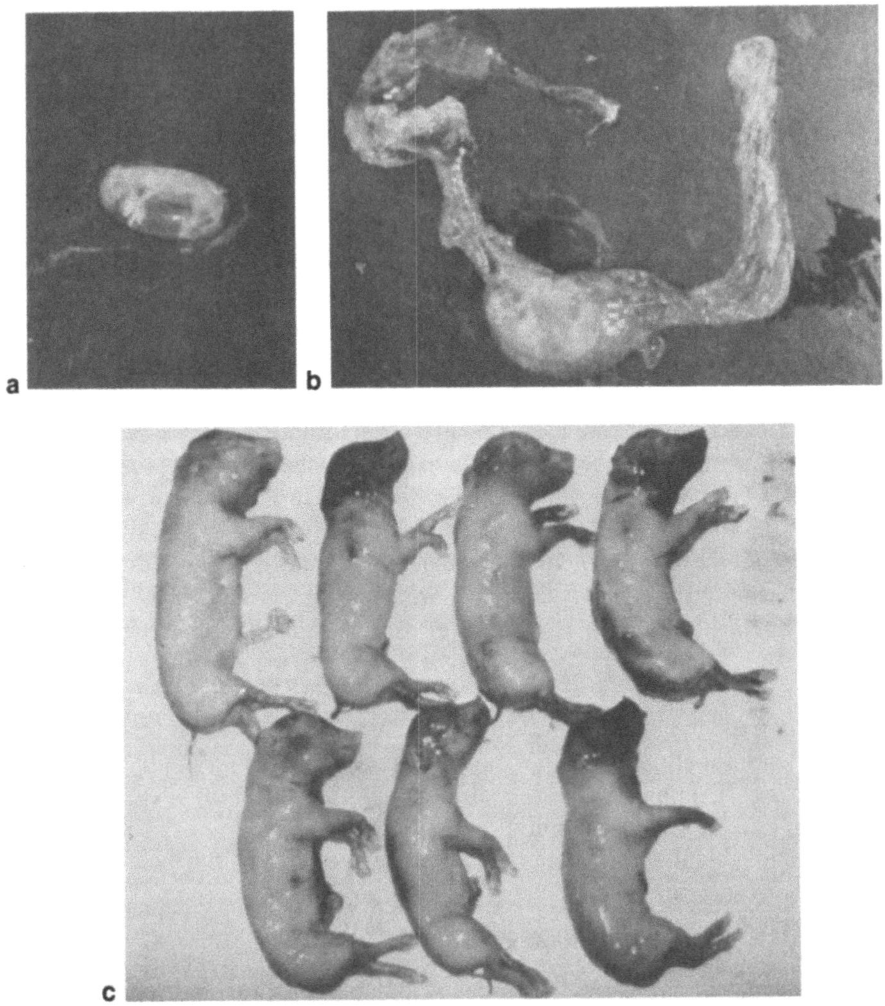

FIGURE 4.2 Development of the fetal pig: (a) 30-day fetus in amnion; (b) 35-day fetus in placenta; (c) 80-day fetuses, where the general appearance is similar to newborn pigs.

yolk stalk which is now greatly enlarged and largely replaces the yolk sac. The caudal end of the digestive tube is the cloaca which ruptures at about this stage, providing the opening to the outside.

The body cavity has not yet divided into the pericardial, pleural, and peritoneal cavities, but there is by now an incomplete division between the liver and heart, the septum transversum. Later this septum fuses with the pleuroperitoneal folds to complete the diaphragm.

The urinary system at this stage is composed of a large network of tubules, the mesonephros, and represents a great mass of tissue in relation to total embryo size, making it a conspicuous organ.

The brain has already reached the five-divisions stage to include the myelencephalon, metencephalon, mesencephalon, diencephalon, and telencephalon as distinguishable entities and the 12 cranial nerves have already developed.

DEVELOPMENT OF THE CIRCULATORY SYSTEM AND BLOOD CELLS

The embryonic development of the pig heart is typical of that of most mammals, but certain features are of special interest in swine production. At day 20, the heart is a tubular organ pumping blood from its posterior and out through its anterior end into a pair of aortas which extend through the length of the developing embryo. The blood is eventually returned to the heart via the common cardinal veins. The single tube eventually develops into the typical four-compartment mammalian heart. However, it is not until later that each compartment is separated by complete septa. Openings remain for some time between the two atria and the two ventricles.

The one between the two atria is the foramen primarum. As it approaches closure, a second opening appears, the interatrial foramen secundum. The septum between the two atria is soon complete except for the foramen secundum. This single opening remains until the time of birth as the foramen ovale, covered by a flap-like tissue to form a valve system between the atria.

Blood from the vena cava flows through the foramen ovale into the left atrium. Meanwhile, the left and right ventricles are separated gradually by the eventual fusing of the ridges along the truncus arteriosus to separate the aortic and pulmonary circulation completely. Along with the increase in the size and capacity of the lungs, as the fetus develops, the blood leaving the right ventricle passes more and more through the lungs and less through the ductus arteriosus. It finally becomes completely nonfunctional and the lungs are able to handle all the blood from the right ventricle. The lungs remain nonfunctional until birth. Oxygenated blood is introduced into the fetal circulation across the placenta.

The umbilical (allantoic) veins become rerouted through the liver at about day 20 and this is accompanied by the rapid enlargement of the liver. The umbilical veins become fused together just before entering the body so that the left umbilical vein takes over the transport of placental blood to the liver. As the vessels through the liver develop, almost all the placental blood soon passes through the liver. Thus, the liver has become the common organ through which all of the extra embryonic circulation and most of the intraembryonic circulation must pass.

The first embryonic blood cells are produced in the yolk sac. However, by the 30th day there is considerable blood-cell formation, including erythrocytes (red blood cells, RBC), within the developing fetus. Later, the kidney and liver take over followed by the lymphoid organs and finally the bone marrow. The changes occurring in blood cell population during prenatal life are summarized in Table 4.2 (Waddill et al. 1962; Wintrobe and Shumacker 1935, 1936).

By day 51, the number of RBC has increased fourfold to about two million per cubic millimeter with a corresponding hemoglobin of about 8 g/100 ml and a hematocrit of about 24. The number of erythrocytes and the hematocrit and hemoglobin concentration continue to increase throughout the remainder of the fetal period (Dittmer and Grebe 1959).

At 30 days most erythrocytes are nucleated, but later the proportion rapidly diminishes and at term almost all are mature cells. Male and female fetuses are similar in RBC morphology and in blood hemoglobin concentration.

The total white blood cell (leukocyte) concentration increases throughout the

TABLE 4.2 Changes in Blood Cell Population and Serum Protein Concentrations during Fetal Life

Blood constituent	*Day after conception*				
	30	*51*	*72*	*93*	*Newborn*
Red blood cells, million/cu mm	0.6	2.1	3.4	4.0	5.6
Hemoglobin, g/100 ml	6.4	8.7	8.6	—	11.7
Hematocrit	20	24	30	32	38
Mean corpuscular volume, cu micra	—	117	91	80	67
Mean corpuscular hemoglobin, picograms	—	31	26	22	21
Leukocytes per cu mm	—	1,430	2,230	2,770	6,270
Total serum protein, g/100 ml	—	2.8	2.3	2.5	2.9
Albumin, % of total protein	—	23	22	19	21
Alpha globulin, % of total protein	—	34	40	50	48
Beta globulin, % of total protein	—	35	27	23	20
Gamma globulin + fibrinogen, % of total protein	—	9	10	9	10

Source: Waddill et al. 1962; Wintrobe and Shumacker 1935, 1936.

embryonic period. At term, males appear to have higher counts than females, and purebreds higher than crossbreds.

At 50 days the total serum protein level is just under 3 g/100 ml consisting of about one-third each of α- and γ-globulins, 9–10% β-globulin, and 20–25% albumin. The concentration of total serum protein then declines somewhat, followed by a gradual rise to term. There are apparently no fetal sex or breed differences either in total serum protein concentration or in proportions of the fractions. A comparison of serum protein values at term with values for older pigs demonstrates the relative physiological immaturity in this respect in newborn pigs. The α- and γ-globulin fractions are bolstered largely by ingestion of colostral proteins shortly after birth. The synthesis of serum proteins by the young pig is not maximum until several weeks after birth.

PLACENTA AND FETAL MEMBRANES (Fig. 4.3)

The placenta of the pig is termed *epitheliochorial.* The extra-embryonic membranes can be separated from the uterus with no damage to or destruction of the uterine epithelium and, unlike in other species, the blood must pass through the chorion epithelium to reach the fetus. The following six tissues separate the maternal and fetal blood in the pig: maternal tissues composed of (1) uterine endothelium, (2) uterine connective tissue, and (3) uterine epithelium, and the fetal chorion composed of the (1) trophoblast, (2) connective tissue layer, and (3) endothelium.

The transfer of large organic molecules across the placenta is limited in the pig so that prenatal passive immunity, provided by maternal blood supply, is almost

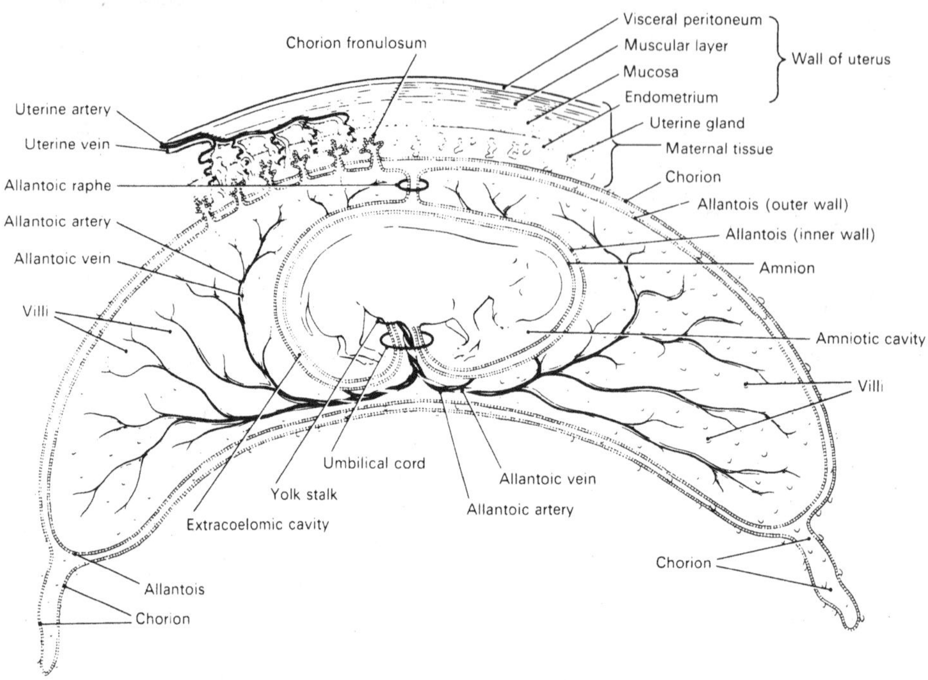

FIGURE 4.3 Semidiagrammatic view of fetal membranes and placenta. From Harrison 1971.

completely unavailable to the fetus (Hemmings and Brambell 1961). The blood of the newborn pig is therefore almost devoid of immune globulins (gamma globulins).

Each embryo has separate covering membranes, except for identical twins. The incidence of identical twins in swine is not known, but is probably quite low. Thus, after implantation, each fetus lives within its own protective membranes and depends on the maternal blood circulation flowing across the six-layered epitheliochorial placenta for its growth and development.

Several membranes are associated with the developing pig fetus. These extra-embryonic membranes are (1) the yolk sac, (2) the amnion, (3) the serosa, and (4) the allantois.

1. **Yolk sac.** It serves mainly in very early embryonic life by carrying nutrients to the embryo. The surface of the yolk sac comes in contact with the uterine epithelium, making transfer of nutrients and oxygen from the uterus to the embryo very efficient. The thick network of blood vessels in the yolk sac which carry these nutrients to the developing embryo is termed the vitelline circulation. This function of the yolk sac disappears as the allantois develops and the yolk sac eventually becomes a shriveled vestige.
2. **Amnion.** This is a single-layered membrane that surrounds the embryo and when completed is filled with fluid which engulfs the fetus and protects it against mechanical injury and adhesions between the membrane and the body surface. The amnion first begins to form at about one week, and by the 17th or 18th day the amnion is complete and the fetus is completely surrounded with amniotic fluid. It is attached where the body opens ventrally in the region of the yolk

stalk, and the opening eventually becomes the umbilical ring, through which nutrients pass via the umbilical artery and umbilical vein.

3. **Serosa (trophoderm or somatopleure).** This membrane has several names, depending on what stage of embryonic development is involved, but the term serosa is used here. The serosa encloses the other extra-embryonic cavities with their respective surrounding membranes and, together with the splanchnopleure of the allantois, represents the chorion, the fetal part of the placenta.
4. **Allantois.** This is a single-layered membrane that arises from the hind gut. It rapidly enlarges and dilates, except at the point where it leaves the body cavity (belly-stalk) where it becomes a constricted connection with the hind gut. The allantois has an abundant blood supply and its rapid growth and vascularity results in a rapid increase in its relative importance as a vehicle for supplying nutrients and oxygen to the fetus. It supplants the yolk sac in this function and eventually occupies an enormous space within the cavity of the serosa and becomes fused with it to form the chorion.

The chorion furnishes a double-layered contact directly with the uterine epithelium and provides the means of nutrient exchange from dam to fetus and excretory exchange from fetus to dam.

The development of the fetus is accompanied, as an important counterpart, by the growth and development of the surrounding membranes and changes in the uterine epithelium. This increase in size and mass of the membranes involves storage of nutrients obtained from maternal circulation, so the net positive balance of nutrient intake and outgo of the pregnant sow represents that amount required for embryonic growth plus that required for the membranes and for the uterus itself. The relative increase in mass of the membranes and of the fetus during the first few weeks of development heavily favors the former. Most of the growth of fetal membranes is completed by day 70 of gestation (Warnick 1928; Pomeroy 1960) while the fetus has its greatest increase in weight after this time. These relative changes in mass of the fetal membranes, fluids, and empty uterus are depicted in Table 4.3 (Moustgaard 1962). Changes in length and weight of the fetus to term are given in Table 4.4 (Ullrey et al. 1965).

TABLE 4.3 Relative Weights of Extra-embryonic Membranes, Their Fluids, and the Uterus at Various Stages of Fetal Development

Days after conception	*No. fetuses*	*Membranes*		*Fluids*		*Empty uterus*	
		g	*% of 47 day value*	*g*	*% of 47 day value*	*g*	*% of nonpregnant uterus*
47	12	800		1,350		1,300	
63	11	2,100	263	5,050	374	2,450	189
66	10	1,570		4,220		2,450	
81	12	2,550	319	5,650	419	2,600	200
86	11	2,150		2,800		2,550	
96	10	2,500	313	2,250	207	3,411	265
102	10	2,500		1,250		3,200	
108	9	2,500	313	1,890	140	3,770	290

Source: Absolute weights of membranes, fluids and empty uterus taken from Moustgaard 1962.

TABLE 4.4 Length and Weight of the Pig Fetus at Various Ages

Days after conception	*Length*[a]		*Weight*	
	cm	*Multiple of 30-day length*	*g*	*Multiple of 30-day wt*
30	2.5 ± 0.3	—	1.5 ± 0.05	—
51	9.8 ± 1.0	3.9	49.8 ± 1.4	33.2
72	16.3 ± 2.0	6.5	220.5 ± 7.3	147.0
93	22.9 ± 2.0	9.2	616.9 ± 15.0	411.3
114	29.4 ± 8.6	11.8	1,040.9 ± 42.7[b]	693.9

[a] Thirty days measured from crown to rump; 51 days and older measured from most anterior portion of frontal crest to anus.

[b] This mean was obtained from first litter dams and represents a mean approximately 100 g less than expected for litters from mature dams.

Source: Ullrey et al. 1965.

The maximum volume of the amniotic and allantoic fluids is highly variable but may reach several hundred milliliters for each fetus at about 65–80 days followed by a constant decline to term. The concentration of cations and anions decreases whereas that of urea and creatinine increases with time. The allantoic fluid is produced first by secretion of the allantoic membranes but later by metabolic products of the fetus as well (McCance and Dickerson 1957).

Compiled data (McCance and Dickerson 1957; Economou-Mavrou and McCance 1958; McCance and Widdowson 1960) on composition of allantoic and amniotic fluid are given in Table 4.5.

TABLE 4.5 Composition of Amniotic and Allantoic Fluids at Various Fetal Ages (per Liter of Fluid)

Constituents	*Age*						
	20 days	*41–46 days*		*55–65 days*		*90 days*	
	Allantoic	*Amniotic*	*Allantoic*	*Amniotic*	*Allantoic*	*Amniotic*	*Allantoic*
No. of samples	32	13	16	5	5	—	—
Volume, ml	5	26	110	60	350	—	—
Total osmolar conc. mmoles	250	290	120	300	90	270	240
Urea, mmoles	3	4	8	5	10	—	—
Creatinine, mg	12	8	37	3	—	—	—
Inorganic phosphorus, mmoles	2	1	2	2	1	—	1
Chlorine, meq	70	104	30	90	20	8	75
Sodium, meq	115	125	13	120	14	110	4
Potassium, meq	14	19	8	9	6	8	8
Magnesium, meq	1.5	1.5	1.0	—	0.5	—	5.0
Calcium, meq	3	5	19	—	15	—	35

Sources: McCance and Dickerson 1957; McCance and Widdowson 1960; Economou-Mavrou and McCance 1958.

The uterus itself also increases in weight throughout pregnancy. Its empty weight increases from approximately 0.5–1.5 kg in the nonpregnant female to 3.0–4.0 kg at term, depending on the number and size of the fetuses. The ratio of placental weight to fetal weight at parturition is about 1 : 10.

GROWTH OF THE FETUS

The number of developing fetuses is inversely related to the ultimate size of individual fetuses at term. The size of individual fetuses is directly related to the size of the placental (chorionic) membranes associated with them. This is probably related to the degree of the vascular connection to the fetus and in turn to the amounts of nutrients available for deposition in the fetus. However, the greater weight of fetuses at the ends of the uterine horn than in the middle and at the ovarian end than at the cervical end cannot be explained on the basis of vascular architecture of the uterus (Perry and Rowell 1969).

Fetal growth is under control of hormones and growth factors, but the exact role of each of these factors is not completely understood in the pig. For example, growth hormone, which is essential for postnatal growth, appears to have no direct role in fetal growth; level of growth hormone circulating in blood of the pig fetus is high but its level is not associated with rate of fetal growth. The importance of insulin, insulin-like growth factors (somatomedins), and other endocrine products in fetal growth remain uncertain (Spencer 1986), but the known effects of protein and energy restriction of the sow during gestation on fetal growth and birth weight (Pond et al. 1987) may operate through changes in the maternal and fetal endocrine systems.

The size of the developing fetus increases tremendously during the final half of the prenatal period, when the bulk of nutrients are stored. Changes in the length and weight of the fetus and in weight of the placenta during the period from conception to birth are shown in Fig. 4.4 (Mountgaard 1962). The *relative* rate of increase in size is greatest in the very young fetus and declines gradually in late prenatal and in postnatal life, even though the greatest daily *absolute* increase in size occurs during late prenatal life and in postnatal life. The comparative rate of growth of various organs during prenatal life is summarized in Table 4.6 (Pomeroy 1960; Ullrey et al. 1965; Lowrey 1911–1912).

Viable and strong young have been known to weigh as little as 400 g (14 oz) and more than 2000 g (70 oz) at birth after a normal gestation period. The size of individual pigs varies inversely with the number of pigs in the litter, and directly with the size of their extra-embryonic membranes. The amounts of the major nutrients (protein, energy, and minerals) deposited in the fetus increase by just more than 3% per day, so that the absolute amounts of each are extremely large during the terminal stages.

The gross chemical composition (water, lipid, protein, ash) of the pig fetus at various stages of development is shown in Table 4.7 (Salmon-Legagneur 1965; Curtis et al. 1967). Changes in water, nitrogen, and major mineral concentration of the fetus from day 51 to term are shown in Table 4.8 (Pomeroy 1960). It is of interest to note the wide variation in chemical composition among newborn pigs. Yet, considering fetuses within litters, very little variation was encountered in percent-

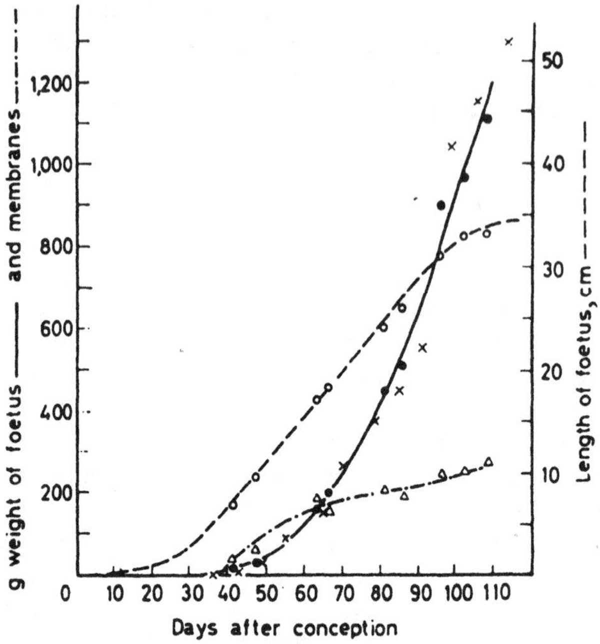

FIGURE 4.4 Length (from end of snout to base of tail) and weight of foetuses and weight of membranes during pregnancy. X indicate American results, ●, ○, and △ indicate Danish results.

TABLE 4.6 Relative Weights of Various Organs at Various Stages of Development

Organ	Organ wt (% of body weight)							
	Prenatal Crown-rump length, mm						Postnatal	
	15[a]	18	25	58	84	97	Newborn	Adult
Brain	8	9	7	7	5	—	4	0.09
Spinal cord	—	1.9	1.3	0.8	0.4	—	0.33	0.04
Eyeballs	—	0.2	0.4	0.9	1.15	—	0.4	0.01
Heart	4.6	3.5	—	1.4	—	0.9	1.0	0.32
Lungs[b]	—	—	0.6	2.9	3.5	—	2.1	0.7
Liver	—	12	16	15	10	3.2	3.1	1.4
Kidneys	—	—	0.3	2.6	2.4	1.1	1.0	0.25
Spleen	—	—	—	0.04	0.05	0.17	0.17	0.13
Pancreas	—	—	—	0.11	0.10	—	0.16	0.14
Thymus	—	—	—	0.1	0.1	—	0.37	0.0001
Adrenals	—	—	—	0.13	0.07	—	0.02	0.005
Stomach + intestine	—	0.3	1.2	2.9	2.8	—	5.9	6.8
Thyroid	—	—	—	0.045	0.035	—	0.025	0.005

[a] Approximately 24 days old.
[b] Right is larger than left by ratio of 1.4:1.
Sources: Pomeroy 1960; Lowrey 1911; Ullrey et al. 1965.

TABLE 4.7 Gross Chemical Composition of the Pig Fetus at Various Ages

Fetal age (days)	*Water*	*Percent of fresh wt*		
		Lipid	*Protein*	*Ash*
30	94.7	0.5	3.6	0.9
60	89.5	0.9	6.2	1.9
100	85.3	1.3	9.1	3.1
107	83.6	1.4	9.7	3.2
Term	80.2(73.5–83.8)	0.6(0.2–1.0)	10.9(5.4–13.0)	4.0(3.2–5.3)

Sources: Salmon-Legagneur 1965 (values for fetal pigs); Curtis et al. 1967 (values for 72 newborn pigs from 12 litters, with ranges in parentheses).

age of mineral composition between the largest and smallest pig (Nielsen 1975). Gestation length apparently has no effect on birth weight, body composition of the newborn pig or thermostability at birth (Curtis et al. 1969).

ADAPTATIONS AT BIRTH

The sudden need for the lung to begin functioning at birth is met by the presence of the foramen ovale and the ductus arteriosus as they remain anatomically open at birth. The anatomical closure of the ductus arteriosus is not complete until 6–8 weeks after birth, although it becomes functionally closed at birth. The foramen ovale also ceases to function within a few days after birth, though it may not actually close until much later.

When the umbilical cord is severed at birth, the umbilical vein and artery involute as they have no further function. There is a great danger of hemorrhaging

TABLE 4.8 Mineral Composition of the Pig Fetus at Various Ages

Constituents	*Fetal age (days)*[a]				
	51	*74*	*97*	*108*	*113*
Body weight, g	55.6	234	823	937	1,127
Water, %	89	89	84	81	81
Total nitrogen, %	1.15	1.07	1.43	1.64	1.84
Calcium, mg/100 g	218	440	800	854	1,070
Magnesium, mg/100 g	14.5	13.5	22.7	27.6	27.6
Phosphorus, mg/100 g	209	297	490	540	628
Sodium, mg/100 g	202	206	202	176	214
Potassium, mg/100 g	185	158	191	214	191
Nitrogen–potassium ratio	6.2	6.8	7.5	7.7	9.6

[a] In fetuses ranging from largest to smallest in litter, very little difference was observed in percent composition of each component in any stage.

Source: Pomeroy 1960.

through the umbilical vessels at birth. It is important to make sure that blood flow through them has ceased. This can be done by pinching the cord with blunt scissors or by tying a knot in the cord close to the umbilicus.

Several hours often elapse between the birth of the first and the last pig. This long delay provides the possibility of suffocation should the umbilical cord be severed sometime before birth. The injection of oxytocin into the sow early in parturition in an effort to hasten parturition has given variable results. Most swine producers who have used oxytocin for this purpose can cite success in individual cases. Apparently, time and amount injected are important.

Breed differences exist in the incidence of stillbirths. The frequency of stillbirths may range from less than 3% to more than 8% among different breeds handled under similar conditions. Stillborn pigs are smaller, on the average, than live littermates at birth, so that probably other causative factors in addition to suffocation are involved. It seems possible that poor adaptation of the blood flow through the heart and lungs associated with closure of the foramen ovale and ductus arteriosus at birth are responsible for some stillborn pigs, but this has not been established. Certainly one of the opportunities for improved efficiency in pig production lies in reducing this high frequency of stillbirths among apparently normal pigs. Experience of veterinary practitioners with hysterectomy-derived pigs suggest that virtually all pigs except mummified fetuses are alive at 112 days.

REFERENCES

Altman, P. L., and Dittmer, D. S. 1968. Metabolism. Fed. Am. Soc. of Exper. Biol., Bethesda. Md.

Book, S. A., and Bustad, L. K. 1974. The fetal and neonatal pig in biomedical research. J. Anim. Sci. *38*, 997.

Brinster, R. L. 1974. Embryo development. J. Anim. Sci. *38*, 1003.

Curtis, S. E., Heidenreich, C. J., and Martin, T. G. 1967. J. Anim. Sci. *26*, 749.

Curtis, S. E., Rogler, J. C., and Martin, T. G. 1969. J. Anim. Sci. *29*, 335.

Dawes, G. S. 1968. Foetal and Neonatal Physiology. Yearbook Medical Publishers, Chicago.

Dittmer, D. S., and Grebe, R. M. 1959. Handbook of Circulation. W. B. Saunders, Philadelphia.

Economou-Mavrou, C., and McCance, R. A. 1958. Biochem. J. *68*, 573.

Gilbert, S. G. 1966. Pictorial Anatomy of the Fetal Pig, 2nd edition. University of Washington Press, Seattle.

Goldstein, S. R. 1926. Anat. Rec. *34*, 25.

Green, W. W., and Winters, L. M. 1946. J. Morphol. *78*, 305.

Hafez, E. S. E. 1975. The Mammalian Fetus. Thomas Publishers, Springfield, Ill.

Harrison, B. M. 1971. Embryology of the Chick and Pig. Wm. C. Brown, Dubuque, Iowa.

Hemmings, W. A., and Brambell, F. W. R. 1961. Br. Med. Bull. *17*, 96.

Heuser, C. H. 1927. Carnegie Inst. Contrib. to Embryol. *19*, 229.

Lowrey, L. G. 1911–1912. Am. J. Anat. *12*, 107.

Marrable, A. W. 1971. The Embryonic Pig: A Chronological Account. Pitman, N.Y.

McCance, R. A., and Dickerson, J. W. T. 1957. J. Embryol. Exp. Morphol. *5*, 43.

McCance, R. A., and Widdowson, E. M. 1960. J. Physiol. *151*, 484.

Moustgaard, J. 1962. *In* Nutrition of Pigs and Poultry (ed., J. T. Morgan and D. Lewis), p. 191. Butterworth's, London.

Nielsen, H. E. 1975. *In* Festskrift til Hjalmar Clausen (ed., H. Staun) p. 165. Danske Landhusholdingsseiskab, Copenhagen, Denmark.

Patten, B. M. 1948. Embryology of the Pig, 3rd ed. McGraw-Hill, New York.

Perry, J. S., and Rowell, J. G. 1969. J. Reprod. Fert. *19*, 527.

Pomeroy, R. W. 1960. J. Agric. Sci. *54*, 31.

Pond, W. G., Yen, J. T., and Mersmann, H. J. 1987. Growth *51*, 355.

Salmon-Legagneur, E. 1965. Chemical composition of the pig foetus. Ph.D. Thesis, Faculty of Science, University of Paris.

Spencer, G. S. G. 1986. *In* Swine in Biomedical Research (ed., M. E. Tumbleson), p. 1205. Plenum Press, N. Y.

Tumbleson, M. E. (ed.) 1986. Swine in Biomedical Research. Plenum Press, N. Y.

Ulberg, L. C., and Rampacek, G. B. 1974. Embryonic and fetal development: uterine components and influence. J. Anim. Sci. *38*, 1013.

Ullrey, D. E., Sprague, J. I., Becker, D. E., and Miller, E. R. 1965. J. Anim. Sci. *24*, 711.

Waddill, D. G., Ullrey, D. E., Miller, E. R., Sprague, J. I., Alexander, E. A., and Hoefer, J. A. 1962. J. Anim. Sci. *21*, 583.

Walker, W. F., Jr. 1964. Dissection of the Fetal Pig. W. H. Freeman, San Francisco.

Warwick, B. L. 1928. J. Morphol. *46*, 59.

Wintrobe, M. M., and Shumacker, H. B., Jr. 1935. J. Clin. Invest. *11*, 837.

Wintrobe, M. M., and Shumacker, H. B., Jr. 1936. Amer. J. Anat. *58*, 313.

5

Postnatal Growth and Development

Various aspects of growth are covered in other chapters, but because it is such an important topic, it is discussed more directly here. The characteristic S-shaped growth curve of all biological systems, whether it be an increase in weight gain of a plant or animal or increase in number of yeast cells in a population (Brody 1945) is shown in Fig. 5.1. The term growth, as applied to animal production, is often considered to be synonymous with increase in body weight of the animal from conception to maturity. Brody (1945) has defined growth as "the constructive or assimilatory synthesis of one substance at the expense of another (nutrient) which undergoes dissimilation." The energetic efficiency of growth as just defined is, therefore, the energy stored in the animal per unit of energy consumed. The pig is especially efficient in growth, in comparison with other farm animals. Detailed discussion of the energetic efficiency of growth in meat-producing animals has been provided by Webster (1980). The complex metabolic and endocrine aspects of tissue and animal growth are exciting areas of animal biology and have been reviewed by Trenkle and Marple (1983) and in a symposium, Concepts of Animal Growth (American Society of Animal Science, 1988).

METHODS OF EXPRESSING GROWTH

Growth, as indicated by increase in body weight, can be expressed as absolute gain in a given period of time or as relative gain (or percentage gain). The absolute growth rate

$$\frac{\text{larger weight} - \text{smaller weight}}{\text{final time} - \text{initial time}}$$

can be misleading from the standpoint that the average value may not represent the actual value at any one point in time. For example, a pig whose average daily weight gain from birth to slaughter was 500 g (17.5 oz),

$$\frac{91 \text{ kg} - 1 \text{ kg},}{180 \text{ days}}$$

might never have gained exactly 500 g on any one day (see S-shaped growth curve in Fig. 5.1). The absolute gain the day after conception might have been a few micrograms, followed by a gain of 50 g (1.75 oz) on the first day after birth, and 1000 g (35 oz) on the last day before slaughter. Nevertheless, from a practical standpoint, the absolute average daily gain to slaughter is an important measure of growth as used in the broad sense.

Relative growth rate

$$\frac{\text{larger weight} - \text{smaller weight}}{\text{smaller weight}}$$

declines steadily as the animal approaches maturity.

The growth rate of individual parts of the pig is not identical. McMeekan (1940) showed, by dissection and carcass studies, that the head and shoulders reach mature size before the posterior parts of the body. That is, early in life, the head and shoulders represent a higher proportion of the total body weight than they do later. Similarly, individual organs and tissues grow at different rates, and indeed, individual cell types within a single organ grow at different rates. The changes in weight of various tissues of pigs of two genotypes representing different degrees of carcass fatness at 12, 17, and 24 weeks of age are illustrated in Table 5.1 (Tess 1981). Note that the relative weight of all organs declines with increasing age and body weight.

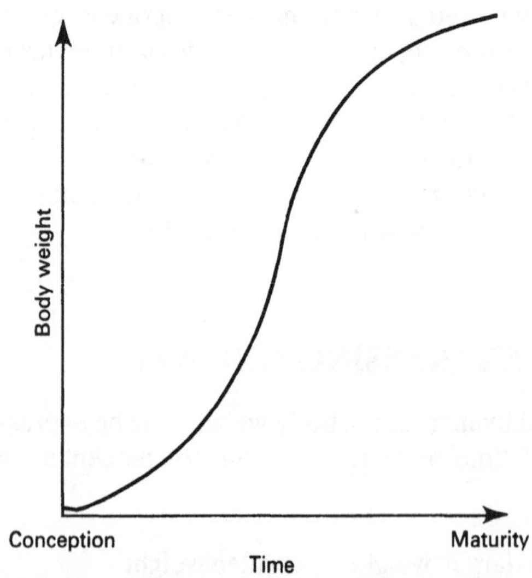

FIGURE 5.1 S-shaped growth curve typical of all biological systems. From Brody 1945.

TABLE 5.1 Means for Organ Weights Expressed as Percentages of Empty Body Weight in Genetically Obese and Contemporary Crossbred Pigs

	Age, weeks		
	10	*17*	*24*
Liver[a]			
Obese	3.08	2.08	1.44
Contemporary	3.20	2.01	1.41
Stomach			
Obese	1.16	0.87	0.67
Contemporary	1.17	0.80	0.61
Small intestine			
Obese	3.97	2.09[a]	1.06
Contemporary	3.67	1.68[b]	1.02
Large intestine			
Obese	2.16	1.79	1.50
Contemporary	2.38	1.71	1.69
Spleen			
Obese	0.21	0.13[a]	0.10[b]
Contemporary	0.23	0.17[b]	0.15[a]
Heart			
Obese	0.52	0.35[a]	0.25[a]
Contemporary	0.57	0.46[b]	0.35[b]
Lungs			
Obese	1.21[a]	0.76	0.45
Contemporary	1.54[b]	0.85	0.56
Kidneys			
Obese	0.61[a]	0.40[a]	0.31[a]
Contemporary	0.51[b]	0.35[b]	0.26[b]
Total viscera and blood			
Obese	19.01	14.03	11.20
Contemporary	19.15	13.69	11.06

[a, b]Means in the same pair with different letters are significantly different ($P < .05$).

Source: Tess 1981.

HYPERPLASIA AND HYPERTROPHY

Growth of tissues and organs and of the whole organism (the pig, in this discussion) occurs in two phases: increase in *number* of cells (*hyperplasia*) and increase in *size* of cells (*hypertrophy*). Immediately after conception, most growth is by hyperplasia. During the prenatal life and early postnatal life, the two processes occur concurrently. Finally, at some point in postnatal life, cell division ceases (except in some tissues) and adult size of specific organs is attained. This is illustrated diagrammatically in Fig. 5.2. Nervous tissue and muscle tissue perhaps fit that concept most closely. Other tissues, such as the epithelium of skin and digestive tract, and the blood have continuous cell turnover throughout life. For example, the life of the RBC of the pig is about 70 days; that of the intestinal mucosal cell is only a few days. For those tissues that reach mature size by proceeding through the

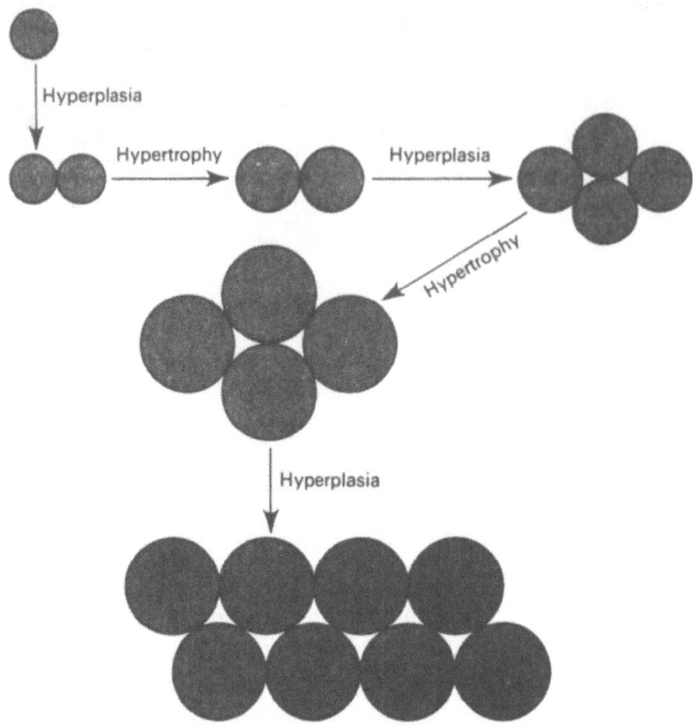

FIGURE 5.2 Diagram of growth by hyperplasia and hypertrophy.

three phases (hyperplasia, hyperplasia-hypertrophy, and hypertrophy) the ultimate size is presumably determined genetically; that is, the upper limit imposed by genetics cannot be exceeded, although environmental factors can have an effect in reducing ultimate size. There is considerable research effort underway in several laboratories at the present time to determine possible effects of nutritional regimes imposed in early life on subsequent growth of body parts of animals. Work of this kind with pigs suggests permanent effects of prenatal malnutrition on postnatal growth and development. The exact nature of this "stunting" is not known, but probably is associated with such factors as reduced cell number,[1] induced by endocrine and other factors that limit gene expression.

The time sequence of cell proliferation and cell growth in individual tissues and organs of pigs is not well mapped. It seems probable that most cell division (hyperplasia) in muscle is complete at birth or shortly thereafter and that most postnatal body growth is associated with an increase in cell size (hypertrophy). Studies of number and cross-sectional diameter of individual muscle cells taken from the loin eye (longissimus) muscle of pigs suggest differences among individuals in *cell number* in this tissue, but they indicate that most of the increase in

[1] Estimates of cell number and cell size can be obtained by measuring nucleic acid concentrations of tissues. The nucleic acid DNA (deoxyribonucleic acid) is found only in cell nuclei and remains constant throughout life of the cell. The nucleic acid RNA (ribonucleic acid) varies with the size of the cell. Therefore DNA, RNA, and the DNA : RNA ratio can be used to estimate cell number and size, except in muscle and other multinucleated tissues.

cross-sectional area of the loin eye (longissimus muscle) as the pig grows is related primarily to an increase in the diameter of individual muscle fibers rather than to an increase in number of fibers. From birth to 25 days of age, fiber diameter increases 100%; but from 100 to 125 days of age, the increase is only 10%. This means that about 95% of maximum diameter is achieved by 150 days of age (Chrystall et al. 1969). Subsequent increases in muscle cross-sectional area, therefore, mainly reflect increases in intramuscular fat and connective tissue. There is a high correlation between cross-sectional area of the loin eye of a pig at slaughter and individual muscle fiber diameter. During the late stages of growth (82–110-kg live weight) the loin eye increases linearly at a rate of about 0.14 cm^2/kg live weight (Miller and Wheat 1970). Net accretion of muscle protein in growing pigs is the result of the balance between protein synthesis and protein degradation, both of which are processes requiring energy. Theoretically, muscle protein accretion should be maximized by increasing protein synthesis or decreasing protein degradation or both. Research now underway should lead to effective ways of maximizing efficiency of lean tissue growth by manipulation of these processes.

The ontogeny of fat cell hyperplasia is also not clear. It is considered by some that the fat cell number is established prenatally in most animals. If hyperplasia does stop at or before birth in the pig, it should be possible, by biopsy of fat deposits in the newborn pig, to predict whether or not a particular individual is likely to become very fat if allowed free access to food in later life. Several groups of scientists (Mersmann et al. 1975, 1981; Steffen et al. 1981; Hausman et al. 1981, 1982; Etherton, 1982; Wright and Hausman 1990) are working to unlock the secrets of the fattening process in swine. Factors affecting rates of lipogenesis (fat formation) and lipolysis (fat breakdown) are under investigation; more complete knowledge of these fundamental processes will allow appropriate adjustments in production practices to improve efficiency of pork production and to meet consumer demands. The significance of these relationships from the standpoint of the animal breeder is discussed briefly in Chapter 6. Exciting research now underway in several laboratories (Dayton and Allen 1987; Reeds et al. 1986; Reeds, 1987) on the regulation of muscle protein synthesis and degradation rates promises to provide new knowledge applicable to lean pork production.

Growth of tissues and organs whose cells are continuously replaced, such as liver and intestinal mucosa, is responsive to diet intake and diet composition. The rapid turnover of cells in these organs and the high energy cost of this cellular-metabolism results in a disproportionately high contribution of these visceral organs to total energy expenditure. Thus, the whole animal energy expenditure is greatly affected by the rate of growth of individual organs and tissues and the rate of cell turnover of each. In growing pigs, visceral organs (stomach, small and large intestine, and pancreas) make up less than 10% of body weight but account for more than 40% of whole body energy expenditure (Koong et al. 1983).

It is well known that hormones (thyroid hormone, growth hormone and growth factors such as insulin-like growth factors (IGF-1 and IGF-II) affect growth of young animals. Newer knowledge of endocrinology and biochemistry shows that these effects of the endocrine system on growth are mediated through effects on enzyme systems that control anabolic and catabolic reactions of nutrients and their metabolites. Thus, we see that growth of the pig, in keeping with the broad definition of growth cited earlier (Brody 1945), proceeds from a fertilized zygote to a mature adult through a process of cell hyperplasia and hypertrophy (accompanied by differentiation of cells, tissues, and organs) under the influence of genetically and environmentally controlled hormone and enzyme systems within the body.

BODY WEIGHT GAIN AS A MEASURE OF GROWTH

In practical swine production, the most meaningful and useful measure of growth is increase in body weight gain. Table 5.2 (Lowrey 1911) illustrates the rapid increase in absolute and relative body weight from conception to slaughter at 90 kg (160 days postnatal age). Growth curves for contemporary pigs representing diverse genetic and environmental backgrounds can be expected to conform to the general S-shaped curve common to all biological systems. A typical growth curve (computer-simulated, Pomar et al. 1990) over the period from 64 to 154 days of age for crossbred contemporary male, female, and castrated male pigs fed *ad libitum* is shown in Fig. 5.3. It should be recognized that these curves are derived from limited numbers of animals of specific genetic background and raised under specific environmental conditions. Genetics as well as environment will affect the shape of the curve, but the examples shown in Fig. 5.3 are useful guidelines.

BIRTH WEIGHT AND LITTER SIZE

The average birth weight of all improved domestic pigs is probably very near 1.2 kg (smaller for "miniature" pigs and many indigenous local strains). A close association exists between birth weight and survival. The frequency of pigs farrowed over the weight range 400 to 2000 g and the relationship between body weight and stillbirths is shown in Fig. 5.4. First-born piglets tend to have larger birth weights than last-born within the same litter.

An increase in litter size (number of live pigs) at birth reduces the chance of survival slightly, owing to smaller individual pigs in large litters (an inverse relationship between total litter weight and weight of individual pigs) (Friend and

TABLE 5.2 Absolute and Relative Weight of the Pig at Various Stages of Prenatal and Postnatal Growth

Stage of development	*Weight (g)*	*No. of times ovum wt*
Ovum	0.000003	
21 days	0.25	83,000
112 days (term)	1,000	333,000,000
135 days (21 days postnatal)	5,000	1,667,000,000
274 days (160 days postnatal)	90,000	30,000,000,000

Source: Lowrey 1911.

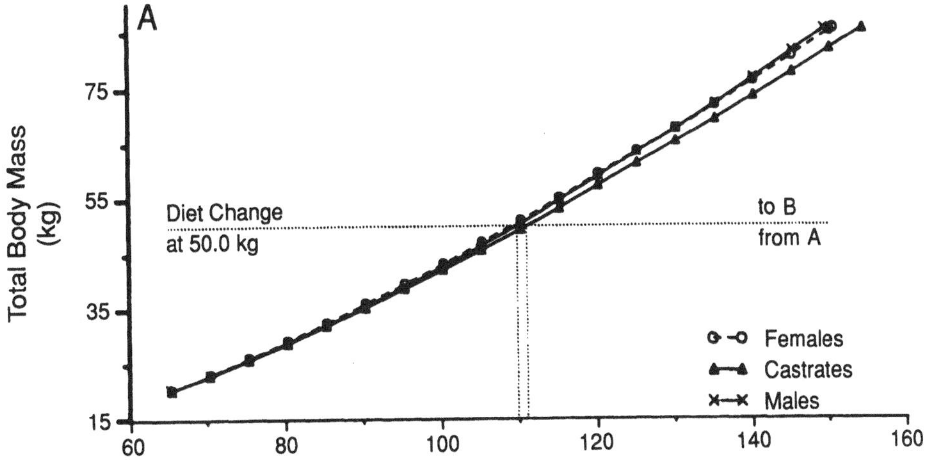

FIGURE 5.3 Empty body weight curves (live body weight minus gastrointestinal tract contents) of contemporary crossbred female, male, and castrated male pigs from age 64 to 154 days as generated by computer simulation. From Pomar et al. 1991.

Cunningham 1966). Increased mortality in smaller pigs is depicted in Danish data (Table 5.3).

The physiological reasons for the lower survival rate of smaller pigs are not completely known. The phenomenon is probably related to the genetic ability of the individual pig to withstand extrauterine stress, to differences in the placental supply of nutrients, to other spacial or endocrine differences in the environment, and to the relative physiological immaturity of the newborn pig. Body fat stores are small and glucose synthesis from other substrates is limited, forcing the newborn pig to depend largely on carbohydrate stores and immediate dietary energy intake for survival (Mersmann 1974). The survival rate of small pigs can be improved greatly by raising them in individual cages, thus eliminating the stress of competition with larger littermates (Lecce 1971; England 1974).

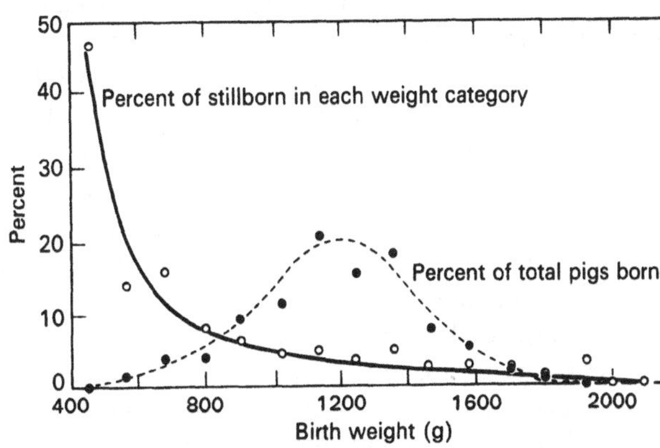

FIGURE 5.4 The relationship between body weight and the incidence of stillbirths. From Indiana Agricultural Experiment Station, Bull. 413, Purdue University.

TABLE 5.3 Influence of Birth Weight on Mortality

Weight intervals, kg at birth	*No. of pigs*	*Mortality % of liveborn*
Up to 0.90 kg	123	56.1
0.91–1.20 kg	416	20.4
1.21–1.40 kg	550	11.3
1.41–1.60 kg	637	9.1
1.61–1.80 kg	431	6.5
1.81–2.00 kg	159	6.9
Above 2.00 kg	46	8.7

Source: Nielsen 1975.

Holub (1968) reviewed the functional changes that occur in early postnatal life of the pig and emphasized the low body energy reserves and immature thermoregulatory and circulatory systems as important factors in neonatal mortality.

If the inherited capacity to attain a large birth weight is not different between individual pigs representing litters of different sizes, there should be a possibility of producing large litters with large individual pigs. This might be achieved by altering the growth pattern of the placenta, the transfer of nutrients across it, or some related aspect of the intrauterine environment. Whether this can be done experimentally or on a practical basis, remains to be seen. The maximum total litter birth weight one can expect to attain at the present time appears to be about 20 kg (44 lb.).

A high proportion of stillbirths occurs during the last third of farrowing (Friend and Cunningham 1966). The inverse relationship between litter size and individual pig birth weight has been observed repeatedly (England 1974). Third and subsequent litters tend to have more and larger pigs than do first and second litters. This is at least partly due to the smaller size of first-litter gilts as compared to mature sows. The average litter size reaches a maximum at the third litter and then plateaus until the sixth or seventh litter, followed by a decline.

By far the greatest rate of mortality in young pigs is during the first day after birth (13% including stillbirths in a study by Cox 1962) with smaller losses during the second and third days (3%), and less during the fourth day (2%) followed by a gradual decline to 0.02% during the period 28 to 154 days of age. This clearly shows the effect of age on mortality rate. In a report from Denmark (Neilsen et al. 1974), stillbirths accounted for 6.1% of pigs born; while deaths due to starvation, trauma, and gastrointestinal diseases accounted for 3.5, 4.0, and 2.0%, respectively, of pigs born; and miscellaneous causes, including infectious diseases and malformations, were responsible for the remainder of the 22.3% total preweaning mortality. This loss is typical of that experienced in the U.S. swine industry today and represents a challenge to the swine producer for improved management practices.

It is important to recognize that prenatal nutrition affects the size and viability of the newborn pig. This is discussed further in Chapter 8.

Several physiological parameters are of importance in the development of the young pig. Each of the following has economic implications in commercial swine production.

BLOOD GLUCOSE

The newborn pig has a very limited supply of readily available energy stored as liver glycogen and, to a lesser extent, heart and skeletal muscle glycogen. The fat content of the body is less than 1% by weight at birth. Because of this small energy reserve, the pig must nurse shortly following birth, or the blood glucose level falls dramatically. The normal value of 100 mg of glucose/100 mL of blood at birth may drop to 10 mg/100 mL or less within the first two days in the fasted newborn pig. When this low blood glucose is reached, lethargy, coma, and death result, unless the condition is promptly corrected by nursing or by oral or intravenous glucose administration. Morrill (1952) completed a series of detailed studies designed to characterize baby pig mortality. The following knowledge was gained from this work.

Providing distilled water or physiological saline *ad libitum* to fasted pigs fails to delay the hypoglycemia but may delay dehydration and aid in renal excretion of nitrogenous wastes and in temperature regulation.

Fasting also results in reduced body temperature (less than 34.5°) vs. 39°C (normal) during the first 2 or 3 days. The hypoglycemia and reduced body temperature are accompanied by a loss of 30% or more of body weight and finally death. The liver loses proportionately more weight than the remainder of the body during fasting. The environmental temperature during fasting affects the rate of decline in blood glucose concentration. Exposure of fasting newborn pigs to low temperatures (<21°C) results in a rapid drop in blood glucose level so that within 24 hours coma and death result. Pigs fasted in the same way at a temperature of above 27°C survived without coma for 2 to 3 days in a controlled experiment.

Hypoglycemic coma can occur even among pigs allowed to nurse. These cases are usually associated with a cold or drafty environment in combination with an inadequate milk supply. Injection of glucose reverses hypolycemic coma and the related symptoms if the condition is of short duration, but coma of 1 hour or longer causes irreversible damage and death.

Carbohydrates other than glucose (such as sucrose, fructose, and galactose) are not utilized by the baby pig for energy when given intravenously. Resistance to fasting hypoglycemia greatly increases as the pig gets older. This resistance to hypoglycemia after the first week of postnatal life is probably related to the amount of reserve energy stored before fasting, as well as to metabolic changes associated with development.

Wide individual differences exist in the ability of pigs to metabolize an oral load of glucose. A normal blood glucose concentration of 60–80 mg/100 mL of blood in weanling pigs shows a sharp rise to 200 mg/100 mL 1 to 2 hours after administration of such a dose in some animals, but only a slight rise in others. This is mostly associated with individual differences, although young animals tend to have a higher glucose tolerance (lower peak blood glucose concentration and faster removal after dosage). Animals with high glucose tolerance tend to have higher muscle glycogen levels.

BODY TEMPERATURE CONTROL AND CLIMATIC PHYSIOLOGY

For a detailed treatment of the climatic physiology of the pig, the student is referred to a book by Mount (1968) and to the Environmental Physiology section of Chapter 10 of this book. The present discussion covers only the salient features of the subject pertaining to development.

The newborn pig is not able to control body temperature efficiently. This works against the survival of newborn pigs exposed to a stressful environment or a limited milk supply. The poor response to low environmental temperatures would be expected because of the newborn pig's sparse hair coat and limited subcutaneous fat. Small pigs are less resistant to cold than are larger pigs because of their greater relative surface area. Maintenance of body temperature in the presence of reduced environmental temperatures requires a high level of heat production (high metabolic rate). Below the critical temperature (that environmental temperature below which an increase in metabolic rate is necessary to maintain body temperature), the newborn pig, in fact, is able to increase heat production (as measured by oxygen consumption) to a surprisingly efficient extent while maintaining body temperature essentially constant. The metabolic rate (oxygen consumption) increases considerably during the first day after birth.

There is an increase in body temperature from about 38°C at birth to 39°C later, so that the observed increase in oxygen consumption is not surprising.

The newborns of wild pigs, which are found roaming free in the southern United States, are more cold resistant than newborn domestic pigs during the first 2 days postpartum. This advantage in thermostability is partly due to a higher oxygen consumption during exposure to cold in wild than in domestic piglets, and partly to a more dense hair cover in the former (Foley et al. 1971).

Heat loss in a cold environment may be lessened by skin vasoconstriction and also by the fact that individual pigs in a litter huddle together, reducing their effective surface area and, therefore, their rate of heat loss (Mount 1964). Weight is of greater importance than age in affecting heat production after the first few days. Development of chemical thermoregulation consists of a period of poor thermo-

TABLE 5.4 Effect of Body Weight and Ambient Temperature on Metabolic Rate (kcal of Heat Produced per Day) in Young Pigs

Body weight (kg)	*Ambient temperature (°C)* 15	20	25	30
	Metabolic rate (kcal/pig/day)			
4.1	530	430	370	330
5.9	700	680	630	610
8.2	990	920	890	880
10.0	1,220	1,160	1,140	1,150
11.4	1,450	1,400	1,400	1,420

Source: Cairnie and Pullar 1957.

regulation (first 6 days), a period of gradual improvement (days 7 to 20), and a fully developed system (after 20 days). Table 5.4 shows the effect of environmental temperature and body weight on metabolic rate (Cairnie and Pullar 1957). As ambient temperature falls, metabolic rate increases to maintain body temperature within normal limits. The increase in energy expenditure in a cold environment becomes smaller as the animal becomes larger, owing to decreased relative surface area and increased insulation with subcutaneous fat.

The baby pig can rebound quickly to warming and feeding following exposure to near freezing temperatures for several hours, even when rectal temperature drops below 33°C and the pig has entered coma.

DIGESTION AND ABSORPTION OF FOOD

Colostrum and Immune Antibodies

The colostrum of the sow is very high in gamma globulin and its associated antibodies, so that the newborn pig is provided with an immediate passive immunity if allowed to nurse. The gamma globulin content of the colostrum gradually declines but considerable amounts persist for several days. During the first 24 hours the intestinal mucosa is permeable to almost any protein materials, including infectious agents. Following the ingestion of colostrum (or milk), the ability of the pig to absorb intact proteins is lost ("gut closure"). This is due partly to a developmental response of the intestinal tract, as evidenced by histological studies (Payne and Marsh 1962a, 1962b) and partly to the presence of a trypsin inhibitor present in sow colostrum (Laskowski et al. 1957). Colostrum-acquired antibodies persist in the serum for 6 weeks or longer after birth. A typical serum protein profile of a newborn colostrum-deprived pig as compared to the mature profile is given in Table 5.5 (Lecce and Matrone 1960). Active antibody production in the young pig begins at about 3 weeks of age, but production is small until 4 or 5 weeks. It is generally considered that 3 weeks of age is the most critical immunological period for the pig, since colostrum-acquired antibodies (passive immunity) are low and active antibody production (active immunity) has not yet attained full development. There is an inverse relationship between the amount of colostral immune

TABLE 5.5 Protein Profile of Newborn Colostrum-deprived Pig Serum as Compared to That of Mature Pig Serum

Serum protein profile	*Newborn*	*Mature*
Total protein, g/100 ml	2.8	6.3
Percentage of total protein:		
Albumin	12	40
Alpha globulin	74	22
Beta globulin	14	20
Gamma globulin	NM[a]	18

[a] NM = not measurable.
Source: Lecce and Matrone 1960.

globulins (IgG) ingested and absorbed during the first 12 hours postnatally and the level of synthesis of the immune globulin by the piglet at age 8 weeks (Butler et al. 1986). This inhibition of active antibody production by passively acquired maternal antibodies is of fundamental importance in immunology and is the subject of active research (Tumbleson 1986) in the pig. Depending on the disease or agent involved, the pig's immunity to it is controlled to varying degrees by passively and actively acquired antibodies. Feeding of immunoglobulins removed from blood serum to colostrum-deprived low birth weight pigs improves survival markedly when given for the first 3 weeks after birth, compared with the improvement achieved when given for only the first day (Whiting et al. 1983). These immunoglobulins are not absorbed after "gut closure," but do protect the pig from infection by pathogenic microbes in the gastrointestinal tract.

Recent work underway at Children's Nutrition Research Center, U.S. Department of Agriculture, Agricultural Research Service, illustrates the importance of factors in colostrum and milk that stimulate growth of intestinal mucosal cells in newborn pigs. For example, insulin (Shulman, 1990) and lactoferrin (Nichols et al. 1990) in colostrum are effective in this regard. It seems likely that this unfolding knowledge will have practical significance in pork production.

SPECIAL CONSIDERATIONS RELATED TO DIGESTIVE SYSTEM DEVELOPMENT

The newborn pig adapts well to a liquid diet for the first few days after birth, and can be trained to survive on dry feed beginning at a few days of age. Synthetic milk or cow's milk can be used to replace sow's milk, even for the newborn pig, and can be supplied either from a bottle and nipple or from a bowl. Fig. 5.5a (Lecce 1971) shows an automated feeding system for metered dispensing of liquid milk replacers on a preset time schedule to a large number of individually caged pigs. Fig. 5.5b and 5.5c (P. J. Reeds and M. Fiorotto, personal correspondence, 1990) show a similar liquid feeding system allowing control of timing and amount dispensed to individual pigs.

Pigs that are suckling their sows normally nurse about every hour around the clock. When early-weaned pigs are fed manually, labor requirements usually limit the feeding frequency to three or four times daily. Pigs fed as infrequently as twice daily convert feed to body weight and gain more efficiently than littermates fed 24 times daily at hourly intervals (Braude et al. 1970), but their feed intake is reduced and slower body weight gains are obtained. More frequent feeding allows more rapid growth, but there is a danger of digestive upsets and death if intake at a single feeding is excessive. Therefore, it is advisable to limit the food offered to the amount that is consumed in a few minutes.

It is advisable to keep artificially-reared pigs separated to avoid injury and continued agitation caused by persistent suckling of ears and nuzzling of other pigs when kept together.

The utilization of nutrients depends on the ability of the animal to release them from the feed in a form available for absorption. For example, amino acids must be released from intact protein, the fatty acids from triglycerides, glucose from the disaccharide, lactose from milk, or from other carbohydrates in other feeds. Each

FIGURE 5.5 (a) Automated feeding device for baby pigs (from Lecce, 1971); (b) view of animals at feeding (courtesy of P. J. Reeds and M. Fiorotto, CNRC, Houston, TX); (c) internal view of the device showing the arrangement of the feeding system.

major nutrient (carbohydrate, fat, protein) will be discussed briefly, with respect to its utilization by pigs of various ages.

Carbohydrates. The newborn pig readily absorbs glucose, but cannot digest sucrose or polysaccharides efficiently. Lactose is efficiently utilized by the newborn pig, but becomes less well utilized in the older pig because of the decline in secretion of lactase, which splits the disaccharide lactose into its component monosaccharides, glucose and galactose. Lactase is an adaptive enzyme. Therefore, when lactose is fed to finishing pigs, most can utilize it efficiently. Some finishing pigs are able to grow well on diets containing up to 40% whey. Galactose is converted to glucose as a source of energy. The young pig is not able to utilize xylose or fructose.

Utilization of complex carbohydrates (polysaccharides) increases with age in pigs and can be improved by injection of glucocorticoid hormones in the 6- to 14-day-old pig, presumably by the increase in pancreatic amylase activity induced by the hormone (Chapple et al. 1989).

Unlike the ruminant, the young pig lacks the capacity to utilize large amounts of some of the complex carbohydrates found in forages (cellulose). The pig does not possess the necessary digestive enzymes to hydrolyze these materials. There is some microbial cellulase activity in the cecum of pigs, but this is not an important nutritional contribution to the host, as it has been shown that cecectomy results in no apparent reduction in nutrient utilization by growing pigs. Feeding large amounts of cellulose to growing pigs increases the size of the stomach, cecum, and colon, increases the rate of food passage through the tract, and has a laxative effect. Microbial degradation of fiber in the large intestine (cecum and colon) of the pig generates volatile fatty acids (VFA) which may provide up to 30% of the energy requirements for growing pigs (Varel 1987). The number of cellulose-splitting bacteria in the pig large intestine can be increased by continuous feeding of high fiber diets (Varel and Pond 1985) and is probably increased as the young pig grows; the longer gastrointestinal tract associated with growth accompanies greater ingesta transit time which permits an extended exposure of the cellulose-degrading microflora to the ingesta.

Fat. The pig is able to hydrolyze and absorb large amounts of fat very efficiently from birth, as would be expected, since approximately one-third of the dry matter of sow's milk is fat. Most of the fat hydrolysis occurs in the small intestine, in the presence of substantial secretion of pancreatic lipases. The apparent absorbability of dietary fat is usually more than 80% in pigs of all ages. Increased chain length of fatty acids decreases absorbability in young pigs as does high unsaturation, but this effect tends to disappear with age. In dry milk replacers for early weaned pigs, lard and beef tallow are utilized with approximately equal efficiency, while absorbability of corn oil and other unsaturated fats is slightly more. Medium-chain triglycerides are absorbed efficiently even in the absence of high lipose activity and are transported to the liver via the portal blood. In addition to lipases, dietary lipids are prepared for absorption by the emulsification of fats and reduction in particle size through the action of bile secretions. Bile secretion is active even in the newborn pig.

Protein. The hydrolysis of protein to its constituent amino acids is inhibited during the first 36 to 48 hours by the presence of a trypsin inhibitor in sow colostrum (Laskowski et al. 1957). Subsequently "gut closure" occurs and proteins must be hydrolyzed before they can be used. Apparent absorbability of milk protein by the newborn pig is very high (92–95%). There is a distinctly higher efficiency of digestibility of casein than of soybean protein (95 vs. 89%) by the very young pig. The secretions of pepsin by the stomach and proteinases from the intestinal mucosa and pancreas are low at birth and gradually increase during the first 8 weeks. The inferior performance of pigs less than 3-weeks old that are fed soybean protein and other plant proteins, rather than casein, probably is related to the low level of proteinase secretions by the very young pig or to the presence of inhibitors in plant proteins. Soybean protein can replace part of the dried skim milk in diets of pigs 2 days to 2 weeks of age without adversely affecting growth or protein utilization (Veum et al. 1986). The difference in utilization of milk protein and plant proteins disappears by about 5 weeks of age.

By about 5 weeks of age, the pig can utilize a wide variety of foods other than milk. The lack of response generally obtained by experimental digestive enzyme supplementation of nonmilk-containing diets for pigs weaned before 3 weeks of age suggests that enzyme secretions may not be limiting factors in utilization of the feed, or that unknown factors in development of digestive capacity are over-riding the beneficial effects of the supplemental enzymes.

RATE OF PASSAGE OF FEED RESIDUES

In the 4-week-old pig fed synthetic milk, the rate of passage of food residues through the alimentary tract varies from 12 hours with soybean protein to 24 hours with casein as the protein source. By 10 weeks of age, approximately 35 hours are required, and there is no effect of protein source. The ratio of dry feed to water plays little, if any, part in 22- to 90-kg pigs nor does the digestibility of the feed. Dry matter content of the feces is remarkably constant over a wide range of feed dry matter, although high fiber diets are associated with a higher water content of colon contents (Pond et al. 1987). Pelleted feed passes through faster than nonpelleted feed (Seerley et al. 1962) and high fiber faster than low fiber feed (Kass et al. 1980). Pigs with a shorter digestive tract have a shorter transit time compared with pigs having a longer tract (Varel et al. 1988). Diurnal variation exists in the composition of feces; nitrogen and ash are in higher concentrations in the A.M. and nonnitrogenous organic matter is in a higher concentration in the P.M., probably due to differences in rate of passage of various feed fractions through the gastrointestinal tract. This phenomenon probably has little or no practical importance in commercial swine production.

INTESTINAL MICROFLORA

Even though the pig depends mostly on its own digestive secretions to prepare its food for absorption, the normal gastrointestinal tract contains a wide variety of microorganisms within 24 hours after birth (Fuller 1962; Willingale and Briggs 1955). A typical microbial population in the feed, feces, and various segments of the gastrointestinal tract of growing pigs is shown in Table 5.6 (Horvath et al. 1958). Diets high in fiber produce an increase in number of fecal cellulolytic microbes. Varel and Pond (1985) found 23.3×10^8 cellulolytic microbes in the feces of sows fed a corn-soybean meal control diet compared with 45.1×10^8 in sows fed 40% alfalfa. The adaptation occurs within about 2 weeks following introduction of the high fiber diet.

With the demonstration that dietary antibiotics stimulate growth of pigs, much attention has been given to alterations in the intestinal microflora that might account for this response. One such study by Larson and Hill (1960) showed that the metabolic activities of the microflora of the ileum are markedly reduced by chlortetracycline supplementation in young pigs. Urea-splitting bacteria are reduced in the large intestine of pigs fed antibiotics (Varel et al. 1987). The inhibition of urease-producing microflora by antibiotics has also been suggested by Visek (1968, 1978) as responsible for the growth stimulation obtained by feeding antibi-

TABLE 5.6 Microflora of Feed, Gastrointestinal Contents, and Feces in 90-kg Pigs[a]

	Coliforms	*Enterococci*	*Lactobacilli*	*Yeasts*	*Molds*
Feed	5.4	4.9	3.0	2.8	3.3
Stomach	7.7	6.8	7.9	4.6	3.5
Small intestine	9.3	6.7	7.8	4.6	3.5
Large intestine	9.2	8.0	8.2	4.4	4.0
Feces	9.9	7.6	8.5	4.1	4.0

[a] Average log of the number of viable cells per gram of dry matter; a mean of 8 animals (except for enterococci with mean of 7).

otics. The growth stimulation by antibiotics has also been suggested to be related to the inhibition of intestinal microflora that reduce nutrient utilization. It has been observed that the thickness of the intestinal lining is reduced by antibiotics. Yen et al. (1987) showed that pigs fed antibiotics have lower small-intestinal weight and suggested that the growth-promoting effects may be related to this phenomenon. Antibiotics may favor nutrient absorption and decrease energy requirement for maintenance of the gastrointestinal tissue, thereby freeing more energy for use in edible tissue growth.

The significance of the microorganisms in the alimentary tract of the pig and the mechanism of antibiotics in promoting growth await further study. Controlling factors must be at work in maintaining a suitable balance of microbes to prevent the overwhelming of the host by the several species that include pathogenic strains but which are ever present as normal inhabitants.

BLOOD AND CIRCULATION

Blood volume of the pig expressed as a percentage of body weight, decreases steadily with age (Bush et al. 1955; Linderkamp et al. 1981; Talbot and Swenson 1970). The ingestion and absorption of milk during the first few hours after birth tend to increase the total volume and probably contribute to the decline in hematocrit and blood hemoglobin concentration that occur in piglets during the first few days after birth. The mean blood volume of a newborn pig is approximately 80–90 mL/kg body weight and increases quickly after nursing to 95–100 mL/kg, where it remains for a few days before gradually declining, so that—at 2 or 3 weeks of age—it is 70–75 mL/kg body weight. Mature obese swine weighing more than 275 kg have a blood volume of only 30–35 mL/kg of weight. There is a gradual decline between these two extremes as the pig becomes larger and fatter.

Systolic blood pressure increases linearly with body weight from birth to maturity (50–60 mm Hg at birth, 140–160 m Hg or more at maturity) (Engelhardt 1966). Diastolic pressure also rises with increased body weight but is 30–45 mm Hg lower than systolic pressure.

Heart rate has been reported in various studies reviewed by Stone and Sawyer (1966) to be as low as 60 and as high as 200 beats per minute. Excitement increases heart rate 40–50% which could account for some of the higher values reported.

Heart rate declines as body weight and age increase. At birth, a heart rate of 200 beats per minute is normal, with a gradual decline to 80 or below at maturity. The current knowledge of heart rate, blood pressure, and cardiac output in swine has been thoroughly reviewed by Hannon (1986). Cardiac output is approximately 150 mL/kg body weight per minute in pigs from 7 to 50 kg body weight, and with a steady decline to approximately 30–40 mL/kg at maturity. Iron-deficiency anemia increases cardiac output severalfold. The inability of the pig to withstand severe exertion is probably related not only to its limited ability to dissipate heat efficiently, due to sparse sweat glands and large subcutaneous fat insulation, but also to its relatively small heart, small blood volume, and unstable circulatory physiology. Detailed information on all aspects of cardiovascular physiology in the pig during rest or exercise, much of which has practical application in commercial swine production, has been reviewed by Hannon and other authors in *Swine in Biomedical Research* (Tumbleson 1986). There is wide variation in heart size among individuals (Wegner 1970). This may be related to differences in susceptibility to stress. Domestication and selection have probably been factors in the characteristic cardiovascular physiology of present breeds of pigs, as wild pigs have a higher blood volume, higher oxygen-carrying capacity of the blood, and better ability to increase blood volume and hemoglobin in response to exercise than do domestic swine.

HEMATOLOGY

Cells

Red blood cells (RBC) (erythrocytes). Hematopoiesis begins during fetal life. The pig RBC is a biconcave disc of approximately 5–6 μ in diameter. The amounts of blood platelets and leukocytes and the changes in hematocrit (packed cell volume as percentage of total blood volume), in the young pig during the first 2 weeks after birth, are shown in Table 5.7 (Lie 1968). Normal values for RBC number, hemoglobin, and indices for RBC size and hemoglobin concentrations for pigs ranging in age from newborn to adults have been summarized by Schmidt and Tumbleson (1986).

The bone marrow is the major site of erythrocyte and leukocyte formation. The RBC during maturation becomes smaller and extrudes its nucleus; its cytoplasm changes from basophilic to eosinophilic and finally takes on the characteristic red color of the mature erythrocyte.

During the suckling period, the pig must depend on a source of supplemental iron for hemoglobin synthesis since poor placental transfer limits body reserves at birth, and milk iron is deficient. Iron deficiency in the young pig results in microcytic hypochromic anemia. Hemoglobin and hematocrit (packed cell volume) are both good indices of iron nutrition. The life of the RBC in pigs is approximately 70 days (Jensen et al. 1956; Withrow and Bell 1969). The RBC lifespan may be shortened by external factors but age-dependent destruction is not subject to environment. Formation may be impaired in various nutritional deficiencies. These are described in Chapter 8.

TABLE 5.7 Platelets, Leukocytes, and Packed RBC Volume of Pigs from Birth to 2 Weeks of Age

Day after birth	*Platelets (thousand/cu mm)*[a]	*Leukocytes (thousand/cu mm)*[b]	*Packed RBC volume (%)*[c]
0	337	16.9	40.9
1	276	15.3	36.1
2	241	9.4	32.9
3	279	10.2	34.0
4	309	11.8	32.2
5	378	11.6	32.4
6	410	10.2	33.0
7	465	10.8	31.5
8	535	9.1	34.2
9	525	8.0	33.9
10	578	9.4	34.4
11	495	8.3	35.0
12	553	8.6	36.8
13	532	8.0	38.5
14	527	9.4	36.1

[a] 51 to 141 observations at each age.
[b] 26 to 106 observations at each age.
[c] 21 to 95 observations at each age.
Source: Lie 1968.

White Blood Cells (Leukocytes)

The white blood cells are fewer in number than the RBC, but with a wide normal range and a lower average concentration in the newborn animal. Infections and stress increase the leukocyte count. Lymphocytes make up the largest fraction of the leukocyte population (40 to 60%), while neutrophils account for about 25 to 45%; eosinophils, 5%; monocytes and basophils, about 2% of the total (Schmidt and Tumbleson 1986). Leukocyte counts are important in clinical diagnosis of disease.

Serum Proteins

The newborn pig is poorly supplied with blood serum proteins until the first colostrum is ingested. Before nursing, the total plasma proteins are only about 2.4 g/100 ml of blood. Since fibrinogen, which is converted to fibrin when blood clots, is present only as a component in the blood clotting mechanism and represents only 3 or 4% of the total serum proteins at birth, it is more meaningful and convenient to discuss the blood proteins in terms of serum proteins (minus fibrin).

Serum proteins, while initially provided in large amounts by colostrum, are soon produced in ample quantities by the liver and to a lesser extent by the spleen and bone marrow. A moderate deficiency of a single amino acid or of total protein in the diet results in a decline in serum protein. Even in extreme and prolonged protein deficiency, the pig is able to maintain serum gamma globulin at about the same absolute level, while albumin is reduced to almost zero. The ratio of albumin to globulin is, therefore, a good index of protein nutrition.

By 8 or 10 weeks, the total serum protein level plateaus at approximately 7 to 8 g/100 mL where it remains until, and during, mature life.

Many physiological studies with swine have resulted in the recording of normal parameters for various blood constituents. Tabular data of this kind have been reported (Bustad et al. 1966; Tumbleson and Schmidt 1986). Tabular summaries of these and other reports are given by Pond and Maner (1984) and Tumbleson and Schmidt (1986). The values represent a variety of ages, breeds, and types of pigs.

The pig has at least 14 blood group systems (Andresen 1962; Moustgaard and Hesselholt 1966) similar to blood groups in man. The blood type of pigs is detected by means of blood typing reagents. It is potentially possible to characterize individual pigs on the basis of a blood type, as this is a heritable and permanent character of that individual. Though the ability to identify such traits is often predicted to be of value, whether or not these traits themselves may eventually have a place in selection of animals in commercial swine breeding remains to be seen.

DENTITION

The dental pattern of the pig is illustrated in Fig. 5.6 (Jump and Weaver 1966). At birth the pig has eight teeth, including four incisors and four canines. The full complement of temporary teeth is 28, and consists of 12 incisors (three pairs on upper and lower jaws), four canine (one pair on upper and lower jaws), and 12 molars (three pairs on upper and lower jaws). The permanent teeth erupt at varying intervals from 3 or 4 months (first molar) to 18 or 20 months (third molar). During

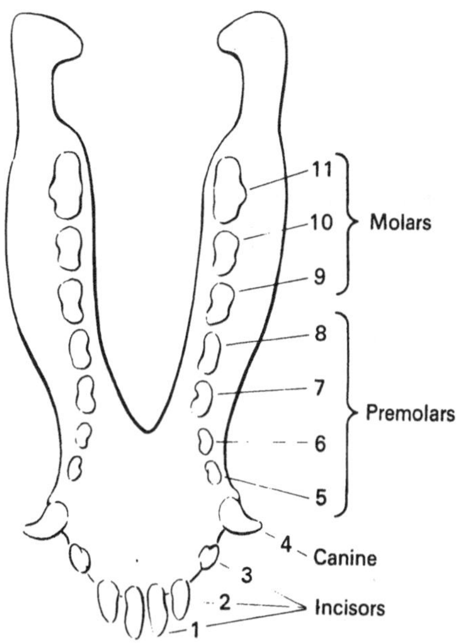

FIGURE 5.6 Dental pattern of the pig.

the preceding time, the dentition shows a mixture of deciduous and permanent teeth. The mature dental pattern is three pairs of incisors, one pair of canines, four pairs of premolars, and three pairs of molars, upper and lower, for a total of 44 teeth present by about 24 to 27 months of age (Gier 1986).

The eight teeth present at birth are very sharp and have been descriptively termed needle teeth. It is common commercial practice to clip these temporary teeth at the gum line shortly after birth to avoid injury to other pigs and to the nipples of the dam during suckling. The canines of the adult grow very long and protrude out of the mouth as tusks. In the male, it is common practice to clip them to avoid injury to humans, or to other pigs.

CHANGES IN BODY COMPOSITION WITH TIME

Rapid changes occur in relative and absolute weights of organs from a few days before birth, to a few days after birth (Brooks and Davis 1969). The stomach, liver, and kidneys increase in weight proportionately faster than the whole body immediately after birth. Nitrogen-free extract (presumably largely glycogen) concentration of liver and carcass increases sharply during the last 10 days before birth and then declines rapidly during the first few hours after birth. This probably reflects the importance of a source of available energy from fetal stores during the early postnatal period, until the pig suckles. The high efficiency of fat synthesis by the newborn pig is indicated by the twofold increase in carcass fat during the first 48 hours after birth. The sharp rise in concentration of linoleic acid in liver and carcass during this same period suggests that transfer of linoleic acid to the pig is more efficient via milk than via the placenta. The changes in weight of several organs from 98 days after conception (approximately 2 weeks before birth) to 6 days after birth are illustrated in Table 5.8.

The newborn pig is much lower in fat content and higher in water than most other newborn mammals. The changes that occur in gross composition and in

TABLE 5.8 Organ Weight Changes in Late Prenatal and Early Postnatal Life

Organ	*Days from birth*						
	−16	*−12*	*−8*	*−4*	*+2*	*+4*	*+6*
Whole body, kg	0.89	1.11	1.18	1.35	1.45	1.71	1.96
Carcass (minus internal organs), kg	0.40	0.52	0.54	0.64	0.72	0.81	0.93
Liver, g	22.8	27.0	28.5	37.5	37.8	51.8	64.3
Stomach (empty), g	6.0	6.8	7.0	6.8	8.5	10.5	12.5
Heart, g	7.3	9.0	10.0	9.8	10.5	12.8	14.0
Kidneys, g	10.0	10.0	9.5	10.3	12.5	12.8	15.0
Spleen, g	2,8	2.0	2.8	1.1	2.0	3.0	4.0
Brain, g	22.8	24.5	25.5	29.3	32.8	35.3	36.8
Lungs, g	28.9	31.5	32.5	22.3	20.3	25.8	29.0

Source: Brooks and Davis 1969.

TABLE 5.9 Body Composition of Pigs During Early Postnatal Life

Constituent	*Postnatal period (age, days)*				
	0	*2*	*7*	*14*	*28*
Livewt, g	1,520	1,815	3,221	5,563	9,928
Empty livewt, g	1,450	1,741	3,044	5,284	9,651
Fat-free tissue, g	1,432	1,701	2,738	4,488	7,888
Fat, g	18	40	306	796	1,763
Fat, %	1.1	—	—	—	—
Water, g	1,198	1,398	2,207	3,557	6,138
Water, %	84.1	—	—	—	—
Protein, g	174	237	437	770	1,427
Protein, %	11.3	—	—	—	—
Ash, g	60.4	66.8	94.1	160.8	323.9
Calcium, g	16.7	17.9	24.1	41.7	87.6
Phosphorus, g	9.3	10.8	16.3	28.1	56.3
Potassium, g	2.8	3.8	6.8	11.5	19.9
Sodium, g	2.8	3.1	4.3	6.9	12.3
Magnesium, g	0.42	0.48	0.83	1.53	3.01
Iron, mg	54	49	77	118	153
Zinc, mg	26	33	60	114	236
Copper, mg	3.9	5.4	10.9	18.6	23.2
Manganese, mg	0.69	0.75	0.70	0.98	1.94

Source: Manners and McCrea 1963

mineral element content during the first 4 weeks after birth, are summarized in Table 5.9 (Manners and McCrea 1963). It is clear that rate of fat deposition exceeds rate of deposition of other organic and inorganic components.

These changes in proportions of ash, fat, protein, and water with increased age and weight for pigs of three genotypes (contemporary, lean, and obese) are depicted in Fig. 5.7 from the data of Tess (1981). Boars are leaner than females at a given weight during growth, and both sexes are leaner than castrated males at the same weight. Pomar et al. (1990) depicts simulated changes in daily body protein accretion in female, male, and male castrate pigs from 64 to 154 days of age. Note that intact males continue to increase in daily protein accretion beyond 154 days, while females and male castrates plateau, females higher than castrates, at about 110 days of age. According to the model generated by Pomar et al. (1990), mature size, in terms of total protein mass and daily protein accretion (Fig. 5.9) is attained by about 2 years of age (daily protein accretion equals zero) for both females and males. Fat accretion would be expected to continue to increase beyond the age of 2 years if feed intake is not restricted, resulting in a steady increase in body weight.

As fat deposition increases with increased body weight, the percentage of saturated acids tends to increase while unsaturated acids tend to decrease (Table 5.10) (Sink et al. 1964). Fatty acid composition of the depot fat is highly responsive to diet. Unsaturated fat, such as safflower oil, results in a soft depot fat, whereas a saturated fat such as hydrogenated coconut oil results in a firm depot fat when fed to growing pigs (Table 5.11) (Babatunde et al. 1968).

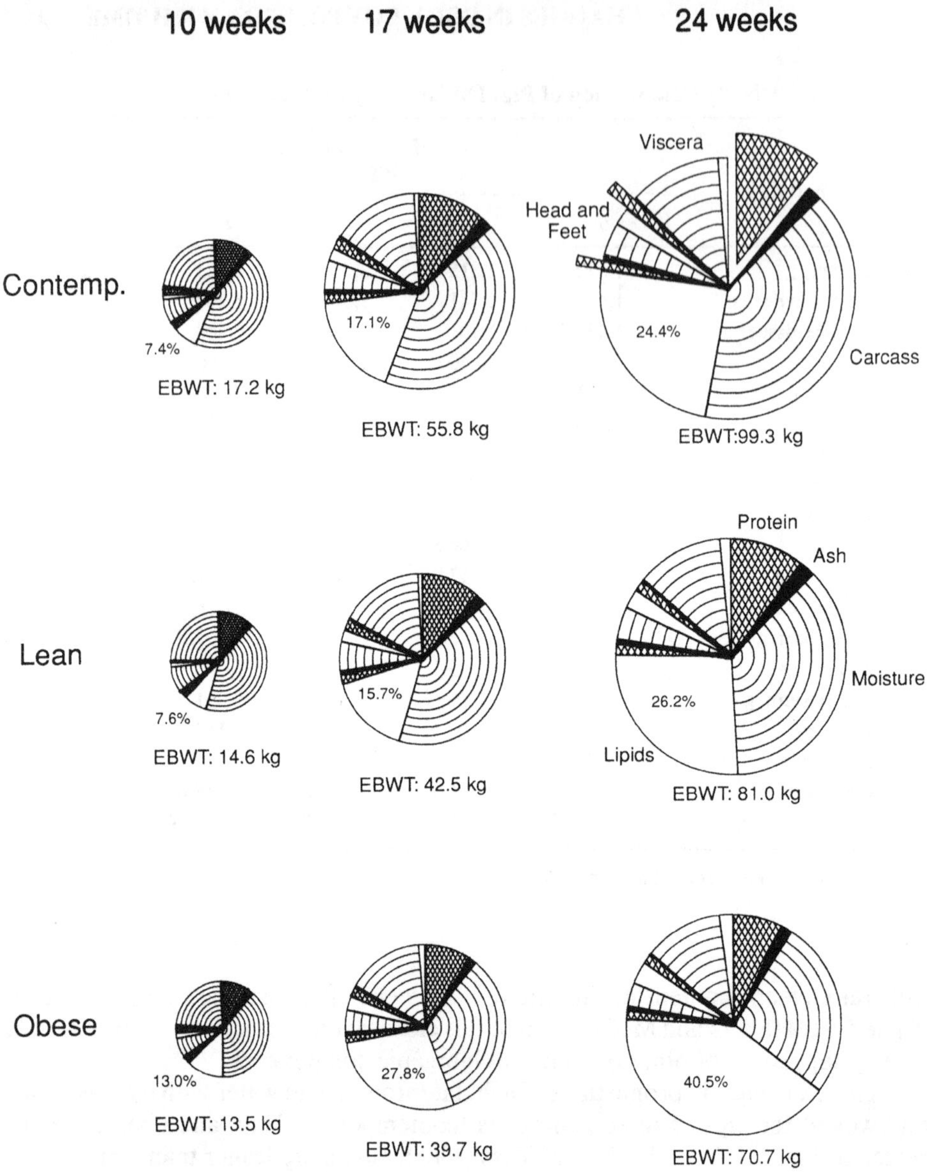

FIGURE 5.7 **Changes in body composition with increasing age in pigs of different genetic background. Percentages shown are for carcass lipids as a fraction of empty body weight (EBWT).**

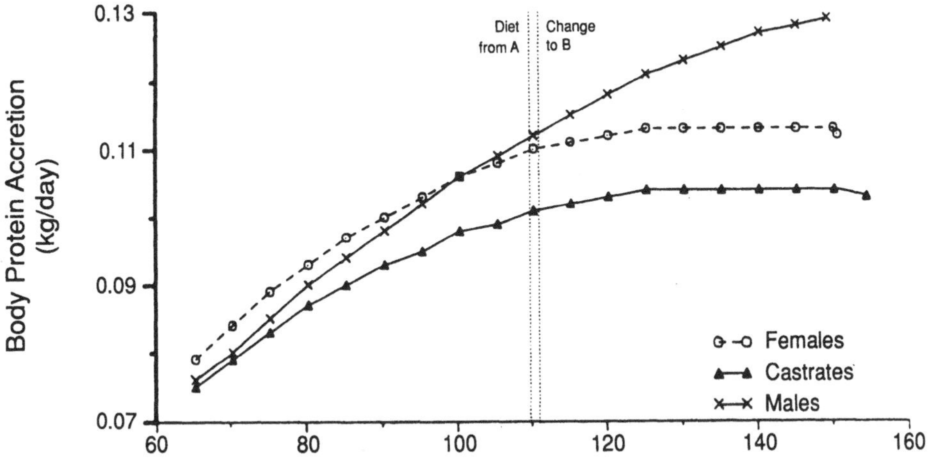

FIGURE 5.8 Changes in daily body protein accretion in female, male, and male castrate pigs, as generated from computer simulation (Pomar et al. 1991), show the greater lean tissue growth rate of intact males than of castrates and the intermediate position of females.

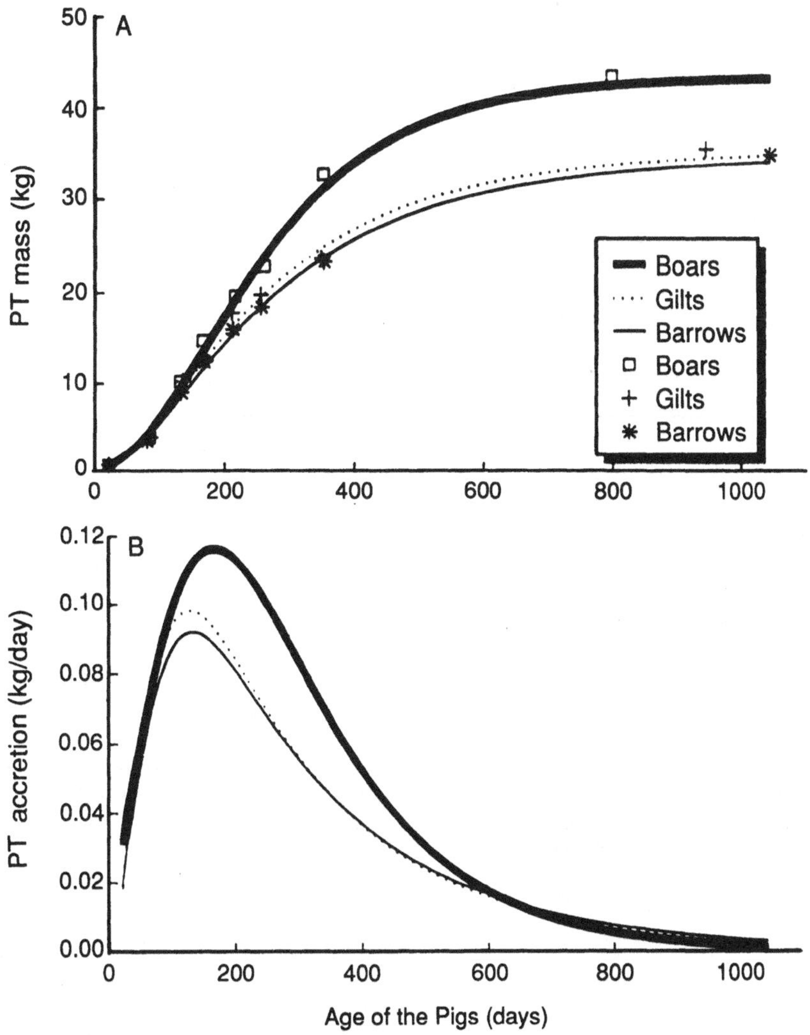

FIGURE 5.9 PT (protein) mass (upper panel) and daily PT (protein) accretion (lower panel), generated from computer simulation (Pomar et al. 1991) suggest that mature size, based on protein accretion, of boar, barrow, or gilt pigs is reached by about two years of age.

TABLE 5.10 Fatty Acid Composition of Depot Fat of Growing Swine as Affected by Body Weight

Body wt (kg)	Fatty acid composition (%)						
	Saturated fatty acids						
	$C_{12:0}$	$C_{14:0}$	$C_{15:0}$	$C_{16:0}$	$C_{17:0}$	$C_{18:0}$	Total
34	0.34	1.75	0.09	22.25	0.52	9.79	34.89
59	0.25	1.72	0.05	22.69	0.46	10.67	35.91
86	0.21	1.98	0.03	23.81	0.23	10.93	37.26
	Unsaturated fatty acids						
	$C_{16:1}$	$C_{17:1}$	$C_{18:1}$	$C_{18:2}$	$C_{18:3}$	$C_{20:4}$	Total
34	4.87	0.75	44.05	14.47	0.96	0.16	65.11
59	4.00	0.54	45.79	12.71	1.06	0.07	64.09
86	3.68	0.26	45.20	12.80	0.76	0.08	62.74

Source: Sink et al. 1964.

TABLE 5.11 Effect of Diet on Fatty Acid Composition of Depot Fat of Growing Pigs

Fatty acid	Diet[a]			
	Fat-free	3% safflower oil	3% hydrogenated coconut oil	Maize–soybean meal
	Percent of total fatty acids			
$C_{10:0}$	0.13	0.20	0.18	0.11
$C_{12:0}$	0.10	0.16	0.63	0.12
$C_{14:0}$	1.79	2.00	3.85	—
$C_{16:0}$	18.71	15.60	20.99	—
$C_{16:1}$	16.85	13.01	16.47	12.68
$C_{18:0+18:1}$	60.16	50.58	55.93	55.76
$C_{18:2}$	1.28	17.15	0.93	8.32
$C_{18:3}$	0.18	0.09	0.09	0.09
$C_{20:1}$	0.98	0.56	0.76	1.08
$C_{20:3}$	0.17	0.46	0.11	0.49
$C_{20:4}$	0.07	0.19	0.07	0.23

[a] Amount of linoleic acid ($C_{18:2}$) in diet was 0, 2.4, 0.06 and 0.8% for fat-free, 3% safflower, 3% hydrogenated coconut oil and maize-soybean meal diets, respectively.
Source: Babatunde et al. 1968.

REFERENCES

American Society of Animal Science. 1988. J. Anim. Sci. *66* (Suppl. 3), 1–97.

Andresen, E. 1962. Annu. N.Y. Acad. Sci. *97*, 203.

Babatunde, G. M., Pond, W. G., Walker, E. F., Jr., and Chapman, P. 1968. J. Anim. Sci. *27*, 1290.

Braude, R., Mitchell, K. G., Newport, M. J., and Porter, J. W. G. 1970. Brit. J. Nutr. *24*, 501.

Braude, R., Mitchell, K. G., and Suffolk, S. F. 1969. J. Inst. Anim. Tech. *20*, 43.

Brody, S. 1945. Bioenergetics and Growth. Hafner, N.Y.

Brooks, C. C., and Davis, J. W. 1969. J. Anim. Sci. *29*, 325.

Brown, H., Speer, V. C., Quinn, L. Y., Hays, V. W., and Catron, D. V. 1961. J. Anim. Sci. *20*, 323.

Bush, J. A., Jensen, A. N., Cartwright, G. E., and Wintrobe, M. M. 1955. Am. J. Physiol. *181*, 9.

Bustad, L. K., McClellan, R. O., and Burns, M. P. 1966. Swine in Biomedical Research. Frayn, Seattle.

Butler, J. E., Klobassa, F., Werhahn, E., and Canbier, J. C. 1986. *In* Swine in Biomedical Research (ed., M. E. Tumbleson), p. 1883. Plenum Press, N.Y.

Cairnie, A. B., and Pullar, J. D. 1957. J. Physiol. *139*, 15.

Chapple, R. P., Cuaron, J. A., and Easter, R. A. 1989. J. Anim. Sci. *67*, 2985.

Chrystall, B. B., Zobrisky, S. E., and Bailey, M. E. 1969. Growth *33*, 351.

Cox, D. F. 1962. Iowa State University Res. Bull. *500*, 339.

Dayton, W. R., and Allen, R. E. 1987. J. Anim. Sci. *65*(Suppl. 2), 1.

England, D. C. 1974. J. Anim. Sci. *38*, 1045.

Engelhardt, W. 1966. *In* Swine in Biomedical Research (eds., L. K. Bustad, R. O. McClellan, and M. P. Burns). Frayn, Seattle.

Etherton, T. 1982. J. Anim. Sci. *54*, 58.

Foley, C. W., Seerley, R. W., Hansen, W. J., and Curtiss, S. E. 1971. J. Anim. Sci. *32*, 926.

Friend, D. W., and Cunningham, H. M. 1966. Can. J. Comp. Med. Vet. Sci. *30*, 179.

Fuller, R. 1962. Studies on the gut flora of the pig and chick. Ph.D. Thesis. University of Reading, England.

Gier, R. E. 1986. *In* Swine in Biomedical Research (ed., M. E. Tumbleson), p. 633. Plenum Press, N.Y.

Hannon, J. P. 1985. *In* Swine in Biomedical Research (ed., M. E. Tumbleson), p. 1341. Plenum Press, N.Y.

Hausman, G. J., and Martin, R. J. 1982. J. Anim. Sci. *54*, 1286.

Hausman, G. J., Campion, D. R., McNamara, J. P., Richardson, R. L., and Martin, R. J. 1981. J. Anim. Sci. *53*, 1634.

Holub, A. 1968. Functional Periodization of the Early Postnatal Period in Piglets. Acta Universitatis Agriculturae, Fakulta Veterinarni. Brno, Czechoslovakia (English summary), 125 pp.

Horvath, D. J., Seeley, H. W., Warner, R. G., and Loosli, J. K. 1958. J. Anim. Sci. *17*, 714.

Jensen, W. N., Bush, J. A., Ashenbrucker, H., Cartwright, G. E., and Wintrobe, M. M. 1956. J. Exp. Biol. Med. *103*, 145.

Jump, E. B., and Weaver, M. E. 1966. *In* Swine in Biomedical Research (eds., L. K. Bustad, R. O. McClellan, and M. P. Burns), p. 543. Frayn, Seattle. 1963. Res. Vet. Sci. *4*, 145.

Kass, M. L., Vansoest, P. J., and Pond, W. G. 1980. J. Anim. Sci. *50*, 192.

Koong, L. J., Nienaber, J. A., and Mersmann, H. J. 1983. J. Nutr. *113*, 1626.

Larson, N. L., and Hill, E. G. 1960. J. Bacteriol. *80*, 188.

Laskowski, M., Kassell, B., and Hagerty, G. 1957. Biochem. Biophys. Acta *24*, 300.

Lecce, J. G. 1971. J. Anim. Sci. *33*, 47.

Lecce, J. G., and Matrone, G. 1960. J. Nutr. *70*, 13.

Lie, H. 1968. Acta Vet. Scand. *9*, 105.

Linderkamp, O., Betke, K., Guntner, M., Jap, G. N., Riegel, K. P., and Walser, K. 1981. Pediat. Res. *15*, 53.

Lowrey, L. G. 1911. Am. J. Anat. *12*, 107.

Manners, M. J., and McCrea, M. R. 1963. Br. J. Nutr. *17*, 495.

McMeekan, C. P. 1940. J. Agric. Sci. *30*, 301, 511.

Mersmann, H. J. 1974. J. Anim. Sci. *38*, 1022.

Mersmann, H. J., Allen, C. D., Chai, E. Y., Brown, L. J., and Fogg, T. J. 1981. J. Anim. Sci. *52*, 1298.

Mersmann, H. J., Goodman, J. R., and Brown, L. J. 1975. J. Lipid Res. *16*, 269.

Michel, M., and François, A. C. 1956. Compt. Rend. *242*, 1770.

Miller, L. R., and Wheat, J. D. 1970. J. Anim. Sci. *30*, 732.

Morrill, C. C. 1952. Am. J. Vet. Res. *13*, 164, 322, 325, 327.

Mount, L. E. 1964. J. Physiol. *170*, 286.

Mount, L. E. 1968. Climatic Physiology of the Pig. Williams Wilkins, Baltimore.

Moustgaard, J., and Hesselholt, M. 1966. *In* Swine in Biomedical Research (eds., L. K. Bustad, R. O. McClellan, and M. P. Burns), p. 25. Frayn, Seattle.

Nichols, B. L., McKee, K. S., and Huebers, H. A. 1990. Pediat. Res. *27*, 525.

Nielsen, H. E. 1975. *In* Festskrift til Hjalmer Clausen (ed., H. Staun), pp. 165–181. Danske Landhusholdingsselskab, Copenhagen, Denmark.

Nielson, N. C., Christensen, K., Bille, N., and Larsen, J. L. 1974. Nord. Veterinaermed. *26*, 137.

Payne, L. C., and Marsh, C. L. 1962a. J. Nutr. *76*, 151.

Payne, L. C., and Marsh, C. L. 1962b. Fed. Proc. *21*, 909.

Pomar, C., Harris, D. L., and Minvielle, F. 1991. J. Anim. Sci. *69:* In press.

Pond, W. G., and Maner, J. H. 1984. Swine Production and Nutrition. AVI, Westport, Conn.

Pond, W. G., Roberts, S. J., Dunn, J. A., and Willman, J. P. 1960. J. Anim. Sci. *19*, 881.

Reeds, P. J., Hay, S. M., Dorwood, P. M., and Palmer, R. M. 1986. Brit. J. Nutr. *56*, 249.

Reeds, P. J. 1987. Anim. Prod. *45,* 149.

Schmidt, D. A., and Tumbleson, M. E. 1986. *In* Swine in Biomedical Research (ed., M. E. Tumbleson) p. 767. Plenum Press, N.Y.

Shulman, R. 1990. Pediat. Res. *28,* 171.

Sink, J. D., Watkins, J. L., Ziegler, J. H., and Miller, R. C. 1964. J. Anim. Sci. *23,* 121.

Steffen, D. G., Arakelian, M. C., Philley, G., Brown, L. J., and Mersmann, H. J. 1981. J. Anim. Sci. *52,* 1306.

Stone, H. L., and Sawyer, D. C. 1966. *In* Swine in Biomedical Research. (eds., L. K. Bustad, R. P. McClellan, and M. P. Burns), p. 411. Frayn, Seattle.

Talbot, R. B., and Swenson, M. J. 1970. Am. J. Physiol. *218,* 1141.

Tess, M. W. 1981. Simulated effects of genetic change upon life-cycle production efficiency of swine. Ph.D. Thesis, University of Nebraska, Lincoln.

Trenkle, A. H., and Marple, D. N. 1983. J. Anim. Sci. *57* (Suppl. 2), 273.

Tumbleson, M. E. 1986. Swine in Biomedical Research, vol. 3, pp. 1811–1961, Plenum Press, N.Y.

Tumbleson, M. E., and Schmidt, D. A. 1986. *In* Swine in Biomedical Research (ed., M. E. Tumbleson), p. 787. Plenum Press, N.Y.

Varel, V. H. 1987. J. Anim. Sci. *65,* 488.

Varel, V. H., and Pond, W. G. 1985. J. Appl. Environm. Microbiol. *49,* 858.

Varel, V. H., Pond, W. G., and Jung, H. G. 1988. J. Anim. Sci. *66,* 707.

Varel, V. H., Robinson, I. M., and Pond, W. G. 1987. Appl. Environm. Microbiol. *53,* 2009.

Vestal, C. M. 1936. Ind. Agric. Exp. Stn. Bull. 413.

Veum, T. L., Zamora, R. G., and Sherry, M. P. 1986. *In* Swine in Biomedical Research (ed., M. E. Tumbleson), p. 1113. Plenum Press, N.Y.

Visek, W. J. 1968. J. Dairy Sci. *51,* 1.

Visek, W. J. 1978. J. Anim. Sci. *46,* 1447.

Webster, A. J. F. 1980. The Energetic Efficiency of Growth. Livestock Prod. Sci. *7,* 243.

Wegner, W. 1970. Fleischwirtschaft *50,* 72.

Whiting, R., Owen, B. D., Elliot, J. I., and Beames, R. M. 1983. Can. J. Anim. Sci. *63,* 993.

Wilbur, R. D., Catron, D. V., Quinn, L. Y., Speer, V. C., and Hays, V. W. 1960. J. Nutr. *71,* 168.

Willingale, J. M., and Briggs, C. A. E. 1955. J. Appl. Bacteriol. *18,* 284.

Withrow, G., and Bell, M. C. 1969. J. Anim. Sci. *28,* 240.

Wright, J. T., and Hausman, G. J. 1990. J. Anim. Sci. *68,* 1170.

Yen, J. T., Nienaber, J. A., and Pond, W. G. 1987. J. Anim. Sci. *65,* 1243.

6

Genetics and Breeding Systems

RELATION BETWEEN BREEDING AND PRODUCTION

In Chapter 3, it was stated that the genetic aspects of the production system set both the limits and potential for performance of the system. The breeding stock input into the system introduces potentials for animal performance but these are not usually exceeded (experimental lean growth promotants such as ractopamine and somatotropins do increase lean growth potential above that expressed without their use). In this chapter, we will examine alternative breeding systems to provide that input, tools available to the breeder and producer to improve the potential for that input, and the evolution of genetic improvement tools from the traditional showring-based visual appraisal approach to modern computer-based performance testing and genetic evaluation procedures.

To provide a framework for this discussion, we shall introduce the concept of the pyramid as illustrated in Figure 6.1. This illustration sketches the main strata of the industry structure relating to the flow of breeding stock and animals through the functions of selection, multiplication, crossing and reproduction, production, processing, and consumption. The top layers represent different generations in the movement from elite breeding herds to production herds. The bottom layers involve the flow of production animals and pork products from producer to processor to consumer. The shape of the pyramid symbolizes that, with the reproductive rate of swine, there can be a smaller number of animals in a limited number of herds practicing intense selection, which contribute breeding stock through multiplier herds into a considerably larger number of production herds that produce pork products that move through the processors to a very large number of consumers. For improvement of all crossbreeding systems, there needs to be a flow of genetic change from the top of the pyramid to the bottom, analogous to water

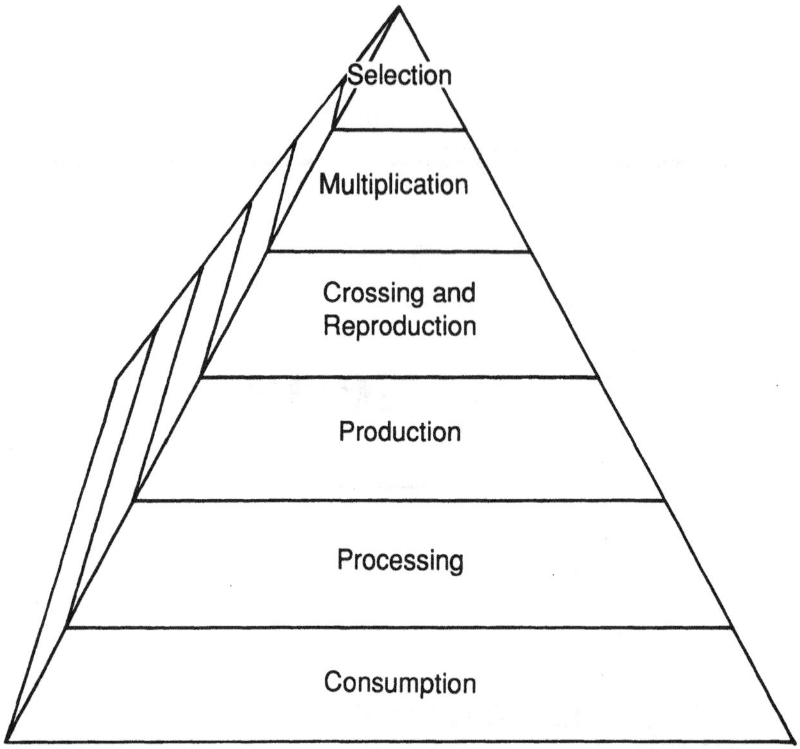

FIGURE 6.1 The major sequential aspects of comprehensive livestock production systems where genetic improvements should flow from the top of the pyramid to the base.

from a fountain at the top flowing down a pyramid. To extend this analogy, genetic selection is the fountain at the top that produces the genetic change. Because the different crossbreeding systems involve different amounts and sexes of breeding stock flowing down the pyramid, we will be relating the different systems described in later sections back to this pyramid concept.

Traditional Purebreeding, Pedigrees, and the Showring

A review of past practices as they evolved to the present will help the reader understand and appreciate the need for and value of modern methodologies. For the first half of the twentieth century, as in past centuries, purebreeding was the predominant breeding system, with merchandizing based on relative merits of the alternative breeds. New breeds were occasionally developed by innovative breeders and groups of breeders from crosses of previous breeds, but the emphasis was on forming and maintaining a new and distinctive group of animals that conformed to some standard of type. Uniform color patterns, ear shapes, and body type were strived for, so that the breed would be distinguished by this "trademark." Inbreeding was often used to establish uniformity for this "trademark."

After breeds became recognized as distinctive, herd books usually were established and records of pedigree became evidence to verify that an animal was truly a member of its breed. The term "purebred" implied that animals without pedigrees were impure. Purebred societies of breeders for each breed were usually formed to promote the breed and the showring became a place to extol the virtues of each breed and to instruct breeders and producers about the "ideal type" for that breed. Prominence was given to pedigrees including ancestors with noted reputations.

For many centuries, fairs and shows have provided an opportunity for livestock producers and breeders to exhibit and compare their animals. Having animals that were judged closest to the breed ideal gave social prestige and established reputations for the breeders and their animals, which, in turn, led to increased sales. Thus, a strong heritage of visual appraisal of live animals as the basis for improvement was developed that continues today.

Most experienced livestock men can make accurate observations about an animal's health, size, skeletal structure, muscling, and fatness from its visual appearance. These abilities are a source of pride to stockmen. Often in the marketing of livestock, visual appearance is all that is available to establish value. This need and pride of ability for visual evaluation, coupled with the heritage of fairs and shows, has led to exaggerated views of these abilities. Terms like "the eye of the master," "form determines function," and "the art of breeding" prevail in early writings on breeding techniques. Other phrases such as "like produces like or the likeness of some ancestor" and "breed the best to the best" guided the practices of breeding in all classes of livestock. Many of these expressions are attributed to Robert Bakewell, a British livestock breeder of the 1700s (Lush 1945). Another phrase attributed to Bakewell was that "inbreeding produces prepotency and refinement." It should be realized that selection and breeding of animals started soon after the domestication of animals but, until this century, was done without clear understanding of the modes of inheritance.

Corn and Poultry Breeding Precedents

A new direction for breeding approaches started in the 1920s and 1930s in the midwestern United States with the development, marketing, and adoption by farmers of hybrid seed corn. Using the tools of inbreeding, selection, and crossbreeding, corn breeders developed hybrid varieties of inbred lines that exhibited performance and production vastly superior to previously used corn breeding techniques that were analogous to purebreeding in the animal species. Hybrid corn not only increased the production of the primary feedstuff for pork production, but established the understanding that systematic crossbreeding could lead to improved performance.

Poultry, for both layer stock and broiler stock, followed corn in use of crossbreeding techniques. Some companies, which had been successful in developing superior hybrid corn varieties, began breeding poultry. They tried to imitate the hybrid corn approach by developing inbred lines to later be crossed. Other poultry breeders chose to adopt the strain-cross approach by crossing non-inbred breeds or strains, avoiding the deleterious effects of inbreeding on the performance of their grandparent stocks.

Swine production also quickly moved to crossbreeding in the 1940s and 1950s, but not to specific terminal crosses as in poultry, but to rotational crosses. These alternative breeding systems will be elaborated in a later section of this chapter. For now, we will note that the recognition of the practicality of exploiting heterosis in crossing inbred lines of corn quickly led to similar recognitions in poultry and swine, with crossbreeding practices quickly following. Purebred breeds were still important, but as interacting components of either form of crossbreeding, rather than as separate products of breeding.

Early hybrid corn varieties involved four inbred lines used as grandparents so that both of the parents were hybrids as well as the commercial crop. Thus, the parents also exhibited heterosis for performance and the poor performance of inbred lines was at the grandparent and earlier generations. In the 1960s, a combination of better performing inbred lines and the producer's willingness to pay larger prices for seed corn made practical the production and sale of single-cross seed corn. But about the same time, those poultry breeders who had attempted to use inbreeding to develop lines for crossing, moved away from that approach to adopt the techniques of selection and crossing of non-inbred strains. Even at the grandparent level, the lower performance of the inbreds made costs of production of parent stock excessively expensive. In the 1940s and 1950s, much research was conducted in midwestern states participating in the Regional Swine Breeding Laboratory concerning the use of inbreeding as a breeding tool, but without real evidence as to its practicality. Since swine have a considerably lower initial level of reproduction, it is unlikely that inbreeding can be practically used as a part of swine breeding, even though it has been used historically. Inbreeding reduces performance in many traits and, especially, those concerned with reproduction. With the usual size of breeding herds, selection strong enough to offset the deleterious effects of inbreeding is not possible. Reproduction is so important in swine production systems that even moderate reductions cannot be tolerated.

Dairy and Beef Breeding Precedents

The dairy industry led the way of the livestock industries in being the first to record performance in an objective, quantitative, and logical manner. This was done through milk recording and butterfat testing activities of the Dairy Herd Improvement Associations (DHIA). The early purpose was to provide management information for the owners of the cows being tested. But soon, this became the basis for culling cows of inferior—and selecting cows of superior—performance. In midcentury, when artificial insemination (AI) became the preferred method of reproduction, the DHIA data base was available for evaluating the genetic merit of the numerous sires being used in many herds. Statistical techniques developed by researchers were enhanced by the rapid development of electronic computer capabilities. Leadership in these developments was by C. R. Henderson and L. D. Van Vleck and their students at Cornell University. National Sire Evaluations have been used for several years to evaluate the merit of the frozen semen available from bulls in several AI studs. Purchase and choice of semen form the major breeding decisions for dairy producers. The tested daughters of a sire located in several dairy herds provided a progeny test of his genetic capabilities. These procedures have recently evolved to the simultaneous evaluation of dairy sires and cows all across the United States in a large analysis done every six months. This analysis is done by the Agricultural Research Service of the USDA on a supercomputer at Cornell University. Genetic evaluation and the use of artificial insemination are very synergistic in their impact upon dairy breeding. Use of bulls in matings across several herds has increased the accuracy of genetic evaluation for performance traits and genetic evaluation has identified those bulls worthy of being utilized in many herds.

Because of the superior milk production performance of the Holstein breed and because of the difficulties of implementing systematic crossbreeding programs in species with low reproductive rate, very little crossbreeding is done in dairy production in the United States or other countries. Although culling and selection of cows continues as an important part of herd improvement, the selection decisions with major impact involve the production and purchase of bull semen.

In recent decades, crossbreeding has been frequently used in the beef industry. This has been stimulated by the introduction of exotic breeds into the United States. To increase numbers, these breeds have been top-crossed onto existing breeds. Systematically producing three or four way terminal crosses has not seemed practical, but much rotational crossbreeding has been done, often in a less than systematic manner. Sometimes efforts have been made to develop new breeds from crosses of two or more existing breeds.

Performance recording within progressive purebred beef herds has led to the use of genetic evaluation tools similar to those in dairy. Performance recording started in some herds in the late 1950s. But the desire for increased growth rates in the traditional breeds stimulated the trends. The use of AI provided the opportunity to make greater use of superior genetic material. The use of bulls across many herds made possible genetic evaluations across herds as well as within herds. The statistical tools of dairy sire evaluation were adopted and extended. The major difference was that the important growth traits could be observed on young bulls so that direct performance information facilitated evaluation before production of semen and the resulting progeny test. But, in beef as in dairy, sire evaluation has been important. R. L. Willham of Iowa State, along with others, has assisted the breed organizations in adopting these modern tools of quantitative genetics. In recent years, the simultaneous evaluation of sires and young animals has been achieved. Most beef breed associations offer computerized performance recording services with interim evaluations for within-herd selection. This is followed by periodic across-herd evaluations of sires and young bulls. The latter are done on larger computers, usually twice a year, and provide the basis for the production and use of semen from bulls with greater expected genetic capabilities.

The swine breeding industry seems directed towards using systematic breed-crossing procedures similar to strain-crossing in poultry coupled with genetic evaluation procedures similar to the beef industry. Genetic evaluation and selection similar to that of poultry to improve their breeds or strains is sometimes done by corporate swine breeders, but is sometimes executed by swine breed associations for their participating members as in beef. Because of limitations of techniques, artificial insemination (AI) is not used as much in swine as in dairy and beef, but is used by some and usage is expected to increase. In Appendix 1, breed associations, corporate breeders, and AI organizations are listed with descriptions and photos of their genetic products, plus summaries of procedures they use.

After reviewing the economic, genetic, and statistical principles, the specifics of crossbreeding and selection procedures currently utilized in swine will be reviewed.

Bioeconomic Objectives

During the showring era of swine breeding, the objective of breeding efforts was towards an "ideal type" for the breed with emphasis on the visual characteristics. The expectation was that "form determines function," which is partially true. A modern text on livestock judging (Hunsley et al. 1978) states it this way:

> The characteristics in body *form* that give rise to variations in *function*, such as early maturity, thrift, activity, and feeding qualities, are depth, width, length, strength of topline, and soundness of feet and legs. There are, of course, many qualities in swine which influence *function* that are not apparent in *form*. [Italics added for emphasis]

It has been hoped by advocates of visual appraisal that easily observed characteristics of *form* "indicate" the more important, but difficult to observe, physiological *functions*. It has become increasingly apparent that the qualities which influence function that are *not* apparent in form greatly exceed those functions that *are* apparent in form. Bone structure, as viewed in the feet and legs and body shape, do affect activity through the ability to walk and might influence the size of the body cavity. However, the internal organs, the circulatory system, the endocrine system, the reproductive system, the nervous system, and the genetic controls on protein synthesis and other cellular and body functions are just not visually observable. The reader should review the traits of animal performance outlined as important to the production system in Chapter 3 and critically assess how well a producer or breeder should expect to determine those characteristics from looking at the animal. Most of the characteristics which are of true economic importance cannot be seen on or through the skin. Thus, modern swine breeding has evolved to recognize the necessity of records of animal performance, followed by statistical analyses of those records for economically important traits, as a more sound and effective basis for herd and breed improvement. Visual appraisal of feet, legs, and udders will continue to have a role, but it must be kept in a realistic perspective concerning its benefit. Visual appraisal without performance records has become antiquated as a breeding technique. The modern view recognizes that it is preferable for *function* (recorded performance) to dictate the best *form* (body shape and structure).

The term "bioeconomic objective" has been coined to represent an economically based mathematical formula of recorded biological traits which defines the real breeding objective for modern genetic evaluation. This formula should evaluate the differences in genetic control of animal performance that impact the pork production system. Alternative formulae that are currently used as selection criteria in a national program for the United States will be presented in a later section. These mathematical functions are chosen to guide the expected improvement of selected breeding stock to increase the economic efficiency of the production system. Such functions have replaced the outmoded "ideal type" as the breeding objective.

IMPACT OF GENETICS AND STATISTICS

In the first half of the twentieth century, two areas of science developed which have had tremendous impact on breeding in all livestock species, including swine. The first of these was genetics, which contributed an understanding of how "like produces like." The second was statistics, which provided the opportunity to describe more precisely the tendency for "like produces like" and, then, to more precisely predict future performance. Coupled with a bioeconomic objective, the breeder now has the opportunity to more accurately identify the "best" of each sex and thus "breed the best to the best" to a degree not possible in the past.

Mendelian Genetics

Early in the twentieth century, the understanding developed that the inheritance of characteristics in plants and animals was through genes located on the chromosomes in the nucleus of the cells. In most cells of most tissues, the chromosomes and thus, the genes, occur in pairs and are termed diploid. The genes occupy specific locations, called loci, on the chromosomes and affect specific characteristics. The form of the genes can change due to infrequent mutations and alternative forms at a locus are called alleles. When the two genes of a locus of an individual are the same, the individual is termed homozygous for that locus. When the pair of genes are different alleles, the individual is said to be heterozygous. Often in heterozygotes, one allele is expressed as phenotypes (observed characteristics) that are the same as those for homozygotes for that allele. Such alleles are said to be dominant to the other allele, which is said to be recessive. In cell division (called mitosis), as occurs in most tissue growth, both pairs of chromosomes and the alleles they carry are duplicated into the new cells.

When the sex cells (sperm and ovum in animals, called *gametes*) are formed, the paired chromosomes separate into two kinds of resulting cells. Thus, the resulting cells have only one-half the number of chromosomes of other cells and are called *haploid*. In this reduction division (called *meiosis*), sometimes the chromosomes recombine (called *crossing-over*) so that a new chromosome may have different parts from each of the individual chromosomes of the pair in the animal producing sex cells. Different pairs of chromosomes segregate independently giving unrestricted recombination. But, normally only one of each pair of genes and loci is present in the resulting gametes. With fertilization, the resulting fertilized ovum (called *zygote*) again has paired chromosomes with one member of each pair coming from each parent. This pairing and separation and re-pairing of chromosomes and the genes and alleles carried on them, with the recombination, leads to genetic similarities between parents and offspring, but also leads to genetic variability of different offspring from the same parents. Each offspring carries a sample half of the genes from each parent. Thus, "like produces like," but the tendency as expressed in many traits is less than perfect.

This understanding of the genetic mechanism of inheritance is usually termed *Mendelian genetics* after Gregor Mendel, who in 1865 first described the phenomenon for segregating characters in the pea plant, even though the role of the chromosomes of the cell nuclei as the location of the genetic factors was not understood at that time. The mechanism is easier to follow through successive generations for contrasting characters such as those of two flower colors, of tall vs. dwarf, of skin color, etc. when there are just two or three contrasting forms controlled by the alleles at one locus. As the mechanism of inheritance for most contrasting forms was found to be the same for most plant and animal species, it became evident that the same mechanism was also involved in the inheritance of continuously varying characters such as weight, height, yield, etc., that do not categorize into distinct classes. Because of the physiological complexity of these characters, it seemed likely that many alleles at many loci on several chromosomes must influence their expression. Because of the likely number of genetic factors involved, it seemed futile to try to identify and follow the individual alleles and loci involved for large populations of animals. However, knowledge concerning the basic mechanism of inheritance not only led to greater understanding of past techniques of breeding but suggested new methodologies.

Biochemical Basis for Genetics

During the second half of the twentieth century, the understanding of inheritance increased through the identification of DNA (deoxyribonucleic acid) as the genetic material of the chromosomes in the cells. These biochemical compounds control most cellular and body functions by directing the synthesis of amino acids into proteins. These proteins include numerous enzymes and hormones which control all the body functions. And, of course, muscle protein is recognized as the primary nutritive constituent of pork. Because so many of the life functions involve protein synthesis or are controlled by proteins, this mechanism is the avenue of genetic control over variation in nearly all body functions. Much exciting progress has been made in elucidating the genetic code for directing synthesis of specific proteins. By following the complex genetic code, it is now possible to trace segregation of genetic factors previously not observable as they are transmitted from ancestor to parent to offspring to descendent. But, on the other hand, the complexity of the processes has also been emphasized, because many DNA combinations—in many different sequences—code for many different proteins involved in all the body processes. Much research effort is being directed towards locating and identifying simple genetic codes for major proteins. Such major genes, if they exist, could be more directly manipulated than in past and current practices.

In particular, the growth hormone (somatotropin) is recognized as a protein of major importance to body processes in the agriculturally important animals. Success has been obtained in isolating genes directing its synthesis from different species and relocating them into the *E. coli* bacteria for artificial, but biological, production of the hormone. Also, these genes have been transferred from one species to another. The potential for this approach to modifying genetic control has stimulated much research in the area but, at this point, has not led to practical procedures for swine breeding. Another related area of research interest is injecting or implanting growth hormone, artifically produced by transformed *E. coli,* into growing-finishing animals to enhance their lean growth potential above that achieved by traditional breeding techniques.

These two sections are inadequate to impart to the student all the knowledge of genetics needed for the basis of swine production and, especially, swine breeding. The reader is encouraged to pursue this area in basic genetics texts giving more complete coverage.

Statistical Tools

Another area of science of fundamental importance to swine breeding and production is statistics. This area provides procedures to measure and describe differences and variability in characteristics that can be represented mathematically. Since many of the aspects of swine production as presented in Chapter 3 are of a quantitative form—amounts, weights, prices, rates, and percentages—it should be evident that mathematics and statistics are necessary to describe and compare alternatives. This is especially true for the breeding segment of the industry.

Space does not allow review of all the statistical procedures of value to swine breeders. We will review some of the key measures of variability and covariation necessary for the discussion in the next few sections. The reader is referred to any of several statistical texts (e.g., Steel and Torrie 1980) for a more thorough review of these concepts and the formulae for calculating their estimates. All of the measures to be presented here pertain to the distribution of the values for a quantitative characteristic of individuals in a population, e.g., the weights of a group of pigs. Pertinent population parameters are:

1. **Mean (or Average).** A measure of the central point in the distribution of observed values, e.g., a group of pigs at 160 days of age might have a mean weight of 100 kg. Common symbols are μ for a population and $\overline{x}$ for a sample. Most individuals in the population would have values close to the mean.
2. **Standard Deviation.** A measure of dispersion of values about the mean. For most characteristics, two-thirds of the population would have observed values between one standard deviation below the mean and one standard deviation above the mean. Ninety-five percent of a population will occur within two standard deviations of the mean. For example, the standard deviation of weights of pigs at 160 days of age might be 10 kg. Thus, 66.6% of individuals would have values from 90 to 110 kg and 95% from 80 to 120 kg. Common symbols are σ for a population and s for a sample.
3. **Variance.** Another measure of variation which is the square of the standard deviation. An alternative definition for variance is the average squared deviation from the mean. For the example used in the standard deviation definition above, the variance would be 10 kg^2, which equals 100 kg^2.
4. **Simple Correlation.** A measure of association of two variables across the population. The possible values for a correlation are between -1 and $+1$, with a

value of 0 indicating no association. For example, the correlation between length and weight of a pig might be 0.7, indicating a strong, but not perfect, association with large values of length associated with large weights and small with small. When large values of one variable tend to be associated with small values of the other, and vice versa, the correlation value will be negative. A common symbol is the letter r, possibly with subscripts indicating the variables involved.

5. **Simple Regression.** Another measure of association is expressed as the amount of change in one variable associated with change in another variable. For example, in a group of pigs varying in both age and weight, a regression coefficient for weight on age of 0.6 kg per day would indicate that, on the average, older pigs tend to be 0.6 kg heavier for each one day increase in age. A common symbol is the letter b with subscripts indicating variables. This coefficient can be used to predict deviations in one variable associated with deviations in another. The prediction will be closer to accurate if the correlation between the two traits is close to either plus one or minus one.
6. **Coefficient of Determination.** Another measure of association representing the fraction of the variation in one trait that is associated with variation in other traits. This quantity is usually symbolized by R^2, which is equal to r^2, the square of the simple correlation, when only one other variable is involved. Thus, for the length and weight example, $R^2 = r^2 = (.7)^2 = 0.49$ of the variation in weight would be associated with variation in length. When more than one associated variable is involved, R^2 can be termed a coefficient of multiple determination.
7. **Multiple Regression.** When the association is between one variable and several other variables, the relationships may be described by coefficients of partial regression, which also describe the change in the dependent variable for each unit change in each of the variables. These partial regression coefficients may differ from the corresponding simple regression coefficients. For example, the relation between the weight of a pig and the weights of its sire and its dam could be described by two partial regressions, one for offspring on sire and another for offspring on dam. Such regressions could predict the combined change in the next generation relative to the current from selecting sires and dams deviating from average.

There are considerably greater considerations in the methodology of statistics, but these few will suffice to describe the basic concepts of quantitative genetics in the next section. Two popular textbooks on statistics are Huntsberger and Billingsley (1986) and Steele and Torrie (1980).

Quantitative Genetics and Variation

Most of the characteristics or traits of concern in swine breeding exhibit continuous variation describable by the tools of statistics, but are influenced by genetic factors too numerous to be categorized before now as being Mendelian. Approaches to genetics and, specifically, breeding for these traits are termed "quantitative genetics." A fundamental parameter for each trait is its heritability. This describes the fraction of the variation in a trait that is associated with the average (or additive) effects of the genes in its makeup. The higher the heritability the greater the degree of genetic control upon the trait. Lowly heritable traits have lesser additive genetic influence and greater nonadditive or nongenetic influences. The symbol used for heritability is h^2. This is analogous to R^2, the coefficient of determination. Heritability can be considered the coefficient of genetic determination of phenotypic variation. Phenotypic value is a term applied to the observed value for a trait for an individual. Genotype refers to the set of genes possessed by an individual. Genotypic value for a specific trait is the conceptual expected average of many phenotypes of individuals of the same genotype. The deviation of the phenotypic value from the genotypic value is considered an environmental deviation due to nongenetic factors.

Each allele is presumed to have an average effect across all the genotypes and phenotypes in which it can occur in a population. The sum of the average effects of all the alleles in a genotype is termed the additive genetic value for the genotype. The genotypic value will differ from the additive genetic value due to interactions among the alleles. Variation due to nonadditive effects of the pairs of alleles at a locus are termed dominance variation. Nonadditive effects between alleles at different loci result in epistatic variation. Neither the genotypic values nor the additive genetic values can be observed directly for individuals. But these are conceptual quantities whose presence can be inferred from the correlations between relatives for specific traits. The correlation between random sires and their offspring is expected to be a bit greater than one-half of the heritability for that trait. A definition of heritability mathematically equivalent to that given above is the ratio of the additive genetic variance to the phenotypic variance for the trait. The amount greater than one-half h^2 is due to the epistatic variation common between sire and offspring. The one-half results from the nature of genetic inheritance, with one-half of the genes from the sire being transmitted to the offspring. The correlation between random dams and their offspring would have similar expectations but might have some additional association due to maternal influences other than through the transmitted genes. The regression of offspring on parents would have similar expectations as the corresponding correlations. Paternal half-sibs would have a correlation of one-fourth the heritability plus some association due to epistatic effects.

Thus, the concept of heritability provides a modern quantification of the tendency expressed in the old breeding expression that "like produces like." The higher the heritability, the greater the tendency. Heritability values might be between 0 and 1. Few traits are controlled exactly by genetics and have heritabilities close to one. Traits related to growth and carcass composition in most populations of swine have heritabilities from 0.4 to 0.6. Traits relating to the reproduction process are less precisely controlled by additive genetic effects with heritabilities from 0.1 to 0.3. Evidence from inbreeding experiments suggests that the heritability of reproductive traits is smaller because of considerable nonadditive genetic variation, especially dominance variance. The greater the heritability, the greater the change in the population when animals are selected on that trait to become parents.

Inbreeding and Heterosis

The term "heterosis" or "hybrid vigor" has been coined to describe the usual greater vigor, fertility, fecundity, and longevity of hybrid or crossbred individuals relative to their parents. The phenomenon has been noted in most plant and animal species. The greater vigor seemingly is associated with the greater heterozygosity of such individuals. Since, at most loci, the heterozygotes express the dominant allele, and, in many cases, the homozygous recessive is undesirable or defective, increased heterozygosity is favorably related to better performance for many traits and, especially for traits relating to reproduction.

Inbreeding is the mating together of related individuals. Since related individuals will tend to have received the same alleles at a locus from their common ancestors, matings between relatives will produce an increased frequency of homozygotes and a decrease in heterozygotes across all loci. Thus, inbred progeny will express more homozygous recessive and less dominant phenotypes. Since dominance is often in the favorable direction, usually inbreeding results in inferior average performance for most traits. In many ways, inbreeding is the opposite of heterosis. This tendency has been noted in most civilizations across species and is the basis for taboos against matings between relatives.

Let's return to our consideration of hybrid corn to assess why inbreeding has been a desirable avenue for developing the lines for crossing. Remember that the normal reproductive rate (progeny per parent) is considerably greater for corn and most plant species than for animal species. Also, inbreeding can be by self-fertilization which is considerably more intense than full-sibbing or parent-offspring matings in animals. With intense inbreeding to produce large numbers of progeny, strong selection pressure can be applied to eliminate the undesirable recessives that are produced. Thus, inbreeding makes selection more effective against undesirable alleles. But in swine, with a much lower inherent rate of reproduction, the selection pressure is seldom great enough to eliminate all the undesirable variation that will be uncovered even with milder inbreeding. Thus, even selected inbreds are likely to exhibit poor reproductive and livability performance. The consensus among most breeders and geneticists is that inbreeding should be avoided in swine. The slight increase in effectiveness of selection with inbreeding seems inevitably offset by the reduced reproductive performance, which leads to reduced selection pressure and increased costs of producing parent stock.

For further elaboration of the principles of quantitative genetics directed towards undergraduate university students, see Pirchner (1983). A more advanced presentation directed towards graduate students is by Falconer (1989).

SYSTEMATIC SWINE BREEDING

To integrate the tools, including both crossbreeding and selection, available to the animal breeder, Harris et al. (1984) summarized nine steps involved in designing an effective animal breeding program. The perspective that should motivate the development of a breeding program is that its results should improve the efficiency of the production system to which it contributes. The steps—with short explanations of each, as pertinent to swine breeding—are as follows:

1. *Describe the production system(s).* In this description include the environments and marketing situations that are to be targets of the breeding program. Specify products, purposes, geographical, and climatic areas of interest, and emphasize benefit and cost considerations. This description should include the normal life cycle of parents and production animals specifying ages for breeding, timing of offspring production, expected fertility and fecundity, ages or weight for weaning, intensive feeding, marketing and slaughter, and all other relevant and important points in the life cycle. The nature of feed, labor, land, buildings, and equipment requirements and the corresponding costs in the various stages, is necessary to fully describe the system. In other words, in this step, make specific the perspective outlined in Chapter 3 as pertains to the pork production systems of the potential breeding stock customers.

2. *Formulate the objective of the system.* The bioeconomic objective should be determined as a mathematical function or set of functions that describe the contributions of various aspects of the system (especially the genetic aspects) to its productive efficiency. This function may simultaneously evaluate the impact of alternative management and nutrition schemes on the system. The focus should be upon the specific segment of the production industry affected by the specific breeding program being planned; in other words, the breeder's intended customers. In this step, the description in words from step 1 is converted into a more exact mathematical form to quantify the net impact of genetic and other changes in the efficiency of the system(s). This formulation will be modified in step 6 to reflect how the breeding stock will contribute to traits of the sires, dams, and progeny. Recently, Stewart et al. (1990) have done this formulation for typical midwestern U.S. swine production.

3. *Choose breeding system and breeds.* The choice of breeding system (specific crossbreeding structure) will influence the performance characteristics of the production units and the costs of producing the production animals through the expansion system of step 8 below. The breeds or strains and their arrangements in the parentage need to be specified. The actual performance characteristics of production animals and of their parents are determined by the breeds or strains chosen for positions in the specified breeding system. The relative emphasis in selection (step 6) on traits of parental performance and progeny performance will change depending on the intended use of a specific breed. The alternative breeding systems appropriate for swine will be discussed in the next section.

4. *Estimate selection parameters and economic weights.* Decades of research with swine have been directed at evaluating heritabilities, genetic and phenotypic correlations, and other parameters pertinent to predicting responses to selection for alternative selection criteria and schemes. These have recently been summarized by Einstein et al. (1989) and include the traits pertinent to steps 3 or 6. Specific values for the economic weights were also developed by Stewart et al. (1990) in conjunction with developing mathematical functions for the objective.

5. *Design animal evaluation system.* Subject to constraints of reproduction and longevity, an animal evaluation system should be designed to measure and record relevant characteristics at appropriate times and to include these in the selection criteria in step 6. This testing system involves specifying the life cycle of test animals (along with the life cycle of near relatives such as sibs or progeny) with measurements to be taken before points of selection and breeding.

Most programs in swine include growth rate and backfat thickness measures taken at or near market weight such as 105 kg (or 230 lb). This allows selection of young animals for retaining in the herd or for sale shortly thereafter. Because of the importance of feed costs, voluntary feed intake is recorded in some corporate breeding programs and test stations; but, alternatively, it can be predicted from the association with growth rate and fatness.

The primary traits of reproduction that are normally recorded for each litter are number born alive, 21-day litter weights and number weaned and are considered to be characteristics of the dam, but do involve some influences of the sire. Selection criteria (step 6) from these traits allow further selection of parent females and their sires for deciding to continue in the breeding herd for each breeder. These traits do not include all traits deemed as important characteristics in Chapter 3, but are those traits which are practical to record prior to selection and have been shown to have reasonably high heritabilities. Other traits that might be recorded in future programs would be indicators of mature size, sexual maturity, fertility, and rebreeding interval for both male and female breeding stock.

6. *Develop selection criteria.* Each selection criterion should be developed as an appropriately weighted mathematical function of the relevant direct and indicator traits, including traits observed for relatives of the candidates for selection. Only observations recorded before that selection point in the cycle can be used. An optimum selection index should be used in preference to simpler criteria. The decision among the selection criteria possible at a specific point in the life cycle should maximize selection response for the objective of step 2 with the economic weights of step 4 and recorded data of step 5. Usually, the selection criteria will include observations on those traits of the bioeconomic objective function (step 2) that have been observed on the candidates for selection or their relatives prior to the specific selection point. The selection criteria will differ depending upon how the breeding is anticipated to be used in the breeding system—to produce terminal sires, parent females, or both. Information from parents and sibs, and more remote relatives, can increase the accuracy of the selection. Information from the progeny can greatly increase the accuracy of the selection criteria but the gathering of such information usually postpones selection until the animal is considerably older.

The phenotype of an animal is not a fully accurate indicator of its additive genetic value; the additive genetic value for related individuals will tend to be similar. Thus, the information from close relatives can increase the accuracy of evaluation relative to just using direct data on each individual, but because different individuals may have differing number of relatives of each type, the development of the optimum selection criteria for each individual in a herd including information from collateral relatives and ancestors is quite complex. Fortunately, the statistical methodology has been developed to the point that, with appropriate computerized calculations, optimum selection criteria (indexes) can be calculated for each animal in a herd, even utilizing information from relatives in other herds. This facilitates selection of the animals ranking highest on the criteria to maximize improvement. The alternative selection criteria developed for STAGES (Swine Testing and Genetic Evaluation System), a national program implemented by the eight U.S. swine breeds, will be presented in a later section.

7. *Design matings for selected animals.* Include in this portion of the design the numbers or proportions of candidates to be selected at each point for the life cycle of each sex of each breed or strain, keeping in mind that an adequate number of selected males and females should be available for mating at the specified breeding seasons for reproducing. Inherent in these decisions are the specification of the mating ratio of females to males and the number of breeding seasons over which selected individuals might be used. However, opportunities for culling previously selected animals should be allowed if progeny information shows that they are not as superior as initially evaluated. As stated earlier, inbreeding should be avoided in swine. A primary goal of designing breeding population sizes should be to maintain a population large enough to sustain sufficient genetic variability for long-term response to selection. Adequate population size is also necessary for stable responses and for supporting the expansion system specified in step 8. For herds within a breed, a greater effective population size is achieved through the purchase of a portion of the selected replacements from other herds with similar selection objectives.

8. *Design system for expansion.* Design a system so that genetic improvements developed in the testing, selection, and mating system can be disseminated into the production system(s) effectively and economically. This link between the herds practicing intense selection and the production herds was implied in the concept of the pyramid introduced in the first section of this chapter. Genetic improvements can be transferred and expanded concurrent with crossing of the improved strains or breeds in order to produce male and female parent-breeding stock or by using artificial insemination (AI) in order to sire production animals. Because testing and selection of breeding stock can be costly, coordination with an expansion and crossing program can prorate the cost of the selection program over more sales animals. When selection is intense in the primary herd, many unselected but superior performing animals will be available that can contribute genetic improvement into multiplication and crossing herds to produce breeding stock to be parents of production herds.

9. *Compare alternative combined programs.* Sometimes the final decision concerning the breeding system and breeds should be deferred from step 3 until this step. In this way, the final decision can be made not only on the merits of the breeding system as a specific part of a total production system, but also on how that purebreeding or crossbreeding system interacts with steps 5 to 8, allowing for improvement of the production system through selection. For example, the specialized selection for terminal crosses (see below) might yield selection responses different from those expected for general purpose selection for input into rotational systems.

CROSSBREEDING SYSTEMS AND THE PYRAMID

When purebreeding systems contributed directly to early production systems, it was through the sale of purebred boars to producers who had a commercial sow herd of similar breeding, but unregistered. The breeders were those who registered their animals with the breed society and participated in breed activities. Increased sales could be accomplished by convincing a producer of the superiority of a different herd or breed and "top-crossing" his sow herd with boars of the different breed or herd. In this form of breeding, the producer kept his replacement gilts from the young finishing animals to replace older sows being culled. Thus, he only purchased replacement boars from breeders. As a result, the flow of genetic improvement down the pyramid was only through the producer's purchase of registered purebred boars from breeders. The multiplication layer in the pyramid existed only in the form of lesser reputation breeders who purchased breeding stock from breeders with greater reputation. Crossing only occurred through top-crossing as a producer changed breeds.

Rotational Crossing

As the merits of crossbreeding became evident, the swine industry easily adapted the approach by the producers rotating through two, three, or four preferred breeds to provide replacement boars and saving the varied crossbred gilt progeny to become parents. With this form, the magnitude of heterosis is not as great, either in the sow herd and in the progeny, as other breeding systems. The advantage is in simplicity. The producer still produces his own replacement gilts and only purchases boars. The pyramid concept is similar, except that the different sides of the pyramid represent the different breeds in the system, with only boars moving between strata in the pyramid. Introducing boars from different breeds and, thus, different breeders can lead to problems of introducing pathogens from different herds into the producer herds. Some breeders have become breeders for three or four breeds so that they can sell boars, with relatively consistent pathogen status, to a particular producer every year. The sow herd and the progeny can become quite variable in color pattern and ear shape with this system, but this has become accepted. However, the variability in performance from rotating diverse breeds is not desirable.

Under this system, the three or four breeds involved really need to be similar in all traits to reduce variation in performance and, thus, in facility utilization over the cycle. For example, a breed that is superior in growth performance but inferior in maternal characteristics might produce good growing-finishing pigs the first year. But when replacement gilts are saved from those pigs, the sow herd might decline in reproductive performance.

Terminal Crossing

The terms, terminal cross or specific cross, are used for breeding systems more akin to corn and poultry breeding approaches. Because maternal characteristics are more important in parent females, it is advantageous to use first-cross or F_1 females that are superior for maternal performance. These can be produced from purebred boars of one breed mated to purebred females of another breed. If both of these breeds are superior in reproduction and lactation traits and there is heterosis between them, the F_1 gilts produced by crossing these breeds are likely to be superior and more uniform in performance than the variable gilts produced from a rotational system of the same three or four breeds. These F_1 gilts and sows can then be mated to a third breed that has superiority in growth and carcass characteristics but might even have deficiencies in the reproduction and lactation traits. Considering the combination of breeding herd performance and growing-finishing-carcass performance, it is likely that the best three-way cross from three breeds will give animal performance superior to the average performance from the fluctuating breed combinations in a rotational system.

Another specific or terminal cross would occur with the parent boars also being F_1 or first-crosses. F_1 boars from crosses of two breeds often exhibit heterosis for traits such as early sexual maturity, vigor, aggressiveness, and longevity. But such terminal crosses usually require four breeds or strains, and both of the breeds or strains producing the parent boars need to be superior in the growth-finishing-carcass characteristics.

For terminal crosses, the three or four sides of the pyramid can symbolize the three or four base breeds with breeding stock of both sexes moving down the multiplication strata for combining at the crossing strata to give the specific terminal cross for the production strata.

In the 1970s, some U.S. corporations began breeding swine with major efforts at developing, producing and merchandizing F_1 boars and F_1 gilts to be used by producers in their breeding herds to produce terminal cross (four-way cross) progeny for growing and finishing. The motivation for this system is its proprietary nature, in that the breeding corporation does not market its improved strains and the producers repeatedly purchase replacement breeding stock from the corporation. This structure is highly similar to that of corporate genetic activities in the corn and poultry industries. Pathogen control programs are usually an integral part of the multiplication and distribution system through affiliated herds.

Many producers, especially some of the larger ones, have become breeding stock customers of these corporations. This fact has partially resulted from the corporate breeders being able to provide larger numbers of replacement breeding stock all at the same time. Some purebred breeders have responded to this competition by moving into the breeding and merchandizing of F_1 gilts, and even F_1 boars, in addition to the merchandizing of purebred boars of two or three breeds. Sometimes the production of F_1 gilts is on separate, but organizationally associated, farms.

The trends described in the above characterization can be related to the pyramid concept with the recognition that some corporate breeders are now participating in all three top layers of the pyramid of Fig. 6-1. Part of the effectiveness of such organizational structure is due to a more coordinated dissemination of genetic improvements from selection into superior terminal crosses (step 8). In other words, the organized structure implied by the pyramid concept is fully achieved. However, most of these corporations also offer F_1 boars for use in rotational breeding systems as alternative products.

Most of these modern efforts utilize performance testing, data recording, and genetic evaluation techniques to make the selection at the top of the pyramid more effective and more directed towards the intended use in the terminal cross by the producer-customer in the middle layer of the pyramid. With such specific terminal cross systems, the selection in the individual breeds and strains can be specific to the characteristics of most importance to the manner in which the strain contributes to the performance of the production system. Thus, breeds and strains can be developed for specialized roles relative to being bred for general purposes in rotational systems.

Rota-terminal Crossing

Rotational breeding systems are seen to have some advantages in simplicity in that the producer saves replacement gilts from market animals and only purchases replacement boars for introduction into the herd. The advantages of terminal systems are the potential superior uniform heterotic performance and more specialized selection. A compromise between the two is called *rota-terminal crossing*. In this system, most market animals are produced by boars from specialized sire strains or F_1's, but the producer produces his replacement gilts from a two-breed rotation in a sub-breeding herd. Thus, only boars need be introduced, but three breeds or strains are necessary—one as the terminal sire and two maternal purpose breeds for rotation in a small sub-herd that produces both its own crossbred replacement gilts, plus crossbred gilts for input to the larger sub-herd. These latter gilts produce most of the market animals from matings to terminal sires. Because of the complexity of maintaining two sub-herds on the same farm, this approach is more often used on larger production farms. In this approach, boars from the component breeds are the primary route of disseminating genetic improvement down the pyramid.

These alternative systems have been reviewed in Ahlschwede et al. (1987) in Pork Industry Handbook 39. The reader is referred to that publication, where eight alternative crossbreeding systems were compared through simulation with a computer model of animal and herd performance reflecting breed and heterosis differences upon conception rate, pigs marketed per litter, days to market, fat thickness, and feed-to-gain ratio. Comparing the eight alternative systems, the three best systems were:

1. Four breed terminal system with Hampshire × Duroc F_1 males mated to Yorkshire × Landrace F_1 females for $94.76 net per litter average.
2. Rota-terminal system with Hampshire × Duroc F_1 males mated to gilt/sows from a three breed rotation of Chester White, Landrace and Yorkshire with $90.63 net per litter average.
3. Rota-terminal system with Hampshire × Duroc F_1 males mated to gilt/sows from a two-breed rotation of Yorkshire and Landrace with $88.37 net per litter average.

Among these eight simulated systems, the crossbreeding system with the lowest net value per litter of $59.27 was a three-breed rotational system involving Hampshire, Duroc, and Yorkshire. Such a system has frequently been used in the past. Terminal and rota-terminal systems seem preferable in economic terms, although rotational systems continue to be used in smaller herds due to their simplicity of implementation.

However, Ahlschwede, et al. (1987) emphasized that breeding herds differ, breeds and herds change over time, and market costs and price incentives change; these changes can cause shifts in the relative evaluations of these alternatives. Also, only systems involving Chester White, Duroc, Hampshire, Landrace, and Yorkshire breeds were simulated.

GENETIC IMPROVEMENT BY SELECTION

Selection of superior phenotypes in an effort to obtain superior genotypes has long been the practice in swine breeding. To the degree that phenotypes have been indicative of the desired genotypes and to the degree that these selected genotypes pass on their superiority to their descendants in the forthcoming generations, such techniques have been effective. The concept of heritability provides a measure of both these degrees. When the objective was to change the visual-appearing form, those visual-appearances, especially for skeletal size and arrangements, were heritable, meaning that transmissible genetic factors influenced their expression to a moderate degree. Often the desired objectives shifted, and breeders responded by changing body form in the direction of the new conceptualization of ideal body form.

Now, as the objective of swine breeding is focusing more on economically important functions, selection is still the appropriate tool. But, the phenotypes need to be objective measures of the functioning of the animals for commercially-relevant purposes. Therefore, data recording of performance that indicates efficiency of the reproductive processes and efficiency of the growth process converting feed intake into lean muscle are now the primary phenotypes that serve as the basis for selection. The focus is on improving the phenotypes of the crossbred descendants rather than just improving the purebred performance.

Performance Recording

The shift from visual appraisal to recorded performance data in swine began in the 1950s with the establishment of central test stations in several states for evaluating hogs for growth rate, fatness, and feed conversion. In early efforts, backfat thickness was measured by a metal ruler probe (Hazel and Kline 1952), but this approach was later replaced by electronic ultrasonic devices. Purebred breeders entered test pens (generally sets of sibs, usually boars) for testing and evaluation. Often, a sib barrow was included in the pen and slaughtered for carcass evaluation. Similar tests have been done earlier in Europe, especially in Denmark. This has been an effective tool for identifying commercially relevant performance. Unfortunately, in the sales at the end of the tests, the better performing boars often were sold into commercial herds, rather than back into purebred herds for breed improvement. The number of boars tested at test stations is only a small fraction of those needed by the industry. Often, test stations were used as merchandizing tools rather than as herd and breed improvement tools. But performance recording and evaluation has also come to be used in the nucleus herds of the corporate breeders and progressive purebred breeders. Sometimes this is done without the recording of feed intake, since considerable labor is required and research has shown that feed intake may be fairly well predicted from body size, growth rate, and fatness.

The other aspect of performance recording is the reproduction and lactation data or, as it is frequently termed, sow productivity. In the process of identification for registration, breeders have long recorded the number of pigs born alive. Following research at Ohio State University (Irvin and Swiger 1984), a Sow Productivity Index was developed to evaluate and compare sows through a function of number born alive (NBA) and the weight of nursed litter at 21 days of age (LW21). Standardization of litter size by cross-fostering is recommended and the recorded LW21 is mathematically adjusted for number after transfer (NAT). The SPI values for each parity are then combined into a predictor of the breeding value of that combined traits. These two values, NBA and LW21, give measures of the gestation and lactation functions of reproducing sows. The American Yorkshire Club has included SPI index values on the pedigree registration certificates for several years. This was a first step in the United States of bringing pedigree information and performance information into a synergistic relationship.

Numerous additional traits can be evaluated by appropriate testing and data recording. But major emphasis in most programs is on growth rate, backfat thickness, number born alive, and litter weaning weight and number as indicators of the genetically controlled physiological processes of lean growth rate, tendency to over-fatten, gestation, and lactation. Undoubtedly, as the swine breeding industry becomes more sophisticated, additional traits that are both economically important and under, at least moderate, genetic control will become parts of performance testing programs.

In the next few sections, we will develop how recorded performance information should be used to guide effective selection decisions by the breeder.

Components of Selection Response

A fundamental formula will now be presented to relate selection response to four components that simultaneously determine the rate of progress. This is:

$$\begin{pmatrix}\text{Rate of}\\ \text{Genetic}\\ \text{Improvement}\end{pmatrix} = \frac{\begin{pmatrix}\text{Intensity of}\\ \text{Selection}\end{pmatrix}\begin{pmatrix}\text{Accuracy of}\\ \text{Selection}\end{pmatrix}\begin{pmatrix}\text{Genetic Variability}\\ \text{for Objective}\end{pmatrix}}{\text{Generation Interval}}$$

As emphasized earlier, the objective of selection should be genetic change that will improve the economic efficiency of pork production systems. Genetic variability for this objective includes the performance differences under genetic control in those traits contributing to system efficiency. This is the raw material for the selection process to produce genetic improvement. In statistical terms, it is the standard deviation of additive genetic values for the composite bioeconomic objective, including several component traits that reflect the contributions to the profitability or economic efficiency of the pork production system. The other three terms determine how efficiently that raw material is utilized. Intensity of selection is related to how small a proportion of the tested population is selected and used to produce the next generation or cycle of selection, with smaller proportions giving higher intensities. But the smallness of the selected proportion is limited by the reproductive rate if the size of the breeding herd is to be maintained. Generation interval is the average age of the parents when the next generation is produced.

This is determined by the age when selected animals start producing offspring in the herd and how long they continue to be used to reproduce.

The remaining term in the numerator, accuracy of selection, involves several considerations. Statistically, it is the correlation between the selection criterion and the bioeconomic objective function. The selection criterion is a function of the traits observed on candidates for selection that is used to rank the candidates. From this ranking, the proportion with greatest values is selected. The traits in the selection criterion may not be precisely the same as in the bioeconomic objective function, but, for high accuracy, they will usually be similar. One possible exception is that survival or longevity of performance is an important part of the objective but cannot be directly included in an early-in-life selection criterion. The weighting of the traits in the bioeconomic objective will reflect the economic importance of that trait. In the selection criterion, the economic importance of the trait will be modified by the heritability of the trait and the interrelationships between traits. Greatest selection emphasis should be given to those traits with large economic importance, greater variability, higher heritability, and positive correlations with other traits. Selection emphasis should be reduced for traits with lower economic importance, lower variability, lower heritability, and negative correlations with other traits.

Since intensity, accuracy, and generation interval may differ between the sexes, the above formula needs to be expanded to:

$$\begin{pmatrix}\text{Rate of}\\ \text{Genetic}\\ \text{Improvement}\end{pmatrix} = \frac{\left[\begin{pmatrix}\text{Selection}\\ \text{Intensity}\\ \text{for Males}\end{pmatrix}\begin{pmatrix}\text{Selection}\\ \text{Accuracy}\\ \text{for Males}\end{pmatrix} + \begin{pmatrix}\text{Selection}\\ \text{Intensity}\\ \text{for Females}\end{pmatrix}\begin{pmatrix}\text{Selection}\\ \text{Accuracy}\\ \text{for Females}\end{pmatrix}\right]\begin{pmatrix}\text{Generic}\\ \text{Variability}\\ \text{for Objective}\end{pmatrix}}{\begin{pmatrix}\text{Generation}\\ \text{Interval}\\ \text{for Males}\end{pmatrix} + \begin{pmatrix}\text{Generation}\\ \text{Interval}\\ \text{for Females}\end{pmatrix}}$$

For greatest effectiveness of selection, the terms in the numerator need to be large while those in the denominator need to be small. But compromises are often necessary in designing a selection program to achieve a maximum rate of genetic improvement.

Often, the amount of information on different candidates may not be the same. All traits may not have been measured on all individuals. When measures on relatives are included to increase accuracy, some individuals may have more full or half sibs, cousins, etc., than others. Thus, the procedures become increasingly complex to include this variable information. Also, difficulties are encountered in accounting for contemporary group, sex, parity, and other extraneous effects. Statistical procedures have been developed and are available to breeders in the form of computer programs to do the complex calculations. These will be described in the next section. But the basic importance of intensity, accuracy, and generation interval in manipulating variabilities as the raw material for genetic improvement continues through these computerized procedures.

Much research and development of statistical procedures has gone into modern genetic evaluation of swine breeding stock. The basic principles of index selection were presented by Hazel (1943) and elaborated by Henderson (1963).

Procedures of handling unequal numbers of records, irregular relationship patterns and the incorporation of relatives in different contemporary groups and herds were consolidated by Henderson (1973) into a statistical approach termed Best Linear Unbiased Prediction (BLUP). These procedures can be combined with the concepts of bioeconomic objectives initiated by Dickerson (1970) and Harris (1970) to provide the foundation for sophisticated genetic evaluation on modern computers. A major set of procedures directed towards selection in swine is described in the next section.

SWINE TESTING AND GENETIC EVALUATION SYSTEM (STAGES)

In this section, a comprehensive genetic evaluation procedure utilizing recorded performance data from purebred swine will be reviewed. This procedure was developed by researchers from Purdue University and the Agricultural Research Service of USDA (Harris et al. 1990) and has been implemented as a computerized data processing service by the eight breed associations of the National Association of Swine Records (see Appendix 1, "Purebred Swine Breed Associations"). This collection of procedures provides for evaluation of purebred swine both across contemporary groups within herds and across herds of purebred swine of the same breed, if those herds are "connected." Connectedness comes from having an adequate number of related animals in the different herds. The form of the procedures is such that the results describe the expected genetic progress from the selection that the breeder practices. A noteworthy point in the development and implementation occurred in 1990 with the execution of the first across-herd sire evaluation for the American Yorkshire Club. Such analyses will follow for other breeds as their data bases expand to a size to support such an evaluation.

Some specific states have alternative computerized procedures. An early progressive program was implemented in Nebraska in association with specific pathogen free (SPF) control programs. There also are available computer software packages for evaluations on microcomputers. Thus, they can be performed with microcomputers on the farms of breeders. These alternatives only provide comparisons within that single herd and do not fully utilize all the indirect genetic information coming from the performance data on relatives of the candidate in other herds or evaluate the long-term progress resulting from the selection.

The statistical procedures incorporated into the Swine Testing and Genetic Evaluation System (STAGES) computer software facilitate full use of the information from relatives and make optimum statistical adjustment for the nongenetic differences between contemporary groups and between herds. Some of the corporate swine breeders utilize similar BLUP-based computer programs to support their selection decisions to genetically improve the breeding stock they offer for sale (see Appendix 1, "Corporate Breeding Programs"). Although some differences in approach were included, the intended utility of these procedures was patterned after successful genetic evaluation procedures in the beef cattle breeds in the United States.

Decision–Support System for Breeders

The STAGES computer software and the genetic evaluations they produce have been described as a decision-support system (Harris et al. 1990). This means that the evaluations support the selection decisions that a breeder must make as he strives to improve his herd. Figure 6.2 presents the more important decisions. First the breeder needs to have a breeding (bioeconomic) objective. STAGES offers three alternative objectives (and corresponding selection indexes) depending on the expected use in crossing systems by the purchasers of the breeder's breeding stock. Other major decisions involve the selection of replacement boars and gilts to become parents of future contemporary groups. Those animals not selected for use within the breeder's herd are likely to be offered for sale to other breeders or to producers to use in crossing systems. These customers might be other breeders or F_1 gilt producers. The value and, thus, the selling price should be related to their expected performance and their descendants' expected performance in the customers' production herds. The genetic evaluations for the important traits and the selection criteria provide a sound genetic and economic basis for the buyer and seller to negotiate the selling price.

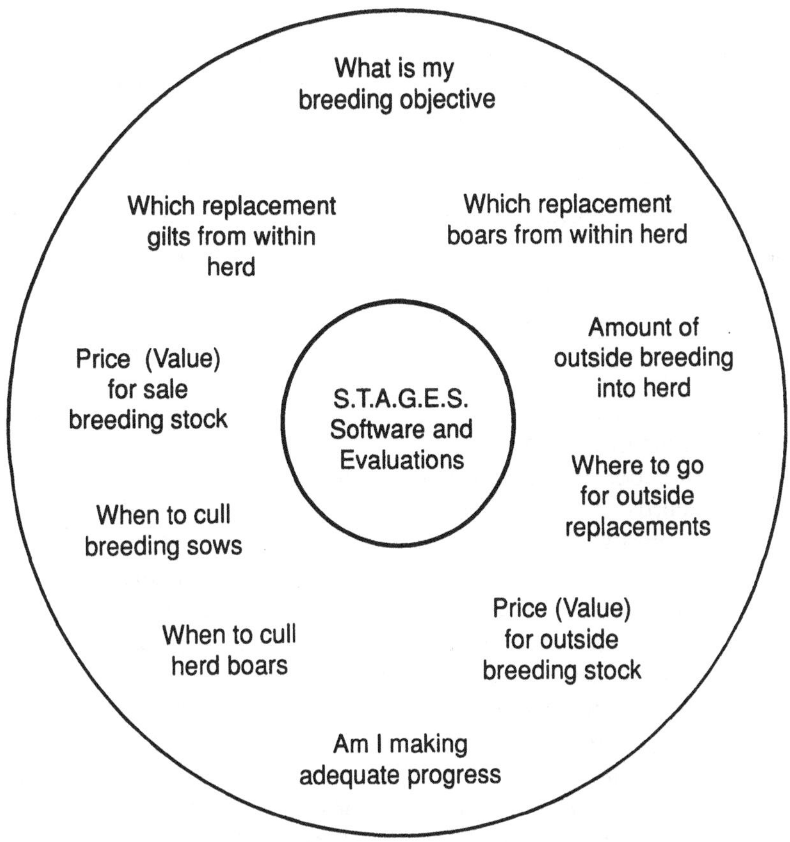

FIGURE 6.2 Decision-support *system* for purebred breeders.

Most breeders should not select all their replacements from within their own herds. If they do, they will likely encounter inbreeding problems when all the members of their herd become related to each other. Therefore, they must decide how much outside breeding to introduce into their herd, who to purchase the outside replacements from, and how much to pay. Breeders need to purchase outside stock that is as good as, or preferably better than, what they can select within their herd and pay for replacements in relation to how much the stock will improve their herd. How much outside breeding depends on both avoiding inbreeding and on the genetic level of the breeder's animals relative to those in other herds. Genetic evaluations across herds provide the comparisons to guide those decisions. This is important because, even if the breeder is making accurate selections within his herd, mistakes in purchasing outside replacements can disrupt his selection program. This is especially hazardous and risky when purchasing and heavily using outside boars in small herds.

The answer concerning when to cull herd boars and breeding sows has to be when they can be replaced by young animals with superior genetic evaluations for the appropriate index. To keep generation interval (in the denominator of previously presented formulae) short to increase the rate of genetic improvement, most animals in an elite breeding herd practicing intense selection should not be kept in the herd for their full reproductive life. An exception can be made if they continue to have genetic evaluations superior to possible replacements. Since STAGES evaluations are comparable with different amounts of information, they provide a basis for comparing animals with several litters of progeny with young animals just coming off a growth test.

Without going into the statistical details of the complex calculations of STAGES, the resulting values can be described as the multiple regression of the unknown true progeny deviation upon the observed values for the individual and his relatives. Alternatively, the calculations provide the expected progeny deviations (EPD) which have the smallest variance of the deviations of predicted values from the conceptual true progeny values. With more observations on the individuals or their relatives, this variance will be smaller indicating predictions closer to the true values. Thus, the accuracy of selection will be greater when there is more information upon which to base the calculations. Greatest accuracy (coefficient of multiple determination) can be achieved when the animal being evaluated has produced large numbers of progeny whose traits have been observed and recorded in addition to records on itself, ancestors and collateral relatives. But the potential for accurate selection with progeny testing is offset by a greater generation interval, so that systematic progeny testing may actually lessen the rate of genetic improvement. For traits observed on individuals of both sexes and with moderate heritabilities, the most effective selection is accomplished prior to breeding age with use of records on the individual, its sibs, its parents and ancestors, and other collateral relatives. If a boar is selected for heavy use in the herd from this information, a progeny test will inevitably occur as performance of his progeny are recorded. The information on these offspring is of greater value as selection information on themselves and their sibs as they approach breeding age.

The traits largely determined by the parent female, such as number born alive, number weaned, and litter weight prior to weaning, cannot be directly measured on male candidates for selection. Therefore, it becomes necessary to initially predict a boar's true genetic potential from observations on his dam, his sisters, and more remote female relatives. Later, observations on his female progeny will increase the accuracy of his evaluation. However, progeny information is only likely to be obtained on animals with good evaluations early in life for the growth traits and the reproductive traits based on female ancestors and collateral relatives.

The STAGES expected progeny deviations (EPD) are predictions of how the progeny of an animal are expected to perform relative to those of an average animal. They are expressed as progeny deviations even if the animal being evaluated has not yet had progeny. This form of presentation is pertinent, because the purchaser of breeding stock is expecting to benefit from the differences in progeny performance and should be willing to pay more for greater expected performance. This combining of information from relatives also involves the heritabilities of the traits, as a measure of expected similarities between relatives, as well as the genetic and phenotypic correlations between traits.

Bio-economic (Breeding) Objective

In addition to combining the information from several related individuals, the information from the different pertinent traits is combined together into selection criteria or indexes in STAGES. These indexes have been chosen to reflect the relative economic importance of differences in each trait as they impact the economic efficiency of the pork production system of the breeding stock purchaser. Because there are different ways that breeding stock from different breeds may be used in a breeding system, alternative selection indexes are presented in the computer output. These were obtained by first determining the mathematical function describing the contribution of the traits to the profitability of the production system (step 2 of the "Systematic Swine Breeding" section), determining the economic weights for those traits (step 4), and putting those together with the recorded data (step 5) into the appropriate selection criteria (step 6).

First, we will consider the growth and fatness traits expressed in the period from weaning to market. Stewart et al. (1990) reviewed the average costs and returns and determined the four fundamental economic values presented at the top of Table 6.1. In conjunction with these, the four assumed means for current crossbred commercial performance are presented in the bottom half of the table. A

TABLE 6.1 Economic Constants: Growth and Fatness

Base Market Price	$0.45 per lb
Backfat Penalty	10.5% per inch above 0.75
Cost Feed	$0.07 per lb
Cost Growing-Finishing Space (labor and facilities)	$0.17 per pig per day
Assumed Means (Crossbred Commercial)	
Market Weight	230 lb
Days 15# to 230#	150 d
Feed 15# to 230#	215*3.5 = 752.5 lb
Backfat	1 in.

post-weaning index (PWI) is presented below representing the income from a 230-lb market hog minus the costs of producing that animal.

$$\begin{aligned} PWI = {} & 230.0 * (.974 + EPD_{Vlu}) * 0.45 \\ & - 150. + EPD_{days}) * 0.17 \\ & - (752.5 + EPD_{feed}) * 0.07 - \mu_{PWI} \end{aligned}$$

where * indicates multiplication and/is read as "per". The income term of the function involves an EPD for value as a function of the EPD for backfat reflecting a 10.5% discount for fatness above 0.75 inch (and a premium for lower backfat). The second term relates the labor and facility cost to the days (indicating growth rate) from 15 lb weaning to 230 lb marketing. The third term predicts progeny feed costs for that period with an EPD for feed being a function of the EPD for growth rate and backfat. As a linear function of EPDs, PWI is also an EPD. To present in the form of an index centered around 100, the PWI is converted to a terminal sire index (TSI).[1]

$$\begin{aligned} TSI &= 100.0 + 10 * PWI \\ &= 100.0 + \$/10 \text{ pigs} \end{aligned}$$

The units of PWI is dollars per pig and the units of TSI is dollars per ten pigs. TSI is the appropriate index for selection within a breed or herd primarily merchandizing boars to use as terminal sires in a three-way cross. Multiplication of the deviations from 100 by one-tenth of the expected number of progeny from a boar would give a prediction of the boar's contribution to the profitability of the customer's production system.

Table 6.2 presents economic weights and assumed means for the breeding sow herd where the reproduction and lactation traits are expressed. A function for sow productivity index (SPI)[2] is presented below to combine the EPD for traits NBA (number born alive), NW (number weaned), and LW21 (21-day litter weight) with an intermediate EPD of days to weaning (DW).

$$\begin{aligned} SPI = {} & (8.5 + EPD_{NW}) * 22.45 \\ & + (15.50 - 0.33) * EPD_{NBA} \\ & - (\mu_{DW} + EPD_{DW}) * [1.50 + 0.035 * (8.5 + EPD_{NW})] \\ & - \mu_{SPI} + 100.0 = 100.0 + \$/\text{litter} \end{aligned}$$

where

$$EPD_{DW} = \frac{(8.5 + EPD_{NW})\,(15.0 - 3.0) * 21.0}{(8.5 * 15.0 + EPD_{LW21}) - (8.5 + EPD_{NW}) * 3.0}$$

with

$$\mu_{DW} \cong 21.0$$

[1] TSI deviations from 100.0 predict differences in net value per 10 pigs.

[2] SPI deviations from 100.0 predict differences in net value up to weaning for a litter of pigs.

TABLE 6.2 Economic Constants: Reproduction and Lactation

Value (cost) of weaned pig (15#)	\$22.45/pig
Value (cost) of baby pig	\$15.50/pig
Costs of handling baby pig	\$ 0.33/pig
Costs for farrowing house (labor and facilities)	\$ 1.50/sow/day
Extra costs per pig weaned (feed and labor)	\$ 0.035/pig/aay
Assumed Means (Crossbred Commercial)	
Number after transfer = Number born alive	10 pigs/sow
Number weaned	8.5 pigs/sow
Wt at birth	3 lb/pig
Wt at weaning	15 lb/pig

This index function predicts the value of a weaned litter at 15-lb, plus the value of transferred pigs at farrowing, minus the cost in days and extra feed and labor for producing that litter. This index is not really a recommended selection criterion but is the only index calculated by STAGES when the post-weaning traits are not recorded. Breeders concerned primarily with herd improvement to produce F_1 gilts should use the MLI below.

When growth and fatness traits are observed and recorded in addition to the reproduction and lactation traits, two additional functions (and indexes) are calculated by STAGES.

Maternal Line Index

$$\begin{aligned} MLI &= (8.5 + EPD_{NW}) * 0.5 * EPD_{PWI} \\ &\quad + (15.50 - 0.33) * EPD_{NBA} \\ &\quad - (21.0 + EPD_{DW}) * (1.50 + 0.035(8.5 + EPD_{NW})) \\ &\quad - \mu_{MLI} + 100.0 = 100.0 + \$/\text{litter} \end{aligned}$$

General Purpose Index

$$\begin{aligned} GPI &= MLI + (8.5 + EPD_{NW}) * EPD_{PWI} \\ &\quad - \mu_{GPI} + 100.0 = 100.0 + \$/\text{litter} \end{aligned}$$

A maternal line index (MLI)[3] is the appropriate index for breeding stock used to produce F_1 gilts. It appropriately combines the five traits reflecting the contribution to the performance of both the sow herd and growing-finishing pigs. The general purpose index (GPI)[3] combines the same traits but gives greater emphasis on the post-weaning traits as is appropriate for boars going into a rotational breeding system.

The breeder needs to choose the index most pertinent for the majority of his customers to direct his herd improvement. But the other index evaluations are available foı customers whose intended use of breeding stock may depart from the primary objective of the breeder. The breeder will probably not select solely on his preferred index. Some selection on structural soundness and avoidance of too many selected individuals from the same families will likely be used to avoid problems in breeding performance and later inbreeding. With the across-herd

[3] GPI and MLI deviations from 100.0 predict differences in net value for a litter of pigs.

evaluations, most breeders will search for superior breeding stock from other breeders to improve his herd and avoid inbreeding.

Documenting Genetic Trend

Together, the STAGES computer software providing within- and across-herd evaluations provides the breeder modern quantitative genetic technology to accelerate his efforts at genetically improving his herd towards commercially relevant objectives.

Figure 6.3 illustrates how the STAGES evaluations separate the genetic changes in the boars and sows in a breeding herd from the nongenetic trends and fluctuations. The symbols S and D reflect the mean sire and dam EPDs for a herd for each contemporary group, whereas M represents the nongenetic environmental and management changes. Thus, STAGES not only provides support for the selection efforts of a breeder, it provides evaluation of both how much improvement he is making and how well his stock performs relative to other breeders.

Progressive breeders have added recording of performance data to their breeding procedures and are utilizing STAGES or similar computerized procedures to gain full benefit of these efforts. Breeders not utilizing such technology and only relying on "eye-ball" techniques are not expected to survive as profit-making suppliers of breeding stock to modern pork production systems.

Elaboration on many of the points in this chapter can be found in the seven fact sheets of the "Breeding and Genetics" section of the Pork Industry Handbook (see

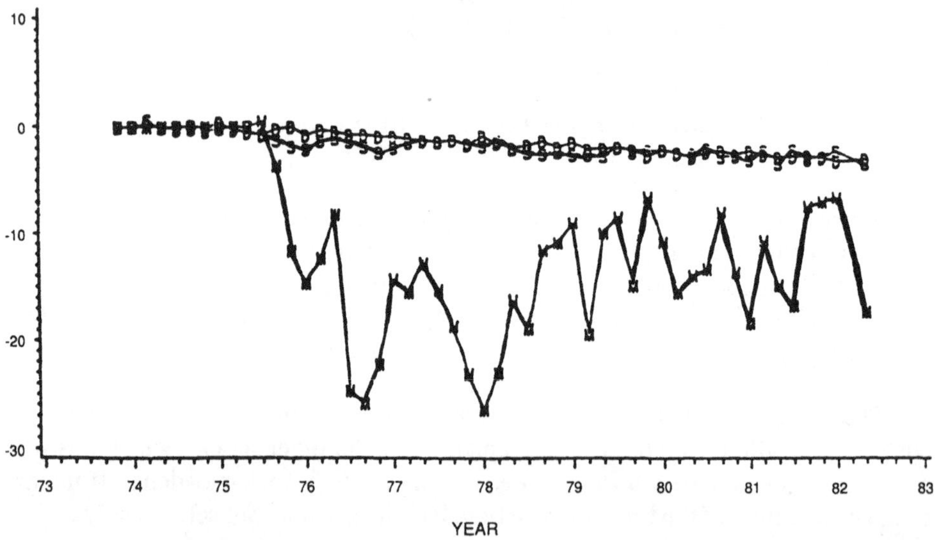

FIGURE 6.3 Plot of STAGES-calculated mean expected progeny deviations (EPD) for days to 220 lb for contemporary groups in one herd, from 1973 through 1982, illustrating genetic trend from selection practiced. S are sire EPD values, D are mean dam EPD values, and M are deviations of observed mean from overall mean plus S and D values. Taken from Harris et al. 1990.

Appendix 5a). Thirteen other similar fact sheets are available through the National Swine Improvement Federation (see Appendix 5b).

Merchandizing of Breeding Stock

Progeny deviations for the relevant production traits and indexes are precisely what the producer is desiring when he purchases replacement breeding stock. Thus, the EPDs and indexes for tested animals give the best objective measures of the value of breeding animals. As a result, these evaluations are not only of value to the breeder as he selects breeding stock to improve his purebred herd, they are of value to the producers as they purchase breeding stock to improve their crossbred herds.

The buyer should ask to study the breeder's genetic evaluations on animals for sale, realizing that the breeder will be keeping the best animals for further improvement. The producer should try to purchase the next best animals on the index appropriate for his use (TSI, GPI, or MLI). But he should expect to have to pay more for the "next best" animals. Payment-for-value is the motivating force that encourages improvement for most industries. In recent years, more accurate payment-for-value of slaughter animals has pushed producers to make a greater effort at controlling fatness. More accurate payment-for-value with premiums for breeding stock with documented evidence of likely superior production performance of their progeny will stimulate breeders to compete for those premiums. Payment-for-value can only occur if there is an objective, credible measure of that value. Performance testing and genetic evaluation provide that objective measure. The STAGES procedures—with documentation of genetic trend and across-herd referencing—provide the most credible and comparable measures. Of course, the accuracy of the evaluations will depend on the accuracy of the recorded data. Since the individual breeder is responsible for data recording, the breeder is responsible for the credibility of his or her records. But the breeder has always had the burden for the credibility of his or her animal's appearance and performance.

PERSISTENT MYTHS

Some concepts from the past heritage of swine breeding continue to persist in the minds of breeders not fully conversant with the research results of recent decades. At the risk of ending this chapter on a negative note, we need to examine some of the quotations from old animal breeding principles that were presented early in this chapter. Like most myths, these ideas do have some basis, but some have been exaggerated beyond their true worth and, thus, have slowed the adoption of better procedures.

Robert Bakewell, the British breeder of the 1700s, was quite correct in his recognition that "like produces like" and the appropriate breeding technique is to "breed the best to the best." However, his view that "inbreeding produces prepotency and refinement" while partially correct, can be misleading. Inbreeding does lead to prepotency, as evidenced by highly uniform single cross hybrid corn produced from highly inbred lines. The "refinement" noted by Bakewell seemingly was the reduction in vigor, now usually termed "inbreeding depression." This lack of vigor is expressed in reduced performance, especially for the reproduction traits, and this is not practical for modern pork production systems.

"Form determines function" was a myth that was greatly exaggerated. Some aspects of form (body shape and visual appearance) do indicate function (performance), but so much of the functioning of swine and other classes of livestock are either not indicated or are poorly indicated by "form," the modern view is that strong selection for performance (function) should be practiced and the results of this will determine what is the best "form."

Coupled with the "form determines function" view was the myth that the "eye of the master" was the appropriate breeding tool. This myth was exaggerated by breeders who promoted the idea that they had some mystical ability for breeding better livestock. Although computerized statistical calculations may seem as mystical to the uninitiated, they do have a real scientific basis.

It has been stated that the "art of breeding" practiced by livestock breeders of past centuries evolved into the "science of breeding" in the first half of the twentieth century. It seems that the "technology of breeding" coming from those scientific developments is going to fully replace the "art of breeding" by the beginning of the twenty-first century. The beneficiaries of these trends will simultaneously be the progressive breeders adopting the technology, the commercial pork producers who purchase their breeding stock, and ultimately, the consumer who purchases lower cost, more nutritious lean pork products.

REFERENCES

Ahlschwede, W. T., Christians, C. J., Johnson, R. K., and Robison, O. W. 1987. Pork Industry Handbook 39.

Cooperative Extension Service. 1976. Pork Industry Handbook.

Dickerson, G. E. 1970. J. Anim. Sci. *30,* 849.

Einstein, M. E., Stewart, T. S., Lofgren, D. L., and Schinckel, A. P. 1988. Purdue Agric. Exp. Sta. Bull. no. 545. 47 pgs.

Falconer, D. S. 1989. Introduction to Quantitative Genetics 3rd ed. Ronald Press, New York.

Harris, D. L. 1970. J. Anim. Sci. *30,* 860.

Harris, D. L., Lofgren, D. L., Stewart, T. S., Schinckel, A. P., and Einstein, M. E. 1990. S.T.A.G.E.S. National Decision Support System for U.S. Swine Breeders. Proc. Ann. Mtg. Natl. Swine Improvement Federation, Toronto, Canada.

Harris, D. L., Stewart, T. S., and Arboledo, C. 1984. Animal Breeding Programs: Systematic approach to their design. USDA-ARS-AAT-NC-8.

Hazel, L. N. 1943. Genetics *28,* 476.

Hazel, L. N., and Kline, E. A. 1952. J. Anim. Sci. *11,* 313.

Henderson, C. R. 1963. *In* Statistical Genetics and Plant Breeding (eds., W. D. Hanson and H. F. Robinson) NAS-NRC Publ. *982,* 225.

Henderson, C. R. 1973. *In* Proc. Symp. Anim. Breed. Genet. Symp. in honor of Dr. Jay L. Lush, pp. 10–14. ASAS, Champaign, Ill.

Hunsley, R. E., Beeson, W. M., and Nordby, J. F. 1978. Livestock Judging, Selection, and Evaluation, 2nd ed. Interstate Printers and Publishers, Danville, Ill.

Huntsberger, David H., and Billingsley, P. 1986. Elements of Statistical Inference, 6th ed. Wm. C. Brown, Dubuque, Iowa.

Irvin, K. M., and Swiger, L. A. 1984. J. Anim. Sci. *58*, 1144.

Lush, J. L. 1945. Animal Breeding Plans. (3rd ed.) Iowa State University Press, Ames.

National Swine Improvement Federation. 1987. Swine Genetics Fact Sheets.

Pirchner, Franz. 1983. Population Genetics in Animal Breeding, 2nd ed. W. H. Freeman, San Francisco.

Steele, R. G. D., and Torrie, J. H. 1980. Principles and Procedures of Statistics: A Biometrical Approach. McGraw-Hill, New York.

Stewart, T. S., Bache, D. H., Harris, D. L., Einstein, M. E., Lofgren, D. L., and Schinckel, A. P. 1990. J. Anim. Breed. Genet. *107*, 340.

7

Reproduction and Lactation

A high reproductive rate is essential in a successful purebred or commercial swine enterprise. A knowledge of the normal anatomy and physiology of the female and male reproductive tract is fundamental to the achievement of a high reproductive rate of swine. In this chapter, the following topics are discussed: anatomy of the female and male reproductive tracts, puberty, the estrous cycle, ovulation, factors affecting ovulation rate and time of estrus, movement and survival of sperm in the female reproductive tract, conception rate, duration of gestation, uterine capacity, parturition, male puberty and sexual behavior, composition of semen, sperm survival and storage, artificial insemination, and potentially important techniques, though presently in limited use, such as embryo transfer, artificial control of the estrous cycle, superovulation, and storage of sperm by freezing. Each of the above subjects is of significance in the total picture of reproduction. Overviews of current knowledge of male and female reproduction in swine are recommended (Dziuk 1977; Dziuk and Bellows 1983; Bazer and First 1983; Amann and Schanbacher 1983; Kirkwood and Aherne 1985; Christenson 1986).

Milk ingestion, particularly colostrum, is important for the survival of the newborn pig. It is possible to wean pigs at birth, but sow's milk is still considered preferable for the suckling-age pig. A knowledge of the yield and composition of sow's milk is needed as the basis for making decisions about sow and piglet feeding and management (Lecce 1986; England 1986).

THE FEMALE

Reproductive physiology of the sow has been reviewed thoroughly by Boda (1959) and more recently specific advances and new technologies have been addressed

(Christenson 1986; Tumbleson 1986) that will be useful for students interested in greater depth of coverage than can be considered here.

ANATOMY OF THE REPRODUCTIVE TRACT

The reproductive tract of the sow includes the ovaries (right and left), the oviducts (fallopian or uterine tubes), uterus, consisting of the body and two uterine horns, cervix, the vagina, and the vulva. A normal reproductive tract is illustrated in Fig. 7.1. The uterine horns are long and tortuous to accommodate numerous developing fetuses. In the mature sow, they may be 1 m or more long when extended, while the body of the uterus is about 5-cm long, and the cervix and vagina each about 10-cm long. The muscular nature of the cervix is also shown in Fig. 7.1. In natural mating, the penis penetrates the first few cervical folds or beyond. The ovaries are lobular, owing to follicles in varying stages of development, and vary in weight from 3 to 7 g and in diameter from 2 to 4 cm. There may be 10 to 25 individual mature follicles each 8 to 12 mm in diameter. The length and weight of the uterus of pre- and

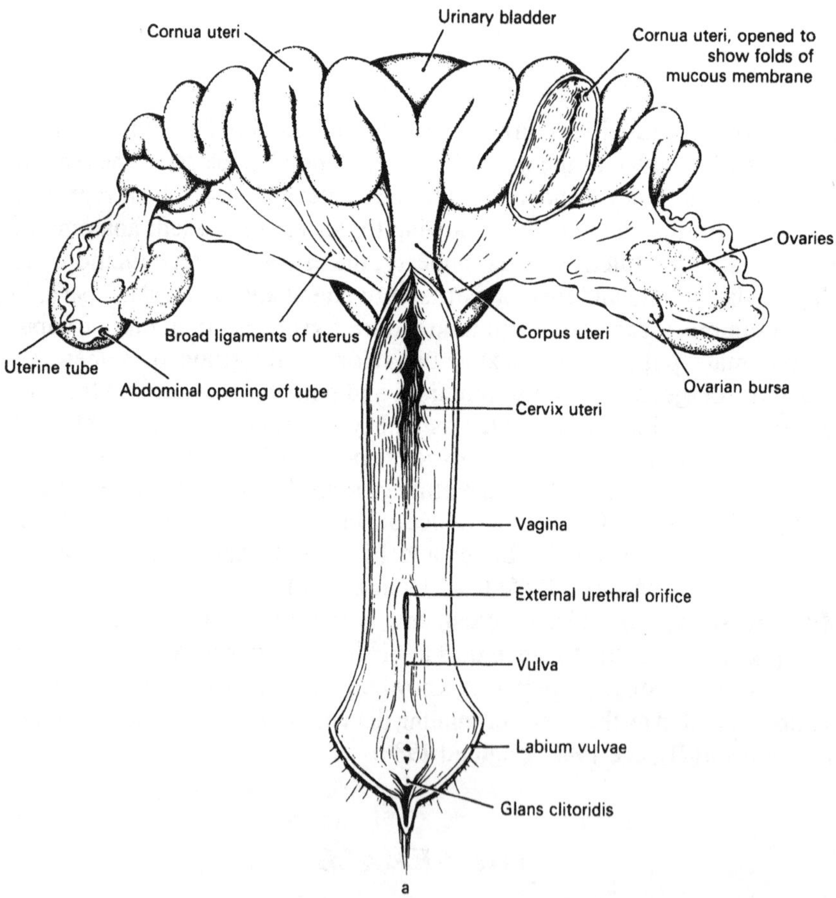

FIGURE 7.1 Normal reproductive tract of the female pig, showing muscular folds of cervix. From Sisson and Grossman 1938.

postpubertal gilts is shown in Table 7.1 (Wu and Dziuk 1988). A normal ovary during early pregnancy is shown in Fig. 7.2. The fallopian tubes are 15 to 30 cm long with a large ovarian end; the vagina length averages about 30 cm.

PUBERTY

Puberty (onset of estrous cycle) occurs at 6 or 7 months of age in most breeds, although in some Chinese breeds it occurs at less than 4 months. The number of ova released per estrus increases gradually over the first several estrous cycles. There is a large range even among normal animals kept under adequate husbandry conditions. Breed is an important factor in the onset of first estrus. Christenson and Ford (1979) reported that the percentage of Landrace gilts that showed regular estrous cycles at 6 months of age was higher than that of Duroc, Hampshire, Large White, and Yorkshire gilts. Puberty is also influenced by season of year (Christenson 1981), social environment, and confinement (Christenson 1981; Christenson and Ford 1979; Rampacek et al. 1981). The percentage of gilts cycling at 9 months of age is higher during the winter than during the summer season. Gilts raised in confinement often show delayed puberty and a higher incidence of behavioral anestrus (silent estrus). Behavioral anestrus is influenced by breed and season of year (higher incidence in summer) as well as by confinement. Inbreeding delays onset of first estrus by several weeks (Warnick 1951), and it is retarded by excessive fatness (Self et al. 1955). Relocation or exposure to a boar consistently advances puberty (Zimmerman et al. 1974). Introduction of a boar to gilts 165–190 days of age that have been previously isolated from boars seems to induce estrus. The adrenal gland may facilitate the onset of puberty through increases in glucocorticoid production, but this action is not obligatory (Killian et al. 1987). Although age at puberty is several months, the number of ova available for release in the course of a lifetime, determined in prenatal life as oogenesis, is complete by day 100 after conception (Block and Erickson 1968).

ESTROUS CYCLE

The estrous cycle length is generally discussed in terms of the estrous periods since it is the important phase from the standpoint of planning of breeding dates. The length of the cycle in gilts and sows (onset of one estrus to the onset of the next)

TABLE 7.1 Length and Weight of the Pig Uterus[a]

	Puberty	*Postpuberty*
Weight, g	36 (n = 40)	161 (n = 12)
Length, cm[b]	40 (n = 72)	120 (n – 12)

[a] Pre- and postpubertal gilts were 5 to 6 months of age when measurements were taken.

[b] Length of uterine horn (left and right selected randomly for measurement). Coefficient of variation for length of uterine horns is about 30%. Measurements must be made under uniform conditions of storage and handling to give meaningful results.

Source: Wu and Dziuk (1988).

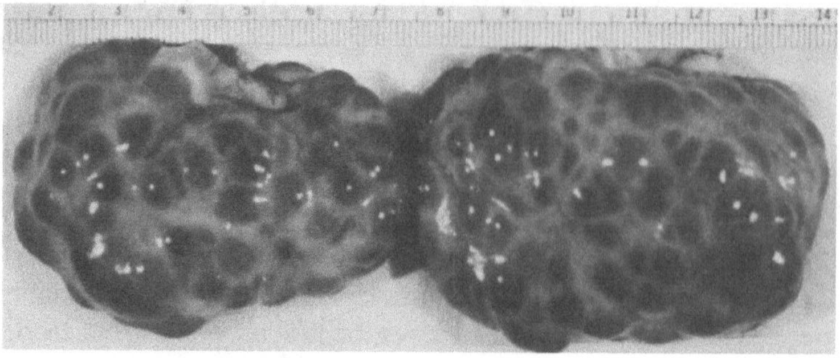

FIGURE 7.2 Normal ovaries from sow at 35 days pregnancy. Note numerous follicles over entire ovary surface. Scale in centimeters. Photo courtesy of R. K. Christenson.

averages approximately 21 days (the range is about 18–24 days). The length of the cycle does not seem to be related to the duration of estrus. Ovulation without external manifestations of estrus (silent estrus) creates problems for the swine manager. Gilts reared in total confinement have a high incidence of abnormalities in the estrous cycle, including delayed puberty and behavioral anestrus ("silent estrus"). Christenson (1981) reported differences in breed, housing (confinement increases), and season (hot weather increases) in these phenomena. The underlying physiological bases for these observations are not clearly understood, but the results illustrate the ineffectiveness of pen mating behaviorally anestrous gilts. This phenomenon is of concern in attempts at synchronization of estrus by administering progestational agents.

The cycle can be divided into proestrus, estrus, and diestrus. During proestrus (1–3 days), females are alert to the approach of the boar, will mount other females and accept mounting by diestrus females, but will not tolerate mounting by a boar. During estrus, the swollen vulva and vaginal discharge first observed during late proestrus are accompanied by restlessness (fence-walking, agitation), mounting of other animals, frequent mounting by other females, and frequent urination. When pressure is applied to the back or rump, the animal takes on a rigid stance and has a characteristic ear carriage. The external signs of estrus may last for a period of 3 to 4 days, or longer, but the female will mate only for a period of 2 to 3 days. These behavioral patterns are reviewed by Hemsworth (1985).

Sex pheromones synthesized in the testes of the boar and transferred to the salivary gland, from which they are secreted during premating behavior, are probably important in enhancing receptivity of the female to the boar. The steroid 5-androst-6-en-3-ol is one such hormone.

Attempts have been made to relate vaginal smears to stage of estrus to utilize more effectively artificial insemination. These attempts have been largely unsuccessful.

ARTIFICIAL CONTROL OF THE ESTROUS CYCLE

Since the estrous cycle is under endocrine control, it is possible to alter it by administering exogenous hormones. Several ways are available, including parenteral gonadotrophin administration after weaning of a follicle-stimulating material (pregnant mare serum gonadotrophin, PMSG) or PMSG followed 56–72 hours later with human chorionic gonadotrophin (HCG), an ovulating hormone; parenteral gonadotrophin administration to the cycling female of PMSG followed by HCG 96 hours later (dosage level is critical; improper levels result in a high proportion of gilts failing to show estrus; dosages of 1200 IU of PMSG and 500–1000 IU of HCG have given good results); oral administration of estrogens followed by progesterone derivatives or of a progestogen, allyl trenbolone (Davis et al. 1979), to the cycling female; oral administration of a dithiocarbamoylhydrazine derivative that affects the hypothalamus so as to inhibit ovulation (this drug, sometimes called methallibure or ICI 33828, has received widespread attention as a means of controlling estrus without problems with cystic ovarian follicles, but it produces teratogenic effects when ingested during pregnancy, so probably will not find commercial use); injection of 10 mg of prostaglandin $F_2\alpha(PGF_2\alpha)$ on days 13 and 14 after inducing accessory corpora lutea with PMSG-HCG administration in cycling sows (since PGF_α is not luteolytic in the pig until day 12 of the estrous cycle, it does not offer a practical means of synchronizing estrus when used alone, according to Hansel and Convey 1983). As more is learned about the physiological effects of prostaglandins, further refinement can be expected in ovulation control methods.

Orally active progestational compounds and nonsteroid inhibitors of anterior pituitary hormone activity hold promise for application in controlled estrus programs to be used in conjunction with artificial insemination. This also offers a practical means of implementing a closely planned farrowing schedule, whereby groups of sows farrow together. The application is foremost with first-litter gilts, as the estrous cycle can be controlled effectively in older sows by a systematic weaning program.

OVULATION

Ovulation occurs late in estrus, usually on the second day after the onset of estrus. There seem to be genetic differences in the length of the estrous period and in the time at which ovulation occurs. Ovulation is delayed with increases in the duration of estrus (Burger 1952). The time span for all ova to be released in a given estrus may be up to at least 7 hours (Burger 1952; Ito et al. 1959). Ova released early are fertilized sooner and may have lower embryonic mortality than those released toward the end of the 7 hour period.

The pig ovum is easily visible under a low-power microscope (120 to 170 μ diameter). Figure 7.3 shows an unfertilized ovum and a series of three stages of fertilization. It is possible to retrieve ova from the flushed oviduct by scanning the fluid collected in a petri dish under low magnification. The ova are not motile but are moved down the oviduct by its peristaltic action. When the number of ova released from the two ovaries is unequal at a given estrus, there is a tendency for fertilized ova to migrate to the uterine horn corresponding to the ovary that released the fewest ova. Unilateral ovariectomy, in which all ova originate from one ovary, clearly illustrates the phenomenon of ova migration to result in roughly equal distribution of embryos in both uterine horns.

Although the mechanism of ovulation is not completely understood, it is closely governed by delicate interactions among hormones of the pituitary, ovary, and uterus. Plasma progesterone concentration rises during the first 14 to 15 days

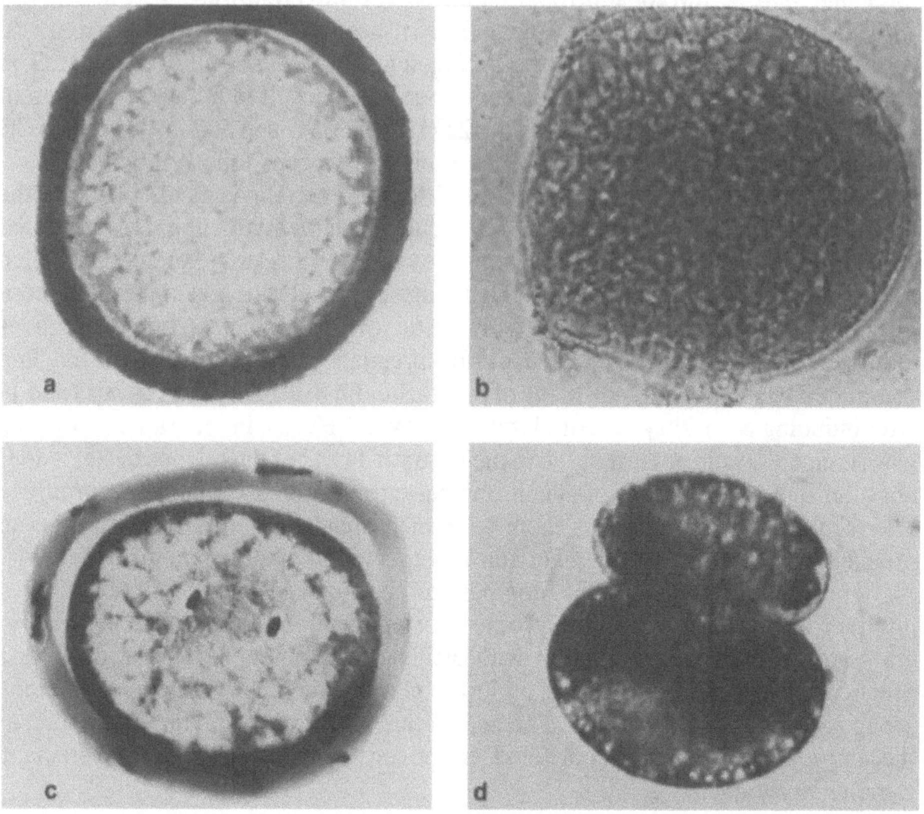

FIGURE 7.3 Ova under low power magnification: (a) Unfertilized ovum. From Thibault, Annals Zootech. *8*, 1959. (b) Ovum with spermatozoa penetrating zona pellucida (lower left) and nucleus on periphery of ovum (bottom). Photo courtesy of R. K. Christenson, Roman L. Hruska U.S. Meat Animal Research Center, Clay Center, Nebraska. (c) Fertilized ovum in first division. From Thibault, Annals Zootech. 8, 1959. (d) Fertilized ovum with two blastomeres. Photo courtesy of R. K. Christenson and R. R. Maurer, Roman L. Hruska U.S. Meat Animal Research Center, Clay Center, Nebraska.

of the estrous cycle, then declines precipitously in the course of a 48-hour period and remains low for 4 or 5 days to the next ovulation. Follicle-stimulating hormone (FSH) and luteinizing hormone (LH) contents of the pituitary are related to the stage of the estrous cycle. Follicle-stimulating hormone increases and LH decreases during pregnancy. Corpora lutea persist during pregnancy and ovulation is prevented. The signal for maintenance of the corpora lutea apparently is mediated by some local uterine effect or by the presence of the embryo. This signal is initiated at about day 12 of pregnancy. The pituitary appears to be essential for extended life of the corpora lutea but not for their formation as ovulation and corpus luteum formation occur in hypophysectomized pigs. The uterus and pituitary apparently regulate the life span of the corpus luteum systemically by both glands as well as locally by the uterus. During pregnancy, progesterone derivatives excreted in the urine increase during the first 54 days, and then decline until term. This is a rough index of blood progesterone levels. Levels of plasma progesterone and LH are not closely related to number of embryos present during pregnancy.

Superovulation can be induced by injection of gonadotrophins.

Fertilization occurs within 6 to 10 hours after natural mating and may occur within 2 hours after artificial insemination. It can be defined as the penetration by a sperm of the outer covering (zona pellucida) of the ovum. Normally, a single sperm enters the ovum but occasionally polyspermy does occur, especially if ovulation has taken place considerably earlier.

Ova remain fertile as long as 24 hours after release so that artificial insemination with fresh semen the day following ovulation can result in conception. However, insemination should be done 6 to 10 hours before ovulation for highest conception rates. Age of sperm at the time of insemination affects embryo survival. Storage of boar semen before insemination results not only in reduced fertilization but in increased early embryonic mortality. The first division occurs very shortly after the time of fertilization.

Two to three days before the onset of estrus the follicle begins to enlarge. The nucleus at this time moves toward the periphery where it undergoes mitosis and the polar body is expelled. At the time of rupture, the granulosa remains intact. The granulosa cells then undergo hypertrophy and are changed to the luteal cells of the mature corpus luteum. By the seventh day after the onset of estrus, the corpus luteum is fully developed. About day 15 of the cycle, regression of the corpora lutea begins unless conception has taken place; if it has, the corpora lutea are maintained. The progesterone content of the corpora lutea increases to about day 14 when it declines if conception has not occurred or persists if it has occurred.

After ovulation, the epithelial cells of the oviduct and uterus rapidly hypertrophy and become high columnar-type cells by the end of the first week following ovulation. About 10 days after ovulation, the columnar epithelium persists and the invasion of eosinophilic leukocytes is at its peak. If conception does not occur, the surface epithelium again reverts to low columnar after day 10 and the eosinophilic leukocytes disappear.

In approximately 21 days the cycle repeats itself. The vaginal epithelium increases in height at estrus and then progressively decreases to a minimum at day 12 to 16 of the cycle. The superficial layers slough beginning at about the fourth day and continuing to day 16. Leukocytes invade the epithelium during late diestrus and reach a maximum just after estrus.

A phenomenon called *arborization of the cervical mucus* occurs when mucus is allowed to dry on an unstained slide. This consists of the appearance of a crystalline pattern of leafy projections resembling fern leaves. Although in humans this occurs at a specific phase of the estrous cycle, in sows it is not predictable enough to be of value in conjunction with artificial insemination. Change in pH of the vaginal smear has also been used as a means of attempting to identify estrus in sows, but, again, results have not been adequate for predictive use.

FACTORS AFFECTING OVULATION RATE AND TIME OF ESTRUS

The number of ova released at a single estrus gradually increases through the first four or five estrous cycles. Therefore, gilts bred at the first estrus have a smaller average litter size than those bred at a subsequent estrus. Because of increased ovulation rate with progressive maturity, first-litter gilts usually farrow fewer pigs than they do subsequently.

Generally, the litter size at birth represents only about 60–70% of the number of ova originally released. In mature sows the average total number of ova released at a single estrus (as measured by counting corpora lutea) is greater than that of first-litter gilts (approximately 16–20 vs. 10–15). Breed differences exist in ovulation rate and in prolificacy. Ovulation rate within a herd can be increased by selection (Zimmerman and Cunningham 1975; Johnson et al. 1984). An increase in the energy (glucose) intake during the 2 weeks prior to breeding ("flushing") is associated with an increase to two to three in total number of ova released. This has also been observed in sows fed increased energy just prior to postlactational estrus. Extreme restrictions in feed intake during the breeding season do not entirely inhibit estrus and, in fact, it is difficult to interfere with the estrous cycle by underfeeding except through its delaying effect on age at puberty as discussed earlier. Ovulation rate does not seem to be affected by continuous light, unlike the effect for the hen. High environmental temperatures (33°C) reduce ovulation rate, but have no effect on estrous cycle length.

Sows do not normally ovulate during lactation, although they often show estrus during the first day or two after parturition. Estrus usually occurs within a few days after the pigs are weaned. When pigs are weaned at 5 weeks of age, estrus usually occurs in less than a week; when pigs are weaned earlier, there is usually a longer interval. The initiation of estrus and ovulation after weaning appears to be associated with changes in production of the gonadotrophin-releasing factor from the hypothalamus of the sow (Cox and Britt 1982).

Estrus and ovulation can be induced during weeks 3–6 of lactation by injection of 1000–1500 IU of pregnant mare serum gonadotrophin (PMSG) followed by 500 or 1000 IU human chorionic gonadotrophin (HCG) 72–96 hours later (Hausler et al. 1980). Hodson et al. (1981) found that best results were obtained with multiparous sows treated with PMSG 25 days or more postpartum.

However, when the pigs are removed from the sow at birth or within 2 days after farrowing, most sows will come into estrus within 2 weeks and the ova are usually fertile.

EMBRYO TRANSFER

Transplantation of pig embryos from the uterus of one female to that of another has been achieved repeatedly (Hancock and Hovell 1962; Smidt et al. 1965; Dziuk et al. 1964, 1968; Hunter et al. 1967). Thus, the groundwork was laid for refinement in conditions and techniques that would allow the routine use of embryo transfer. This technique could have special use as a means of increasing the number of progeny obtained from genetically superior females, by using common or inferior recipient mothers to incubate the developing fetuses. In addition, embryo transfer experiments provide a unique approach to gaining a greater understanding of other aspects of reproductive efficiency in swine (Dziuk and Bellows 1983; Polge 1982).

Another option with potential eventual application is the manipulation of embryonic cells in culture, which can then be cloned[1] (the production of genetically identical individuals) and inserted into the embryo.

[1] The nucleus from a single cell of a developing embryo is removed and replaced by another nucleus from a multicellular embryo. A single embryo can be used to harvest several nuclei.

MOVEMENT AND SURVIVAL OF SPERM IN THE FEMALE REPRODUCTIVE TRACT

Studies of sperm transport in the female reproductive tract have established that sperm reach the uterotubular junction within 10 minutes but that passage into the oviducts may not occur for more than 1½ hours after mating. There appears to be no difference in the number of spermatozoa recovered from the oviducts of sows inseminated with live or dead spermatozoa or in the number of sperm in the oviducts of sows inseminated during estrus or on day 7 postestrus. Spermatozoa can maintain viability in the oviduct for 25–30 hours or longer. More than half of the spermatozoa have disappeared from the uterus within 15 minutes after insemination and almost all are gone after 24 hours. Motility of spermatozoa in the uterus is drastically reduced after 2 hours, and phagocytosis has been observed at 8 hours.

Sperm age has an effect on fertilization rate. Semen stored for 54 hours after collection produced a lower rate of fertilization in artificially inseminated gilts than semen stored for 6 hours. The number of sperm in the zona pellucida of 3-day cleaved eggs is also significantly lowered with older sperm (First et al. 1963).

CONCEPTION RATE

Conception rate (percentage of matings, either artificial or natural, that result in development of viable fetuses) may refer to first service or to first and second services or to total services. An acceptable first service conception rate for gilts and sows is 70%. Failures to attain 100% conception are due to many factors. Some of these factors are abnormalities in the anatomy of the female reproductive tract, cystic follicles, the improper timing of mating, and infertility of the boar. Environmental temperature exerts an effect on conception rate in swine. Cold stress on the day of mating (−25°C) does not interfere with conception (Swierstra 1970), but heat stress as indicated by summer versus winter conception rates clearly reduces reproductive efficiency (Teague 1970; Omtvedt et al. 1971).

In swine, it has been generally considered that conception is an all-or-nothing phenomenon since several billion sperm are competing to fertilize fewer than 25 ova. However, there is evidence that smaller litters may be associated with marginal fertility in the boar. Double mating of the sow (two services 24 hours apart during estrus) is associated with a higher conception rate and with one or two extra pigs per litter, probably because the duration of ovulation may be several hours.

DURATION OF GESTATION

The gestation period of swine averages about 114 days, but ranges 111 to 119 days (Allen and Stewart 1983). There are small genetic differences. There is no important effect of litter size or of first pregnancy vs. subsequent pregnancy on gestation length.

PRENATAL DEATHS: UTERINE CAPACITY AND LITTER SIZE

It is common to have a 30–40% loss of developing young between the first and 114th day of gestation. More than half of these losses occur during the first 25 days (before or shortly after implantation). Mummified fetuses also account for a surprisingly large total. These represent deaths that occur after about day 40 of gestation and are differentiated from stillborn pigs by partial resorption. Typical sequence and proportion of losses are illustrated diagrammatically in Fig. 7.4.

Litter size depends on the number of ova produced, on the percentage of fertilization, and on the number of prenatal deaths and stillbirths.

It has been suggested that early embryonic mortality may be due to an excess secretion of progesterone by the corpora lutea. More recent data (Archibong et al. 1987) suggest that a lower circulating progesterone to estrogen ratio may be associated with a higher rate of embryo mortality in gilts mated at pubertal estrus.

The high embryo mortality rate in swine has also been proposed as being related to uterine crowding. However, negative results were obtained by ligating portions of the uterine horns to provide uterine spaces of varying capacity (Dziuk 1968). Uterine crowding may affect fetal survival when the number of fetuses exceeds about 14 but it is probably not an important factor until after day 30 of gestation (Dhindsa and Dziuk 1968b; Fenton et al. 1970). From embryo transfer studies, in which color inheritance was used as a code for identifying fetuses in late gestation, evidence has been presented to suggest that the barrier imposed by uterine capacity must be removed somehow before a marked increase in reproductive rate of sows can be realized (Bazer et al. 1969).

The concept of uterine capacity as a limiting factor in litter size at birth has been refined in recent research. Wu et al. (1988) reported that the length of uterine horns was related to the incidence of mummified fetuses (fetal deaths occurring late enough in pregnancy so that resorption of the dead fetus did not occur) and that a longer uterus had greater space per fetus, a larger number of live fetuses, and a lower incidence of mummified fetuses. Leymaster et al. (1986), Huang et al. (1987)

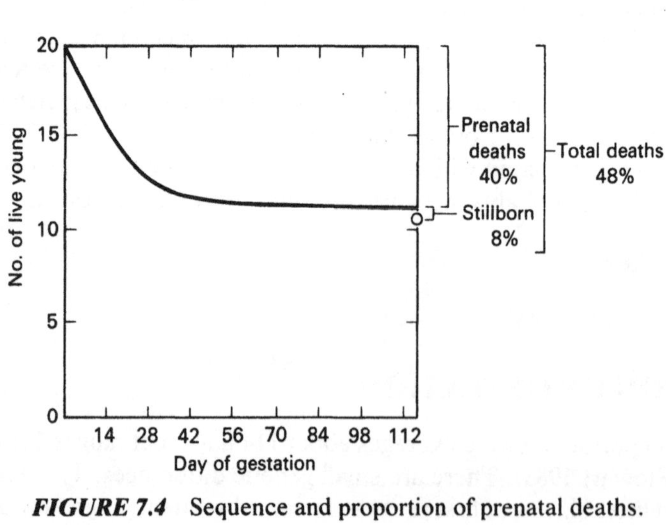

FIGURE 7.4 Sequence and proportion of prenatal deaths.

and Christenson et al. (1987) have addressed the concept of separate effects of ovulation rate and uterine capacity on ultimate litter size in swine by systematic experiments involving unilateral hysterectomy-ovariectomy (UHO). The research should provide information on the relative contribution of ovulation and uterine factors in controlling litter size and the extent to which gilt selection for increased uterine capacity might increase average litter size in a population. Currently, 30–40% of ova shed are not represented by live pigs at birth (Pope and First 1985), and greater knowledge of causative factors in this inefficient process is needed to provide practical means of improving fetal survival.

A large environmental effect on embryonic mortality is related to energy intake. A high plane of nutrition immediately following breeding results in an increase in embryonic mortality. The metabolic basis for this is not known. Reciprocal embryo transfer studies using gilts fed high or low energy diets have shown that embryos from gilts of each group are equally viable (Bazer et al. 1968). Therefore, it appears that the controlling factor is some uterine factor rather than any effect of the embryo per se. Although a marked decline in body weight occurs in gilts subjected to severe energy restriction throughout gestation (Pond and Mersmann 1988) or inanition in midgestation, fetal survival rate is unaffected even though piglet birth weight is reduced when feed restriction extends over two-thirds or all of gestation (Pond et al. 1985; Pond and Mersmann 1988). Hard and Anderson (1982) have found that uterine blood flow is maintained at a high level during late pregnancy of energy-restricted gilts despite changes in plasma volume. They have suggested that this special adaptation allows normal fetal development during prolonged maternal starvation.

The incidence of stillbirths is greater with large litters than with small ones. Virtually all pigs, except mummified fetuses, are alive at 112 days in the experience of veterinarians involved with hysterectomy-derived pigs. The simultaneous occurrence of stillbirths and mummified fetuses does not have an influence on frequency of stillbirths. In one study (Pond et al. 1960), only 29 out of 70 litters that contained stillbirths contained mummified fetuses. Single mummified fetuses occurred more frequently than two or more in a litter. Mummified fetuses were more prevalent in large than in small litters.

Normal 25-day embryos are approximately 20 mm long and by 40 days they have reached approximately 40 mm (Gossett and Sorenson 1959b). Most mummified fetuses are at least 50-mm long and often nearly full term, indicating that total resorption probably occurs when death precedes the 40th day.

PARTURITION

The duration of parturition may vary from 1 to 2 to as many as 24 hours in a normal pregnancy. Attempts to hasten parturition by injection of oxytocin are not always successful, probably because dosage level and timing of the injection are critical. Progesterone in maternal blood plasma declines before the onset of natural parturition. Since corpora lutea are the primary source of progesterone for maintenance of pregnancy, cessation of their function is necessary for initiation of parturition. Prostaglandin $F_{2}\alpha(PGF_{2}\alpha)$ is a natural luteolytic agent in swine at term, producing a decline in progesterone and a release of relaxin from the corpora lutea. Exogenous $PGF_{2}\alpha$ administration results in loss of corpora luteal function and a drop in plasma progesterone and has been found to be effective in inducing rapid parturition in sows (Nara and First 1981).

When an appropriate amount of $PGF_{2}\alpha$ or one of its analogs is administered intramuscularly on day 112 of gestation, parturition can be expected to begin 20 to 28 hours later in about 90% of the sows (First et al. 1982; Welp and Holtz 1985). This phenomenon provides a practical tool for adjusting parturition time to the convenience of the producer, rather than allowing farrowing to be spread randomly over a 24-hour period.

Parturition can be delayed for several days beyond the normal gestation by subcutaneous administration of progesterone (Minar and Schilling 1976; Wilson et al. 1981) or a progesterone derivative, 6-methyl-17-acetoxy-progesterone (MAP) (Nellor et al. 1975). However, the delayed parturition may be associated with abnormal patterns of plasma progesterone, estrogen, and relaxin, which in turn may contribute to the high incidence of stillbirths in delayed parturition. Further research is needed to eliminate these increased stillbirths and thereby to create the conditions under which artificially delayed parturition can be used as a management tool for controlling the time of farrowing.

Just before parturition, the sow shows general uneasiness and attempts to make a nest with the bedding if available. The presence of colostrum in the mammary glands is a signal of imminent parturition. The birth of the first pig is usually accompanied by the placental membranes but occasionally one or more pigs are born before any placenta is expelled. The umbilical cord is usually still attached when the pig is expelled but breaks soon afterward as the pig struggles to stand and explore its new environment. Presentation may be with the front feet first or the hind feet first and there appears to be no chronological pattern within a given litter. Occasionally, lack of uterine motility or the inability of a large pig to pass the pelvic region necessitates assistance, but usually this is unnecessary. Occasionally cesarean section becomes necessary, or the failure of some of the placental membranes to be expelled results in uterine infection. This can be avoided or alleviated by oxytocin injection and antibiotic administration. Within the first few hours after the completion of parturition, the sow will be mobile and will accept water and feed. Pigs born earliest soon find their way to a nipple and may suckle from several glands before the last pig is born.

THE MALE

ANATOMY OF THE REPRODUCTIVE TRACT

The anatomy of the reproductive organs of the boar was thoroughly described by Mackenzie et al. (1938). A normal reproductive tract is shown in Fig. 7.5. The size and weight of the primary and accessory reproductive organs are shown in Table 7.2 (Blom 1962). Differentiation and development of the testes, anatomical description of the adult testes, spermatogenesis, endocrine control of testicular function, and epididymal structure and function have been reviewed (Amann and Schanbacher 1983), but detailed summary of this information is beyond the scope of this book.

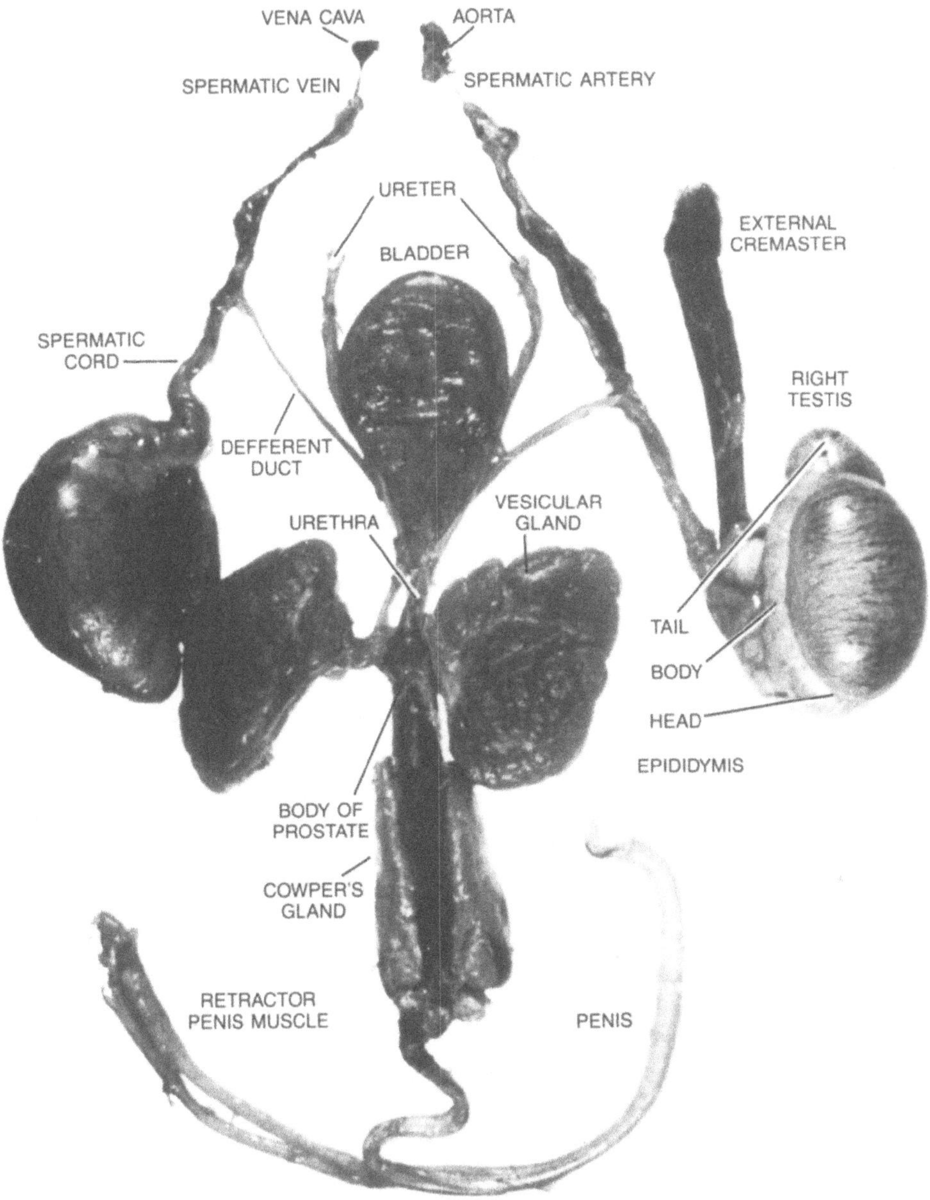

FIGURE 7.5 Normal reproductive tract of the boar. Photo courtesy of R. K. Christenson, Roman L. Hruska U.S. Meat Animal Research Center, Clay Center, Nebraska.

All the typical accessory organs of the mammal are present. The testes are carried in the external body cavity in the scrotum, where they become located after migration out of the body cavity by approximately the 100th day of prenatal life. The testes, covered by a tough capsule (tunica albuginea), are carried in a near vertical position with the head of the epididymis on the ventral end. The tail of the epididymis forms a cap over the dorsal end of the testis. In normal castration

TABLE 7.2 Size of Boar Reproductive Organs

Organ	*Size*
Testis plus epididymis	13 × 7 × 7 cm 250–300 g (single)
Epididymal duct length	55–65 m
Vas deferens	25–30 cm long × 2 mm outside diameter
Seminal vesicle	13 × 7 × 5 cm
Prostate	3 × 3 × 1 cm (15–25 g)
Cowper's gland	17 × 3 × 3 cm
Penis	50–75 cm long 2 cm diameter (erect)

Source: Blom 1962.

procedure, the epididymis and the attached tunica vaginalis are removed with the testes. Stored sperm leave the epididymis via the vas deferens. The seminal vesicle, Cowper's (bulbourethral) gland, and prostate gland all contribute to semen volume. The semen leaves the reproductive tract through the urethra and the penis at ejaculation.

A microscopic section through the testis (Fig. 7.6) shows the relative proportions of seminiferous tubules and interstitial cells. The interstitial tissue in boars is more plentiful than in most other animals. The epididymis consists of a long, highly convoluted tube which increases in diameter as it reaches the point of transition into the vas deferens. On each side of the pelvic urethra lie the seminal vesicles. They cover the neck of the bladder, the Cowper's gland, the prostate, and the vas deferens. The openings from the vas deferens and the seminal vesicles into the urethra represent four separate openings located in close proximity to each other. The Cowper's glands lie parallel to the urethra and are attached to it on the ventral side. The prostate gland is a multilobular gland composed of the body (bulb) attached on the dorsal surface of the urethra and of the pars disseminata attached to the body of the gland itself. The pars disseminata of the prostate is conspicuous at its anterior end because of its yellow color.

PUBERTY AND SEXUAL BEHAVIOR

Sexual maturity in the boar is a gradual process in which sperm production and sexual desire begin concurrently in increasing intensity at about 4 months of age (even younger in some breeds of Chinese pigs). However, it is generally considered that boars should not be used in service before 6–8 months of age. Semen has been collected as early as 5 months of age, but volume and percentage of normal sperm are less than in semen from mature boars. Extending daylight by artificial light may hasten onset of sexual interest and libido in young boars but does not appear to affect semen quality (Berger et al. 1980). Sexual behavior of males has been

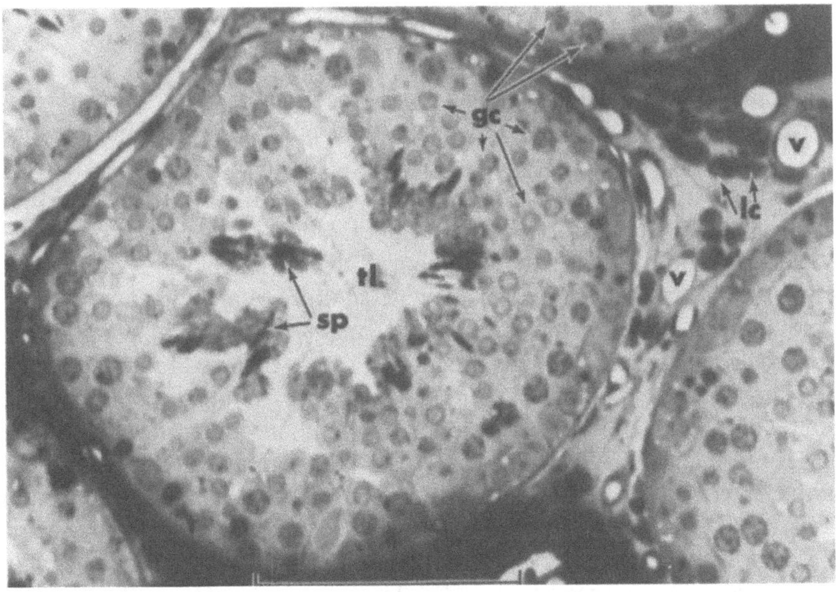

FIGURE 7.6 Microscopic appearance of boar testes showing interstitial tissue with Leydig cells, blood and lymph vessels and semeniferous tubules with spermatozoa in various stages of development. Key: sp = elongated spermatids; tl = tubules lumen; gc = germ cells in different stages of development; lc = Leydig cells; v = blood and lymph vessels; and bar = 100 microns. Photo courtesy of D. D. Lunstra, Roman L. Hruska U.S. Meat Animal Research Center, Clay Center, Nebraska.

described by Hafez et al. (1963), and by Houpt and Wolski (1982). It includes phases of courtship, mounting, and intromission and ejaculation. Mature boars should not be allowed together if not accustomed to being penned together, as the fighting instinct will result in injury and death to one or both boars and possible injury to attendants. When boars are raised together, there is no problem in maintaining them in the same pen at maturity. Shy breeders or those introduced to a strange environment will often serve better when the female to be bred is brought to them rather than vice versa. The duration of copulation is 3 to 20 minutes or longer. Competition among boars for a female in estrus usually results in an establishment of order of dominance among two or more boars that influences the frequency of siring of litters by a particular boar. There are apparently genetic differences in pattern and intensity of sexual behavior.

Mature size is not reached until more than one year of age so that increases in total semen and sperm production accompany both the development of sexual maturity and the increase in body size until about one year of age. The size of the testis tends to increase proportionately with total body size and is directly related to total sperm output. Severe restriction of feed intake delays puberty. It results in a reduction in the diameter of the seminiferous tubules of the testis as well as a reduction in its total growth. The effect is reversible by returning the animal to a normal level of intake.

ARTIFICIAL INSEMINATION AND SEMEN PRODUCTION

The practice of artificial insemination on a commercial basis exists in some countries of the world. For economic reasons the practice, even though technically feasible, has not achieved widespread use in North and South America. Two problems that still exist are difficulty in detecting estrus in the gilt and the poor survival of boar sperm stored beyond about 3 days. The first problem can be alleviated by keeping a vasectomized boar in the presence of the gilts to be bred. The volume of total fluid (semen) inseminated and the total number of motile sperm are both important factors in successful insemination. The total number of motile sperm required to result in a high conception rate is believed to be in the order of three billion and the minimum volume about 50 ml. Assuming 60% motility, this would require a total of five billion sperm in 50 ml of inseminating fluid. Since boar semen is more concentrated than this, it is possible to use a diluent. None of the diluents (extenders) in use to date will promote viability of sperm beyond a few days of storage at 5° to 15°C. Freezing usually kills boar sperm, but research has resulted in viable litters of pigs from semen frozen under special conditions (Graham et al. 1971; Johnson et al. 1981). This is a technical breakthrough that will allow eventual commercial application. When conditions are favorable, the rate of conception with artificial insemination is equal to that obtained by natural service (Self 1959; Hancock 1960). The development of extenders and of thawing procedures for frozen semen (Pursel 1983) allows storage of semen over longer periods of time and will encourage the use of outstanding sires on a commercial basis in artificial insemination. The optimum time for insemination of fresh, whole semen is 12–16 hours before ovulation (Dziuk 1970); in contrast, maximum fertility with frozen-thawed spermatozoa is obtained when pigs are inseminated at about the time of ovulation (Pursel 1980).

The total volume of semen obtained in a single ejaculate is affected by the size of the boar and the frequency of semen collection. In a study involving Yorkshire and Chester White boars (Turkheimer et al. 1958), from which semen was collected weekly, the total volume per ejaculate for boars 18–33 weeks of age was 125 ml, whereas that of boars 47–52 weeks of age was 184 ml. Therefore, one can expect a sizable increase in total semen volume between initial semen collection at puberty and at 1 year of age. In boars housed together, there is a tendency toward pederasty which makes it advisable to house boars individually for utilization of semen production for either artificial insemination or natural service.

The total volume per ejaculate is decreased, but the total sperm collected is increased if collections are daily rather than less frequent. Motility is apparently not affected appreciably. In commercial production, it is common to use a mature boar in natural service to breed three sows in a single day. The total semen production of a mature boar is, therefore, probably considerably greater than the amount measured during semen collection studies. Total semen production (as measured by number of fertile matings) remains at a high level for several years after the boar reaches sexual maturity. Boars are often kept in heavy service to an age of 5 or 6 years, or more, and are usually disposed of for other reasons, such as an ill temper (which represents a danger to the safety of those handling the boar) or physical disability, rather than a failure in semen production.

Low ambient temperature (−15° to 20°C) does not interfere with testicular development, sperm production, or semen quality in healthy boars (Swierstra 1970). On the other hand, acute heat stress (exposure to environmental temperatures of 33.3°C for 72 hours) reduces total number of sperm per ejaculate, reduces sperm motility, increases abnormal sperm per ejaculate, and reduces pregnancy rates of gilts inseminated with semen from boar postexposure (Christenson et al. 1972). Detrimental effects on semen quality and fertility were first evident at 2–3 weeks postexposure, continued to decline until 4.5 weeks, and then returned gradually to pre-exposure conditions by 8–9 weeks. Likewise, chronic heat stress (exposure to an environmental temperature of 34.5°C for 8 hours and 31°C for 16 hours during each 24-hour period for 11 weeks) reduces daily sperm production by nearly one-half and reduces the conception rate significantly (Wettemann et al. 1979). Furthermore, the reduced sperm motility associated with heat stress persists for at least 5 weeks after the heat stress is removed. Results of such studies under controlled laboratory conditions simulating temperature fluctuations that normally occur in many environments in which swine are produced suggests the importance of providing a means of artificial cooling for boars during summer months. Overhead misting or sprinkling systems or temperature-modified housing may prove to be a worthwhile investment for boars in many environments.

In large mature boars, it is often possible to collect up to a liter or more of semen at a single ejaculate. This represents a tremendous fertility potential for outstanding boars that remain in service for any length of time. However, more common sperm production figures and the number of services that might be expected from them are shown in Table 7.3 (Foote et al. 1959; Kennelly and Foote 1964). Only about two-thirds of the total sperm produced are actually obtained

TABLE 7.3 Characteristics of a Typical Single Ejaculate of Boar Semen[a]

Criterion	*Amount*
Total volume, ml	300
Strained volume, ml	210
Gelatinous volume, ml	90
Spermatozoa per ml of strained semen	200×10^6
Total spermatozoa per ejaculate	42×10^9
Motile sperm, %	70
Total motile sperm per ejaculate	29.4×10^9
Number of services possible	10

[a] Minimum volume of fluid and number of motile sperm needed for 1 service: 3×10^9 motile sperm in 50 ml of fluid.

Maximum number of services from above ejaculate:

$$\frac{29.4 \times 10^9}{3 \times 10^9} = 9.8 = \text{approximately } 10$$

500 ml of fluid needed for 10 services
210 ml of semen + 290 ml of extender = 500 ml of diluted semen for 10 services = 50 ml per service.

Source: Foote et al. 1959; Kennelly and Foote 1964.

when semen is collected by an artificial vagina because of masturbation, resorption of sperm before ejaculation, losses in the urine, and adherence to surfaces of collection equipment (Kennelly and Foote 1964).

COMPOSITION OF SEMEN

The composition of boar semen is summarized in Table 7.4 (Westmoreland et al. 1967; White 1958) and that of seminal vesicle secretion in Table 7.5 (Mann 1954, 1959). The total composition is a reflection of the secretions from the secondary sex glands as well as from the testes. The composition of a single ejaculate is conveniently divided into three components: (1) pre-sperm fraction, often high in bacterial contamination and containing mainly secretions of the accessory sex glands; (2) sperm-rich fraction containing the largest concentration of sperm and contributing the largest volume; (3) the post-sperm fraction consisting mainly of gelatinous material. The three fractions can be collected in separate containers. In practice, it is common to collect the entire ejaculate in one container and strain the semen through a sterile cheesecloth to separate the gelatinous mass. This is to avoid problems with clogging of the inseminating tube and to obtain a sample of uniform sperm concentration. Generally, the higher the sperm concentration, the lighter the color of the semen (Polge 1956b).

SPERM SURVIVAL AND STORAGE

Removal of any one accessory gland apparently does not alter libido nor does it change the activity of the remaining accessory glands or the process of ejaculation,

TABLE 7.4 Chemical Composition of Boar Semen (per 100 ml of fluid)

Constituent	*Mean*[a] *(per 100 ml of fluid)*
Water, g	95(94–98)
Sodium, mg	650(290–850)
Potassium, mg	240(80–390)
Calcium, mg	5(2–6)
Magnesium, mg	11(5–14)
Chloride, mg	330(260–430)
Zinc, μg/ml	31.8(7.8–78)[b]
Fructose, mg	13(3–50)
Sorbitol, mg	12(6–18)
Citric acid, mg	130(30–330)
Inositol, mg	530(380–630)
Glycerylphorphoryl choline (GPC), mg	(110–240)
Ergothioneine, mg	(6–23)
Protein, g	(3–7)

[a] Mean, with range in parentheses.
[b] Westmoreland et al. 1967.

Source: White 1958.

TABLE 7.5 Composition of Seminal Vesicle Secretions in Boars (per Liter of Fluid)

Constituent	*Seminal vesicle secretion*[a]
Water (g)	814
Total N (g)	15.5
Sodium (meq)	30
Potassium (meq)	55
Chloride (meq)	3
Magnesium (meq)	—
Calcium (meq)	6
Citric acid (mmole)	30
Total phosphorus (mmole)	10
Inorg. phosphorus (mmole)	1
Acid sol. phosphorus (mmole)	9

[a] The concentration of nitrogen, citric acid and calcium is higher than in whole semen; that of chloride, sodium and total phosphorus is much lower; that of other listed constituents is similar.

Source: Mann 1954, 1959.

but it does cause a proportional reduction in semen volume. Fertility apparently is not reduced when the seminal vesicles and Cowper's glands are removed and, in fact, the number of abnormal sperm may be reduced and the duration of sperm motility increased. Regeneration of tissue from completely removed glands does not take place. Motility is possible both in aerobic and anaerobic conditions; in the presence of oxygen, and spermatozoa oxidize lactic acid and in its absence they utilize glycolizable sugar, mainly fructose.

The viability of boar sperm after ejaculation is only a few days. Unlike bull semen, the semen from boars cannot be readily frozen without loss of motility and fertility. Much effort has been exerted to determine the conditions required to maintain motility of boar semen by various storage procedures. Conception has been accomplished in gilts inseminated with frozen boar semen at the USDA (Pursel and Johnson 1971; Johnson et al. 1981), in Scandinavia (Crabo and Einarsson 1971), and at the University of Minnesota (Graham et al. 1971). A variety of extenders have been tried, including solutions of sodium sulfate and potassium sulfate, calcium chloride, peptone, egg yolk-phosphate mixtures, cow's milk, glucose tartrate, and antibacterial agents such as penicillin and streptomycin. In view of the high concentration of inositol, citrate, glucose, and glycine in semen, each of these has also been used. Citrate and phosphate extenders, which are capable of prolonging the life of bull sperm, are toxic to boar sperm under the same conditions.

Boar spermatozoa are subject to "cold shock" when cooled suddenly to 6°C or less storage. Attempts to prevent the loss of potassium and other ions from sperm exposed to quick cooling by addition of substances such as glycerol and dimethylsulfoxide (DMSO) to the diluent have been unsuccessful.

Boar semen exhibits anabiosis—that is, quick loss of motility and settling out on short-term storage—but reactivation is possible by warming to 37°C and shaking in the presence of oxygen. Therefore, it is necessary to take this into consideration in studies of boar sperm motility. As scientists continue to search for improved methods of freezing boar spermatozoa for long periods of storage, they are obtaining greater knowledge of sperm physiology, which will be of use in application of artificial insemination in commercial swine production in the future.

Modern genetic evaluation procedures (Chapter 6) are expected to stimulate the desire for improved commercial application at artificial insemination techniques. As truly superior boars are identified, the opportunity to extend their use to more sows and more herds will motivate their owners and potential customers to make the effort to utilize this potential tool. Present practical techniques limit the utilization relative to dairy and beef industries, but the potential utilization and benefits are similar if techniques can be improved.

LACTATION AND MILK YIELD

Milk yield is more difficult to measure directly in the sow than in the cow, since oxytocin must be injected to induce milk letdown when the milk is obtained by hand or by machine (Hartman and Pond 1960). It can be estimated by weighing the pigs immediately before and after nursing (Speer and Cox 1984), by machine-milking, by hand-milking of the sow at predetermined intervals after oxytocin injection, or by use of deuterium oxide as a tracer to estimate milk consumption via estimation of body water turnover in nursing pigs (Pettigrew et al. 1985). The peak daily yield is reached during the fourth or fifth week (Hartman and Pond 1960; Allen and Lasley 1960). By the ninth or tenth week, daily production is negligible. The average daily yield over an 8-week lactation may range from 5 to 8 kg or more (Smith 1952, 1959; Salmon-Legagneur 1958; Hartman et al. 1962), depending on breed, body size, nutrition, and other management factors. Extrapolation of values obtained by the "weigh-suckle-weigh" procedure to daily production from nine consecutive hourly measurements showed a mean of 5.9 kg daily milk yield from 175 sows sampled at 14 and 20 days of lactation (Speer and Cox 1984). Comparable yields were reported by Noblet and Etienne (1989). Milk production efficiency in swine (milk energy per unit feed energy) is very high (45%) (Bowland 1966). Environmental temperature, number of young suckled, body fatness of the sow at the beginning of lactation, and level of energy intake during lactation all affect the apparent gross efficiency of milk production.

Lactation curves representative of older data reported in the literature are shown in Fig. 7.7. More recent data based on measurements taken in early lactation (Speer and Cox 1984; Mahan et al. 1971; Noblet and Etienne 1989) appear to fit these curves reasonably well. The daily milk yield tends to be directly related to the number of pigs sucking, but inversely related to the yield per pig. That is, as litter size increases, total milk yield increases, but not sufficiently to accommodate maximal milk intake by each pig. By about week 4 of lactation, daily pig nutrient requirements begin to exceed the amount supplied by milk so supplemental feed must be made available. The relationship between milk supply and pig nutrient demands during the first 6 weeks of postnatal life is illustrated in Fig. 7.8.

Individual pigs identify themselves with a particular teat during the first few days of life and jealously protect their access to it. In small litters it is common for a pig to claim more than one nipple. On the other hand, in large litters, the weakest

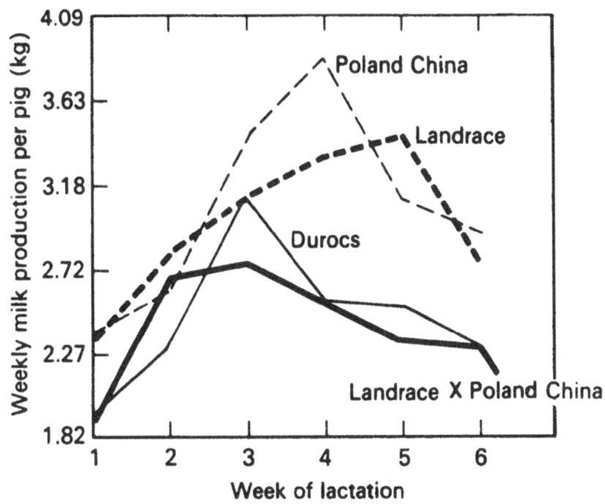

FIGURE 7.7 Lactation curves for sows. From Allen and Lasley 1960.

pigs may starve because they cannot compete for a nipple. The dam will normally not accept pigs other than her own, unless transfer of orphan pigs is made during the first few days of lactation. However, when two or more sows and litters are penned together, it is not uncommon for some pigs to nurse more than one sow. When pigs are introduced to a foster dam at greater than 2 or 3 days of age, it is advisable to disguise odors and other cues by allowing the adopted pig to mingle with the others before introduction to the foster dam. This procedure allows a means of equalizing the size of litters if litters of disparate size are born during a short time interval.

The normal nursing behavior pattern is for a uniform time interval between each feeding over a 24-hour period. The average nursing interval is less than 1 hour, day and night, so that the suckling pig normally receives more than 24 feedings daily. Each feeding ritual consists of only a few minutes, with the actual suckling phase lasting less than a minute.

Nursing frequency tends to decrease with the advance of lactation as indicated

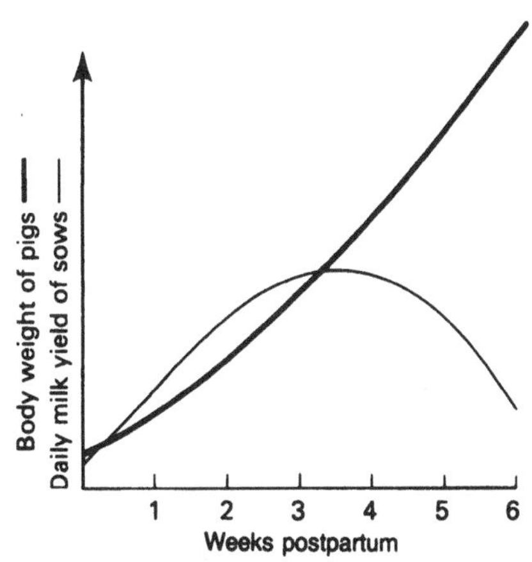

FIGURE 7.8 Milk supply in relation to pig nutrient needs to 6 weeks.

in Table 7.6, which summarizes the results of a study of the nursing interval using time-lapse cinematography (Hartman et al. 1962).

COMPOSITION

The gross composition of sow milk is highly variable. These data are summarized in Table 7.7. The large variation is probably due partly to genetics, stage of lactation, prepartum diet, and other environmental factors.

The composition of colostrum during the first day is considerably different from that of later milk, as shown in Table 7.8 (Perrin 1955). The gradual change to milk occurs in the course of a 2- to 3-day transition period. Colostrum is higher in percentage of total solids and in protein than milk, but lower in ash, fat, and lactose. Data (Klaver et al. 1981) obtained with Dutch Landrace sows agree well with earlier published data on sow milk yield and composition. Milk production at day 5 of lactation averaged 4.55 kg daily; energy concentration was 1.202 kcal/kg and dry matter, ash, protein, fat, and lactose were; 19.2, 0.74, 6.41, 6.88, and 5.16%, respectively. Such values are similar to those reported earlier for different breeds of swine (Bowland 1966). Sow condition affects milk production in early lactation; a high level of intake by thin sows may reduce fat content without increasing total milk production (Klaver et al. 1981). In normal sows, high feed intake in early lactation increases milk production and tends to prevent excessive lactation weight loss. Considerable evidence (Boyd et al. 1978a; Seerley et al. 1978a,b) shows increased pig survival by feeding the sow a high fat diet during late gestation and early lactation. This effect appears to be the result of increased milk fat which provides a greater piglet calorie intake during the critical perinatal period (Pettigrew 1981). Variations in the fat content of sow milk due to genetic background, diet composition, or level of intake (Van Kempen et al. 1985) result in piglets suckling sows with equal milk production, but receiving different total energy and gaining at slightly different rates on the same milk intake (Noblet and Etienne 1989).

The higher protein content of colostrum is due to its high immune globulin content. The ingestion of colostrum is especially critical to the survival of the newborn pig as this represents the chief source of immunity to disease during early life. The gamma globulin fraction is nearly one-half of total colostral protein at parturition and declines gradually during the first 24 hours, then plateaus (Table

TABLE 7.6 Nursing Interval of Litters through 6 Weeks of Lactation

Time	Week						Average
	1	2	3	4	5	6	
Day	48[a]	54	51	58	54	56	53
Night	39	46	53	52	56	60	51
Average	44	50	52	55	55	58	52

[a] Minutes from initial phase of one nursing to initial phase of next.

Source: Hartman et al. 1962.

TABLE 7.7 Fat, Protein, Lactose, Vitamin, Calcium, Phosphorus, and Ash Content of Swine Colostrum and Milk

Constituents	*Milk*		*Colostrum*
	Unweighted mean	*Range of means*	*Range of means*
Total solids, %	19.4	17.1–25.8	22.0–33.1
Fat, %	7.2	3.5–10.5	2.7–7.7
Protein, %	6.1	4.4–9.7	9.9–22.6
Lactose, %	4.8	2.0–6.0	2.0–7.5
Ash, %	0.96	0.78–1.30	0.59–0.99
Calcium, %	0.21	0.12–0.35	0.05–0.08
Phosphorus, %	0.14	0.10–0.19	0.08–0.11
Vitamin A, μg/100 ml	—	15–255	44–144
Vitamin D, IU/100 ml	10	—	—
Vitamin E, mg/100 ml	0.14	—	—
Vitamin C, mg/100 ml	14.6	11.0–24.6	—
Thiamin, μg/100 ml	65	60–71	56–97
Riboflavin, μg/100 ml	355	137–820	45–650
Pantothenic acid, μg/100 ml	405	190–568	130–680
Niacinamide, μg/100 ml	741	430–903	165–167
Vitamin B_6, μg/100 ml	20.0	—	2.5
Biotin, μg/100 ml	1.4	—	5.3
Vitamin B_{12}, μg/100 ml	0.14	—	0.15
Folic acid, μg/100 ml	390	—	—

Source: Bowland 1966.

7.9; Bourne et al. 1969). Gamma globulins are absorbed intact by the newborn pig but are not absorbed in appreciable proportions after about 36 hours (Westrom et al. 1986).

Other differences (Salmon-Legagneur 1965) that exist between colostrum and milk are these: (1) Phospholipids, unsaturated fatty acids, and vitamins A, C, and E

TABLE 7.8 Changes in Gross Composition of Swine Colostrum and Milk with Time following Parturition

Constituents	*Time after parturition*								
	Hours								*Weeks*
	0	*3*	*6*	*9*	*12*	*15–24*	*27–48*	*72–120*	*2–8*
Total solids, %	30.2	28.7	26.6	23.6	20.8	19.6	21.2	21.8	21.2
Fat, %	7.2	7.3	7.8	7.8	7.2	7.7	9.5	10.4	9.3
Protein, %	18.9	17.5	15.2	11.7	10.2	7.2	6.9	6.8	6.2
Lactose, %	2.5	2.7	2.9	3.0	3.4	3.7	4.0	4.6	4.8
Ash, %	0.63	0.62	0.62	0.63	0.63	0.66	0.72	0.77	0.95
Calcium, %	0.05	0.04	0.05	0.05	0.06	0.07	0.11	0.16	0.25
Phosphorus, %	0.11	0.11	0.11	0.11	0.11	0.12	0.13	0.14	0.15

Source: Perrin 1955.

TABLE 7.9 Concentrations (g/100 ml) of Certain Proteins in Sow Colostrum and Milk

Hour postpartum	*Total protein*	*Beta globulin*	*Gamma globulin*
0	19.6	3.7	8.6
1	18.8	3.4	8.2
2	15.7	3.0	6.8
3	14.3	2.7	6.0
4	12.7	2.6	5.2
6	10.5	2.1	3.9
8	8.6	1.9	2.9
11	6.1	1.5	2.0
13	5.1	1.2	1.6
16	4.3	1.0	1.2
24	4.1	1.0	0.9
Day postpartum			
2	3.5	0.9	0.7
7	3.3	0.9	0.5
14	3.3	0.9	0.5
21	3.3	0.9	0.5
28	3.3	0.9	0.5

Source: Bourne 1969.

and thiamin are all higher in colostrum. (2) Nonprotein nitrogen, pantothenic acid, and niacin are lower in colostrum. (3) The amino acid distribution of colostrum is different from that of milk, although the difference probably is not of practical importance.

The diet fed during gestation and lactation also has an effect on the levels of other nutrients in the milk. Diets fortified with large quantities of some of these nutrients increase milk levels by increased transfer to the mammary gland. Thus, milk constituents are derived partly from synthesis within the mammary gland (lactose, fatty acids, some of the milk protein fractions) and partly by filtration or active transport from the blood.

There appears to be little variation among glands on the same sow in the composition of the milk produced (Pond et al. 1962), with the exception of the percentage of fat, which seems to be more variable. The level of vitamins in milk is related both to dietary intake during lactation, and to the amounts stored in liver and other tissues at the onset of lactation. For example, a large liver storage of vitamin A could temporarily mask the effect of a dietary insufficiency of vitamin A during lactation.

The same applies for some of the inorganic elements, such as, calcium and phosphorus, but milk volume probably suffers before striking changes occur in concentration in the milk.

The content of several mineral elements in swine milk is summarized in Table 7.10 (Guegen and Salmon-Legagneur 1959; Pond et al. 1965; Plumlee et al. 1956). Most mineral levels of milk appear to be resistant to dietary levels, while zinc (Earle and Stevenson 1965; Pond and Jones 1964) and manganese (Plumlee et al.

TABLE 7.10 Mineral Content of Swine Milk

Element	*Content*
Calcium	2.13 g/kg
Phosphorus	1.54 g/kg
Potassium	1.00 g/kg
Sodium	0.34 g/kg
Magnesium	0.20 g/kg
Chloride	1.0 g/kg
Iron	1.33 mg/liter
Zinc	4.94 mg/kg
Manganese	0.38 mg/kg (dry basis)

Sources: Perrin et al. 1955; Guegen, Salmon-Legagneur 1959; Pond et al. 1965; Plumlee et al. 1956.

1956) can be increased in the milk by increasing dietary intake. Iron and copper, unlike zinc and manganese, are not appreciably increased in the milk by feeding (Veum et al. 1965) or even by injection (Pond et al. 1961). As a result, milk is characteristically deficient in iron and copper regardless of gestation or lactation diet.

The fatty acid composition of the milk is sensitive to changes in the composition of the dietary fat. This relationship is shown in Table 7.11 (de Man and Bowland 1963; Tollerz and Lindberg 1965). The unsaturated fatty acids, linoleic and linolenic, are more closely related to the dietary levels than are the other fatty acids.

There is no evidence to indicate the mechanism of milk fat secretion in swine is different from that of other mammals (Keenan et al. 1970).

TABLE 7.11 Fatty Acid Composition of Swine Colostrum and Milk

Fatty acid	*Colostrum*	*Standard milk*	*6.3% cottonseed oil[a] milk*	*6.3% linseed oil[a] milk*
Capric ($C_{10:0}$)	—	0.2	—	—
Lauric ($C_{12:0}$)	—	0.3	—	—
Myristic ($C_{14:0}$)	1.4	3.3	3.2	2.9
Palmitic ($C_{16:0}$)	22.5	30.3	28.3	27.8
Palmitoleic ($C_{16.1}$)	5.0	9.9	6.6	8.1
Stearic ($C_{18:0}$)	5.7	4.0	5.4	3.2
Oleic ($C_{18:1}$)	41.7	35.3	34.3	27.1
Linoleic ($C_{18:2}$)	20.9	13.0	20.5	11.8
Linolenic ($C_{18:3}$)	2.4	2.5	0.9	18.3
Unidentified	0.4	1.2	—	—

[a] Cottonseed oil contains 51.9% $C_{18:2}$, 0.8% $C_{18:3}$; iinseed oil contains 14.1% $C_{18:2}$, 58.2% $C_{18:3}$.

Sources: de Man and Bowland 1963; Tollerz and Lindberg 1965.

MAMMARY GLANDS

The number of mammary glands varies from 6 to 20 (3–10 pairs) or more, arranged in two parallel rows, one on each side of the ventral median line and extending from the pectoral to the inguinal region.

There may be an even or an odd number on each side and in total; frequency distribution of even or odd is about 60 : 40. Each gland is separate from adjacent glands in secretory tissue.

Most breeds have a mean of approximately 12–14 nipples in females and a slightly higher number in males. Some Chinese breeds have more. A relatively low heritability coefficient of 0.10–0.20 (Enfield and Rempel 1961, Allen et al. 1959) suggests that it is impractical to select for high nipple number. Most breed associations require that individual animals have a minimum of 12 functional nipples for purebred registration.

A normal nipple has two external duct openings. Supernumerary (nonfunctional) nipples are often present between normal ones, usually between the third and fourth pairs. Most of these are underdeveloped, even though they may be in association with a functional gland. Rudimentary nipples also sometimes occur on the posterior end of one or both rows of normal nipples. "Inverted" nipples occur occasionally. This is a situation in which the nipple fails to project out from the surface of the gland, making it impossible to suckle.

The blood supply to the mammary glands comes from two branches of the arterial system; the common carotid artery supplies the anterior glands and a branch of the abdominal aorta supplies the posterior glands. However, there is considerable anastomosis, and some of the blood supplying the pectoral (anterior) glands may pass posteriorly to the posterior glands.

The lymphatic system of the mammary glands of swine is more complex than that of other farm animals. The details of structure have been described by Turner (1939).

In animals with red or black skin, there is pigmentation within the nipple. Presence of this pigment in the nipples of slaughter animals has led to the term "seedy belly" in the bacon trade. A red pigment in the nipples of some females of all breeds is found at estrus and is referred to as "red seedy cut" in carcasses from slaughter pigs. The mammary system appears early in prenatal life and, at birth, the nipples are easily visible. By 5 months of age, a well-developed duct system is formed; and by puberty (6 months), the extent of development is still more advanced. During early pregnancy, there is a further increase in the duct system; and by midpregnancy, the lobules and alveoli are fully developed. Secretion into the lumen of the ducts and alveoli begins by the 75th day of pregnancy; by 90 days, the enlargement of the gland is easily visible. By about 114 days (a few hours before parturition), the glands are full of colostrum. Lactation begins when the first pig is born.

REFERENCES

Allen, A. D., and Lasley, J. F. 1960. J. Anim. Sci. *19*, 150.

Allen, A. D., Tribble, L. F., and Lasley, J. F. 1959. Mo. Agric. Exp. Stn. Res. Bull. 694.

Allen, M. A., and Stewart, T. S. 1983. Agric. Systems *10*, 193.

Amann, R. P., and Schanbacher, B. D. 1983. J. Anim. Sci. *57* (Suppl. 2), 380.

Archibong, A. E., England, D. C., and Stormshak, F. 1987. J. Anim. Sci. *64*, 474.

Bazer, F. W., and First, N. L. 1983. J. Anim. Sci. *57* (Suppl. 2), 425.

Bazer, F. W., Clawson, A. J., Robison, O. W., and Ulberg, L. C. 1969. J. Reprod. Fert. *18*, 121.

Bazer, F. W., Clawson, A. J., Robison, O. W., Vincent, C. K., and Ulberg, L. C. 1968. J. Anim. Sci. *27*, 1021.

Berger, T., Mahone, J. P., Svoboda, G. S., Metz, K. W., and Clegg, E. D. 1980. J. Anim. Sci. *51*, 672.

Block, J. L., and Erickson, B. J. 1968. Anat. Rec. *161*, 45.

Blom, E. 1962. *In* Reproduction in Farm Animals (ed., E. S. E. Hafez). Lea & Febiger, Philadelphia.

Boda, J. M. 1959. *In* Reproduction in Domestic Animals (eds., H. H. Cole and P. T. Cupps). Academic Press, New York.

Bourne, F. J. 1969. Anim. Prod. *11*, 337.

Bowland, J. P. 1966. *In* Swine in Biomedical Research (eds., L. K. Bustad, R. G. McClellan, and M. P. Burns). Frayn, Seattle.

Boyd, R. D., Moser, B. D., Peo, E. R., Jr., and Cunningham, P. J. 1978a. J. Anim. Sci. *47*, 874.

Boyd, R. D., Moser, B. D., Peo, E. R., Jr., and Cunningham, P. J. 1978b. J. Anim. Sci. *47*, 833.

Brinkley, H. J., Wickersham, E. W., First, N. L., and Casida, L. E. 1964. Endocrinology *74*, 462.

Christenson, R. K. 1981. J. Anim. Sci. *52*, 821.

Christenson, R. K. 1986. J. Anim. Sci. *63*, 1280.

Christenson, R. K., and Ford, J. J. 1979. J. Anim. Sci. *49*, 743.

Christenson, R. K., Leymaster, K. A., and Young, L. D. 1987. J. Anim. Sci. *65*, 738.

Christenson, R. K., Teague, H. S., Grifo, A. P., Jr., and Roller, W. L. 1972. The effect of high environmental temperature on the boar. Ohio Research Summary *61*(4), 19. Dept. Anim. Sci., Ohio State University, Columbus.

Cox, N. M., and Britt, J. H. 1982. Biol. Reproduction *27*, 70.

Crabo, B., and Einarsson, S. 1971. Acta Vet. Scand. *12*, 1.

Crighton, D. B., and Lamming, G. E. 1969. J. Endocrinology *43*, 507.

Davis, D. L., Knight, J. W., Killian, D. B., and Day, B. N. 1979. J. Anim. Sci. *49*, 1506.

De Man, J. M., and Bowland, J. P. 1963. J. Dairy Res. *30*, 339.

Dhindsa, D. S., and Dziuk, P. J. 1968a. J. Anim. Sci. *27*, 122.

Dhindsa, D. S., and Dziuk, P. J. 1968b. J. Anim. Sci. *27*, 668.

Dziuk, P. J. 1968. J. Anim. Sci. *27*, 673.

Dziuk, P. J. 1970. J. Reprod. Fertil. *22*, 277.

Dziuk, P. J. 1977. Reproduction in pigs. *In* Reproduction in Domestic Animals (eds., H. H. Cole and P. T. Cupps), pp. 455–474. Academic Press, New York.

Dziuk, P. J., and Bellows, R. A. 1983. J. Anim. Sci. *57* (Suppl. 2), 355.

Dziuk, P. J., Polge, C., and Rowson, L. E. 1964. J. Anim. Sci. *23*, 37.

Earle, I. P., and Stevenson, J. W. 1965. J. Anim. Sci. *24*, 325.

Enfield, F. D. and Rempel, W. E. 1961. J. Anim. Sci. *20*, 876.

England, D. C. 1986. J. Anim. Sci. *63*, 1297.

Fenton, F. R., Bazer, F. W., Robison, W., and Ulberg, L. C. 1970. J. Anim. Sci. *31*, 104.

First, N. L., Lohse, J. K., and Nara, B. S. 1982. *In* Control of Pig Reproduction (eds., D. J. A. Cole and G. R. Foxcroft), p. 311. Butterworths, London.

First, N. L., Stratman, F. W., and Casida, L. E. 1963. J. Anim. Sci. *22*, 135.

Foote, R. H., Young, D. C., Turkheimer, A. R., and Bratton, R. W. 1959. *In* Colloquium on Reproduction and A. I. of the Pig, p. 27. Institut National de la Recherche Agronomique, Paris.

Garbers, D. L., and First, N. L. 1969. J. Reprod. Fertil. *20*, 465.

Graham, E. F., Rajamannan, A. H. J., Schmehl, M. K. L., Maki-Laurila, M., and Bower, R. E. 1971. A. I. Digest *19*, 6 (June).

Guegen, L., and Salmon-Legagneur, E. 1959. C. R. Séances Acad. Sci. *249*, 784.

Guthrie, H. O. 1979. J. Anim. Sci. *49*, 158.

Hafez, E. S. E., Sumption, L. J., and Jakway, J. S. 1963. *In* Behavior of Domestic Animals (ed., E. S. Hafez), p. 334. Williams & Wilkins, Baltimore.

Hancock, J. L. 1960. Stud. Fertil. *9*, 146.

Hancock, J. L., and Hovell, G. J. R. 1962. J. Reprod. Fertil. *4*, 195.

Hansel, W., and Convey, E. M. 1983. J. Anim. Sci. *57* (Suppl. 2), 404.

Hard, D. L., and Anderson, L. L. 1982. Biol. Reprod. *27*, 79.

Hartman, D. A., and Pond, W. G. 1960. J. Anim. Sci. *19*, 780.

Hartman, D. A., Ludwick, T. M., and Wilson, R. F. 1962. J. Anim. Sci. *21*, 883.

Hausler, C. L., Hodson, H. H., Jr., Kao, D. C., Kinney, T. J., Rauwolf, V. A., and Strack, L. E. 1980. J. Anim. Sci. *50*, 773.

Hemsworth, P. H. 1985. J. Anim. Sci. *61* (Suppl. 3), 75.

Hodson, H. H., Jr., Hausler, L. L., Snyder, D. H., Wilkens, M. A., and Arthur, R. D. 1981. J. Anim. Sci. *52*, 688.

Houpt, K. A., and Wolski, I. R. 1982. Domestic Animal Behavior for Veterinarians and Animal Scientists, pp. 1–356. Iowa State University Press, Ames.

Huang, Y.-T., Johnson, R. K., and Eckardt, G. R. 1987. J. Anim. Sci. *65*, 1298.

Hunter, R. H. F., Polge, C., and Rowson, L. E. 1967. J. Reprod. Fertil. *14*, 501.

Jensen, A. H., Yen, J. T., Gehring, M. M., Baker, D. H., Becker, D. E., and Harmon, B. G. 1970. J. Anim. Sci. *31*, 745.

Johnson, L. A., Aalbers, J. G., Willems, J. T., and Sybesma, W. 1981. J. Anim. Sci. *52*, 1130.

Johnson, R. K., Zimmerman, D. R., and Kittok, R. J. 1984. Livestock prod. Sci. *11*, 541.

Keenan, T. W., King, J. L., and Colenbrander, V. F. 1970. J. Anim. Sci. *30*, 806.

Kennelly, J. J., and Foote, R. H. 1964. J. Anim. Sci. *23*, 160.

Kenneth, J. H. 1953. Gestation Periods. A table and bibliography. Tech. Communications, The Commonwealth Bureau of Animal Breeding, Genetics no. 5, Farnham House, Farnham Royal, Slough, Bucks., England.

Killian, D. B., Kiesling, D. O., Wulff, F. P., and Stewart, A. N. V. 1987. J. Anim. Sci. *64*, 231.

Kirkpatrick, R. L., Lauderdale, J. W., First, N. L., Hauser, E. R., and Casida, L. E. 1965. J. Anim. Sci. *24*, 1104.

Kirkwood, R. N., and Aherne, F. X. 1985. J. Anim. Sci. *60*, 1518.

Klaver, J., van Kempen, G. J. M., de Lange, P. G. B., Verstegen, M. W. A., and Boer, H. 1981. J. Anim. Sci. *52*, 1091.

Lasley, J. F., Day, B. N., and Mayer, D. T. 1963. J. Anim. Sci. *22*, 422.

Lecce, J. G. 1986. J. Anim. Sci. *63*, 1307.

Leymaster, K. A., Christenson, R. K., and Young, L. D. 1986. Proc. 3rd World Congress on Genetics Applied to Livestock Production. *11*, 209.

MacKenzie, F. F., Miller, J. C., and Bauguess, L. C. 1938. Mo. Agric. Exp. Stn. Res. Bull. 279.

Mahan, D. C., Becker, D. E., Norton, H. W., and Jensen, A. H. 1971. J. Anim. Sci. *33*, 35.

Mann, T. 1954. The Biochemistry of Semen. Methuen, London.

Mann, T. 1959. *In* Reproduction in Domestic Animals (eds., H. H. Cole and P. T. Cupps). Academic Press, New York.

Minar, V. M., and Schilling, E. 1976. Dtsch. Tieraerztl. Wochenschr. *77*, 428.

Nara, B. S., and First, N. L. 1981. J. Anim. Sci. *52*, 1360.

Nellor, J. W., Daniels, R. W., Hoefer, J. A., Wildt, D. E., and Dukelow, W. 1975. Theriogenology *4*, 23.

Noblet, J., and Etienne, M. 1989. J. Anim. Sci. *67*, 3352.

Omtvedt, I. T., Nelson, R. E., Edwards, R. L., Stephens, D. F., and Turman, E. J. 1971. J. Anim. Sci. *31*, 312.

Perrin, D. R. 1955. J. Dairy Res. *22*, 103.

Pettigrew, J. E., Jr. 1981. J. Anim. Sci. *53*, 107.

Pettigrew, J. E., Sower, A. F., Cornelius, S. G., and Moser, R. L. 1985. Can. J. Anim. Sci. *65*, 989.

Plumlee, M. P., Thrasher, D. M., Beeson, W. M., Andrews, F. N., and Parker, H. E. 1956. J. Anim. Sci. *15*, 352.

Polge, C. 1965. Vet. Rec. *77*, 232.

Polge, C. 1968. Proc. 6th Int. Congr. Anim. Reprod. (Abstr. 294).

Polge, C. 1982. *In* Control of Pig Reproduction (eds., Foxcroft and Cole) p. 277. Butterworths, London.

Pond, W. G., and Jones, J. R. 1964. J. Anim. Sci. *23*, 1057.

Pond, W. G., Lowrey, R. S., Maner, J. H., and Loosli, J. K. 1961. J. Anim. Sci. *20*, 747.

Pond, W. G., and Mersmann, H. J. 1988. J. Nutr. *118*, 1223.

Pond, W. G., Mersmann, H. J., and Yen, J. T. 1985. J. Nutr. *115*, 179.

Pond, W. G., Roberts, S. J., Dunn, J. A., and Willman, J. P. 1960. J. Anim. Sci. *19*, 881.

Pond, W. G., VanVleck, L. D., and Hartman, D. A. 1962. J. Anim. Sci. *21*, 293.

Pond, W. G., Veum, T. L., and Lazar, V. A. 1965. J. Anim. Sci. *24*, 668.

Pope, W. F., and First, N. L. 1985. Theriogenology *23*, 91.

Pursel, V. G. 1983. J. Anim. Sci. *56,* 1161.

Pursel, V. G. 1980. J. Anim. Sci. *51* (Suppl. 1), 316.

Pursel, V. G., and Johnson, L. A. 1971. Agric. Res. Service, USDA ARS *44,* 227.

Rampacek, G. B., Kraeling, R. R., and Kiser, T. E. 1981. Theriogenology *15,* 1491.

Rook, J. A. F., and Witter, R. C. 1963. C. R. Acad. Sci. (Paris) *256,* 1372.

Salmon-Legagneur, E. 1958. Ann. Zootech. *2,* 143.

Salmon-Legagneur, E. 1965. Ann. Zootech. *14,* 1.

Seerley, R. W., Griffin, F. M., and McCampbell, H. C. 1978a. J. Anim. Sci. *46,* 1009.

Seerley, R. W., Maxwell, J. S., and McCampbell, H. C. 1978b. J. Anim. Sci. *47,* 1114.

Self, H. L. 1959. *In* Colloquium on Reproduction and A.I. of the Pig. Institut National de la Recherche Agronomique, p. 121. Paris.

Self, H. L., Grummer, R. H., and Casida, L. E. 1955. J. Anim. Sci. *14,* 573.

Sisson, S., and Grossman, V. D. 1938. The Anatomy of Domestic Animals, W. B. Saunders, Philadelphia.

Smidt, D., Steinbach, J., and Scheven, B. 1965. J. Reprod. Fertil. *10,* 153.

Smith, D. M. 1952. N. Z. J. Sci. Technol. *A34,* 65.

Smith, D. M. 1959. N. Z. J. Agric. Res. *2,* 1057.

Speer, V. C., and Cox, D. F. 1984. J. Anim. Sci. *59,* 1281.

Swierstra, E. E. 1970. *In* Effect of Disease and Stress on Reproductive Efficiency in Swine, p. 8. Proc. Symposium, Extension Service, Univ. of Nebraska, College of Agric., Lincoln, and USDA, Washington, D.C.

Teague, H. S. 1970. *In* Effect of Disease and Stress on Reproductive Efficiency in Swine, p. 21. Proc. Symposium, Extension Service, Univ. of Nebraska, College of Agric., Lincoln, and USDA, Washington, D.C.

Tollerz, G., and Lindberg, P. 1965. Acta Vet. Scand. *6,* 118.

Tumbleson, M. E. 1986. Swine in Biomedical Research. Plenum Press, N.Y.

Turkheimer, A. R., Young, D. G., and Foote, R. H. 1958. Cornell Vet. *58,* 291.

Turner, C. W. 1939. The Comparative Anatomy of the Mammary Glands. University Coop. Store, Univ. of Missouri, Columbia.

Van Kempen, G. J. M., Geerse, C., Verstegen, M. W. A., and Mesu, J. 1985. Neth. J. Agric. Sci. *33,* 23.

Veum, T. L., Gallo, J. T., Pond, W. G., Vanvleck, L. D., and Loosli, J. K. 1965. J. Anim. Sci. *24,* 1169.

Warnick, A. C., Wiggins, E. L., Casida, L. E., Grummer, R. H., and Chapman, A. B. 1951. J. Anim. Sci. *10,* 479.

Welp, C., and Holtz, W. 1985. Anim. Reprod. Sci. *8,* 171.

Westmoreland, N., First, N. L., and Hoekstra, W. G. 1967. J. Reprod. Fertil. *13,* 223.

Westrom, B. R., Karlsson, B. W., Ekstrom, G., Svendsen, J., and Svendsen, L. S. 1986. *In* Swine in Biomedical Research (ed., M. E. Tumbleson), p. 1297. Plenum Press, New York.

Wettemann, R. P., Wells, M. E., and Johnson, R. K. 1979. J. Anim. Sci. *49,* 1501.

White, I. G. 1958. Anim. Breed. Abstr. *26,* 109.

Wilson, M. E., Edgerton, L. A., Cromwell, G. L., and Stahly, T. S. 1981. J. Anim. Sci. *52,* 323.

Wu, M. C., and Dziuk, P. J. 1988. J. Anim. Sci. *66,* 1712.

Wu, M. C., Hentzel, M. D., and Dziuk, P. J. 1988. J. Anim. Sci. *66,* 3202.

Young, D. C. 1958. Sperm production, sperm preservation, heat detection and artificial insemination in swine. M. S. Thesis. Cornell Univ., Ithaca.

Zimmerman, D. R., Carlson, R., and Lantz, B. 1974. J. Anim. Sci. *39,* 230.

Zimmerman, D. R., and Cunningham, P. J. 1975. J. Anim. Sci. *40,* 61.

8

Nutrition and Feed Formulation

Feed represents 55–85% of the total cost of commercial swine production, depending mainly on the relative costs of feed, labor, and housing in a particular situation. For this reason, it is highly important that economical as well as nutritionally balanced diets are provided during all phases of the life cycle. The economics of swine feeding are largely dependent on local conditions of feedstuff availability and competition for the same foodstuff for use by either humans or other animals. The pig has a digestive system with limited ability to utilize large quantities of forage so that, like the chicken, it is in direct competition with man for available food supplies. The degree of competition is related to cultural differences in food preferences. For example, wheat and potatoes are not usually fed to swine in the United States as the demand for human consumption holds the price too high, but in many other parts of the world these crops are commonly fed to swine. Similar relationships exist for other crops in other parts of the world.

PRACTICAL SWINE FEEDING

Practical swine feeding consists of meeting biological needs with appropriate combinations of feed ingredients and pure sources of limiting nutrients to yield an economical and nutritionally balanced diet. The nutrient requirements of swine are similar within each phase of the life cycle except for genetic differences in utilization or quantitative requirements for individual nutrients and for environmental effects such as ambient temperature, altitude, and season. Therefore, specific nutrients and signs of deficiency are discussed first, quantitative nutrient requirements for each phase of the life cycle are discussed next, and finally, practical aspects of diet formulation and feeding are referred to, but not addressed in detail.

Since available feedstuffs are quite different in different parts of the world, practical feeding must take this into account. The nutrient composition and feeding value of the common feedstuffs available for formulation of swine diets in temperate and tropical environments are discussed in detail in the 1984 edition of *Swine Feeding and Nutrition* (Pond and Maner 1984). Average composition of some common feed ingredients for swine is given by the National Research Council (1982A) and in the National Research Council (1988) publication on nutrient requirements of swine.

NUTRIENTS REQUIRED BY SWINE

A nutrient, for the purpose of this discussion, is any chemical entity required by the animal to meet metabolic needs. It can be supplied by feeding, or in some cases by parenteral administration. Quantitative requirements and clinical signs of deficiency are given in publications by the National Research Council (1988) and by the Agricultural Research Council (1981). Practical feeding recommendations and current knowledge of digestion and absorption of nutrients by swine are reviewed by Low and Partridge (1980) and provided in several textbooks (Cunha 1977; Morrison 1956; Krider et al. 1981; Pond and Houpt 1978). In addition, the effects of environment on nutrient requirements have been reviewed (National Research Council 1982B). Literature on early research in swine nutrition can be found in *Swine Production in Temperate and Tropical Environments* (Pond and Maner 1974) and *Swine Production and Nutrition* (Pond and Maner 1984) and technical aspects of swine nutrition are found in Miller, Ullrey, and Lewis (1991).

The nutrients known to be required by swine follow:

Water. Water is of course necessary.

Fatty acids. Linoleic acid and perhaps arachidonic acid are required and some fat (0.06%) is needed for absorption of fat-soluble vitamins.

Vitamins. Vitamins include the fat soluble—A, D, E, and K; and the water soluble—biotin, choline (replaceable by methionine), folacin, niacin (nicotinic acid), pantothenic acid, riboflavin (vitamin B_2), thiamin (vitamin B_1), vitamin B_{12}, and vitamin B_6 (pyridoxine, pyridoxamine, or pyridoxal).

Protein. Proteins are probably not required as such, but they are necessary as a source of essential amino acids and N for nonessential amino acid synthesis.

Amino acids. Amino acids include arginine (for growth only), histidine, isoleucine, leucine, lysine, methionine (50% replaceable by cystine), phenylalanine (30% replaceable by tyrosine), threonine, tryptophan, and valine.

Inorganic elements. Inorganic elements include calcium, chlorine, cobalt (as component of vitamin B_{12}), copper, iodine, iron, magnesium, manganese, sodium, phosphorus, potassium, selenium, sulfur (as component of sulfate and of thiamin, methionine, cystine, and several other organic compounds), and zinc.

Carbohydrates. There is no proof that carbohydrates are required as a source of energy, but since all natural feedstuffs contain some carbohydrate, it becomes an academic question. Certainly, by far the greatest source of energy commonly available in swine feeding is carbohydrate. If a dietary requirement exists, it is as glucose, since the ultimate utilization of energy by the cell is as glucose from dietary intake or from glucose formation from amino acids and fat.

Several of the vitamins are required only under specialized conditions. Biotin deficiency is normally encountered only when large quantities of raw egg white are fed. This contains a protein, avidin, which ties up biotin in the gastrointestinal tract. Recent field reports indicate that biotin deficiency may occur under special conditions of confinement feeding in commercial production or when high levels of antimicrobial agents are fed. Vitamin K deficiency has been observed in pigs kept in wire- or slotted-floor pens which prevent coprophagy. Apparently sufficient vitamin K is synthesized by microflora in the lower intestinal tract to meet the metabolic requirement of the pig when given access to the feces. As confinement feeding of swine on slotted floors becomes more common, it behooves scientists and swine producers to be alert to deficiencies of vitamins and other nutrients induced by prevention of coprophagy.

A series of fact sheets are available from The National Pork Industry Handbook (available from county agricultural extension offices throughout the United States) that deal with practical aspects of swine feeding (PIH 2, PIH 3, PIH 5, PIH 7, PIH 23, PIH 31, PIH 52, PIH 71, PIH 73, PIH 108, and PIH 112). No attempt is made here to duplicate or paraphrase these excellent resources.

CLINICAL SIGNS OF NUTRIENT DEFICIENCIES

Under practical feeding conditions, in which natural feedstuffs are used, only some of the required nutrients are likely to be deficient. These include the following: total protein; the amino acids, lysine and tryptophan and occasionally others; the vitamins A, D, E, riboflavin, pantothenic acid, niacin, choline, vitamin B_{12}; and the minerals calcium, phosphorus, iodine, sodium, chlorine, zinc, and selenium.

A chart adapted from a National Research Council publication (1979) summarizing clinical signs of deficiency of nutrients is shown in Table 8.1 and photographs of selected deficiency signs are in Fig. 8.1. A similar chart of signs of dietary excesses is shown in Table 8.2. Clinical signs of deficiency of some of these nutrients are recorded in several textbooks on nutrition and in the scientific literature. Brief descriptions of typical nutrient deficiency and toxicity signs are given in this book. Gross and microscopic pathology of nutrient deficiencies are illustrated in books by Follis (1958) and King et al. (1976).

The preparation of tables of nutrient requirements as done by the National Research Council (1988) and the Agricultural Research Council (1981) for use by feed manufacturers, swine producers, and scientists requires compilation and evaluation of results of research obtained under a wide variety of experimental conditions and environments. A diet that is nutritionally adequate for one population of pigs might not be entirely adequate for another population or for the same population under different environmental conditions. Therefore, a clear distinction must be made between minimum requirements and minimum recommended allow-

TABLE 8.1 Signs of Nutrient Deficiencies

Nutrient	*Signs of nutrient deficiency*	
	Clinical	*Subclinical*
Energy	Weakness, low body temperature, loss of weight, coma, and death	Hypoglycemia Loss of subcutaneous fat Elevated hematocrit and serum cholesterol Reduced blood glucose, calcium, and sodium
Protein: amino acid	Impaired growth Unthriftiness Reduced resistance to bacterial infection	Kwashiorkor-like signs in baby pigs, including reduced serum protein and serum albumin, anemia, gross edema, and increased lipid liver concentration
Fat: linoleic acid	Scaly dermatitis may appear	Small gall bladder Elevated triene/tetraene in tissue lipids
Vitamin A	Incoordination Lordosis Paralysis of rear limbs Night blindness Congenital defects	Retarded bone growth Increase in cerebrospinal fluid pressure Degeneration of sciatic and femoral nerves Minimal visual purple Atrophy of epithelial layers of genital tract
Vitamin D	Rickets Osteomalacia Low calcium tetany	Lack of bone calcification and proliferation of epiphyseal cartilage Rib and vertebra fracture Low plasma calcium, magnesium, and inorganic phosphorus levels Elevated serum alkaline phosphatase levels
Vitamin E—selenium	Edema Sudden death	Generalized edema Liver necrosis (hepatosis dietetica) Microangiopathy Cardiac muscle degeneration (mulberry heart) Pale, dystrophic muscle
Vitamin K	Pale newborn pigs with loss of blood from umbilical cord Sudden death following dicoumarin intake	Increased prothrombin time Increased blood-clotting time Internal hemorrhage Anemia due to blood loss
Thiamin	Poor appetite Poor growth Sudden death	Cardiac hypertrophy Bradycardia First and second degree auriculoventricular block Elevated plasma pyruvate
Riboflavin	Slow growth Seborrhea Impaired sow reproductivity	Lens cataracts Increase in neutrophilic leukocytes Birth of weak pigs with skeletal anomalies
Niacin	Poor appetite Poor growth	Necrotic lesions of intestine

TABLE 8.1 (*Continued*)

Nutrient	*Signs of nutrient deficiency*	
	Clinical	*Subclinical*
	Severe diarrhea	
	Dermatitis	
Pantothenic acid	Poor appetite	Inflammation of colon
	Poor growth	Degeneration of sciatic and peripheral nerves
	Diarrhea	
	Unusual gait (goose-stepping)	Reduced blood pantothenic acid level
	Impaired sow reproductivity	Reduced free pantothenic acid level in milk
Vitamin B_6	Poor growth	Microcytic hypochromic anemia
	Epileptic seizures	Elevated serum iron
		Fatty infiltration of liver
		Elevated urinary xanthurenic acid
		Elevated gamma globulin-like blood protein fraction
Vitamin B_{12}	Depressed growth	Reduced serum and tissue B_{12} levels
	Hypersensitivity	
	Reduced sow reproductivity	
Choline	Slow growth	Fatty infiltration of liver
	Reduced litter size	Reduced conception rate
Biotin	Dermatosis	Reduced urinary biotin excretion
	Spasticity of hind legs	
Folacin	Poor growth	Normocytic anemia
	Weakness	
Calcium	Rickets	Lack of bone calcification
	Osteomalacia	Bones easily fractured
	Low calcium tetany	Low plasma calcium level
		Elevated serum inorganic phosphorus and alkaline phosphatase
Phosphorus	Poor growth	Lack of bone calcification
	Rickets	Bones easily fractured
	Osteomalacia	Low serum inorganic phosphorus level
		Elevated serum calcium and alkaline phosphatase
		Enlarged costochondral junction (beading)
Magnesium	Poor growth	Low serum magnesium and calcium
	Stepping syndrome	Reduced bone magnesium
	Weakened carpo-metacarpo-phalangeal and tarso-metatarso-phalangeal joints	
	Tetany	
Potassium	Anorexia	Reduced heart rate
	Rough hair coat	Increased *PR*, *QRS*, and *QT* intervals on electrocardiogram
	Emaciation	
	Ataxia	Reduced serum potassium

(*continued*)

TABLE 8.1 Signs of Nutrient Deficiencies (*Continued*)

Nutrient	*Signs of nutrient deficiency*	
	Clinical	*Subclinical*
Sodium	Poor appetite Low water consumption Unthriftiness	Negative sodium balance Elevated serum potassium Elevated plasma urea nitrogen Reduced chlorine retention
Chlorine	Poor growth	Reduced plasma chlorine Reduced sodium and potassium retention
Iron	Poor growth Rough hair coat Pallor Anoxia	Hypochromic microcytic anemia Enlarged heart and spleen Enlarged fatty liver Ascites Clumping of erythroblastic cells in bone marrow Reduced serum iron and percent transferrin saturation
Copper	Leg weakness Ataxia	Microcytic hypochromic anemia Reduced serum copper and ceruloplasmin Aortic rupture Cardiac hypertrophy
Zinc	Poor growth Poor appetite Parakeratosis	Reduced serum, tissue, and milk zinc Reduced serum albumin–globulin ratio Reduced serum alkaline phosphatase Reduced thymus weight Retarded testicular development Impaired reproductivity of sows
Iodine	Goiter Myxedema Sows farrow weak, hairless pigs	Enlarged hemorrhagic thyroid Hyperplasia of follicular epithelium of thyroid Reduced plasma protein-bound iodine
Manganese	Lameness in growing pigs Increased fat deposition in pregnant gilts with birth of weak pigs with poor sense of balance	Replacement of cancellous bone with fibrous tissue Early closure of distal epiphyseal plate Low serum manganese and alkaline phosphatase Negative manganese balance
Water	Poor appetite Dehydration Loss of body weight Possible salt poisoning Death	Elevated hematocrit Elevated plasma electrolytes Loss of temperature regulation Tissue dehydration

Source: National Research Council 1979.

ances. The minimum requirement can be considered to be that amount of a single nutrient needed to allow normal body function of all animals in a population under ideal conditions. The minimum recommended allowance can be considered the minimum requirement plus some arbitrary increment of that nutrient added as a "safety factor" to allow for increased metabolic needs induced by environmental stressors. This concept is illustrated graphically in Fig. 8.2.

Genetic differences between breeding populations in performance potentials are known to exist. Similarly, genetic variation for performance occurs within populations of similar breeding. Thus, it follows that the nutrient requirements to support the performance potential changes from one genetic population to another or from animal to animal within a population. A challenge for the future is to develop the ability to more precisely describe genetic and environmentally induced differences in potential performance so that feeding programs can be customized to those differences. Superimposed on the differences in requirements is that variability in appetite results in variation in the amount of nutrients consumed between and within genetic populations. The following discussion of individual nutrient deficiency signs pertains to results obtained with diets that failed to meet the minimum requirement of the specified nutrient. The same deficiency signs could be expected in animals in which the diet would have contained the minimum requirement if it had been fed under ideal conditions, but in which the metabolic requirement had been increased by some environmental stress. Examples of such stresses are cited under individual nutrients. From a practical standpoint, a reduction in rate of body weight gain, in efficiency of feed utilization, or in other productive function is the most important and usually the most common clinical sign of a nutrient deficiency. The following descriptions of individual deficiency signs, therefore, emphasize more specific changes.

Protein

Inadequate dietary protein is probably the most frequent nutrient deficiency, primarily because the common energy sources (maize, cereal grains, root and tuber crops) are low in protein, and protein supplements are expensive. The main signs of protein deficiency in growing pigs are slow growth and inefficient feed utilization, accompanied in severe cases by unthriftiness. Blood serum protein is depressed, the liver becomes infiltrated with fat, and in severe deficiency there may be fluid accumulation (edema) in the jowl and scrotum.

Regarding protein requirements for gestation and lactation, it appears that lactation is considerably more critical. Protein deficiency during lactation results in reduced milk yield and in reduced protein concentration of the colostrum but not of milk. The pregnant dam has the ability to partially protect the developing fetus against protein deficiency by using her own protein reserves to meet the needs of the fetus. The National Research Council (1988) recommends a diet containing 12% protein during gestation. That this high level of protein is not required over a single pregnancy seems clear, because opaque-2 maize (containing 9.7% protein and approximately 0.3% lysine) as the only protein source allows normal reproduction, although subsequent lactation requires a higher protein level. Amino acid balance is also apparently of importance, since common maize is inferior to

a

b

c

d

e

f

g

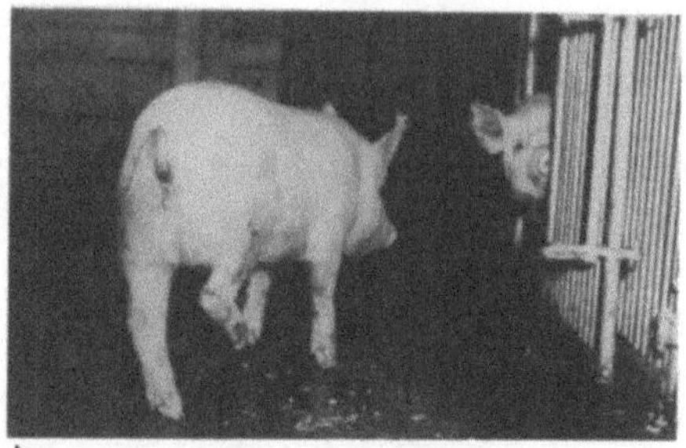

h

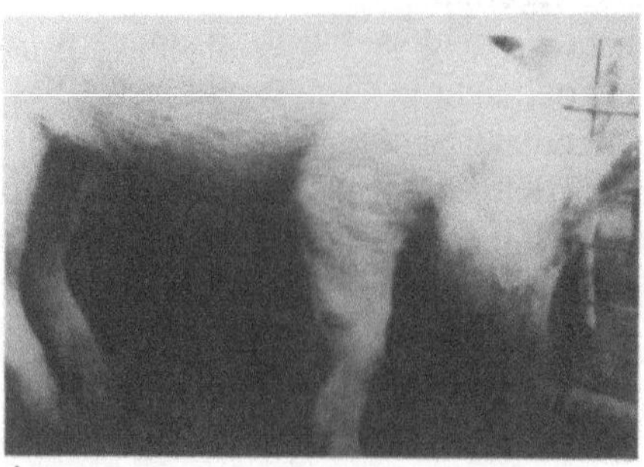

i

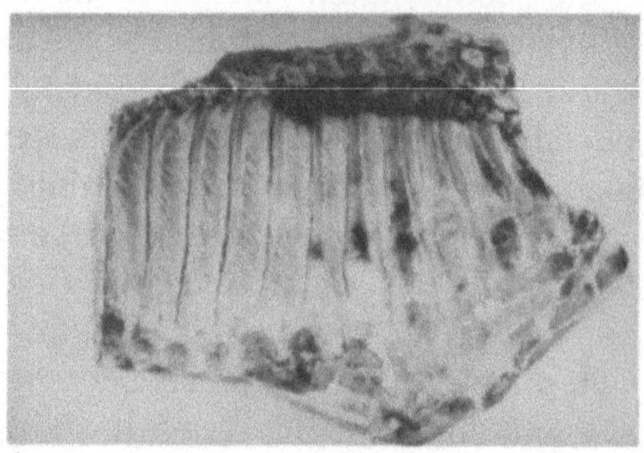

j

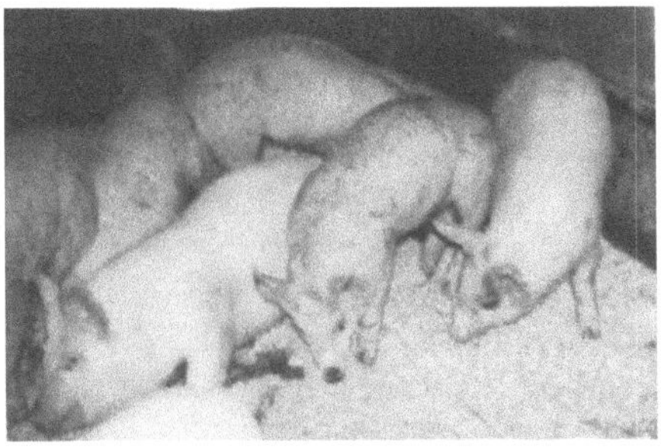

k

l

m

FIGURE 8.1 Pigs and organs of pigs showing clinical signs of nutrient deficiencies. (a) Energy deficiency in adult; note the extreme emaciation. (b) Energy deficiency in growing pig; this pig was weaned at 2 days and fed cow's milk in restricted amounts to maintain body weight for 8 weeks; the pig remains active but stunted in growth. (c) Protein deficiency in growing pig; this pig was weaned at 3 weeks and fed a 3% protein, 23% fat diet for 8 weeks; note the rough hair coat and severe edema; serum protein is depleted and liver is high in fat. (d) Amino acid deficiency (lysine and tryptophan) in growing pigs; note the greater size of pig fed opaque-2 maize, compared with littermate fed common maize. (e) Tryptophan deficiency in growing pigs; note the difference in size between littermates fed diets identical except for tryptophan content. (f) Liver from a growing pig fed a diet deficient in vitamin E and selenium; note the massive liver damage as evidenced by the rough, necrotic appearance. (g) Vitamin A deficiency results in birth of stillborn pigs with skeletal abnormalities and eyelessness. (h) Pantothenic acid deficiency in growing pig; note the peculiar "goose-stepping." (i) Calcium deficiency in growing pig caused by feeding 0.2 calcium; lameness results from subnormal calcification of bones. (j) Ribs from pig fed a calcium-deficient diet; note the multiple fractures. (k) Vitamin D deficiency rickets in growing pigs; these pigs were fed a normal calcium-phosphorus diet, but the diet contained no added vitamin D and the pigs were denied access to sunlight; note the lameness and soreness caused by poor absorption and utilization of calcium (courtesy of J. A. Hoefer, Michigan State University). (l) Zinc deficiency, parakeratosis, in a growing pig; this pig was fed a cassaba-rice bran diet unsupplemented with zinc; note the thickened, encrusted skin. (m) Fat deficiency in growing pig; note the pustules on the skin and loss of hair in deficient pigs in contrast to normal littermates.

TABLE 8.2 Signs of Dietary Excesses

Nutrient	*Toxic dietary level*[a]	*Age*	*Signs of dietary excess*
Calcium	1% (with limited zinc)	Immature	Depressed appetite, reduced rate of gain, parakeratosis
	1% (with adequate zinc and limited phosphorus)	Immature	Reduced rate of gain and reduced bone strength
Copper	300–500 mg/kg (in absence of higher levels of dietary iron and zinc)[b]	Immature	Reduced growth, lower hemoglobin, icterus, and death[c]
Iodine	800 mg/kg	Immature	Depressed feed intake and rate of gain, lowered hemoglobin,[d] and eye lesions
Iron	5000 mg/kg	Immature	Depressed feed intake and rate of gain, reduced serum inorganic phosphorus and femur ash, rickets[e]
Manganese	4000 mg/kg	Immature	Depressed feed intake, reduced growth rate, stiffness, and stilted gait
Selenium	5–8 mg/kg[f]	Immature	Anorexia, hair loss, separation of hoof and skin at coronary band, degenerative changes in liver and kidney
	10 mg/kg	Breeding (sows)	Reduced conception; pigs small, weak, or dead at birth
Sodium chloride and other sodium salts	1–8% (with severe water restriction)	All ages	Nervousness, weakness, staggering, epileptic seizures, paralysis and death
Zinc	2000 mg/kg	Immature	Growth depression, arthritis, hemorrhage in axillary spaces, gastritis, and enteritis
Arsenic	990 mg/kg	Immature	Poor growth, erythema, ataxia, posterior paralysis, quadraplegia, and blindness; myelin degeneration of optic and peripheral nerves
Cadmium	50 mg/kg	Immature	Reduced gain and hematocrit
	150 mg/kg	Immature	Severe depression of gain and hematocrit
	450 mg/kg	Immature	Severe depression of gain and hematocrit and appearance of dermatitis
Cobalt	400 mg/kg	Immature	Anorexia, growth depression, stiff-legged, humped back, incoordination and muscle tremors, anemia[g]
Fluorine:			
Soluble fluorides	100 mg/kg	Mature	Mottled enamel, enamel hypoplasia, softening of teeth, osteomalacia, excessive loss of weight by lactating sows
Rock phosphate F	200 mg/kg	Mature	
Gossypol[h]	200 mg/kg	Immature	Muscular weakness, dyspnea, generalized edema, death; myocarditis, hepatitis, and nephritis

TABLE 8.2 (*Continued*)

Nutrient	*Toxic dietary level*[a]	*Age*	*Signs of dietary excess*
Lead	660 mg/kg	Immature	Squealing as if in pain, diarrhea, salivation, grinding of teeth, depressed appetite, reduced growth rate, muscular tremors, ataxia, increased respiratory rate, decreased heart rate, enlarged carpal joints, impaired vision, clonic seizures, death
Mercury	Single oral dose of 5–15 mg methyl mercury dicyandiamide per kilogram of body weight	Immature	Anorexia, body weight loss, central nervous system depression, weakness, gagging, vomiting, diarrhea, ataxia, cyanosis, muscular tremors, postural and gait abnormalities, polyuria
Nitrate	1800 mg NO_3/kg	Immature	Growth depression, dyspnea and cyanosis, elevated methemoglobin, lymphocytosis, reduced serum vitamin A and E levels
Nitrite	400 mg NO_2/kg	Immature	
Urea	2.5%	Immature	Reduced feed intake and growth rate; increased plasma urea nitrogen level

[a] The toxic dietary levels listed are those that have experimentally produced the signs indicated and are not necessarily minimum toxic or maximum tolerant levels.
[b] In a few instances, a dietary level of 250 mg/kg has resulted in signs of excess.
[c] In some instances, 500 mg/kg of copper has been fed without icterus or death occurring.
[d] Anemia of iodine toxicity alleviated with supplemental iron.
[e] Rickets from excessive dietary iron alleviated by increasing dietary phosphorus.
[f] Selenium toxicity partially alleviated with arsenic.
[g] Cobalt toxicity alleviated by supplemental methionine, iron, zinc, and manganese.
[h] Gossypol toxicity alleviated by increasing dietary iron to equal the weight of free gossypol.

Source: National Research Council 1979.

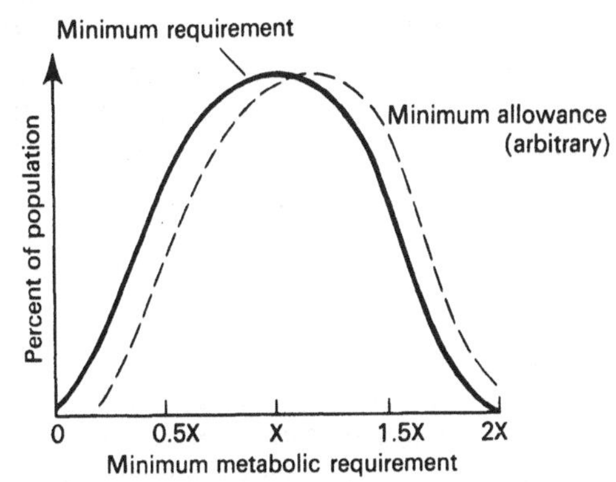

FIGURE 8.2 Minimum requirements vs. recommended allowance of nutrients.

opaque-2 maize and to maize soybean meal diets. Complete deprivation of protein throughout pregnancy may cause the birth of smaller pigs but apparently has no effect on number born or their livability. The long-term effects of protein deficiency on reproductive performance over more than two or three reproductive cycles is not known.

The pig's requirement for protein is, in reality, a requirement for individual amino acids not synthesized by the body (indispensable or essential amino acids) plus a requirement for nonprotein nitrogen that can be used by the animal for synthesis of dispensable amino acids to allow tissue protein synthesis to occur. The National Research Council (1988) has compiled tabular summaries of research findings describing quantitative dietary requirements of indispensable amino acids for growing pigs of various weights and for gestating and lactating sows. For the student of nutrition, a careful review of these tables provides insight into the criteria used to establish requirements and the types of diets used to produce deficient levels of specific amino acids.

Lysine

This is the first or second limiting amino acid in maize and in most of the cereal grains, so that a deficiency is common in pigs that are fed all-plant diets, unless the diet is maize and soybean meal, a combination which provides an excellent balance of amino acids. A deficiency of lysine results in anorexia and an unthrifty appearance in young pigs, but in older pigs a moderate deficiency may result in less protein deposition in the carcass and slightly less efficiency of feed utilization with no effect on weight gain. The requirement for gestation and lactation is apparently less than that for growth, but it is probably important to supplement low protein diets with protein sources high in lysine for optimum reproductive performance.

Tryptophan

Along with lysine, this amino acid is likely to be deficient in diets composed mainly of common maize. The signs of deficiency are slow growth, unthriftiness, and eye cataracts. Tryptophan can be converted by the tissues of the pig to the vitamin, niacin, so it is important to provide adequate niacin to minimize the tryptophan requirement. The hydroxy and keto analogues of tryptophan are utilized to some extent for growth, indicating that either the pig or the microflora of the gastrointestinal tract or both have the ability to add ammonia to form the amino acid. This may have future economic importance in amino acid fortification of swine diets.

Calcium

It is inappropriate to discuss mineral requirements independent of one another because of their interrelationships in metabolism. For example, phosphorus and vitamin D are both related to calcium metabolism, and levels of these nutrients in the diet affect the calcium requirement. An optimum ratio of calcium to phosphorus is between 1 : 1 and 2 : 1, depending on the biological availability of each in the feedstuffs used. When the ratio is less than 1 : 1, even a nearly adequate absolute amount of calcium in the diet may be associated with signs of calcium deficiency. Absorption of calcium from the gastrointestinal tract is increased by dietary vitamin D, decreased by high dietary fat, decreased by acid pH in the contents, decreased by phytin phosphorus, and probably affected by other poorly understood factors.

A deficiency in growing animals results in rickets (crooked legs, spontaneous bone fractures) due to failure of the protein matrix of bone to ossify. The corresponding condition in adults is osteomalacia. A dietary imbalance of calcium to phosphorus (ratio of less than 1 : 1) results in excessive bone resorption (fibrous osteodystrophy) due to secondary nutritional parathyroidism (Brown et al. 1966). The low level of calcium in the blood serum resulting from the calcium—phosphorus imbalance triggers the parathyroid to hypersecrete parathyroid hormone in an attempt to maintain blood serum calcium homeostasis. Calcium is lost from the bone in excessive amounts and replaced by fibrous connective tissue. This is a generalized phenomenon involving the entire skeleton. Clinical signs include distortion and loss of the nasal turbinates sometimes accompanied by a crooked snout and spontaneous fractures of ribs and long bones.

Calcium deficiency in gestation may result in spontaneous fractures of long bones, parturient paresis, or lameness of the dam. The fetus is partially protected by transfer of labile maternal sources of calcium across the placenta.

Phosphorus

A deficiency of phosphorus results in classical rickets. Adequate phosphorus supplementation to pig diets is especially important since there is a wide variation in the biological availability of the phosphorus from various sources. Plant sources of phosphorus (phytin phosphorus) are rather poorly utilized by the pig. Dicalcium phosphate is superior to soft phosphate with colloidal clay (colloidal phosphate) in promoting growth, feed utilization, femur ash, and inorganic serum phosphorus.

Phosphoric acid and di- and monocalcium phosphate are equal as sources of phosphorus while steamed bone meal and defluorinated rock phosphate are second in acceptability followed by Curaçao Island phosphate and colloidal phosphate. An average availability of phosphorus of 50% is believed to be typical for most diets fed to growing pigs (Agricultural Research Council 1981). Dietary phosphorus levels should be calculated on the basis of biologically available phosphorus because of the wide variability encountered in bioavailability of phosphorus in feed resources.

Zinc

The role of zinc in preventing skin lesions (parakeratosis) was first reported by Tucker and Salmon (1955). Since then, it has been established that a zinc deficiency can be precipitated by a high level of dietary calcium. Diets containing more than

1% calcium predispose to zinc-deficiency symptoms which can be prevented by the addition of zinc salts (carbonate, chloride, oxide, or sulfate) to the diet.

Since parakeratosis is not entirely absent even in diets containing less than 1.0% calcium, zinc should be added to most diets for swine. Hand-feeding a wet diet results in a lower incidence of parakeratosis than *ad libitum* feeding of a similar diet in the dry form.

The reduced growth rate and efficiency of feed utilization associated with parakeratosis are not entirely a reflection of food intake as shown by improved weight gains and feed conversion in the presence of added zinc at equalized energy intake. Serum alkaline phosphatase activity and tissue concentration of zinc are decreased in zinc deficiency. Phytic acid present in plant protein depresses zinc absorption and increases the incidence of parakeratosis. For this reason, the zinc requirement of pigs fed an all-plant protein diet is higher than that of pigs fed a milk-protein diet. Sex also affects dietary zinc requirement: females and intact males have a higher requirement than castrated males.

A wide margin of safety exists in zinc intake. Only levels as high as 1000 ppm of zinc (as zinc carbonate) in a maize–soybean meal diet produce toxicity symptoms. The addition of calcium to the diet has no effect on the level of zinc tolerance. Injection of zinc sulfate in parakeratotic animals causes local irritation, although it does relieve skin lesions.

Sodium and Chlorine

These two nutrients are considered together since the major dietary source is common salt (sodium chloride). While the required levels of sodium and chloride have been reported to be approximately 0.10 and 0.13% of the diet, respectively, in practice, it is common to add 0.4% sodium chloride to the diet. Signs of deficiency include anorexia, muscular weakness, and a depraved appetite. An excess of salt can be induced by a diet containing as little as 2% of salt if water intake is restricted. Water restriction prevents normal urinary excretion of excess sodium and results in acute toxicity and death. There is very little danger of salt toxicity if water is not restricted.

Iron

Iron deficiency is common in the newborn pig because placental and mammary transfer are inadequate. Therefore, unless a supplemental source of iron is available to the baby pig, iron deficiency anemia (microcytic hypochromia) results within 2 to 3 weeks. Oral iron can be provided by tablet or drench (ferrous form is more efficiently absorbed than ferric form), but a more efficient way is by intramuscular injection of an iron—dextran complex. A level of approximately 150 mg of iron injected at 2–3 days of age is sufficient to meet the needs until about 3 weeks of age, when dry feed consumption provides adequate iron. Administration of iron to sows either orally or by injection during gestation or lactation does not prevent anemia in the suckling pigs. This reflects the poor placental and mammary transfer of iron. The slight increase in pig hemoglobin that has been observed by feeding sows high levels of iron during gestation is due to fecal contamination rather than metabolic transfer of iron through the milk.

Copper

The copper requirement of pigs is extremely small and symptoms of copper deficiency are obtained only by careful control of evironmental contamination. In copper deficiency, there is microcytic hypochromic anemia, bowing of the legs, and spontaneous bone fractures.

Of far greater practical importance in swine production is the use of high levels of copper in the diet of growing pigs to stimulate growth. Levels of copper far above the metabolic requirement (additions of 125 to 250 ppm of diet) sometimes improve growth.

The mechanism of action is unknown. There is a danger of copper toxicity when dietary level of copper exceeds 200 ppm. The future of copper as a replacement for antibiotics as a growth stimulant for swine remains uncertain but the practice is widely used.

Iodine

Iodine deficiency in swine is manifested mainly by the birth of hairless pigs, often weak or stillborn. The thyroid of the progeny of sows fed iodine-deficient diets is increased in size and decreased in iodine content. The addition of 0.4% iodized salt containing 0.007% iodine (0.035 ppm in diet) to the diet of growing and adult animals prevents all signs of iodine deficiency and results in normal thyroid function.

Vitamin E and Selenium

A deficiency of vitamin E and selenium causes liver necrosis, a yellowish brown discoloration of body fat, and sudden death in young pigs. Whenever both the level of vitamin E and the level of selenium in the diet are marginal, the symptoms may occur. Apparently either one present in an adequate amount prevents the appearance of the lesions.

Common feedstuffs contain several tocopherols, the most biologically available of which is α-tocopherol. It has not always been considered important to be concerned with the level of vitamin E, when maize or cereal grains make up a significant proportion of the diet, but frequent field reports of vitamin E and/or selenium deficiency suggest that some methods of harvesting and storage (heat drying, for example) may destroy vitamin E or make it sufficiently biologically unavailable to precipitate deficiency signs when selenium is also limiting.

Some areas of the United States and other parts of the world are known to have soils of low selenium content (Kubota et al. 1967). Crops grown on these soils, in turn, are low in selenium.

Of practical concern in some locations is selenium toxicity, which occurs in areas where the soil levels and, in turn, plant levels, of selenium are excessive. Toxicity is manifested by poor growth, poor reproduction, and death. The reproductive failure associated with selenium toxicity can be prevented by feeding organic arsenic, which apparently reduces the amount of selenium absorbed from the gastrointestinal tract.

Cobalt

The only known metabolic requirement of cobalt is as a constituent of vitamin B_{12}. Some synthesis of vitamin B_{12} occurs in the lower intestinal tract, but absorption from this part of the tract is limited. Some reports have indicated a beneficial response to high cobalt feeding which would depend on vitamin B_{12} synthesis by the microflora of the gastrointestinal tract and absorption directly or after ingestion by coprophagy.

Sulfur

The requirement for sulfur is as inorganic sulfate and as a component of several organic nutrients, including the amino acids, methionine and cystine, and the vitamin thiamin. The normal diet provides adequate sulfur to meet these needs.

Manganese

The quantitative requirement of the pig is not clear, probably owing to differences among animals in the amounts stored in body tissues. Most feedstuffs contain manganese in amounts generally considered to exceed the dietary requirement. However, because of the widely diverse estimates of requirement, it is desirable to provide supplemental manganese in diet. The toxic level is several times the requirement.

Potassium and Magnesium

These are both required in large amounts, but are present in most common feedstuffs in amounts exceeding nutritional requirements. Therefore, they are not considered to be of practical importance in balancing diets.

Other Elements

Several elements normally present in animal tissues may have metabolic significance, but no requirement has been demonstrated for them in swine. These include aluminum, antimony, arsenic, barium, boron, bromine, cadmium, cesium, fluorine, germanium, lead, lithium, molybdenum, nickel, niobium, radium, rubinium, strontium, tellurium, tin, and zirconium. Their significance in practical swine nutrition remains unknown. Most of them are of more concern because of their toxic properties than as possible required nutrients. Others have been found to be required for specific animal species, but have not yet been shown to be required by swine. Such elements include vanadium, chromium, and silicon (National Research Council 1980; Church and Pond 1988).

Vitamin A

With the exception of yellow maize and forages, most common feedstuffs for swine are low in carotene and vitamin A. Liver storage on an adequate diet prevents immediate appearance of vitamin A deficiency symptoms when animals are fed a deficient diet; however, after tissue stores are depleted, growth is reduced and other deficiency signs appear. These include incoordination, emaciation and loss of weight, paralysis, xerophthalmia, and reproductive failure. In addition, several teratological effects have been noted; including malformations of the brain and spinal cord and of the palate, eyes, limb, heart, lungs, diaphragm, liver, and kidneys.

Since vitamin A is stored in large quantities in the liver, hypervitaminosis A can be produced by excessive feeding. Effects include anomalies of the eyes, lungs, and major blood vessels in progeny of females fed an excess during pregnancy.

The pig can efficiently convert dietary carotene to vitamin A, so that blood levels of carotene are normally low while those of vitamin A are normally high. Thus, pork fat is low in carotene and has a characteristic white color.

Vitamin D

Vitamin D (vitamins D_2 and D_3 are equally utilized by the pig) is closely related to dietary calcium and phosphorus metabolism through its effect on their absorption and utilization for bone metabolism. Vitamin D deficiency results in reduced absorption of dietary calcium and phosphorus. The physiologically active forms of vitamin D are its metabolites, 1-hydroxy, and 1,25 dihydroxy vitamin D and other derivatives. Vitamin D is converted to its active metabolites in the liver and kidneys.

The source of protein, also is related to the vitamin D requirement. The requirement is increased with soybean protein as compared with casein; this may be related to the presence of high levels of phytate in soybean protein. Dietary vitamin D is not required if the pig is allowed exposure to ultraviolet light from the sun which activates 7-dehydrocholesterol to vitamin D_3 which can then be utilized for metabolic purposes. Signs of deficiency include stiffness and lameness even in the presence of adequate calcium, phosphorus, and other minerals. Lameness is followed by appetite failure and growth depression; it can be corrected quickly by injection or feeding of vitamin D. As the trend toward complete confinement of swine throughout the life cycle continues, the importance of adequate dietary vitamin D fortification increases, especially in temperate climates where completely enclosed housing is used.

Riboflavin

Signs of riboflavin deficiency in baby pigs include dermatitis, eye lens cataracts, diarrhea, vomiting, and inflammation of the colon and anal mucosa. Baby pigs from sows fed riboflavin-deficient diets are incoordinated and weak, and mortality rate is high.

Niacin

The niacin present in most cereal grains and maize is not biologically available for the pig, so that it must be added to diets which, on an analytical basis, contain adequate quantities. Niacin deficiency results in unthriftiness, diarrhea, and necrotic enteritis. The lesions in the gastrointestinal tract resemble those produced by a pathogen, *Salmonella choleraesuis*. Since the pig can convert dietary tryptophan to niacin, the level of tryptophan has an influence on the response to added niacin.

Pantothenic Acid

Pantothenic acid deficiency results in a characteristic gait involving elevation of the hind limbs ("goose-stepping"), alopecia, diarrhea, and abscesses and ulcers in the colon. In the adult female deficiency results in the birth of weak pigs or complete reproductive failure.

Vitamin B_{12}

Tissue storage is quite considerable and accounts for the earlier reference to vitamin B_{12} as animal protein factor (APF).

Vitamin B_{12} deficiency results in macrocytic anemia in growing animals and reproductive failure in adults. Because of the trend in many parts of the world toward all-plant diets for swine, the addition of vitamin B_{12} is of special importance in diet formulation, as tissue depletion occurs within a few weeks when a deficient diet is fed.

Vitamin B_1, Vitamin B_6, Folacin, Biotin, Choline, and Vitamin K

None of these is likely to be deficient in practical diets for swine. Vitamin B_1 (thiamin) deficiency, in addition to its adverse effect on normal growth and reproduction, causes bradycardia and damage to the heart. The small quantitative dietary requirement for thiamin and its relatively high tissue storage by pigs account for the infrequency of thiamin deficiency in pigs. The high tissue storage makes pork a good source of thiamin in human nutrition.

A deficiency of vitamin B_6 (pyridoxine, pyridoxal, and pyridoxamine are all active forms) results in microcytic hyperchromic anemia, epileptic-like convulsions, reduced growth, and impaired reproduction.

A folacin deficiency results in macrocytic anemia but a deficiency has never been reported in pigs fed diets composed of common feedstuffs.

Biotin is not usually considered of practical concern as it is normally synthesized in adequate amounts by the microflora of the intestinal tract. However, there have been field reports of deficiency signs under some conditions of confinement. These symptoms are dermatitis, alopecia, and cracked hoofs.

Choline is present in most feedstuffs in excess of the dietary requirement. However, as with other nutrients, the use of certain processed feedstuffs as a major part of the diet may result in suboptimum choline intake. Therefore, a supplemental source should be provided.

Choline deficiency results in fatty infiltration of the liver, kidney lesions, birth of weak and uncoordinated pigs, and reduced growth. Dietary methionine at a level above that needed for tissue protein synthesis allows normal growth in choline-deprived pigs. Therefore, a high protein diet, unless severely deficient in methionine, would probably prevent choline deficiency, but this would not be an economical way to meet the choline requirement.

Vitamin K is normally synthesized in adequate amounts by the microflora of the intestinal tract. However, some field reports have suggested that a hemorrhagic syndrome in growing pigs, causing sudden deaths and subcutaneous hemorrhages, may be the result of vitamin K deficiency that has been induced by consumption of moldy feeds containing a vitamin K antagonist. Therefore, under certain conditions there may be a valid reason for adding vitamin K (menadione) to practical swine diets.

Fat and Fatty Acids

The pig has not been shown to require fat except as a source of possible essential fatty acids and as a vehicle for absorption of fat-soluble vitamins. The effect of dietary unsaturated fatty acids on growth remains controversial. Although early data indicated growth impairment with fat-free diets, more recent data have failed to demonstrate an effect on growth, although skin lesions develop. It is not clear whether linoleic acid deficiency or that of another fatty acid or an imbalance of fatty acids is responsible for the skin lesions.

A level of linoleic acid of 1% of the dietary calories has been suggested as a minimum requirement. Fat deficiency is not likely when diets contain appreciable amounts of maize or the cereal grains. However, when roots and tubers (such as potatoes and cassava) or processed carbohydrates (such as cane sugar) make up a major part of the energy source and when solvent-extracted protein supplements are used exclusively, there is potential danger of skin lesions related to fat (or fatty acid) deficiency.

The addition of fats to diets of early-weaned pigs or lactating sows may be economically advantageous as a means of improving energy intake.

QUANTITATIVE NUTRIENT REQUIREMENTS IN LIFE CYCLE FEEDING

Estimates of quantitative dietary requirements of each nutrient required by swine during each phase of the life cycle have been compiled by several groups of researchers (National Research Council 1988, Agricultural Research Council 1981). The data of the National Research Council are in Tables 8.3, 8.4, 8.5 and 8.6. Requirements for *ad libitum*-fed breeding herd replacements are in Table 8.7 and daily energy and feed requirements for pregnant and lactating gilts and sows are in Tables 8.8 and 8.9. It must be recognized that the values presented must be considered only as guidelines since values were obtained under experimental conditions using, in most studies, better than average physical facilities and management. Such factors as differences in genetic background and climatic or physical environment might be expected to affect the quantitative requirements.

Gestation and Prenatal Nutrition

The nutrient requirements for gestation are influenced by two separate productive functions; first, the needs of maintaining the pregnant animal per se and, second, the provision of an adequate environment and nutrient supply for the developing fetus. The first necessitates more than just meeting the maintenance requirements of an adult, since the endocrinological changes during pregnancy include increased basal metabolic rate and increased size and vascularity of the uterus as well as of the mammary glands in preparation for lactation. It is clear, then, that it is necessary to provide sufficient nutrition of the pregnant sow to allow normal reproduction and mothering ability (including subsequent lactation) and the marketing of normal progeny that have in no way been stunted in postnatal development due to prenatal malnutrition. It is possible, at least with some nutrients, to obtain normal reproduction through one reproductive cycle on a diet that is clearly not adequate for the dam. This possibility has been demonstrated experimentally for protein, and the results clearly show that the dam is able to draw on her own reserves to meet the needs of the fetus. However, such a diet cannot be considered adequate since it does not satisfy the long-term requirements of both dam and fetus.

The nutrition of the fetus depends on the transfer of nutrients across the placenta from the maternal blood. It is clear that adequate fetal nutrition, therefore, depends on adequate levels of circulating nutrients in the blood of the dam. With some nutrients the dam's blood serum concentration is maintained within a fairly normal range by homeostatic processes. For example, calcium can be resorbed from bone if there is insufficient dietary intake to maintain normal blood concentration. In contrast, a deficiency of the B vitamin, riboflavin, for example, causes a drop in blood riboflavin level of the dam because of limited tissue storage, and the amount available for transfer to the fetus is reduced accordingly. Therefore, in discussing fetal nutrition, it is important to specify the nutrient under consideration.

The nutrient requirements for fetal growth are hidden under the umbrella of pregnancy plus fetal nutrient requirements so that, until technical skills in gynecol-

TABLE 8.3 Nutrient Requirements of Swine Allowed Feed Ad Libitum (90 percent dry matter)

	Swine liveweight (kg)				
Intake and performance levels	*1–5*	*5–10*	*10–20*	*20–50*	*50–110*
Expected weight gain, g/day	200	250	450	700	820
Expected feed intake, g/day	250	460	950	1,900	3,110
Expected efficiency, gain/feed	0.800	0.543	0.474	0.368	0.264
Expected efficiency, feed/gain	1.25	1.84	2.11	2.71	3.79
Digestible energy intake, kcal/day	850	1,560	3,230	6,460	10,570
Metabolizable energy intake, kcal/day	805	1,490	3,090	6,200	10,185
Energy concentration, kcal ME/kg diet	3,220	3,240	3,250	3,260	3,275
Protein, %	24	20	18	15	13
	Requirement (% or amount/kg diet)[a]				
Nutrient					
Indispensable amino acids (%)					
Arginine	0.60	0.50	0.40	0.25	0.10
Histidine	0.36	0.31	0.25	0.22	0.18
Isoleucine	0.76	0.65	0.53	0.46	0.38
Leucine	1.00	0.85	0.70	0.60	0.50
Lysine	1.40	1.15	0.95	0.75	0.60
Methionine + cystine	0.68	0.58	0.48	0.41	0.34
Phenylalanine + tyrosine	1.10	0.94	0.77	0.66	0.55
Threonine	0.80	0.68	0.56	0.48	0.40
Tryptophan	0.20	0.17	0.14	0.12	0.10
Valine	0.80	0.68	0.56	0.48	0.40
Linoleic acid (%)	0.1	0.1	0.1	0.1	0.1
Mineral elements					
Calcium, %	0.90	0.80	0.70	0.60	0.50
Phosphorus, total, %	0.70	0.65	0.60	0.50	0.40
Phosphorus, available, %	0.55	0.40	0.32	0.23	0.15
Sodium, %	0.10	0.10	0.10	0.10	0.10
Chlorine, %	0.08	0.08	0.08	0.08	0.08
Magnesium, %	0.04	0.04	0.04	0.04	0.04
Potassium, %	0.30	0.28	0.26	0.23	0.17
Copper, mg	6.0	6.0	5.0	4.0	3.0
Iodine, mg	0.14	0.14	0.14	0.14	0.14
Iron, mg	100	100	80	60	40
Manganese, mg	4.0	4.0	3.0	2.0	2.0
Selenium, mg	0.30	0.30	0.25	0.15	0.10
Zinc, mg	100	100	80	60	50
Vitamins					
Vitamin A, IU	2,200	2,200	1,750	1,300	1,300
Vitamin D, IU	220	220	200	150	150
Vitamin E, IU	16	16	11	11	11
Vitamin K, menadione, mg	0.5	0.5	0.5	0.5	0.5
Biotin, mg	0.08	0.05	0.05	0.05	0.05

(*continued*)

TABLE 8.3 Nutrient Requirements of Swine Allowed Feed Ad Libitum (90 percent dry matter) (*Continued*)

	Swine liveweight (kg)				
Intake and performance levels	*1–5*	*5–10*	*10–20*	*20–50*	*50–110*
Choline, g	0.6	0.5	0.4	0.3	0.3
Folacin, mg	0.3	0.3	0.3	0.3	0.3
Niacin, available, mg	20.0	15.0	12.5	10.0	7.0
Pantothenic acid, mg	12.0	10.0	9.0	8.0	7.0
Riboflavin, mg	4.0	3.5	3.0	2.5	2.0
Thiamin, mg	1.5	1.0	1.0	1.0	1.0
Vitamin B_6, mg	2.0	1.5	1.5	1.0	1.0
Vitamin B_{12}, μg	20.0	17.5	15.0	10.0	5.0

NOTE: The requirements listed are based upon the principles and assumptions described in the text of this publication. Knowledge of nutritional constraints and limitations is important for the proper use of this table.

[a] These requirements are based upon the following types of pigs and diets: 1- to 5-kg pigs, a diet that includes 25 to 75 percent milk products; 5- to 10-kg pigs, a corn-soybean meal diet that includes 5 to 25 percent milk products; 10- to 110-kg pigs, a corn-soybean meal diet. In the corn-soybean meal diets, the corn contains 8.5 percent protein; the soybean meal contains 44 percent.

Source: National Research Council 1988.

TABLE 8.4 Daily Nutrient Intakes and Requirements of Swine Allowed Feed Ad Libitum

	Swine liveweight (kg)				
Intake and performance levels	*1–5*	*5–10*	*10–20*	*20–50*	*50–110*
Expected weight gain, g/day	200	250	450	700	820
Expected feed intake, g/day	250	460	950	1,900	3,110
Expected efficiency, gain/feed	0.800	0.543	0.474	0.368	0.264
Expected efficiency, feed/gain	1.25	1.84	2.11	2.71	3.79
Digestible energy intake, kcal/day	850	1,560	3,230	6,460	10,570
Metabolizable energy intake, kcal/day	805	1,490	3,090	6,200	10,185
Energy concentration, kcal ME/kg diet	3,220	3,240	3,250	3,260	3,275
Protein, g/day	60	92	171	285	404
			Requirement (amount/day)		
Nutrient					
Indispensable amino acids (g)					
Arginine	1.5	2.3	3.8	4.8	3.1
Histidine	0.9	1.4	2.4	4.2	5.6
Isoleucine	1.9	3.0	5.0	8.7	11.8
Leucine	2.5	3.9	6.6	11.4	15.6
Lysine	3.5	5.3	9.0	14.3	18.7
Methionine + cystine	1.7	2.7	4.6	7.8	10.6
Phenylalanine + tyrosine	2.8	4.3	7.3	12.5	17.1
Threonine	2.0	3.1	5.3	9.1	12.4
Tryptophan	0.5	0.8	1.3	2.3	3.1
Valine	2.0	3.1	5.3	9.1	12.4
Linoleic acid (g)	0.3	0.5	1.0	1.9	3.1

TABLE 8.4 (*Continued*)

Mineral elements					
Calcium, g	2.2	3.7	6.6	11.4	15.6
Phosphorus, total, g	1.8	3.0	5.7	9.5	12.4
Phosphorus, available, g	1.4	1.8	3.0	4.4	4.7
Sodium, g	0.2	0.5	1.0	1.9	3.1
Chlorine, g	0.2	0.4	0.8	1.5	2.5
Magnesium, g	0.1	0.2	0.4	0.8	1.2
Potassium, g	0.8	1.3	2.5	4.4	5.3
Copper, mg	1.50	2.76	4.75	7.60	9.33
Iodine, mg	0.04	0.06	0.13	0.27	0.44
Iron, mg	25	46	76	114	124
Manganese, mg	1.00	1.84	2.85	3.80	6.22
Selenium, mg	0.08	0.14	0.24	0.28	0.31
Zinc, mg	25	46	76	114	155
Vitamins					
Vitamin A, IU	550	1,012	1,662	2,470	4,043
Vitamin D, IU	55	101	190	285	466
Vitamin E, IU	4	7	10	21	34
Vitamin K (menadione), mg	0.02	0.02	0.05	0.10	0.16
Biotin, mg	0.02	0.02	0.05	0.10	0.16
Choline, g	0.15	0.23	0.38	0.57	0.93
Folacin, mg	0.08	0.14	0.28	0.57	0.93
Niacin, available, mg	5.00	6.90	11.88	19.00	21.77
Pantothenic acid, mg	3.00	4.60	8.55	15.20	21.77
Riboflavin, mg	1.00	1.61	2.85	4.75	6.22
Thiamin, mg	0.38	0.46	0.95	1.90	3.11
Vitamin B_6, mg	0.50	0.69	1.42	1.90	3.11
Vitamin B_{12}, μg	5.00	8.05	14.25	19.00	15.55

Source: National Research Council 1988.

TABLE 8.5 Nutrient Requirements of Breeding Swine

Intake levels	*Bred gilts, sows, and adult boars*	*Lactating gilts and sows*
Digestible energy, kcal/kg diet	3,340	3,340
Metabolizable energy, kcal/kg diet	3,210	3,210
Crude protein, %	12	13
	Requirement (% or amount/kg diet)[a]	
Nutrient		
Indispensable amino acids (%)		
Arginine	0.00	0.40
Histidine	0.15	0.25
Isoleucine	0.30	0.39
Leucine	0.30	0.48

(*continued*)

TABLE 8.5 *(Continued)*

Intake levels	*Bred gilts, sows, and adult boars*	*Lactating gilts and sows*
Lysine	0.43	0.60
Methionine + cystine	0.23	0.36
Phenylalanine + tyrosine	0.45	0.70
Threonine	0.30	0.43
Tryptophan	0.09	0.12
Valine	0.32	0.60
Linoleic acid (%)	0.1	0.1
Mineral elements		
Calcium, %	0.75	0.75
Phosphorus, total, %	0.60	0.60
Phosphorus, available, %	0.35	0.35
Sodium, %	0.15	0.20
Chlorine, %	0.12	0.16
Magnesium, %	0.04	0.04
Potassium, %	0.20	0.20
Copper, mg	5.00	5.00
Iodine, mg	0.14	0.14
Iron, mg	80.00	80.00
Manganese, mg	10.00	10.00
Selenium, mg	0.15	0.15
Zinc, mg	50.00	50.00
Vitamins		
Vitamin A, IU	4,000	2,000
Vitamin D, IU	200	200
Vitamin E, IU	22	22
Vitamin K (menadione), mg	0.50	0.50
Biotin, mg	0.20	0.20
Choline, g	1.25	1.00
Folacin, mg	0.30	0.30
Niacin, available, mg	10.00	10.00
Pantothenic acid, mg	12.00	12.00
Riboflavin, mg	3.75	3.75
Thiamin, mg	1.00	1.00
Vitamin B_6, mg	1.00	1.00
Vitamin B_{12}, μg	15.00	15.00

NOTE: The requirements listed are based upon the principles and assumptions described in the text of this publication. Knowledge of nutritional constraints and limitations is important for the proper use of this table.

[a] These requirements are based upon corn-soybean meal diets. In the corn-soybean meal diets, the corn contains 8.5 percent protein; the soybean meal contains 44 percent.

TABLE 8.6 Daily Nutrient Intakes and Requirements of Intermediate-Weight Breeding Animals

	Mean gestation or farrowing weight (kg) of:	
	Bred gilts, sows, and adult boars	*Lactating gilts and sows*
Intake and performance levels	*162.5*	*165.0*
Daily feed intake, kg	1.9	5.3
Digestible energy, Mcal/day	6.3	17.7
Metabolizable energy, Mcal/day	6.1	17.0
Crude protein, g/day	228	689
	Requirement (amount/day)	
Nutrients		
Indispensable amino acids (g)		
Arginine	0.0	21.2
Histidine	2.8	13.2
Isoleucine	5.7	20.7
Leucine	5.7	25.4
Lysine	8.2	31.8
Methionine + cystine	4.4	19.1
Phenylalanine + tyrosine	8.6	37.1
Threonine	5.7	22.8
Tryptophan	1.7	6.4
Valine	6.1	31.8
Linoleic acid (g)	1.9	5.3
Mineral elements		
Calcium, g	14.2	39.8
Phosphorus, total, g	11.4	31.8
Phosphorus, available, g	6.6	18.6
Sodium, g	2.8	10.6
Chlorine, g	2.3	8.5
Magnesium, g	0.8	2.1
Potassium, g	3.8	10.6
Copper, mg	9.5	26.5
Iodine, mg	0.3	0.7
Iron, mg	152	424
Manganese, mg	19	53
Selenium, mg	0.3	0.8
Zinc, mg	95	265
Vitamins		
Vitamin A, IU	7,600	10,600
Vitamin D, IU	380	1,060
Vitamin E, IU	42	117
Vitamin K (menadione), mg	1.0	2.6

(continued)

TABLE 8.6 (*Continued*)

	Mean gestation or farrowing weight (kg) of:	
	Bred gilts, sows, and adult boars	*Lactating gilts and sows*
Intake and performance levels	*162.5*	*165.0*
Biotin, mg	0.4	1.1
Choline, g	2.4	5.3
Folacin, mg	0.6	1.6
Niacin, available, mg	19.0	53.0
Pantothenic acid, mg	22.8	63.6
Riboflavin, mg	7.1	19.9
Thiamin, mg	1.9	5.3
Vitamin B_6, mg	1.9	5.3
Vitamin B_{12}, μg	28.5	79.5

Source: National Research Council 1988.

ogy allow the "test tube" rearing of pig fetuses, the question of the nutritive requirement of the fetus per se is, in a sense, academic. Therefore, these requirements are presented as requirements for gestation and are at best imprecise, as illustrated in the foregoing discussion.

Placental transfer of nutrients must occur or there would be no fetal growth. Some nutrients are transferred more efficiently than others. All of the fat-soluble vitamins and B vitamins are readily transferred to the fetus. Intact protein does not pass the placental barrier except in negligible amounts but individual amino acids

TABLE 8.7 Requirements for Several Nutrients of Breeding Herd Replacements Allowed Feed Ad Libitum

	Weight (kg) of:			
	Developing gilts		*Developing boars*	
Intake levels	*20–50*	*50–110*	*20–50*	*50–110*
Energy concentration, kcal ME/kg diet	3,255	3,260	3,240	3,255
Crude protein, %	16	15	18	16
Nutrient[a]				
Lysine, %	0.80	0.70	0.90	0.75
Calcium, %	0.65	0.55	0.70	0.60
Phosphorus, total, %	0.55	0.45	0.60	0.50
Phosphorus, available, %	0.28	0.20	0.33	0.25

[a] Sufficient data are not available to indicate that requirements for other nutrients are different from those in Table 8.6 for animals of these weights.

TABLE 8.8 Daily Energy and Feed Requirements of Pregnant Gilts and Sows

Intake and Performance levels	Weight (kg) of bred gilts and sows at mating[a]		
	120	*140*	*160*
Mean gestation weight, kg[b]	142.5	162.5	182.5
Energy required, Mcal DE/day			
Maintenance[c]	4.53	5.00	5.47
Gestation weight gain[d]	1.29	1.29	1.29
Total	5.82	6.29	6.76
Feed required/day, kg[e]	1.8	1.9	2.0

[a] Requirements are based on a 25-kg maternal weight gain plus 20-kg increase in weight due to the products of conception; the total weight gain is 45 kg.
[b] Mean gestation weight is weight at mating + (total weight gain/2).
[c] The animal's daily maintenance requirement is 110 kcal of DE/kg$^{0.75}$.
[d] The gestation weight gain is 1.10 Mcal of DE/day for maternal weight gain plus 0.19 Mcal of DE/day for conceptus gain.
[e] The feed required/day is based on a corn-soybean meal diet containing 3.34 Mcal of DE/kg.
Source: National Research Council 1988.

do, and these form the main source of tissue protein synthesis by the fetus. Nonprotein nitrogen is used by the fetus to synthesize nonessential amino acids probably as early as 30 days of gestation. Individual fatty acids, but not triglycerides, apparently are transferred across the placenta, but the main source of energy for the fetal pig is glucose transported from maternal blood.

Transfer of water and the inorganic elements to the fetus increases several-fold during the latter part of gestation. The mineral composition of maternal plasma and the products of conception in swine at 110 days of gestation are summarized in Table 8.10. The mineral transfer to the fetus involves several factors which have

TABLE 8.9 Daily Energy and Feed Requirements of Lactating Gilts and Sows

Intake and performance levels	Weight (kg) of lactating gilts and sows at postfarrowing		
	145	*165*	*185*
Milk yield, kg	5.0	6.25	7.5
Energy required, Mcal DE/day			
Maintenance[a]	4.5	5.0	5.5
Milk production[b]	10.0	12.5	15.0
Total	14.5	17.5	20.5
Feed required/day, kg[c]	4.4	5.3	6.1

[a] The animal's daily maintenance requirement is 110 kcal of DE/kg$^{0.75}$.
[b] Milk production requires 2.0 Mcal of DE/kg of milk.
[c] The feed required/day is based on a corn-soybean meal diet containing 3.34 Mcal of DE/kg.
Source: National Research Council 1988.

TABLE 8.10 Mineral Composition of Maternal Plasma and Placenta, Placental Fluids, and Fetus at 110 Days after Conception (per Kilogram Fresh Weight)

Mineral	*Maternal Plasma*	*Placental Fluids*	*Placenta*	*Fetus*
Calcium, g	0.118	0.620	0.589	9.9
Chloride, meq	106	87	88	50
Copper, mg	1.02	0.3	1.6	2.5
Iron, mg	1.15	0.8	30.5	34.8
Magnesium, g	0.026	0.022	0.085	0.3
Phosphorus, g	0.053	0.148	0.620	5.3
Potassium, meq	6	15	20	50
Sodium, meq	145	108	97	92
Sulfur (inorg.), g	0.012	0.037	0.032	1.0
Zinc, mg	0.92	0.4	5.8	6.8

Source: Hansard 1966.

been outlined by Hansard (1966) as follows: (1) availability of the mineral to the dam for transfer; (2) placental type and degree of permeability; (3) physiological age and size of the fetus; (4) size, weight, and charge of the particles being transferred. In general, the greatest placental transfer of minerals occurs during the last one-third of pregnancy. In comparison with the transfer of other mineral elements, the efficiency of iron transport to the fetus is very low and is unresponsive to maternal diet. This inefficient transport of iron across the placenta is not fully understood.

Lactation

The nutrient output in milk during a 5-week lactation period is much greater than the nutrient deposition in fetuses and placental membranes during a 114-day gestation period. Therefore, the nutrient requirements of lactation are much more demanding than those of gestation. While the sow can draw on her own body reserves to some extent for milk production, the extent of this is limited, and complete or partial lactation failure results if nutrient restriction is severe or prolonged. The degree to which reserves of the dam can be drawn upon for lactation when dietary intake is insufficient is related to the specific nutrient concerned. For example, as discussed in the foregoing paragraphs, the transfer of iron into the milk is extremely inefficient so that even a very high dietary concentration does not appreciably increase the concentration in the milk. In general, a deficiency of a particular nutrient in the diet is manifested more by a reduction in total milk production than by a decreased concentration of that nutrient in the milk. The most striking effect of level of intake on total milk production is with energy. If the lactating sow is not allowed *ad libitum* or near *ad libitum* intake especially after the first week, milk production is decreased. Milk production increases as litter size increases, up to the point at which genetic capacity of the sow for milk production is reached. Mature sows raising large litters will often consume 8–10 kg or more of high concentrate diet (3600 kcal of metabolizable energy per kilogram of diet) per day at the peak of lactation. The amount of diet consumed and the changes in body weight during lactation depend on the obesity of the sow at the beginning of lactation. Obese sows tend to consume less feed during lactation and lose more weight than thin sows. The energetic efficiency of lactation is higher when milk is produced by current energy intake than by dependence on body fat depots. Therefore, in practical feeding, the greatest efficiency of energy utilization is achieved by controlled energy intake during gestation to minimize fattening and by full-feeding during lactation to minimize mobilization of depot fats for milk production.

The peak of lactation is reached at 4–5 weeks postpartum so that the quantitative daily nutrient requirement for lactation coincides with this peak. The daily milk production may be as high as 8 kg or more in individual sows nursing large litters at the peak of lactation. Since the caloric density of sow milk is considerably higher than that of cow milk, it represents a large supply of energy to individual pigs during a period of extremely rapid growth.

Quantitative nutrient requirements for lactation, assuming *ad libitum* energy intake, are given in Tables 8.5 and 8.6. It should be recognized that total feed requirements (expressed as units per kilogram of diet) will vary depending on the density of the diet and previous nutritional history of the sow.

Preweaning

The nutrient requirements (except iron) of the suckling pig are met by sow milk during the first 2 to 3 weeks. After this time, the rapid growth of the pig with the concomitant increase in daily nutrient requirement, combined with the decline in milk yield, necessitates the provision of supplemental feed. These relationships are illustrated diagrammatically in Fig. 7.8 (Chapter 7).

A supplemental source of feed (creep feed) must be supplied beginning at 3 weeks of age or earlier to allow maximum weight gain during the suckling period. The protein content, as a percentage of diet, does not have to be as high as that of sow milk, as the energy concentration is less, so that the ratio of calorie to protein is less. The carbohydrate source must be palatable and low in fiber to encourage consumption. For this reason, sucrose, dried molasses, glucose, or some other simple sugar is often added to make up 5 or 10%, or even more, of the diet. Nonnutritive sweeteners such as saccharine or monosodium glutamate are also used, but at a much lower concentration. While energy concentration of the creep diet is not critical, the addition of 5 or 10% fat to the diet has been shown to improve palatability (probably by reducing dustiness) and to encourage early consumption. Thus the quantitative nutrient requirements during the early preweaning period are met mostly by sow milk, but energy becomes limiting as the daily requirements increase in relation to the amount of milk available. The introduction of a dry, well-balanced diet early enough so that the pig consumes sufficient quantities for maximum weight gain is, therefore, a major consideration during the preweaning period, even though the total amount consumed per pig is negligible, compared with the total amount required to reach market weight.

Removal of the pig from the sow earlier than at 5 or 6 weeks of age can be considered early weaning for the purposes of this discussion. The younger the pig at weaning, the more critical are the dietary requirements. Primary specific pathogen-free (SPF) pigs, weaned at birth, are of the earliest possible weaning age. Because of deprivation of colostrum, not only must these pigs be fed a highly fortified diet, but also they must be kept in a sanitary environment and given parenteral antibody protection. The quantitative nutrient requirements of the newborn pig probably coincide closely with the composition of sow milk (for gross composition see Chapter 7). The digestive enzyme capacity of the newborn pig is limited when nonmilk proteins are fed. This demonstrates a difference in protein digestibility and in utilization of absorbed protein between pigs fed casein and those fed soybean protein. The liquid diets as used in this work must be fed at frequent intervals (a minimum of four times daily) to provide adequate intake for normal growth. When they are fed at this frequency, intake must be kept at less than full-feed to minimize problems with diarrhea. The digestive capacity of the young pig is not sufficient to handle large meals at infrequent intervals. Therefore, it is advisable to feed more frequently in smaller quantities per feeding to achieve best results. A commercially available automated feed dispenser that allows feeding more than 100 individually caged pigs 24 feedings per day at 1-hr intervals is shown in Fig. 5.5. Pigs can be weaned to dry diets at a few days of age. This reduces problems with sanitation.

Other experimentally developed automated feeders for baby pigs also have been developed (Lecce 1969; Braude et al. 1969; P. Reeds and M. Fiorotto, personal communication, 1990). A less elaborate feeding system involves individually caged pigs fed four times daily from porcelain feeding bowls. It is important to house pigs individually when not allowed frequent feeding as the strong suckling instinct results in undue suckling and nuzzling of the ears and navels of other pigs when several pigs are kept together. It is important to increase intake gradually, as rapidly as freedom from digestive upsets will permit. Rate of weight gain is closely related to feed intake during this early period. The energy concentration of cow milk is not high enough to support maximum growth rate even when fed at nearly *ad libitum* intake. Therefore, most liquid sow milk replacers contain a higher dry matter and higher fat concentration than cow milk (Schneider and Sarett 1969). Young pigs can survive long periods on a daily energy intake so restricted that weight gain is almost nil. Although the pig has a tendency to compensate by extra rapid growth when it is transferred to a dry diet offered *ad libitum*, the effect of early deprivation on body size is probably never fully overcome (Barnes et al. 1967).

Growing Period (Early Postweaning)

This stage of the life cycle is arbitrarily set as the period from weaning (usually 5 or 6 weeks of age) to the point at which the pig has attained approximately 45-kg body weight. This arbitrarily assigned period represents a stage of nutrient requirements less critical than earlier stages, but more critical than that of the finishing period (from 45 kg to slaughter at 90 kg).

An important phenomenon of weaning is the "post-weaning growth check" that accompanies the transition from sucking to consumption of dry post-weaning diet. The degree and duration of this inflection in the body weight curve is variable, but experimental evidence shows that even in cases in which the pig maintains a constant weight or continues to gain weight slowly shortly after weaning, there is a net loss in body fat, concurrent with sustained tissue protein accretion. There is a considerable amount of interest in, and concern about, this phenomenon and efforts are underway to develop practical means of minimizing or preventing it.

The changes in nutrient requirements, as the pig matures, are related to changes in growth rate and body composition. Fat concentration of the body increases rapidly at the expense of water as the pig matures, and although the percentage of protein remains rather stable, the ratio of calorie to protein in the body increases steadily to market weight. Feeding a protein deficient diet during the growing period results in an increase in this ratio (a fatter carcass); feeding protein in excess of the dietary requirement, on the contrary, does not yield a leaner carcass than that obtained when protein equal to the requirement is supplied. Clausen (1965) advocated high dietary protein for maximum carcass leanness but emphasized that the pig cannot be forced to increase lean meat deposition beyond its genetic capacity. Genetics influences carcass composition, but the effect of genetics on the response to diet remains uncertain. Experimental data with growing pigs suggest that the total protein required for normal performance is less than previously recognized. The addition of certain crystalline amino acids (lysine and methionine), in appropriate amounts and combinations to low protein diets, is now economically feasible, so that when the essential amino acid requirements are met in this way, the total amount of nitrogen needed for normal performance is reduced. Protein and amino acid requirements are usually expressed as a percentage of diet, but since energy concentration of the diet is important in affecting total daily feed intake, the expression of protein and of amino acid requirements as grams per unit of feed energy would be better. A clear-cut effect of energy concentration of the diet of growing pigs on performance can be observed by the addition of fat to the diet. Feed intake is thereby reduced, and if other nutrients are present in marginal or suboptimum amounts, the deficiency is aggravated.

Recent advances in the development of nutrient repartitioning agents such as beta-adrenergic agonists (epinephrine analogues) and anabolic growth promoters such as porcine somatotropin have raised the likelihood that quantitative nutrient requirements for optimum growth will have to be increased to support the enhanced potentials. Research is underway to determine the extent of such effects on nutrient requirements so that full use may be made of the beneficial effects of these agents following their approval for use by regulatory agencies (i.e., Food and Drug Administration) in the United States.

There are other examples of dietary interactions among nutrients. These include calcium and zinc (excess calcium increases the zinc requirement), methionine and choline (methionine can provide labile methyl groups used in metabolism that are normally supplied by choline, when choline is deficient in the diet), leucine and isoleucine (excess leucine inhibits the utilization of isoleucine), and tryptophan and niacin (tryptophan can be converted to niacin in the body). This last example is worthy of further consideration because it has considerable practical significance in diets for growing swine, especially when maize or cereal grains make up a major part of the diet. This is for two reasons: tryptophan is a limiting amino acid in diets of this kind, and niacin, although present, is biologically unavailable. Therefore, the inadequate niacin absorption diverts the already short supply of tryptophan away from use for tissue protein synthesis, and growth depression results (Harmon et al. 1969).

In addition to interrelationships among nutrients, antagonism exists between some nonnutritive constituents and certain nutrients. An example is the relationship between a protein source and quantitative zinc requirement whereby soybean protein increases the quantitative zinc requirement. Subsequent research has demonstrated that phytic acid and perhaps other organic constituents of plant materials are capable of altering the biological availability of zinc and other trace elements for growing pigs. As the trend in swine production moves more toward the use of all-plant diets, it seems likely that other such examples of antagonisms will be encountered. The swine producer, nutritionist, and feed manufacturer should be alert to these occurrences.

In addition to strictly nutritional factors affecting performance of growing swine, the physical form of the diet is important. For example, pelleting some types of diets has an effect on performance. The improved performance may be due to improved utilization of nutrients or to less feed wastage or both. Other factors, including fineness of grind (particle size) and dustiness, contribute to variations in performance. Such factors are discussed in more detail later in this chapter.

Finishing Period (45–90-kg Slaughter Weight)

The same principles and problems in meeting nutrient requirements during the growing period apply during the finishing period. However, the quantitative requirements for nutrients, other than energy, are less (in terms of percentage of diet) during the finishing period. The total daily energy requirement is considerably greater during finishing, not only because of the larger body size, but also because of the higher feed requirement per unit of body weight gain. This correspondence between decreased efficiency and increasing weight is related to the increased fat deposition and the high energy requirement associated with it.

Since energy is quantitatively the most important nutrient during the finishing period, it is important to consider the acceptability of available energy sources. Although the gross energy concentration of common cereal grains and root and tuber crops and plant by-products may be similar, the utilizable energy as well as the palatability varies widely. There are also differences in amino acid pattern that may have an effect when the protein supplement is eliminated or limited. Table 8.11 shows that maximum weight gain is not necessarily associated with maximum carcass leanness (Gallo and Pond 1968). Thus, it is important to consider nutrients in terms of their effects on efficiency of lean meat production, not only in terms of their effects on weight gain.

NUTRITIONAL DISEASES

Marginal dietary levels of one or more nutrients are probably common in swine fed homegrown feeds not fortified with commercial sources of vitamins. Mineral and protein supplements are tailored to meet specific needs. These marginal (subclini-

TABLE 8.11 Effect of Protein Deficiency during the Finishing Period on Carcass Leanness

Constituents	*Composition of diets, %*	
	Maize–soybean meal	*Maize–glucose*
Maize (No. 2 yellow)	71.50	59.00
Glucose	—	34.32
Soybean meal	23.00	—
Decalcium phosphate	2.50	3.00
Ground limestone	1.50	1.25
Salt (trace mineralized)	0.50	0.50
Vitamin zinc supplement	0.50	0.50
Antibiotic supplement	0.50	0.50
Potassium phosphate	—	0.50
L-Lysine-HCl	—	0.37
L-Tryptophan	—	0.06
Total, %	100.00	100.00
Total protein, %	20.50	7.60
Criteria	*Performance*[a]	
Daily gain, g	670.00	680.00
Loin eye area, cm^2	29.40	19.90
Back fat, cm	3.30	3.90
Leaf fat, kg	1.36	1.91
Serum protein, g/100 ml	7.10	5.50

Conclusion: Diets for finishing pigs that provide adequate protein for normal weight gains may not be sufficient for normal carcass leanness.

[a] Five pigs per group; av initial wt, 47 kg; av final wt, 91 kg.

Source: Gallo and Pond 1968.

cal) deficiencies of a general nature are manifested as reduced body weight gain or reduced efficiency of feed utilization, often so slight that they are not even recognized.

However, over and above these marginal deficiencies, usually involving more than one nutrient, there are several specific nutrient deficiencies that manifest themselves by typical clinical signs (National Research Council 1979; Whitehair and Miller 1981; Cunha 1977; Pond and Houpt 1978). As pointed out by Whitehair and Miller (1981), nutritional deficiencies result from not only specific dietary deficiencies, but also from substances that interfere with nutrient utilization, from infections that lower the intake or utilization of nutrients, and from diseases that increase nutritional requirements (for example, infections that cause fever and thereby increase the metabolic rate). Only simple deficiencies of some nutrients of special concern are described here.

Swine are affected by a variety of metabolic disorders of importance in commercial production. Some of these have their counterpart in humans or in other farm animals. The use of swine as the model animal for studying these disorders has been reviewed (Bustad et al. 1966; Pond and Houpt 1978). They include atherosclerosis, thrombosis, osteoporosis, and gastric ulcers. The economic losses to the swine industry as a result of degenerative diseases such as atherosclerosis are probably not great, since in today's husbandry practices the animals are usually slaughtered before they develop the disorder enough to show clinical signs. However, losses due to death or crippling of breeding animals from osteoporosis are significant in commercial herds.

Some metabolic diseases of major economic importance to the swine industry are discussed here.

Zinc Deficiency (Parakeratosis)

The specific clinical sign is a typical skin lesion characterized by excess keratinization of the epidermis accompanied by a brown exudate and alopecia. The lesion resembles that produced by the mange mite (*Sarcoptes scabiei*), but is distinguished from mange by the rapid and dramatic disappearance of the lesion when the diet is supplemented with zinc. The skin lesion is accompanied by reduced growth rate and efficiency of feed utilization, and the serum alkaline phosphatase concentration is depressed. All of these effects are dramatically reversed by zinc supplementation.

Tucker and Salmon (1955) first published evidence that zinc would cure or prevent the lesion and these workers are generally given credit for the discovery, although R. F. Sewell and co-workers at the University of Georgia had previously obtained firm evidence of a relationship between parakeratosis and zinc.

The biological availability of zinc is affected by several dietary factors. A high level of calcium increases the dietary zinc requirement so that supplemental zinc is required to prevent parakeratosis when the diet is high in calcium, but may not be required with lower levels of calcium. Phytic acid (the hexaphosphoric acid ester of inositol), present in high concentrations in soybean meal and other plant proteins, ties up zinc in the gastrointestinal tract and thereby increases the dietary requirement. Parakeratosis can be prevented by the addition of 100 ppm of zinc as an inorganic salt to the diet of growing pigs. The toxic level is not reached until the diet contains more than 1000 ppm of zinc (Brink et al. 1959).

Iron Deficiency (Baby Pig Anemia)

Iron deficiency anemia is of practical importance mainly in suckling pigs as a result of inefficient placental and mammary transfer of iron. The newborn pig has very limited body reserves of iron and must get iron in addition to that obtained from milk to avoid anemia. Pigs allowed access to soil or pasture are exposed to enough iron from environmental contamination to meet their needs, but those born and raised in concrete-floor pens develop anemia within a few days after birth if not given a supplemental source. Oral tablets and pastes or parenteral administration of commercial iron–dextran preparations will cure or prevent anemia. Despite the availability of iron sources, baby pig anemia remains as a major nutritional disease of swine, primarily because many management systems do not ensure that oral sources of iron offered to suckling pigs are actually consumed in sufficient amounts. Prevention is assured by parenteral administration of nontoxic iron–dextran preparations. Iron deficiency is not a serious problem after weaning, because normal feedstuffs usually contain more iron than required.

Selenium and Vitamin E Deficiency (Hepatosis Dietetica, Mulberry Heart Disease)

The exact metabolic interrelationships between selenium and vitamin E are not well understood, but it appears that one can substitute for the other as a dietary component to prevent the liver or heart lesions that result in development of deficiency signs. The deficiency signs are seen almost entirely on postmortem examination, following sudden death of apparently healthy animals. The liver shows massive necrosis and the heart often has grossly visible muscle dystrophy, verifiable by histology. Skeletal muscles may also show dystrophy, although this is not as commonly seen in pigs as in sheep that are deficient in vitamin E. Depot fats may have a brown discoloration.

Although vitamin E (tocopherol) is stored in the liver and other tissues, it is possible to produce deficiency signs by long periods of dietary vitamin E inadequacy. For many years it was considered that swine fed diets composed mainly of maize or cereal grains had no danger of developing vitamin E deficiency. However, in recent years many cases of vitamin E deficiency have been reported in swine fed diets composed of feedstuffs grown over a wide geographic area. D. E. Ullrey and associates at Michigan State University (Ku et al. 1972) have shown a close relationship between body tissue concentration of selenium in pigs and selenium content of the diet fed.

It is now well known from soil surveys (Kubota et al. 1967) that selenium-deficient soils exist in many parts of the world. Crops grown on these soils tend to be low in selenium and when they are fed in diets also deficient in vitamin E, the above syndrome is produced. The reason for the apparent increase in incidence of selenium and vitamin E deficiency in swine in recent years is not clear but may be related to the methods of processing feeds during and following harvest and the increasing practice of confinement rearing with limited access to fresh forages.

Calcium, Phosphorus, or Vitamin D Deficiency (Rickets)

Failure to provide adequate calcium, phosphorus, or vitamin D in the diet of the growing pig results in abnormalities in bone structure induced by inadequate mineralization. A deficiency of any one or all three of these nutrients produces rickets in growing animals. Vitamin D can be supplied as such in the diet or can be synthesized in skin by ultraviolet light activating 7-dehydrocholesterol to form the active vitamin or its precursor (current evidence indicates that the active form of vitamin D is 1,25-dihydroxycholecalciferol). Thus, pigs given exposure to as little as 30 minutes of direct sunlight daily have no need for dietary vitamin D. If adequate vitamin D is supplied either in the diet or by sunlight, dietary calcium and phosphorus utilization are adequate. Therefore, if rickets develops in the presence of adequate vitamin D, dietary calcium and/or phosphorus deficiency is indicated. Symptoms are reduced mineralization of the protein matrix and the outward manifestations are bowing of the legs, lameness, and occasionally a spontaneous fracture of an extremity or of the ribs. Growth rate and efficiency of feed utilization may be reduced. Rickets can be prevented by feeding adequate calcium and phosphorus and by feeding vitamin D or its metabolites or allowing exposure to ultraviolet light.

Calcium–Phosphorus Imbalance (Fibrous Osteodystrophy)

A dietary calcium—phosphorus ratio of less than 1 : 1 may result in excessive bone resorption in relation to deposition with replacement of bone by fibrous tissue (Brown et al. 1966). This is a manifestation of nutritional secondary hyperparathyroidism induced by continued low serum calcium concentration.

The lesion is generalized throughout the skeleton and, in severe cases, results in spontaneous fracture of long bones and of ribs as in rickets. However, the histological picture is different in fibrous osteodystrophy than in rickets. The high phosphorus content of many plant materials probably contributes to the frequent occurrence of this nutritional imbalance. Prevention is achieved by dietary calcium adequacy and by maintenance of a calcium–phosphorus ratio greater than 1 : 1.

Protein Deficiency (Hypoproteinemia, Fatty Liver, Edema, Osteoporosis)

Dietary protein deficiency is most easily produced in young rapidly growing animals. Older animals have a lower protein requirement and are more resistant to deficiency signs. In young pigs fed corn or cereal grain as the only source of protein, body weight gain may be reduced to zero, liver fat is increased, and serum total protein and serum albumin concentrations are severely reduced (Gallo and Pond 1968). Failure of normal formation of protein matrix for ossification of skeletal tissue results in reduced activity of the growth plate of long bones and decreased density and thickness of bone cortices (osteoporosis) (Lowrey et al. 1962). These signs are aggravated when the energy concentration of the diet (and consequently the calorie–protein ratio) is increased by adding fat (Pond et al. 1965). This increased calorie–protein ratio produces edema, evident by puffiness around the umbilical area, neck, and jowl and in the scrotum of males. The syndrome produced by low-protein–low-fat diets (maize alone) resembles the condition of marasmus in human infants whereas that produced by low-protein–high-fat diets resembles kwashiorkor in human infants (Pond et al. 1966). Protein deficiency can be prevented by the addition of a protein supplement to low-protein, high-energy feedstuffs. Some of the new varieties of maize, which contain a superior balance of amino acids and, in some cases, a higher protein content than common varieties, can be used with less protein supplement.

Niacin Deficiency (Necrotic Enteritis)

A suboptimum level of niacin in the diet, in addition to reducing growth rate (as does any B-vitamin deficiency) results in ulceration and necrosis of the epithelial lining of the gastrointestinal tract. These lesions are associated with severe diarrhea, dehydration, and weight loss. The lesions and the clinical signs resemble those produced by a pathogenic bacteria, *Salmonella cholerasuis,* but are not responsive to antibiotics or furazolidone, except in cases in which secondary infections have occurred.

The biological availability of niacin in maize and cereal grains is very low, so that supplementation is necessary even in diets for which analytical data indicate an adequate intake. An intake of niacin in excess of the requirement is merely excreted in the urine. Since there is no evidence for niacin toxicity, deficiency can be prevented by dietary supplementation.

NUTRIENT TOXICITY DISEASES

In the broad sense, any excess of a single nutrient has potential danger to health, including even fat, in which the threat to health is obesity. However, under most conditions only a few nutrients cause serious ill effects if fed continuously in excess. Some of the most common of these maladies are included here.

Hypervitaminosis A

Massive oral doses of vitamin A during the first half of pregnancy result in numerous teratological effects in the offspring of swine (Palludan 1966). These include histological abnormalities in the eyes, and a variety of abnormalities in the heart and circulatory system. There are no data available on hypervitaminosis A in growing pigs, but the observed effects on the pregnant sows in the Palludan experiment suggest that the greatest effect might be discomfort and lameness as a result of skeletal changes in the joints of extremities. Hypervitaminosis A is likely to occur only as a result of experimental or accidental administration of an excess either parenterally or in the feed.

Hypervitaminosis D

Massive doses of vitamin D given parenterally or orally produce excessive calcification. Vitamin D synthesis in skin, resulting from exposure to sunlight, is inversely related to the amount of exposure, so that hypervitaminosis D is not likely as a result of this. As with hypervitaminosis A, the only likely means of producing hypervitaminosis D is by experimental or accidental excesses given parenterally or by errors in feed mixing. Plants containing compounds that mimic or that are active metabolites of vitamin D may also create signs of vitamin D toxicity.

Selenium Toxicity

In some areas of the United States and other parts of the world the selenium content of soil is sufficiently high to result in excessive concentration of selenium in the plant. Swine fed cereal grains or other feeds grown on these soils develop alopecia and crooked hoofs, and weight gains are severely depressed (Wahlstrom et al. 1955). Excesses of dietary selenium in pregnant sows result in reduced birth weights and increased mortality in progeny. The addition of arsanilic acid to high-selenium diets counteracts some of the signs of selenium toxicity, presumably by interfering with absorption of selenium from the gastrointestinal tract.

Molybdenum Toxicity

Pigs are more resistant to molybdenum toxicity than are ruminants. Levels that would be toxic for ruminants have been fed to swine throughout the growing-finishing period with no toxic signs except a slight reduction in average daily gain.

The adverse effect of molybdenum is apparently the result of its ability to tie up copper in the gastrointestinal tract so that it induces a copper deficiency. When larger than adequate amounts of copper are fed, the adverse effects of excessive molybdenum are prevented.

Copper Toxicity

The addition of copper salts to diets for growing swine as a means of stimulating weight gains has become a common practice in some parts of the world. Isolated cases of copper poisoning have been reported from Europe and more frequently from experimental studies in the United States. The main signs of acute toxicity are internal hemorrhages and generalized icterus throughout the body fluids and tissues. Chronic toxicity is manifested by depressed hemoglobin concentration and reduced body weight gain. Sudden death occurs in some affected animals, presumably as a result of hemorrhage and/or impaired liver function. Other dietary factors affect the biological availability of copper. High-protein diets depress absorption; casein induces more severe copper-toxicity symptoms than does soybean meal. The anemia produced by high-copper diets is due to induced iron deficiency probably caused by interference with iron absorption from the gastrointestinal tract (Gipp 1971). More work is needed to elucidate the factors affecting the response to high levels of copper.

Sodium (Sodium Chloride or Salt) Toxicity

Pigs deprived of water and ingesting diets containing 1–2% sodium chloride may show clinical signs of salt toxicity including severe thirst, convulsions, and death. Levels of sodium in blood serum above 160 meq/liter indicate excessive sodium retention. Free access to water allows sodium to be excreted by the kidney and protects against the effects of excess salt ingestion.

Toxicity of Other Minerals

An excess of any required nutrient or ingestion of harmful levels of minerals in the environment but not required by the animal for normal metabolism may result in adverse effects on normal body processes. Tolerance levels for required and toxic levels of minerals for farm animals, including swine, have been described (National Research Council 1980) and signs of toxicity of other nutrients required by swine have been reviewed (Carlson and Lloyd 1981; Church and Pond 1988) but are not discussed here.

METABOLIC DISEASES

Gastric Ulcers

The occurrence of stomach ulcers in growing-finishing pigs has apparently increased substantially in recent years. The reasons for this are not clear but it appears that stress is an important factor. O'Brien (1981) summarized the current knowledge of gastric ulcers in swine and reported a widespread incidence around the world; 5% in United States, 29% in England, 22% in Sweden, and even higher levels in several other countries. The secretion of gastric juice with high acidity and high pepsin concentration increases the susceptibility to ulceration, but the mechanism of the effect of diet is not clear. Heated (gelatinized) maize is associated with a higher incidence of ulcers than unheated maize; in comparison, oats tend to protect against ulcers. An alcohol-soluble substance extracted from oats has been shown to be a potent inhibitor of gastric acid secretion and to protect against ulcers in the pig. The process of ulceration can occur within a 2- or 3-day period. Stress associated with marketing (fasting, long duration of shipment) and crowding increases the incidence. Finely ground feeds increase the incidence. Other dietary components and nutritional factors such as the inclusion of cheese whey, corn starch, and unsaturated fats have been implicated. The extent of economic loss to the industry is difficult to evaluate but probably losses are restricted mainly to deaths among affected pigs, as there is no good evidence that growth rate or feed utilization are adversely affected; presumably deaths are due either to perforation of the stomach wall, resulting in peritonitis, or to such severe ulceration that digestive function is halted. Until more is known concerning etiology and pathogenesis of gastric ulcers, the preventions seem to be avoidance of stresses during growing and at marketing, and avoidance of feeding finely ground diets.

Osteoporosis

Although documentation is not extensive, there is sufficient indirect evidence showing that many of the cases of lameness and spontaneous bone fractures in extremities in older sows result from excessive bone resorption due to suboptimal dietary calcium intake in the course of extended periods of time. The calcium demands during lactation are great, and insufficient dietary calcium during this stage of production may result in excessive bone resorption with no noticeable change in composition or quantity of milk produced. In growing pigs kept in close confinement, or in sows kept continuously in stalls during gestation and in farrowing crates throughout lactation, a potential exists for development of osteoporosis severe enough to produce clinical signs and economic losses through crippling or bone fractures. Present knowledge of etiology and pathogenesis of osteoporosis does not permit solid recommendations, but adequate dietary calcium and at least moderate exercise are probably helpful.

Ketosis

An abnormally high level of ketone bodies in the blood during late gestation or early lactation is, on rare occasion, a problem in swine. Metabolically, this disease is probably identical to or similar to ketosis in dairy cattle, in which it is a much more serious problem. The basic causes of the disease have not been determined although it is now recognized that it involves a defect in energy metabolism or in energy intake relative to current metabolic needs. It is seen most commonly in lactating sows in thin condition and manifests itself by anorexia, body weight loss, and the obvious presence of acetone and other ketones in the expired air. In severe ketosis the animal may become comatose and death may occur following a period of unconsciousness. Treatment is by injection of sodium propionate or of glucose.

This is merely an expediency, as the basic metabolic defect is not corrected. Adequate nutrition, especially during late gestation and in lactation, is important for minimizing the incidence.

Parturient Paresis (Milk Fever)

This disease is related to calcium metabolism and is precipitated by hypocalcemia induced by dietary or metabolic means. Metabolically it is probably identical or similar to the disease as it occurs in dairy cattle, in which it is much more common. It almost always occurs during lactation when large amounts of calcium and other minerals are needed for milk production. The inability of the diet to supply adequate calcium or of the body to mobilize sufficient calcium from bone to maintain serum calcium homeostasis results in a decline in serum calcium concentration to a level low enough to cause the clinical signs of parturient paresis. The signs include anorexia, apathy, ataxia, tetany, unconsciousness, and death. Treatment is to inject a calcium salt such as calcium gluconate intravenously. The response is dramatic. Animals nearly comatose are able to rise and appear normal within a few minutes after calcium administration. Adequate dietary calcium, phosphorus, and vitamin D intake, especially during lactation, minimize the incidence. There are undoubtedly endocrine factors of importance in addition to the nutritional factors. Parathyroid hormone injection can be used as a means of mobilizing bone calcium, but can be considered only as an expediency. There is some evidence that excess calcitonin production from the thyroid gland may be a factor in inhibiting bone resorption to a sufficient degree to induce parturient paresis. There is also the possibility that exercise during gestation acts through the endocrine system to produce an effect on bone mineral mobilization. Research on parturient paresis has been centered on dairy cattle, although similar mechanisms probably control the syndrome in swine. Current knowledge of the disease has been reviewed by Littledike et al. (1981).

Photosensitization

This is a problem in white pigs or those with small amounts of skin pigments. It involves sensitization of the skin to ultraviolet light following the ingestion of plants containing photosensitizing agents. Since both ultraviolet light exposure and ingestion of certain plant materials are needed to cause photosensitization, the problem exists only in swine kept outdoors or otherwise exposed to ultraviolet light. Several legumes, including alfalfa forage and other common legumes as well as buckwheat, contain these photodynamic agents. The symptoms begin approximately one week after the animal begins to consume the diet and include severe reddening, swelling, and exudation of parts of the body exposed to the sun, especially the ears. This swelling and reddening is associated with hemolysis of red blood cells and is followed by tissue necrosis of affected areas. Behavioral changes may also be observed, including staggering and repeated falling.

Osteochondrosis

This is a disturbance in ossification of endochondral tissue and in epiphyses of long bones, especially ulna and femur. The syndrome involves lameness and discomfort and appears to have a higher incidence in pigs raised on concrete or on slatted floors. Lameness develops when a weakened area of the articular-epiphyseal cartilage separates from the underlying bone. Scandinavian work has shown a relationship between visual scoring of locomotion in young pigs and the severity of osteochondrosis, but the causal relationship between the abnormal shape of the articular surfaces and long bones and the development of osteochondrosis is controversial.

The syndrome appears to be noninfectious but attempts to identify nutrient deficiencies or imbalances (vitamins A and D, protein, calcium, phosphorus) as causative factors have been unsuccessful. There is evidence that severity of osteochondrosis has some degree of heritability, so that selection against the syndrome may have some merit.

The endocrine and metabolic bases for osteochondrosis have not been thoroughly studied and it appears that until the syndrome is more completely characterized, the major means of control will be by culling affected animals from breeding and providing comfortable housing and opportunity for exercise.

Rectal and Vaginal Prolapse

Cases of prolapse are sporadic on individual farms and often the cause is obscure. Culling rates for rectal and/or vaginal prolapse in breeding swine have been reported to be as high as 8.9% in a single herd, but more commonly the incidence is less than 1%. Rectal prolapse is most common in pigs 3- to 5-months old, usually at an incidence of less than 1%. The physiological basis for vaginal and rectal prolapses is unclear, although such factors as nutrition (dietary constituents that lead to constipation) drugs (estrogenic substances), genetics (inbreeding may increase incidence), and lack of exercise have been implicated. Treatment must be cost effective; often a simple purse-string suture is sufficient to allow reversal of the prolapse.

Many potentially useful sources of nutrients for swine would remain unused if appropriate processing were not possible. The feeding value of some particular feedstuffs is low because of physical properties (such as hardness of kernel) or because of some chemical property which reduced the biological availability of one or more nutrients (such as heat-labile inhibitors in raw soybeans). Several processing methods, including grinding, pelleting, heating, drying, and oil extraction, have been developed in the feed industry to improve the usefulness of a wide variety of feedstuffs for swine. This chapter is devoted to a discussion of these methods and the beneficial effects of their use.

TYPES OF FEED PROCESSING

Preparation of swine feeds has been drastically modified and in the course of the past decade, the degree of processing has increased from slight to extensive. There has been a definite shift away from the free-choice, whole-grain-plus-supplement feeding to the self-feeding of complete mixed and processed diets. Although many of the changes have been brought about by the development of complete commercial and on-the-farm processing units and the desire to minimize labor requirements through the mechanization of feeding systems, many changes in processing have been utilized in an attempt to improve the efficiency of utilization of the feedstuffs available.

Processing may be defined as a series of actions, operations, motions, and the like that are employed in the accomplishment of a definite end from a definite beginning. An understanding of the reasons for processing of feedstuffs will provide a better basis for evaluation of each process. It is beyond the scope of this chapter to describe in detail each and every processing term but a glossary of processing terms is presented in Appendix 4.

Feed processing is a physical, chemical, thermal, bacterial, or other treatment or alteration of a feed ingredient before it is fed. Feeds are processed for the following reasons: (1) to alter the physical form or particle size, (2) to preserve, (3) to isolate specific parts, (4) to improve palatability, (5) to improve digestibility, (6) to alter nutrient composition, and (7) to detoxify. Table 8.12 (Harris and Crampton 1972) clearly indicates that alteration of particle size or form is the most important consideration given to processing, and that this purpose plus preservation and isolation of specific parts constitute approximately 87% of the motivation for processing.

Modification of a feed may take place during the many processes that occur between harvest and feeding. Some of these modifications may directly alter the feeding value of the products treated; therefore, it is important that the feeder be aware of the processes to which a given feed has been subjected and the effect that each has on the nutritive value of the feed. Details of the effects of processing on the nutrition value of feeds were presented by the National Academy of Sciences (1973).

Hogberg et al. (1980) summarized the current knowledge of the effects of feed processing on the growth rate, feed intake, and efficiency of feed utilization of growing swine. This information is reproduced in Table 8.13. It must be recognized that the direction and magnitude of response to a particular set of processing conditions is never completely predictable, due to differences in pigs, plant growing conditions, crop harvesting times and methods, and plant variety.

TABLE 8.12 Reasons for Processing and Their Related Importance

Purpose of treatment	*Total treatments (%)*
Alteration of form or particle size	40
Preservation	25
Isolation of specific part	22
Improvement of palatability	8
Improvement of digestibility	4
Alteration of nutrient make-up	<1
Detoxification	<1
Total	100

Source: Harris and Crampton 1972.

TABLE 8.13 Effects of Processing Feedstuffs on Pig Performance

Processing method	*Type of grain*	*Growth rate*	*Change in feed intake*	*Feed conversion*	*Comments*
Grinding	Corn	Improved 3–5%	No effect	Improved 3–5%	Medium screen ($\frac{1}{4}$–$\frac{3}{8}$ in.). Generally most acceptable.
	Grain sorghum	Improved	No effect	Improved	Too fine of grind can reduce palatability.
Pelleting	Corn, grain sorghum	Improved 3–6%	Reduced 1–3%	Improved 5–8%	
	Barley, oats, alfalfa, wheat, bran	Improved 3–6%	Reduced 1–3%	Improved 7–10%	Greater improvement with higher fibrous materials.
Paste	Corn	Improved 10–15%	Increased 10–15%	No change	Feed:water ratio of 1.2–1.5:1. Advantage with G-F pigs but not weanlings.
Liquid	Corn	Improved	Increased	Improved	Water:feed ratio of 2.1. Advantages with limit-fed systems. Full-fed levels fail to improve performance.
Roasting	Corn	No change	No change	Slightly improved	Processing cost greater than improvement return.
	Soybeans	No change	Decreased	Improved 4–6%	Necessary for adequate performance. Comparisons made with SBM.
Steam flaking	Corn, grain sorghum	No change	No change	No change	
Micronizing	Corn, grain sorghum	No change	No change	Variable	Availability of amino acids may be reduced.
Extruding	Wheat, grain sorghum	No change	No change	No change	
	Soybeans	No change	Decreased	Improved 4–6%	Comparison made with SBM—necessary for adequate performance.

Source: Hogberg et al. 1980.

DIET FORMULATION

Adequate diet formulation depends on the use of valid measures of nutrient adequacy in the feedstuffs available for use. A chemical characterization of feeds as accomplished by *proximate analysis* is given as follows.

Nitrogenous compounds (analyzed as N in Kjeldahl procedure): These are protein N, nonprotein N, di-, tri-, polypeptides, amino acids, creatine and creatinine, purines, pyrimidines, urea, uric acid, amines, nucleotides, nucleic acids, ammonia, and some water-soluble vitamins.

Fat-soluble compounds (analyzed as ether extract): These are tri-, di-, and monoglycerides; free fatty acids; waxes; phospholipids; glycolipids; cholesterol, vitamin D and other sterols; vitamin A and carotenoids; vitamin E (tocopherols); and vitamin K.

Poorly digested carbohydrates (analyzed as crude fiber or as neutral detergent fiber (cell wall) and acid detergent fiber): These are cellulose and other carbohydrate fractions not readily degraded and lignin-cellulose complexes.

Ash (determined by weight remaining after ignition of fresh sample): This is all inorganic constituents including essential major and trace elements.

Well-digested carbohydrates (calculated by difference): These are hemicellulose, starch, dextran, glycogen, tri-, di-, and monosaccharides, and some water-soluble vitamins.

This is a widely used procedure that incorporates the analysis of protein (nitrogen × 6.25), ash, ether extract, moisture, crude fiber, and—by subtracting the sum of these from 100—the analysis of nitrogen-free extract (NFE). The specific classes of compounds contained in each fraction are indicated. The wide variety of compounds present in each fraction demonstrates the rather imprecise nature of the proximate analysis. It allows description of feeds only according to amounts of major nutrient groups and gives no information on amino acid content (protein quality), mineral composition, vitamin content, or such factors as palatability and toxicity. Also, the determination of crude fiber directly and of NFE (by subtracting the sum of all other components from 100) does not give an accurate prediction of the biological availability of the carbohydrates. It is now recognized that values obtained for crude fiber do not accurately describe the feed as an energy source for the pig, owing to the imprecise chemical characterization of the carbohydrates contained in it. Simpler techniques now available, utilizing acid and neutral detergent fiber analysis, are of greater value in characterizing the complex carbohydrates of plants (Van Soest 1963, 1967; Van Soest and Wine 1967). The cell wall of the plant contains the hemicellulose, cellulose, and lignin that are responsible for the low utilization of forages and seed hulls by swine. Hemicellulose is more digestible than cellulose by swine (Keys et al. 1970), and lignin is essentially nondigestible by swine as well as by ruminants. Thus, the overall digestibility of fibrous feeds by swine depends on the relative amounts of these cell-wall constituents present.

The crude fiber measurement obtained by proximate analysis is not completely without merit as a guide to the energy value of a feed, since levels of more than 5%

in the diet generally reduce growth rates. Therefore, crude fiber content can be used as a general guide in selecting energy constituents for swine diets. The digestibility of crude fiber differs according to source; for example, the fiber of corn cobs has a lower energy value for pigs than fiber of barley.

The energy value of a feed for growth or other productive functions can be determined by using information obtained from the proximate analysis of the feed and of excreta of animals to which it is fed. The apparently digestible energy is computed by measuring the difference between the amount of energy consumed in a specified period of time and the amount of energy excreted in the feces in the same period of time. This is called the total digestible nutrient (TDN) value of the feed. The procedure for determining TDN of a feedstuff is as follows:

1. Determine the nitrogen, ether extract, crude fiber, and NFE content of the dried feed by proximate analysis.
2. Feed known quantities of the feed at constant daily intake for a specified time period (usually 5–10 days).
3. Collect the feces from the animal during the same time period.
4. Determine the nitrogen, ether extract, crude fiber, and NFE content of the dried feces by proximate analysis.
5. Calculate the amount of each nutrient absorbed (feed minus feces).
6. Calculate the percentage of each nutrient absorbed (amount absorbed divided by amount in feed). This is the *digestion coefficient.*
7. Multiply the value obtained for ether extract by 2.25 to account for the higher energy value of fat, which is 2.25 times that of carbohydrates.
8. Multiply the percentage of each nutrient in the diet by the percentage of each nutrient absorbed (digestion coefficient).
9. Add the percentage of each nutrient and divide the sum by 100. This gives TDN as a percentage of diet.

A more direct approach to energy utilization, is to consider *calorie*[1] absorption and utilization. The calories contained in any material can be determined by measurement of the heat produced by complete ignition in a bomb calorimeter.

This *gross* energy content can be expressed as calories per gram of material. Feedstuffs widely different in utilizable energy (such as corn vs. corn cobs) will yield similar values for gross energy. *Digestible* energy can be determined by completely combusting a sample of feed and of feces obtained in the same way as for the determination of TDN just described. In fact, TDN is really an indirect means of estimating digestible energy of a feed, because each nutrient (protein, fat, and carbohydrate) contains a specific approximate calorie concentration. For practical purposes, carbohydrate and protein are both considered to contain 4000 cal/g of pure nutrient,[2] while fat is considered to contain about 9000 cal/g (2.25

[1] A calorie is that amount of heat required to raise the temperature of 1 g of water 1°C (from 15.5° to 16.5°C).

[2] Protein contains about 5000 cal/g, but about 20% is unavailable to the body for use as energy because nitrogen from its metabolism is lost in the urine as urea.

times as much). One gram of TDN contains approximately 4400 digestible calories (1 kg contains 4400 kcal).

Although digestible energy (or TDN) provides useful information about the value of feedstuffs, it does not take into account losses of energy through the urine, expired air, and so on. More refined techniques take these losses into account. The energy remaining for use by the body after these losses are accounted for is *metabolizable* energy. After a meal has been eaten, there is extra heat produced associated with assimilation of the meal. This is called the *heat increment*. The energy remaining for use after the heat increment has been deducted is the *net energy* of the feed and represents the most refined expression of the energy value of a feed. A flow sheet showing these relationships is shown as follows:

Energy value of feeds

Gross energy − fecal losses

↓

Digestible energy − urinary, expired air, and other losses

↓

Metabolizable energy − heat increment

↓

Net energy

Protein Value of Feeds

The protein value is basically the amino acid balance. The importance of proper amounts and proportions of amino acids has been amply demonstrated in the discussion of nutrient requirements of growing and finishing pigs. Methods of assessing amino acid adequacy, aside from analytical determination of the composition of a feedstuff, include the growth trial (measurement of growth of animals full-fed the test diet adequate in all other nutrients), protein digestibility (as described under TDN), nitrogen balance [feed nitrogen minus (fecal + urinary nitrogen)], and body composition studies. The terms PER (protein efficiency ratio) and NPU (net protein utilization) are sometimes used to describe experimentally determined values for protein adequacy of a feedstuff. Detailed descriptions of their derivation and application have been given by Loosli et al. (1977) and Church and Pond (1988). For practical purposes the most indicative index of amino acid adequacy of a feedstuff is the growth trial, as the appetite of the animal and its body weight gain are sensitive to a poorly balanced amino acid mixture.

Modern Feed Formulation

For the formulation of diets of swine that meet the requirements for growth, maintenance, reproduction, and lactation, a wide variety of sources of nutrients, from both plants and animals, is available. Since the nutrient requirements of the pig are similar to those of man, theoretically the feedstuffs used for swine are available only at the expense of food supplies for man. This is potentially true, but many of the feedstuffs used for swine are unsuitable for man or unacceptable to him on esthetic grounds. By-products of the milling industry, the meat-packing industry, the fisheries industry, and the plant oil industry all provide important feeds for swine that are not well suited for direct human use.

The efficient production of pork depends on equating the nutrient requirements of the pig to the available feed supply by using feedstuffs that complement one another in providing all nutrients in proper amounts and proportions at lowest economic cost.

The explosive growth of computer technology has brought linear programming and least-cost diet formulation to the farm. Now, not only commercial feed companies but individual pork producers use computers to capture optimum combinations of available feedstuffs for economical diet formulation. The nutrient composition and other attributes of a vast array of feed resources are used to make intelligent choices in feed formulation, using computer technology to process appropriate inputs of information.

Techniques of computer simulation (described in Chapter 10) are expected to enhance the ability to formulate diets and rations to maximize lean growth to the potential of the genetic population but control wasteful fat production.

CLASSIFICATION OF FEEDSTUFFS

Feedstuffs can be classified on the basis of origin (plant or animal) and major nutrient contributions. An exhaustive catalog of potential feedstuffs for swine is not given here. Rather, the most commonly used feedstuffs and those that seem to show promise of greater future use are described in terms of general characteristics, nutrient contents, and factors limiting their use. The swine industry has tended to be concentrated in areas in which production of these crops is high. The composition of many feedstuffs for swine has been listed by the National Research Council (1982). The feedstuffs listed here include some that are contained therein, but also several that are not. Other tables of nutrient composition of feedstuffs have been prepared by Morrison (1956) and Schneider (1947). Amino acid composition of a wide variety of feeds has been reported by Eggum (1968).

The feedstuffs in common use or of potential value in swine feeding are discussed in detail in the next chapter (energy sources, protein sources, and mineral and vitamin sources). Most of the feeds considered as protein or energy sources also supply significant amounts of minor constituents (vitamins and minerals). Therefore, when diets for swine are being formulated to meet all nutrient requirements, the amounts of supplemental vitamins and minerals needed to fulfill the deficits existing in natural sources may vary from zero to considerable, depending on the nature of the feedstuffs used in the formulation. One must, therefore, consider the contributions of feedstuffs not only in terms of energy and protein but also in terms of their vitamin and mineral content. These considerations are included in the descriptions of individual feedstuffs that follow in Chapter 9. Energy, protein, amino acid, vitamin and mineral content of common ingredients in swine feeds are given in Tables 8-14 and 8-15.

Feedstuffs that are normally considered as energy sources also contribute protein to the diet. The contribution of cereal grains to protein nutrition for man and animals is emphasized in a reference list prepared by Coons (1968) and pub-

TABLE 8.14 Average Composition of Some Feed Ingredients (Excluding Amino Acids) Commonly Used in Swine Diets on an As-Fed Basis

Line No.		International Feed Number[a]	Dry Matter (%)	Energy (kcal/kg)		Protein (%)	Ether Extract (%)	Crude Fiber (%)	Minerals			
				DE	ME				Calcium (%)	Phosphorus (%)	Potassium (%)	Chlorine (%)
01	Alfalfa meal, dehy., 17% protein	1-00-023	92	2580	2270	17.5	2.5	24.1	1.44	0.22	2.40	0.46
02	Barley	4-00-549	89	3086	2870	11.6	1.8	5.1	0.05	0.36	0.48	0.15
03	Barley, Pacific Coast	4-07-939	89	3130	2940	9.0	2.0	6.4	0.05	0.32	0.53	0.15
04	Beans, field (*Vicia faba*)	5-09-262	89	3263	3080	26.0	1.4	8.2	0.14	0.54	1.20	—
05	Beet pulp, dried	4-00-669	91	2866	2345	8.0	0.5	21.0	0.60	0.10	0.21	—
06	Blood meal, spray or ring dried	5-00-381	86	2690	1927	85.0	1.0	0.6	0.30	0.25	0.90	0.27
07	Brewers dried grains	5-02-141	92	1940	1710	25.3	6.2	15.3	0.29	0.52	0.09	0.12
08	Corn, dent yellow, grain	4-02-935	89	3525	3325	8.8	3.8	2.2	0.02	0.28	0.30	0.04
09	Corn and cob meal	4-02-849	85	3086	2500	7.8	3.0	10.0	0.04	0.21	0.45	0.04
10	Corn, gluten feed	5-02-903	90	3307	2400	22.0	2.5	10.0	0.40	0.80	0.57	0.22
11	gluten meal, 41%	5-02-411	91	3230	3069	41.0	2.5	4.0	0.23	0.55	0.31	0.11
12	Corn, distillers grain w/solubles, dehy.	5-02-843	93	3568	3390	27.2	9.0	9.1	0.35	0.95	1.00	0.17
13	Corn, distillers solubles, dehy.	5-02-844	92	3307	2900	28.5	9.0	4.0	0.35	1.33	1.75	0.26
14	Corn, hominy feed	4-02-887	90	3615	3365	10.0	6.9	6.0	0.04	0.50	0.67	0.05
15	Cottonseed meal, mechanical extracted	5-01-609	93	2954	2453	40.9	3.9	12.6	0.17	1.05	1.19	0.04
16	Cottonseed meal, solvent extracted	5-01-619	92	2689	2555	41.4	1.5	11.3	0.15	0.97	1.22	0.03
17	Feather meal	5-03-795	93	2778	2270	86.4	3.3	1.0	0.20	0.80	0.31	—
18	Fish meal, anchovy	5-01-985	92	3086	2450	64.2	10.0	1.0	3.73	2.43	0.90	0.29
19	herring	5-02-000	93	3086	2500	72.3	10.0	0.7	2.29	1.70	1.50	0.90
20	Menhaden	5-02-009	92	2734	2230	60.5	9.4	0.7	5.11	2.88	0.77	0.60
21	Fish solubles, condensed	5-01-969	51	3307	3190	31.5	4.0	0.2	0.30	0.50	1.74	2.65
22	Meat-and-bone meal, 50%	5-09-322	93	2866	2434	50.4	8.6	2.8	10.10	4.96	1.40	0.74
23	Meat meal, 55%	5-09-323	92	2998	2540	54.4	7.1	2.5	8.27	4.10	1.40	0.91
24	Molasses, beet	4-00-668	79	2460	2320	6.1	0.0	0.0	0.13	0.06	4.83	1.30
25	Molasses, cane	4-04-696	74	2469	2343	2.9	0.0	0.0	0.82	0.08	2.38	—
26	Oats	4-03-309	89	2866	2668	11.4	4.2	10.8	0.06	0.27	0.37	0.11
27	Oat groats (dehulled oats)	4-03-331	91	3690	3400	16.0	5.5	3.0	0.07	0.43	0.34	—
28	Peas	5-03-600	90	3527	3200	23.8	1.3	5.5	0.11	0.42	1.02	0.06
29	Peanut meal, expeller	5-03-649	90	3600	3200	45.0	7.3	12.0	0.16	0.55	1.12	0.03
30	Peanut meal, solvent	5-03-650	90	2845	2920	47.0	1.2	13.1	0.20	0.65	1.15	—
31	Rapeseed meal, solvent	5-03-871	94	2998	2670	35.0	1.8	12.4	0.66	1.09	0.80	—
32	Rice bran, solvent	4-03-930	91	3080	2200	12.9	0.6	11.4	0.07	1.50	1.35	0.07
33	Rice, broken	4-03-932	89	2513	2360	8.7	1.7	9.8	0.08	—	—	0.08
34	Rice, polishings	4-03-943	90	3792	3000	12.2	11.0	4.1	0.05	1.31	1.06	0.11
35	Rye, grain	4-04-047	89	3307	2712	12.6	1.8	2.8	0.08	0.30	0.46	—
36	Safflower meal, solvent	5-04-110	91	2960	2435	28.5	0.5	30.6	0.40	1.10	0.80	—
37	Sesame meal, expeller	5-04-220	93	3130	2560	42.0	7.0	6.5	1.99	1.37	1.20	0.06
38	Skim milk, dried	5-01-175	92	3792	3360	33.5	0.9	0.0	1.28	1.02	1.59	0.50
39	Sorghum, grain (Milo)	4-04-444	89	3439	3229	8.9	2.8	2.3	0.03	0.28	0.32	0.09
40	Soybeans, full-fat cooked	5-04-597	90	4056	3540	37.0	18.0	5.5	0.25	0.58	1.61	0.03
41	Soybean meal, dehulled, solvent	5-04-612	90	3860	3485	48.5	1.0	3.9	0.27	0.62	2.02	0.05
42	Soybean meal, expeller	5-04-600	90	3483	2990	42.6	4.0	6.2	0.27	0.61	1.83	0.07
43	Soybean meal, solvent	5-04-604	89	3350	3090	44.0	0.8	7.3	0.29	0.65	2.00	0.05
44	Sunflower meal, dehulled, solvent	5-04-739	93	2998	2605	42.0	2.9	12.2	0.37	1.00	1.00	0.10
45	Wheat bran	4-05-190	90	2513	2320	15.7	4.0	11.0	0.14	1.15	1.19	0.06
46	Wheat shorts	4-05-201	89	3175	2910	16.8	4.2	8.2	0.11	0.76	0.88	0.07
47	Wheat middlings	4-05-205	88	3050	2940	16.0	3.0	7.0	0.12	0.90	0.60	0.03
48	Wheat, hard, red winter	4-05-268	87	3483	3220	14.1	1.9	2.4	0.05	0.37	0.45	0.05
49	Wheat, soft, red winter	4-05-294	86	3659	3416	10.2	1.8	2.4	0.05	0.31	0.40	0.08
50	Whey, dried	4-01-182	93	3439	3190	13.6	0.8	1.3	0.97	0.76	1.05	1.50
51	Whey, low lactose	4-01-186	91	3307	2750	15.5	1.0	0.3	1.95	0.98	3.00	2.10
52	Yeast, brewers dried	7-05-527	93	3135	2707	44.4	1.0	2.7	0.12	1.40	1.70	0.12

lished by USDA. That list was prepared "to facilitate awareness and evaluation of available research data on cereal proteins . . ." and, though not recent, is still pertinent for those deeply involved with feed formulation.

The digestible energy value of some of the most common energy and protein sources in use in North America has been summarized in Table 8.16. It should be noted, first, that a wide variation exists in digestible energy content of feeds which

TABLE 8.14 *(Continued)*

									Vitamins													
Line No.	Magnesium (%)	Sodium (%)	Sulfur (%)	Copper (mg/kg)	Iron (mg/kg)	Manganese (mg/kg)	Selenium (mg/kg)	Zinc (mg/kg)	Biotin (mg/kg)	Choline (mg/kg)	Folacin (mg/kg)	Niacin (mg/kg)	Pantothenic Acid (mg/kg)	Pyridoxine (mg/kg)	Riboflavin (mg/kg)	Thiamin (mg/kg)	Vitamin B_{12} (mg/kg)	Vitamin E (IU/kg)
01	0.26	0.08	0.21	8.2	310	28.0	0.60	17	0.30	1097	6.3	38	28.4	6.5	15.7	3.4	0.004	125
02	0.14	0.04	0.15	7.5	50	8.0	0.10	17	0.08	990	0.5	63	9.2	3.0	1.2	4.0	—	36
03	0.12	0.02	—	7.7	60	16.3	0.10	15	0.15	1034	0.5	48	7.0	2.9	1.6	5.5	—	36
04	0.13	0.80	—	4.1	70	8.4	—	42	0.09	1670	—	22	3.0	—	1.6	5.5	—	1
05	0.27	0.32	0.20	12.5	300	35.0	—	0.7	—	800	—	20	0.8	—	1.1	0.2	—	—
06	0.22	0.33	0.32	8.1	3000	6.4	—	306	0.30	749	0.3	22	1.1	4.4	1.3	0.5	0.440	—
07	0.16	0.15	0.31	21.1	250	37.8	0.70	98	0.96	1723	7.1	29	8.0	0.7	1.4	0.5	—	25
08	0.12	0.02	0.08	3.4	35	5.0	0.04	10	0.11	530	0.2	34	7.5	7.0	1.0	3.5	—	22
09	0.13	0.01	0.18	6.7	70	7.7	0.07	9	0.05	393	0.3	17	4.0	5.0	0.9	—	—	19
10	0.29	0.95	0.22	47.9	460	23.8	0.10	48	0.33	1518	0.3	66	17.0	15.0	2.4	2.0	—	15
11	0.05	0.07	0.40	28.3	400	8.9	1.00	20	0.18	330	0.2	50	10.0	7.9	1.7	0.2	—	20
12	0.35	0.90	0.30	44.7	280	30.0	0.39	80	0.30	3400	0.9	80	11.0	2.2	8.6	3.5	—	40
13	0.64	0.26	0.37	82.7	560	73.7	0.33	85	1.40	4842	1.1	116	21.0	10.0	11.6	6.9	—	55
14	0.24	0.10	0.03	13.3	70	14.5	—	3	0.13	1500	0.3	46	8.0	11.0	2.2	7.9	—	—
15	0.42	0.04	0.40	18.6	160	22.9	0.90	57	0.60	2753	2.7	38	7.7	5.3	4.2	9.7	—	15
16	0.40	0.04	—	17.8	110	20.2	—	—	0.55	2933	2.7	40	9.9	3.0	4.0	7.7	—	15
17	0.20	0.71	—	—	—	21.0	—	—	0.04	891	0.2	27	10.0	—	2.1	0.1	0.600	—
18	0.24	1.10	0.54	9.3	220	9.5	1.36	103	0.23	5100	0.2	135	20.0	4.0	7.1	0.1	0.352	6
19	0.15	0.61	0.69	4.5	80	4.7	1.93	132	0.20	5306	0.5	142	22.0	4.0	9.9	0.1	0.588	17
20	0.16	0.41	0.45	10.8	440	33.0	2.10	147	0.15	3056	1.0	55	9.0	4.0	4.9	0.2	0.150	7
21	0.02	3.10	0.12	44.9	30	14.4	2.00	38	0.18	4028	—	169	35.0	12.2	14.6	5.5	0.347	—
22	1.12	0.72	0.26	1.5	490	14.2	0.25	93	0.14	1996	0.6	46	4.1	12.8	4.4	0.2	0.070	0.8
23	1.13	0.73	0.26	1.5	440	12.3	0.25	103	0.14	2077	0.6	57	5.0	3.0	5.5	0.2	—	0.8
24	0.23	—	0.48	17.7	70	4.7	—	14	0.70	880	—	48	4.0	—	2.1	—	—	4.4
25	0.35	0.90	0.35	59.6	200	42.2	—	—	0.70	660	—	45	39.0	—	2.3	0.9	—	4.4
26	0.16	0.06	0.21	5.9	70	43.2	0.30	1	0.30	1100	0.4	15	29.2	1.0	1.1	6.0	—	20.0
27	0.09	—	0.20	6.4	90	28.6	—	—	0.20	1232	0.3	18	11.0	—	1.3	6.8	—	15.0
28	—	0.04	—	—	50	—	—	30	0.18	642	0.4	17	4.6	1.0	0.8	1.8	—	—
29	0.32	—	0.28	—	—	24.8	—	—	0.39	1640	—	165	46.8	—	5.1	7.1	—	2.9
30	0.40	0.10	—	—	—	29.9	—	—	0.39	1980	—	165	50.6	—	11.0	6.6	—	3.0
31	0.51	0.50	—	7.0	180	43.0	0.98	66	—	6464	—	153	9.0	7.0	3.7	1.7	—	19.1
32	0.95	0.07	0.18	13.0	190	138.0	—	30	4.20	1135	—	293	23.0	14.0	2.5	22.5	—	59.8
33	0.11	0.07	0.06	—	—	18.0	—	17	0.08	800	0.2	46	8.0	—	0.7	—	—	14.5
34	0.65	0.10	0.17	—	160	—	—	—	0.61	1237	—	520	47.0	—	1.8	19.8	—	90.0
35	0.12	0.02	0.15	7.8	100	66.9	—	31	0.60	—	0.6	16	9.2	—	1.5	4.4	—	15.0
36	0.37	0.06	—	10.8	560	19.8	—	44	1.56	2247	0.5	60	43.8	—	11.3	2.8	—	0.9
37	0.86	0.04	0.43	—	—	47.9	—	100	0.34	1690	—	30	6.0	12.5	3.6	2.8	—	—
38	0.11	0.44	0.31	11.5	50	2.0	0.12	40	0.33	1250	0.6	12	33.0	3.9	22.0	3.5	0.010	9.1
39	0.20	0.01	0.09	14.1	40	12.9	—	14	0.09	678	0.2	41	12.0	3.2	1.1	4.0	—	12.0
40	0.21	0.28	0.22	15.8	80	29.8	0.11	16	0.27	2420	3.5	22	15.6	10.8	2.6	6.6	—	0.9
41	0.27	0.34	0.43	36.3	120	27.5	0.10	45	0.32	2850	0.7	22	15.0	5.0	2.9	1.7	—	3.3
42	0.26	0.27	0.33	18.0	140	30.7	0.10	60	0.33	2703	0.5	37	14.0	—	3.7	1.7	—	6.1
43	0.27	0.34	0.43	36.3	120	29.3	0.10	27	0.32	2794	0.5	60	13.3	8.0	2.9	1.7	—	2.1
44	0.75	2.00	—	3.5	30	22.9	—	—	1.45	2894	—	220	10.0	16.0	3.1	—	—	11.0
45	0.52	0.05	0.22	10.2	170	100.0	0.50	95	0.10	980	1.8	321	31.0	7.0	3.1	8.0	—	10.8
46	0.26	0.07	0.23	12.1	100	115.0	0.50	106	0.10	930	1.4	100	17.6	11.0	2.0	19.9	—	29.9
47	0.29	0.60	0.16	4.4	40	43.0	0.80	64	0.10	1100	0.6	53	13.0	9.0	2.2	18.9	—	—
48	0.17	0.04	0.12	10.6	50	62.2	0.06	14	0.04	1090	0.4	56	13.5	3.4	1.4	4.5	—	12.6
49	0.10	0.04	0.12	9.7	40	51.3	0.06	14	0.04	788	0.4	48	11.0	4.0	1.2	4.3	—	13.2
50	0.13	2.00	1.04	40.0	130	6.1	0.06	—	0.34	1980	0.8	10	44.0	4.0	27.1	4.1	0.015	0.2
51	0.25	1.50	—	—	—	14.0	0.06	—	0.64	4392	1.4	19	69.0	4.0	29.9	5.7	0.015	—
52	0.23	0.07	0.38	32.8	120	5.2	1.00	39	1.05	3984	9.9	448	109.0	42.8	37.0	91.8	—	—

NOTE: A blank space indicates that the feed does not contain the particular nutrient.

Source: National Research Council 1988.

are normally considered to be good energy sources for swine (for example, the difference in energy content between maize and oats) and, second, that many of the protein supplements are good sources of digestible energy. Of course, the prices of most protein supplements preclude their use as energy sources.

Modern feed formulation laboratories use highly specialized equipment to

TABLE 8.15 Average Amino Acid Composition of Some Commonly Used Feedstuffs

	International Feed Number[a]	Protein (%)	Arginine (%)	Histidine (%)	Isoleucine (%)	Leucine (%)	Lysine (%)	Methionine (%)	Cystine (%)	Phenylalanine (%)	Tyrosine (%)	Threonine (%)	Tryptophan (%)	Valine (%)
Alfalfa meal, dehy., 17% protein	1-00-023	17.5	0.8	0.3	0.8	1.3	0.73	0.2	0.2	0.8	0.6	0.70	0.28	0.8
Barley	4-00-549	11.6	0.6	0.3	0.5	0.8	0.40	0.2	0.3	0.6	0.3	0.42	0.14	0.6
Barley, Pacific Coast	4-07-939	9.0	0.5	0.2	0.4	0.6	0.29	0.1	0.2	0.5	0.3	0.30	0.12	0.5
Beans, field (*Vicia faba*)	5-09-262	27.4	2.5	0.7	1.1	1.9	1.72	0.2	0.2	1.2	0.7	3.96	0.24	1.2
Beet pulp, dried	4-00-669	8.0	0.3	0.2	0.3	0.6	0.60	0.01	0.01	0.3	0.4	0.40	0.10	0.4
Blood meal, spray or ring dried	5-00-381	85.0	4.1	5.5	1.0	12.7	8.10	1.5	1.5	7.3	3.0	4.90	1.10	9.1
Brewers dried grains	5-02-141	25.3	0.8	0.6	1.4	2.5	0.90	0.6	0.4	1.5	1.2	0.98	0.34	1.7
Corn, dent yellow, grain	4-02-935	8.8	0.5	0.2	0.4	1.1	0.24	0.2	0.2	0.5	0.5	0.39	0.05	0.4
Corn and cob meal	4-02-849	7.8	0.4	0.2	0.4	1.0	0.18	0.1	0.1	0.4	—	0.35	0.07	0.4
Corn, gluten feed	5-02-903	22.0	1.0	0.7	0.7	1.9	0.63	0.5	0.5	0.8	0.6	0.89	0.10	1.0
gluten meal, 41%	5-02-411	40.6	1.4	1.0	2.2	7.2	0.78	1.0	0.7	2.9	1.0	1.40	0.21	2.2
Corn, distillers grain w/solubles, dehy.	5-02-843	27.2	1.0	0.7	1.0	2.6	0.60	0.6	0.3	1.2	0.7	0.92	0.19	1.3
Corn, distillers solubles, dehy.	5-02-844	28.5	1.1	0.7	1.3	2.1	0.90	0.5	0.4	1.3	1.0	1.00	0.30	1.4
Corn, hominy feed	4-02-887	10.0	0.5	0.2	0.4	0.8	0.40	0.1	0.1	0.4	0.5	0.40	0.10	0.5
Cottonseed meal, mechanical extracted	5-01-609	40.9	4.3	1.1	1.6	2.5	1.51	0.6	0.6	2.2	1.1	1.38	0.55	2.0
Cottonseed meal, solvent extracted	5-01-619	41.4	4.6	1.1	1.3	2.4	1.71	0.5	0.6	2.2	1.0	1.32	0.47	1.9
Feather meal	5-03-795	86.4	3.9	0.3	2.7	6.7	1.10	0.4	3.0	2.7	6.3	2.80	0.50	4.6
Fish meal, anchovy	5-01-985	64.2	3.7	1.5	3.0	5.0	5.10	1.9	0.6	2.7	2.2	2.68	0.74	3.4
herring	5-02-000	72.3	4.8	1.7	3.2	5.3	5.70	2.1	0.7	2.8	2.3	3.00	0.81	4.4
Menhaden	5-02-009	60.5	3.8	1.5	2.9	5.0	4.83	1.8	0.6	2.5	2.0	2.50	0.68	3.2
Fish solubles, 50% solids	5-01-969	31.5	1.6	1.6	0.7	1.9	1.73	0.5	0.3	0.9	0.4	0.86	0.31	1.2
Meat-and-bone meal, 50%	5-09-322	50.4	3.6	1.2	1.4	3.2	2.60	0.7	0.3	1.5	0.8	1.50	0.28	2.3
Meat meal, 55%	5-09-323	54.4	3.7	1.3	1.6	3.3	3.00	0.8	0.7	1.7	1.8	1.74	0.36	2.6
Oats	4-03-309	11.4	0.8	0.2	0.5	0.9	0.40	0.2	0.2	0.6	0.5	0.43	0.16	0.7
Oat groats (dehulled oats)	4-03-331	16.0	0.7	0.3	0.5	1.0	0.60	0.2	0.3	0.7	0.9	0.50	0.18	0.7
Peas	5-03-600	23.8	1.4	0.7	1.1	1.8	1.60	0.3	0.2	1.3	—	0.94	0.24	1.3
Peanut meal, expeller	5-03-649	45.0	4.7	1.1	1.8	3.6	1.55	0.4	0.7	2.6	—	1.40	0.46	2.6
Peanut meal, solvent	5-03-650	47.0	4.9	1.2	2.1	3.7	1.76	0.4	0.8	2.8	2.0	1.45	0.48	2.8
Rapeseed meal, solvent	5-03-871	35.0	1.9	1.0	1.3	2.3	2.10	0.7	0.4	1.4	0.8	1.53	0.45	1.8
Rice bran, solvent	4-03-930	12.9	0.9	0.3	0.4	0.9	0.59	0.2	0.1	0.6	0.7	0.48	0.15	0.6
Rice, broken	4-03-932	8.7	0.6	0.2	0.3	0.5	0.24	0.1	0.1	0.3	—	0.27	0.10	0.5
Rice, polishings	4-03-943	12.2	0.8	0.2	0.4	0.8	0.57	0.2	0.1	0.5	0.6	0.40	0.13	0.8
Rye, grain	4-04-047	12.6	0.5	0.3	0.5	0.7	0.49	0.2	0.2	0.6	0.3	0.86	0.12	0.6
Safflower meal, solvent	5-04-110	28.5	3.7	1.0	1.7	2.5	1.30	0.7	0.7	1.9	—	1.35	0.60	2.3
Sesame meal, expeller	5-04-220	42.0	4.2	1.1	2.1	3.3	1.30	1.2	0.6	2.2	2.0	1.65	0.80	2.4
Skim milk, dried	5-01-175	33.5	1.1	0.8	2.2	3.2	2.40	0.9	0.4	1.6	1.1	1.60	0.44	2.3
Sorghum, grain (Milo)	4-04-383	8.9	0.4	0.3	0.5	1.4	0.22	0.1	0.2	0.4	0.4	0.27	0.10	0.5
Soybeans, full-fat cooked	5-04-597	37.0	2.8	0.9	2.0	2.8	2.40	0.5	0.6	1.8	1.2	1.50	0.55	1.8
Soybean meal, dehulled, solvent	5-04-612	48.5	3.7	1.3	2.6	3.8	3.18	0.7	0.7	2.1	2.0	1.91	0.67	2.7
Soybean meal, solvent	5-04-604	44.0	3.3	1.2	2.4	3.5	2.93	0.7	0.7	2.3	1.3	1.81	0.62	2.3
Sunflower meal, dehulled, solvent	5-04-739	42.0	3.3	1.4	2.8	3.9	1.70	0.7	0.7	2.9	1.2	2.13	0.71	3.2
Wheat bran	4-05-190	15.7	1.0	0.3	0.6	0.9	0.59	0.2	0.3	0.5	0.4	0.42	0.30	0.7
Wheat, hard, red winter	4-05-268	14.1	0.6	0.2	0.6	0.9	0.40	0.2	0.3	0.7	0.6	0.37	0.18	0.6
Wheat middlings	4-05-205	16.0	1.8	0.4	0.6	1.1	0.69	0.2	0.3	0.6	0.5	0.49	0.20	0.7
Wheat shorts	4-05-201	16.8	1.2	0.5	0.6	1.1	0.81	0.2	0.3	0.7	0.5	0.61	0.19	0.8
Wheat, soft, red winter	4-05-294	10.2	0.4	0.2	0.4	0.6	0.31	0.2	0.2	0.5	0.4	0.32	0.12	0.4
Whey, dried	4-01-182	12.0	0.3	0.2	0.8	1.2	0.97	0.2	0.3	0.3	0.3	0.89	0.19	0.7
Whey, low lactose	4-01-186	15.5	0.7	0.1	0.3	0.2	1.47	0.6	0.6	0.1	0.2	0.50	0.18	0.3
Yeast, brewers dried	4-05-527	44.4	2.2	1.1	2.1	3.2	3.23	0.7	0.5	1.8	1.5	2.06	0.49	2.3

NOTE: A blank space indicates that the feed does not contain the particular nutrient.

Source: National Research Council. 1988.

determine vitamin, mineral, amino acid, and fatty acid analyses of feedstuffs. These analyses are expensive, but with modern automated equipment it is possible to obtain a complete nutrient analysis on a wide variety of feedstuffs, so that diets can be balanced within a few hours or days after the sample is obtained. As a result, computer formulation of diets, whereby prescribed limitations can be imposed on acceptable levels of nutrients (upper and lower limits) as well as of feedstuffs, has become increasingly convenient. It allows "least-cost" diet formulation which

TABLE 8.16 Digestible and Metabolizable Energy Values of Some Common Feedstuffs for Swine (kcal/kg, as fed)

Feedstuff	*Energy value: DE*	*ME*
Alfalfa meal, dehydrated, 17% protein	2580	2270
Alfalfa meal, dehydrated, 17% protein[a]	2870	2610
Barley	3086	2870
Corn, dent yellow, grain	3525	3325
Corn and cob meal	3086	2500
Cottonseed meal, solvent	2689	2555
Fish meal, herring	3086	2500
Fish meal, Menhaden	2734	2230
Hominy feed[a]	4020	3840
Meat meal, 55% protein	2998	2540
Molasses, cane	2469	2343
Oats	2866	2668
Oat groats (dehulled oats)	3690	3400
Rice[b]	3700	3580
Rice bran[b]	3880	3710
Rice, broken	2513	2360
Rice, polishings	3792	3000
Skim milk, dried	3792	3360
Sorghum, grain, milo	3439	3229
Soybeans, full-fat	4056	3540
Soybean meal, dehulled, solvent	3860	3485
Soybean meal, solvent	3350	3090
Wheat, soft, red winter	3659	3416
Wheat, hard, red winter	3483	3220
Wheat bran	2513	2320
Whey, dried	3439	3190
Yeast, brewers' dried	3135	2707

Source: National Research Council 1979; [a] Stanley and Ewan 1982; [b] Robles and Ewan 1982.

provides for frequent substitutions and alterations in proportion of feedstuffs, depending on current prices. Feed manufacturers and many individual swine producers now balance diets with the use of computers for least-cost formulation. The relatively low cost and easy programming of home computers now available encourages their use on the farm for diet formulation as well as the total business management program. The mechanics of least-cost formulation require that the following information be supplied to the machine: (1) tolerance (upper and lower) of specific ingredients, (2) accurate feedstuff composition data, and (3) current ingredient costs. If these three pieces of information are used, diets that meet all mathematical requirements can be formulated, but the final measure of acceptability is the performance of the animal to which the diet is fed. Least-cost formulation does not account for factors, such as palatability, that may have an important effect on animal response.

The technique of computer simulation mentioned in the previous section will allow predicted body and carcass composition and time of slaughter to be incorporated into the considerations so that ration formulation can be directed to "maximum economic efficiency'' for producing the pork products. In the final analysis, the most sophisticated computer cannot take the place of the animal itself for which the diet was formulated in determining the acceptability of the diet.

REFERENCES

Aricultural Research Council. 1981. The Nutrient Requirements of Pigs. Commonwealth Agricultural Bureaux, Farnham Royal, Slough. U. K.

Barnes, R. H., Moore, A. U., Reid, I. M., and Pond, W. G. 1967. Am. Diet. Assoc. *51*, 34.

Braude, R., Mitchell, K. G., and Suffolk, S. F. 1969. J. Inst. Anim. Technicians *20*, 43.

Brown, W. R., Krook, L., and Pond, W. G. 1966. Cornell Vet., Suppl. 1 *56*, 108.

Bustad, L. K., McClellan, R. A., and Burns, M. P., eds. 1966. Swine in Biomedical Research. Frayn, Seattle.

Carlson, T. L., and Lloyd, W. E. 1981. *In* Diseases of Swine (eds., A. D. Leman et al.) 5th ed., ch. 57, p. 603. Iowa State University Press, Ames.

Church, D. C., and Pond, W. G. 1988. Basic Animal Nutrition and Feeding, 3rd ed. Wiley, New York.

Clausen, H. 1965. World Rev. Anim. Prod. *1*, 28.

Coons, C. M. 1968. Selected References on Cereal Grains in Protein Nutrition. Human and Experimental Animal Studies of Major and Minor Cereals, 1910–1966. ARS 61-5, USDA, Washington, D.C.

Cunha, T. J. 1977. Swine Feeding and Nutrition. Academic Press, New York.

Eggum, B. O. 1968. Aminosyrekoncontration og proteinkvalitet. Stougaards Forlag, Copenhagen, Denmark.

Follis, R. H., Jr. 1958. Deficiency Disease, p. 43. Charles C. Thomas, Springfield, Ill.

Frank, G. R., Bahr, J. M., and Easter, R. A. 1984. J. Anim. Sci. *59*, 1567.

Gallo, J. T., and Pond, W. G. 1968. J. Anim. Sci. *27*, 73.

Gipp, W. F. 1971. The effect of supplemental copper on growth, hematology and iron utilization of swine. Ph.D. Thesis, Cornell University, Ithaca, N.Y.

Hansard, S. L. 1966. *In* Swine in Biomedical Research (eds., K. L. Bustad, R. O. McClellan, and M. P. Burns), p. 79. Frayn, Seattle.

Harmon, B. G., Becker, D. E., Jensen, A. H., and Baker, D. H. 1969. J. Anim. Sci. *28*, 848.

Harris, L. E., and Crampton, E. W. 1972. Symposium on the Effect of Processing on the Nutritive Value of Feeds. NRC-NAS and Univ. Florida Center for Tropical Agric., Gainesville, Fla.

Hogberg, M., Mahan, D., and Seerley, R. 1980. Physical Forms of Feed—Feed Processing for Swine. Pork Industry Handbook PIH 71. Cooperative Extension Service, Purdue Univ., West Lafayette, Ind.

Keys, J. E., Jr., Van Soest, P. J., and Young, E. P. 1970. J. Anim. Sci. *31*, 1172.

King, J. M., Hsu, F. S., Hong, C. B., and Lee, R. C. T. 1976. An Atlas of General Pathology. Pig Research Institute of Taiwan and Joint Commission on Rural Reconstruction, Taipei, Taiwan. 300 pp.

Krider, J. L., Conrad, J. H., and Carroll, W. E. 1981. Swine Production, 5th ed. McGraw-Hill, New York.

Ku, P. K., Ely, W. T., Groce A. W., and Ullrey, D. E. 1972. J. Anim. Sci., *34,* 208.

Kubota, J., Allaway, W. H., Carter, D. L., Corey, E. E., and Lazar, V. A. 1967. J. Agric. Food Chem. *15,* 448.

Lecce, J. G. 1969. J. Anim. Sci. *28,* 27.

Littledike, E. T., Young, J. W., and Beitz, D. C. 1981. J. Dairy Sci. *64,* 1465.

Low, A. G., and Partridge, I. G. 1980. Current concepts of digestion and absorption in pigs. Tech. Bull. 3, Natl. Inst. Res. Dairying, Reading, England.

Lowrey, R. S., Pond, W. G., Krook, L., and Loosli, J. K. 1962. J. Nutr. *78,* 245.

Loosli, J. K., Warner, R. G., Hintz, H. F., and Maynard, L. A. 1977. Animal Nutrition, 7th ed. McGraw-Hill, New York.

Maner, J. H., Pond, W. G., and Loosli, J. K. 1961. J. Anim. Sci. *20,* 614.

Miller, E. R., Ullrey, D. E., and Lewis, A. J. 1991. Swine Nutrition. Plenum Publications, New York.

Morrison, F. B. 1956. Feeds and Feeding, 22nd ed., Morrison, Ithaca, N.Y.

National Academy of Sciences. 1973. Effect of Processing on the Nutritive Value of Feeds. NAS, Washington, D.C. 494 pp.

National Research Council. 1979. Nutrient Requirements of Swine, no. 2. National Acad. Sci., Washington, D.C.

National Research Council. 1988. Nutrient Requirements of Swine, 9th revised ed. National Acad. Press, Washington, D.C.

National Research Council. 1980. Mineral Tolerance of Domestic Animals. Natl. Acad. Sci., Washington, D.C. 577 pp.

National Research Council. 1982A. Joint U.S.–Canadian Tables of Feed Composition Natl. Acad. Sci., Washington, D.C.

National Research Council. 1982B. Effect of Environment on Nutrient Requirements of Domestic Animals, NRS. Natl. Acad. Sci., Washington, D.C.

O'Brien, J. J. 1981. *In* Diseases of Swine (eds., A. D. Leman et al.) 5th ed., ch. 59, p. 632, Iowa State University Press. Ames.

Palludan, B. 1966. *In* Swine in Biomedical Research (eds., L. K. Bustad, R. O. McClellen, and M. P. Burns). Frayn, Seattle.

Pond, W. G., and Houpt, K. A. 1978. Biology of the Pig. Cornell University Press, Ithaca, N. Y. 371 pp.

Pond, W. G., and Maner, J. H. 1974. Swine Production in Temperate and Tropical Environments. W. H. Freeman, San Francisco. 646 pp.

Pond, W. G., and Maner, J. H. 1984. Swine Production and Nutrition, AVI, Westport, Conn. 731 pp.

Pond, W. G., Yen, J. T., Mersmann, H. J., and Maurer, R. R. 1990. Growth, Development, and Aging. *54,* 77.

Robles, A., and Ewan, R. C. 1982. J. Anim. Sci. *54,* 1175.

Schneider, B. H. 1947. Feeds of the World Agric. Exp. Stn., West Virginia University, Morgantown.

Stanley, D. L., and Ewan, R. C. 1982. J. Anim. Sci. *54,* 1175.

Tucker, H. F., and Salmon, W. D. 1955. Proc. Soc. Exp. Biol. Med. *88,* 613.

Van Soest, P. J. 1963. J. Assoc. Off. Agric. Chem. *46,* 829.

Van Soest, P. 1967. J. Anim. Sci. *26,* 119.

Van Soest, P. J., and Wine, R. H. 1967. J. Assoc. Off. Agric. Chem. *50,* 50.

Whitehair, C. K., and Miller, E. R. 1981. *In* Diseases of Swine (eds., A. D. Leman *et al.*) 5th ed., ch. 61, p. 656. Iowa State University Press, Ames, Iowa.

9

Feedstuffs

A broad array of feed resources is available worldwide from which to formulate balanced diets for all stages of the life cycle of the pig. Those interested in entering the literature are referred for early background to Morrison's *Feeds and Feeding* (1956), which enjoyed wide use throughout the first half of the twentieth century, and for more recent detailed coverage to *Swine Production and Nutrition* (Pond and Maner 1984), *Basic Animal Nutrition and Feeding* (Church and Pond 1988), and *Swine Nutrition Guide* (Patience and Thacker 1989). Tabular data on composition of common swine feeds is presented in *Nutrient Requirements of Swine* (NRC, 1988). These tables are reproduced in Chapter 8 of this book.

Feedstuffs can be classified broadly according to their major contribution of nutrients to the total diet. In this chapter, we discuss briefly the following classes of feedstuffs, energy sources, protein sources, and vitamin and mineral sources.

ENERGY SOURCES

Carbohydrates are the most abundant form of energy in plant materials and, as such, are the most widely available sources of energy for swine feeding. In addition to the array of energy feedstuffs discussed here, many by-products of agriculture and industry are produced but have been underutilized as feed resources despite their potential value. Such products, including industrial food processing wastes, nonfood industrial wastes, forest residues, animal wastes, crop residues, and aquatic plants, were featured in a review to encourage their greater exploitation for animal feed use (National Research Council 1983). The future should bring increased use of these valuable resources in pork production.

GRAIN AND GRAIN BY-PRODUCTS

Grains and their by-products are by far the most important sources of carbohydrates. Although the grains and their by-products discussed in this section also supply important amounts of proteins, vitamins, and minerals, they are included here because of their usefulness in supplying available energy to swine of all ages.

Corn (Maize)

Corn (*Zea mays*) is the most common grain used for feeding swine. Although in the United States it is referred to as corn, it is called maize in most parts of the world. It is hereafter called *maize* in this chapter. It is the basis for swine diets throughout the Americas, and in the United States alone about one-half of all maize produced is marketed for feeding to swine. The amounts of the most abundant feed grain sources produced in the United States and the world in 1988 are shown in Table 9.1 (FAO 1988). In the temperate environments of Canada and Europe, maize is commonly replaced by barley, wheat, rye, and triticale.

Although common maize predominates in yield and area planted, plant geneticists, in their search for maize of improved nutritional value, have identified mutants with interesting nutritional qualities. Even though commercially these mutants are of little importance, the nutritional characteristics of several are discussed briefly to indicate their unique potential.

Common Maize

Common maize is an important grain in the food supply of swine. Although maize is commonly used for swine feeding and many times as the only source of high energy feed, it should never be fed as the sole source of protein because of its relatively low protein content and poor balance of essential amino acids. It was known as early as 1914 (Osborne and Mendel 1914) that the quality of maize protein was poor because of a deficiency of several of the essential amino acids, especially lysine and tryptophan. Without adequate supplementation with protein and/or essential amino acids, maize is an inadequate protein source for swine of all ages.

TABLE 9.1 Feed Grains Produced in the United States and the World in 1988 (millions of metric tons)

Grains	*United States*	*World*
Total Cereals	206.5	1,743.0
Maize	125.0	405.5
Wheat[a]	49.3	509.9
Rice[a]	7.2	483.5
Barley	6.3	168.4
Grain Sorghum	14.7	61.8
Oats	3.2	38.8
Rye	0.4	29.6

[a] Most of wheat and rice produced is sold for human consumption; only a small fraction is feed to swine.

Source: FAO Production Yearbook 1988.

The chemical composition of the maize grain has been studied in detail. Extraction of the protein fractions has divided the protein of the endosperm and germ into four fractions; the zein, glutelin, acid-soluble, and residual fractions. The importance of the proportions of these fractions in the grain is obvious from a study of the amino acid composition of each fraction (Table 9.2). The reason behind the poor protein quality of common maize is that the protein, zein, contributes as much as 50% of the total protein of the kernel in most varieties. Zein has been studied extensively and has been shown to contain very low levels of both lysine and tryptophan. Environment and variety have a significant effect on lysine content. Nitrogen fertilization increases protein content and decreases protein quality. Reduced protein quality is associated with an increase in the zein fraction. An increase in the zein fraction which accompanies increases in protein content of maize produces a high-protein maize of lower biological level.

The energy value of maize is generally used as a standard with which other energy sources are compared. Thus, if the relative energy value of maize is taken as 100, the energy value of other available plant energy sources is generally less. This is mainly due to the low fiber content of the maize kernel and the high digestibility of its starch.

White and yellow maize are of similar composition except that yellow maize is high in carotene, a vitamin A precursor. Both are fair sources of vitamin E, but low in vitamin D and the B vitamins. They are very deficient in calcium, and although they contain a reasonably high level of phosphorus, much of it is present in the biologically poor form of phytic acid phosphorus. Trace elements are present only in marginal quantities in maize.

Numerous studies with growing pigs have dealt with crystalline amino acid supplementation of maize protein. Both lysine and tryptophan are severely limiting and neither elicits a favorable response without supplementation with the other; tryptophan is slightly more limiting than lysine. The third and fourth limiting amino

TABLE 9.2 Amino Acid Composition of the Protein Fractions of Normal Maize

	Endosperm			
Amino acid	*Acid-soluble fraction*	*Alcohol-soluble fraction (zein)*	*Alkali-soluble fraction (glutelus)*	*Germ*
Lysine	1.8	0.3	3.6	6.1
Histidine	1.1	1.6	4.2	2.9
Arginine	2.8	2.5	5.3	9.1
Threonine	2.4	3.4	3.4	3.9
Tryptophan		0.1	0.9	
Valine	2.4	4.5	5.6	5.3
Methionine	1.4	2.5	1.1	1.7
Cystine		1.0	3.1	1.0
Isoleucine	3.0	4.2	3.4	3.1
Leucine	10.1	20.3	8.6	6.5
Phenylalanine	3.2	7.1	3.8	4.1

Source: Adapted from Mertz 1963.

acids in maize are threonine and isoleucine, respectively. For the gravid gilt, maize protein is first limiting in lysine and second limiting in tryptophan. Maize diets containing supplemental lysine and tryptophan have been shown to be adequate for swine during gestation.

When soybean meal is used to supplement maize-based diets, it is generally acknowledged that lysine is first limiting. Crystalline lysine is available at economically competitive cost, and can be used to substitute for a portion of the supplemental protein and thereby reduce crude protein concentration of the diet. The requirement for lysine, expressed as a percentage of the diet, is directly related to the dietary protein concentration. The lysine requirement, expressed as a percentage of the diet, decreases by 0.02 percentage units with each 1% reduction in dietary protein level.

Quality Protein Mutant Maize

In 1963 it was determined that the opaque-2 mutant contained much more lysine and tryptophan than normal maize kernels (Mertz 1963). A portion of the change in amino acid composition is attributed to a reduction in the ratio of the prolamines (made up of zein and zein-like proteins) to glutelins. Whereas normal lines of maize contain 50–60% prolamines and 17–18% glutelins, maize containing the opaque-2 gene contain approximately 42% glutelins and only 16% prolamines (zein). Since glutelins are 18 times richer in both lysine and tryptophan than the prolamines, the nutritional advantage is apparent. With the proportion of glutelins increasing at the expense of prolamines, the overall lysine content of the endosperm protein rises from 2.6% in common maize to about 4% in opaque-2 maize. Additionally, tryptophan also increases substantially.

The initial opaque-2 maize work with pigs was conducted by Purdue University workers (Beeson et al. 1966; Pickett 1966). It was shown that growing pigs between 13.8 and 25.7 kg grew 3.6 times faster on opaque-2 maize than on normal maize. Those fed opaque-2 grew at a rate equal to that of pigs fed normal maize and soybean meal balanced to supply the same quantity of crude protein (11.6%) as that found in opaque-2 maize. Finishing pigs with an average initial body weight of 59 kg gained weight 50% faster on opaque-2 maize than on normal maize and about the same rate as pigs fed a 13% protein diet. In subsequent studies at Purdue, these studies were confirmed. It was further shown that nonessential nitrogen, in the form of diammoniun citrate, glutamin, and glycine, added to the normal maize diet was without effect in improving pig performance. The work demonstrated that the higher levels of both lysine and tryptophan present in opaque-2 maize were responsible for the improved protein quality.

The baby pig, because of its requirement for higher levels and quality of protein, provides the most sensitive and critical test for assessing the value of the opaque-2 maize protein. Thirty-five-day-old pigs with an average body weight of 8.9 kg were fed diets in which all the protein was supplied by either common maize or opaque-2 maize (Maner et al. 1969).

The baby pigs fed common maize as the sole source of protein performed very poorly, gained an average of only 21 g/day during the entire 130-day experimental period and required 35.19 kg of feed to produce a kilogram of grain. The very poor balance of essential amino acids in the common maize was associated with a low level of feed consumed (0.74/day). As the experiment progressed, the pigs fed only common maize demonstrated the classical symptoms of protein deficiency and some died a few days after the experiment was terminated at 130 days.

In contrast, the growth rate of pigs fed the opaque-2 hybrid diet was increased to 254 g/day, which was a 12-fold increase when compared to those fed common maize. These pigs consumed 53% more feed but required only 4.43 kg/kg of body weight gain. The performance of these pigs fed on opaque-2 was, however, inferior to that of pigs fed an equal level of protein supplied by the mixture of maize and soybean meal. This inferior performance indicates that, although the level and balance of essential amino acids of the opaque-2 maize are much improved when compared to those of common maize, the proportion and level of these essential amino acids are not adequate to support a level of growth produced by the maize–soybean meal control diet containing an equal level of protein.

Opaque-2 maize has been fed as the only source of protein and energy to gestating sows without altering normal reproduction. When these sows were compared to control sows that had been fed only common maize or maize-soybean meal diets containing from 11.2 to 20% crude protein, it was shown that opaque-2 maize fed during gestation had no effect on number or weight of pigs farrowed. However, number of pigs weaned per litter, litter gain, and total litter weaning weight were superior from gilts fed the opaque-2 maize diets as compared to those fed the normal maize diet (Baker et al. 1970).

Maner and Colombian co-workers (1971) and Mahan et al. (1971) demonstrated that sow lactation rations based on opaque-2 maize supplemented with vitamins and minerals was adequate to support equal baby pig survival and weaning weight but was not adequate to maintain sow weight. Although equal pig numbers and weights were recorded with opaque-2 maize rations and balanced maize-soybean meal diets containing 10 to 18% total protein, lactating sows fed only opaque-2 maize lost significantly more weight than sows fed adequate diets. It appears that a higher level of protein than is supplied by opaque-2 maize is needed for optimum lactation performance, even though its amino acid composition is superior to that of common maize.

Although opaque-2 maize was shown to be nutritionally superior to normal maize, its soft endosperm and low grain density proved to be a serious limitation. Farmers throughout the world failed to accept the new opaque-2 maize varieties because of lower yields and increased susceptibility to insect damage.

When these technical and economic obstacles hindered the spread of opaque-2, scientists at the Centro International de Mejoramiento Maize y Trigo (CIMMYT) utilized gene modifiers to create opaque-2 germplasm with a hard endosperm while maintaining the high-quality protein. In 10 to 12 years breeders at CIMMYT, under the direction of Dr. S. Vasal, have developed quality-protein maize breeding pools and varieties with similar levels of lysine and tryptophan and nutritional quality and with grain characteristics similar to those of normal flint maize. These agronomic improvements and stabilized nutritional value have stimulated renewed interest in these hard-endosperm, high-lysine maizes in both developed and developing countries.

Other Maize Mutants

Other maize mutants, such as floury-2 and waxy maizes, have been studied and their nutritional quality compared to normal and opaque-2 maize. Both are reviewed in *Swine Production and Nutrition* by Pond and Maner (1984).

Grain Sorghum (Milo)

Sorghum (*Sorghum bicolor* L. Moench) is a hardy plant able to grow and produce under a wide range of environmental conditions, particularly in warmer temperatures and tropical regions of the world. In the United States, sorghums are cultivated throughout the Great Plains states from Texas to South Dakota and in the hot, irrigated valleys of southern Arizona and California. Outside of the United States, southern and eastern Asia and sub-Sahara Africa are the principal regions of cultivation, although production in South America has increased significantly during the past 10 years.

Great volumes of research have been published concerning the value of sorghum grain as a feed for swine. The composition and nutritive value of sorghum grain has been summarized by Hulse et al. (1980). It seems clear that maximum gains and feed efficiency from grain sorghums are obtained when the grain is incorporated into a complete mixed feed. Complete mixed rations based on grain sorghum produce faster and more efficient gains than when ground grain sorghum is fed free-choice with a protein supplement. When good quality protein supplement is fed to either growing or finishing pigs, sorghum grain has a feeding value not different from that of maize. However, when poor quality protein such as peanut meal is fed, the value of sorghum grain is only 69.2% that of maize. Replacing as much as 8% of the peanut meal with fish meal improved gains but not to a level equal to those obtained with maize. Sorghum has a feeding value equal to or superior to that of barley.

The great variability in sorghum varieties and hybrids and the lack of accurate physical and chemical description of the materials used make it very difficult to accurately analyze the data available. In more recent research, scientists have attempted to characterize better the materials utilized. Studies have focused on (1) endosperm texture, (2) starch type, (3) endosperm color (type), (4) content of tannins and seed coat color, and (5) amino acid content, availability, and adequacy. Each of these characteristics have been extensively reviewed in *Swine Production and Nutrition* by Pond and Maner (1984) and the reader is directed there for such information.

Grinding and rolling of sorghum to disrupt the intact kernel has consistently improved its nutrient digestibility and feeding value for growing-finishing swine. Processing sorghum also makes it possible to feed complete mixed diets, which results in better pig performance than does the feeding of grain and protein supplement separately. For swine, the grain should be ground to be fine to moderately fine since coarsely ground grain is not as well digested. Successive reduction in sorghum particle size by grinding improves feed efficiency but not daily gains.

Amino Acids

A review of the more recent research published on sorghum grain provides information that could be used to indicate that sorghum grain is inferior, equal, or superior to maize as a feed source for swine. When maize is replaced by sorghum grain in well-balanced diets supplemented with soybean meal, growth rate is generally not affected or only slightly reduced; however, the quantity of feed required per unit of weight gain is consistently increased by 10–15%. A more marked depression in both growth rate and efficiency of feed conversion is generally noted when protein supplements such as cottonseed meal and peanut meal are included in the diet. Lysine is the most limiting and threonine is the second limiting amino acid in sorghum grain for growth. Other studies indicate that tryptophan appears to be third limiting in sorghum and that methionine and isoleucine may be fourth limiting for growing swine.

Grain sorghum is an excellent grain for pigs of all ages if properly fed and properly supplemented to provide adequate quantities of the essential amino acids. Tannin content can reduce the digestibility and utilization of some nutrients; although protein content and quality may not be related to tannin content, tannin in sorghums may alter amino acid availability, as has been demonstrated in the case of methionine. To be efficiently utilized, diets based on sorghum should be supplemented with protein supplements and/or amino acids to provide adequate quantities of all essential amino acids, especially lysine, threonine, tryptophan, isoleucine, and methionine which are limiting in sorghum grain.

Barley

Barley ranks fourth in total production among the cereal grains in the world with an annual production of 168.4 million metric tons (FAO 1988) which represents approximately 9.7 percent of total world cereal production of 1,743 million metric tons. The greatest portion of this production is concentrated in Eastern Europe and the Soviet Union with lesser amounts in Asia and the northern part of North America.

Barley is a commonly utilized grain source in swine rations in countries in northern latitudes where maize and grain sorghum are not grown. A summary of 32 early experiments concluded that barley had a feeding value of 91% that of maize (Morrison 1956). The inferior feeding value of barley for growing-finishing and finishing swine has been shown by a number of researchers. However some studies have clearly demonstrated that barley is equal to maize, sorghum, and wheat for young starter pigs and for growing pigs. This apparent inconsistency appears to be mitigated by a high fiber and low energy content on the one hand, and a high level of crude protein and essential amino acids on the other. Recent research by Newman and associates at Montana State has shown that the feeding value of barley for swine is greatly influenced by the presence of beta glucans; high levels reduce performance of pig fed barley (Graham et al. 1989).

A comparison of the crude protein and essential amino acid content of barley and maize indicates that barley contains more total protein and higher levels of lysine, tryptophan, and the sulfur-containing amino acids, methionine and cystine, and suggests that barley may be superior to maize in its capacity to meet the requirements of the young growing pig for these nutrients. During the early growth period, pig performance is more heavily dependent on level of dietary protein and level and balance of essential amino acids than total energy concentration. It appears, therefore, that caloric concentration of barley based rations is adequate to meet the needs of the young growing pig and that the improved level and balance of limiting amino acids, especially lysine and tryptophan, are able to support pig performance similar to that obtained with complete mixed rations based on maize, wheat, or sorghum. When barley is fed with only vitamins and minerals, substantial improvement in protein utilization and pig performance can be achieved by the simultaneous addition of 0.38% L-lysine and 0.18% L-threonine (Fuller and Livingston 1979; Fuller and Mennie 1979).

Various methods of processing of barley have been attempted to improve its feeding value. The removal of the hull through a process known as "pearling" greatly improves the nutritive value of barley (Dinusson et al. 1956). This process effectively reduces the fiber content and increases the digestible energy. The genetic removal of hulls from barley and the production of hull-less varieties has been shown to effectively improve the feeding value of barley. Most hull-less varieties are superior to isogenic hulled varieties for growing-finishing and finishing swine (Joseph 1924; Gill et al. 1966; Newman et al. 1968; Newman and Eslick 1970; Newman et al. 1980). Hull-less barley contains more digestible energy than its hulled counterpart, 3295 vs. 3133 kcal/kg (Mitchell et al. 1976) and 3398 vs 2962 kcal/kg (Bhatty et al. 1979).

Pelleting of most barley diets improves weight gains and feed intake and apparently efficiency of feed utilization. Increasing the available energy content of barley diets with added fat or with whole, full-fat roasted soybeans has been shown to improve pig performance.

Oats

Oats (*Avena sativa L.*) make up only 2.2% of total world production of cereal grains, with most of this production of 38.8 million metric tons concentrated in the Soviet Union, Europe and North America. Whole oats contain approximately one-third hull and are very high in crude fiber (10.7%) and low in digestible energy (2670 kcal/kg) and are not widely recommended as a good source of carbohydrates for young growing pigs. The gains of growing pigs as well as the efficiency of feed conversion are decreased as the quantity of ground oats is increased in the diet to as much as 30–40% (Jensen et al. 1959a; Meade et al. 1966; Owen and Ridgman 1967; Katz et al. 1973; Wahlstrom et al. 1977). Depression in both growth and feed efficiency is due to the hull portion of the oats (Jensen et al. 1959b) and is mediated by the dilution of utilizable energy of the ration and by lowered feed intake. Similar depression in performance also occurs when oat hulls are included in maize-soybean diets.

Oats vary markedly in test-weight (weight for volume). Lighter-weight oats have a higher proportion of hulls, lower content of digestible and metabolizable energy and lower feeding value.

Oat groats (oats with hull removed), a high grade but expensive feed, tend to be used in prestarter and starter diets, since then the high digestibility and low fiber content are of most benefit. Wahlstrom et al. (1977) fed dehulled, high-protein oats supplemented only with lysine to young pigs and obtained the same growth and feed conversion as from pigs on 17% crude protein, maize-soybean meal starter diets. Research results (Christison and Bell 1980) with a new cultivar ("Terra") of naked oats (*Avena nuda*) indicate that this cultivar is similar to wheat and maize as an energy source for young pigs.

New oat varieties grown today in the United States and Canada are higher in both protein and lysine than normal oats. Robbins et al. (1971) determined the protein and amino acid content of 289 samples of oat groats and suggested that increased protein in oats is not accompanied by decreased lysine concentration in the protein as in many other grains. Selections of new varieties of high-protein oats such as Dal, Hinoats, and Spear, and others now available have shown generally good, but variable results.

Rye

Rye is a cereal not commonly used as a grain for swine feeding in the United States although it makes up a significant proportion of grain for swine in Poland, the Soviet Union and some European countries where it is commonly grown on the lighter soils. Numerous studies conducted over the past three decades have demonstrated the inferior growth promoting properties of rye in swine diets (Bowland 1966; Holzschul and Trautman 1962; Friend and MacIntyre 1969; Hale et al. 1967, Esnoala et al. 1976). High levels of rye as a substitute for maize or barley depresses both gains and feeding efficiency. Esnoala et al. (1976) substituted rye at incremented levels from 0 to 100% for barley and reported a significant linear depression in both feed intake and average daily gains. Palatability appeared to affect feed consumption which at the highest level of substitution was depressed by about 12%. In studies where lower levels of substitution have not depressed growth, feed required per unit of gain has been increased.

Digestibility studies comparing rye and barley (Friend and MacIntyre 1969) clearly demonstrate that lowered digestibility of rye is not the limiting factor, since the apparent digestibility of both the dry matter (88 vs. 84%) and the crude protein (84.5 vs. 82.1%) is higher for rye than for barley and nitrogen retention is similar.

The presence of ergot (*Clauiceps purpurea*) in rye grain has also been studied as a causative agent for the poor performance of pigs fed diets containing rye. The presence of ergot in rye has long been recognized as a problem in its use as a livestock feed. Ergot in rye grain fed to pregnant sows has caused reproductive problems and abortions. Proportions of 0.10 and 0.05% ergot allowed only moderate feed intake and weight gains which were inferior to those of pigs fed ergot-free rations. In addition to reducing weight gains and feed efficiency, 0.10% ergot also significantly reduced nitrogen retention. The ergot used in these studies contained 0.292% alkaloids with a predominance (71%) of ergocristine.

Wieringa (1967) indicated that rye contains a fat-soluble, growth-inhibiting substance, reported to be a mixture of 5-n-alkyl resorcinal with odd-numbered side chains of 15–23 carbon atoms and a small amount of 5-n-alkenyl resorcinal.

Wheat

Wheat and wheat products are utilized extensively for human consumption in almost every major area of the world, so that wheat is generally too expensive to be utilized for swine feeding. Wheat is included in swine diets only when surpluses or grain of a quality not suited for milling are available. Wheat has excellent palatability and is well consumed by swine of all ages.

The crude protein of wheat is of high digestibility and generally high quality. Lysine has been shown to be the first most limiting amino acid in wheat protein; threonine, second; and valine or methionine, third (Ivan and Farrell 1975).

Wheat compares favorably with other grains in energy content. Expressed on a dry matter basis, wheat contains an average of 4430 kcal/kg gross energy, 3820 kcal/kg DE, 3680 kcal/kg ME, 3660 kcal/kg nitrogen-corrected metabolizable energy, and 2450 kcal/kg net energy (Wu and Ewan 1979). When properly supplemented with protein and/or amino acids to supply adequate quantities and proportions of amino acids, wheat can be used as the sole cereal in growing and finishing diets for both meat and bacon pigs without reducing daily gains, feed efficiency, or carcass quality (Cole et al. 1969; Beames 1963; Jensen et al. 1969; Ivan et al. 1975a,b; Allee 1977; Hale et al. 1979; Erickson et al. 1980).

Although there is some disagreement as to the best physical form and particle size for wheat included in swine diets, palatability studies (Hale et al. 1979) showed that pigs consumed seven times more crushed wheat than ground wheat when both were offered simultaneously. When processed, rolled, or crushed wheat has been shown to be equal or superior to ground wheat (Braude et al. 1961; Lawrence 1967; Ivan et al. 1974; Hale et al. 1979). Pelleting of diets containing ground wheat improves both growth rate and feed efficiency when compared with similar diets fed in meal form (Lennon et al. 1972; Erickson et al. 1980).

Triticale

Triticale is a polyploid hybrid cereal derived from an interspecific cross between durum wheat (*Triticum durum*) and rye (*Seclae cereale*). Although naturally occurring wheat-rye hybrids were observed early in the nineteenth century (Leighty 1915), present-day triticale hexaploid hybrids are artificially produced. Many of these hybrids were derived from triticale developed at the University of Manitoba in Canada and improved and extended at CIMMYT in Mexico (Larter 1974).

Interest in triticale as a feed has been generated because of its higher protein and improved amino acid composition as compared to other cereal grains. Although triticale has shown promise as a nutritionally competitive feed grain with maize, wheat, sorghum, and barley in some studies with growing and finishing swine (Bowland 1968; Shimada et al. 1971; Noland et al. 1972; Wilson 1972; Allee and Hines 1972; Martynov and Shtan'ko 1979), several feeding studies with growing and finishing pigs indicated that triticale was inferior to maize, wheat, sorghum, and barley when more than 20% of the grain source was triticale (Longernecker et al. 1969; Stothers 1970; Harrold et al. 1971; Allee and Hines 1972; Stothers and Shebeski 1965; Erickson et al. 1979).

Some inconsistencies in nutritive value reported for triticale are related to the fact that triticale is not a discrete entity. Considerable variation exists among the numerous crosses and varieties that have been developed. Screening tests for protein and lysine conducted at CIMMYT on 2400 lines of triticale showed a mean protein content of 13.4% with a range of 12.9–19.1% and mean lysine content was 3.4% with a range of 2.6–3.9% (Villegas 1973). Erickson et al. (1978) also observed wide variation in amino acid content of four varieties produced during the same season at Michigan State University.

As early as 1965, Stothers and Shebeski (1965) indicated that palatability was a problem with some triticale samples used in swine starter diets. Studies have shown that neither ergot (Bowland 1968) nor trypsin inhibition (Erickson et al. 1979) were responsible for the reduced palatability.

Triticale can be used as a substitute for a portion or all of the grain in swine diets. However, the level of substitution must be based on the chemical composition (protein, lysine, energy, ergot, and trypsin inhibitor) of the specific grain source used, and each source must be properly supplemented to provide nutritionally adequate diets.

Rice

Rice (*Oryza sativa L.*) is the most important food crop in Asia and plays a very important role in the nutrition of millions of people in the Latin American countries. There are some 483 million metric tons produced annually (FAO 1988) and a wide variety of by-products arises from the milling of paddy rice to produce polished rice for human consumption. In particular, the bran and polishings make valuable animal feedstuffs as do the broken grains or brewers' rice.

Large-scale milling consists of many processes. After the hull has been removed from the *paddy rice,* the germ and outer layers are removed and form the product known as *rice bran,* of which there may be more than one grade. After this process, the grains are nearly white but still rather rough, and they are polished until they become *polished rice*. The fine particles resulting from the polishing of the grain are called *rice polishings*. The very small pieces of broken seeds that are removed are separated and sold as *brewers' rice* or *broken rice*. The seed coat layers and embryo (germ) contain more than one-half of the mineral matter of the grain, one-quarter of the protein, practically all of the vitamins, and about three-quarters of the fat (Kik 1956). The oil is largely contained in the germ and about 85% of the entire oil content of the hulled rice goes into the by-product.

Rough Rice or Paddy Rice

Rough rice is sometimes available at a price that makes it economical to use as a feed for growing and finishing swine. Rough rice contains about 10.0% moisture, 8.2% crude protein, 9.2% crude fiber, 1.9% ether extract, 6.5% ash, and 64.2% nitrogen-free extract. The hulls, which contain about 40% crude fiber and 11–19% silica, make up about 25% of the weight of rough rice. The hull portion of the rough rice greatly reduces the energy value of the product.

Research in Colombia showed that when rough rice was substituted for 100% of the maize in a maize-soybean meal diet for growing-finishing pigs (17.9–90 kg), gains were not different but feed required per kilogram of grain was increased by 18%. Economic studies based on these trials (P. Andersen and J. H. Maner, unpublished) indicate that the price of rough rice must not be greater than 75% of that of maize for equal cost of utilization.

Brown and Polished Rice

Brown rice contains more protein, fiber, ether extract, and ash, but less nitrogen-free extract than polished rice. However, in diets supplied with adequate vitamins, minerals, and protein, the feeding value of each rice is similar and not different from that of maize (Bray 1943; Noland 1960). Average daily gains, feed intake, and efficiency of feed conversion are equal to those of maize for growing-finishing pigs. Therefore, for growing-finishing pigs both white rice and brown rice have a feeding value equal to that of maize and can be used as a substitute when prices of both are equal to that of maize. Since the rice grains are very hard, they should be finely ground for most efficient utlization.

Buckwheat

Buckwheat (*Fagopyrum esculentum*) is not a true cereal (Gramineae). It belongs to the family Polygonaceae. Buckwheat is grown primarily for human consumption but it has potential in animal feeding. The average protein content of buckwheat is about 11% and is about 74% digestible. Whereas cereals are low in lysine content, buckwheat has a high content of this amino acid. The average lysine content of buckwheat on an as-fed basis is 0.61% compared with only 0.30–0.45% for either wheat or barley and 0.18% for maize.

Anderson and Bowland (1979) reported the gross energy in buckwheat to be 68% digestible and to have an average digestible energy value of 3077 kcal/kg. The low digestible energy content of buckwheat is due to its woody hull which constitutes 18–20% of the seed. The crude fiber content of buckwheat is about 10.5% and is similar to that of oats.

Feeding high levels of buckwheat to pigs housed outdoors has led to the development of fagopyrism (buckwheat poisoning). Fagopyrism occurs only in light- or white-skinned pigs when exposed to direct sunlight. This condition is a photosensitization of the skin causing itching and dermatitis. Therefore, the inclusion of buckwheat in rations for pigs housed outdoors should be limited. When pigs are housed in total confinement, fagopyrism is not a problem and the inclusion of buckwheat does not have to be limited.

Grain Amaranth

An ancient crop used as human food for centuries in some parts of the world has received recent attention as a potentially valuable feed for pigs. Several cultivars of Amaranthus species are currently under study to identify superior agronomic and nutritional sources for propagation. Feeding studies with laboratory animals indicate that some cultivars of grain amaranth have excellent protein quality and that grain from these cultivars can be used as the sole source of energy and protein for normal growth of nonruminant animals. The future of amaranth as a swine feed appears bright. *Amaranthus cruentus* and *Amaranthus hypochondriacus* cultivars have received the greatest attention as feed and food resources, and selected populations of these species show particular nutritional promise.

Milling By-Products

The manufacture of flour from cereal grains and maize for human use results in several by-products available for feeding to livestock. The seed is covered with high-fiber layers which are also higher in protein and minerals than the inner substance which is mainly starch. The bran fraction represents the outermost layer of the seed and just under this layer is the aleurone layer. The germ, which is at the base of the seed, is high in fat, fat-soluble vitamins, and minerals. Under the aleurone layer, and constituting most of the mass of the seed, is the endosperm composed of thin-walled cells packed with starch intermixed with gluten. In milling, the objective is to remove as much starch and gluten as possible while maximizing the exclusion of the germ, aleurone, and bran fractions. Owing to variability in the degree to which this separation is made, flour by-products that contain a wide range of fiber (bran and aleurone fractions) are produced.

Wheat By-Products

The currently accepted method of labeling wheat flour by-products is based on fiber content. Thus, the wheat flour by-product with more than 9.5% crude fiber is wheat bran; that with 9.5% maximum fiber is middlings or mill run; that with 7% maximum fiber is red dog, white shorts, or white middlings; and that with only 1.5% maximum fiber is feed flour. Although these are the categories established by the Millers' National Federation, extreme variations exist among and between products and in the nomenclature used in the literature. These variations are reflected in results obtained by different researchers and from different samples of the same product.

Wheat bran is highly palatable and has a laxative effect. Its amino acid balance is superior to that of whole wheat and improves with increasing rates of milling (Nehring et al. 1970). It is extremely high in phosphorus, but low in calcium, so that a nutritional imbalance occurs if wheat bran is a major component of the diet. It is high in niacin, but the niacin present is almost completely unavailable. Most of the nutrients and a major part of the potential metabolizable energy of wheat bran are contained in the cells of the aleurone layer. Contrary to what might be expected, the pig is able to completely break down the aleurone layer including the cell walls, although a portion of the carbohydrate and about 10% of the protein are not absorbed and appear in the feces (Saunders et al. 1969, 1974).

Feeding studies with growing and finishing pigs have demonstrated that increasing levels of wheat bran above 40% to 45% decreases both rate of gain and efficiency of feed conversion in both growing and finishing pigs (Forsythe 1979).

Wheat middlings contain 3470 kcal/kg DM of digestible energy and 3340 kcal/kg DM of metabolizable energy (Pals and Ewan 1978), have a high quality protein (Beames and Natoli 1969), and have a higher feeding value than wheat bran. Wheat middlings can replace a portion or all of the wheat and soybean meal or maize and soybean meal in diets for growing-finishing pigs with only small reductions in pig performance.

Wheat shorts have been used in growing pig diets with variable results. Patience (1977a,b) observed that wheat shorts could replace all of the maize and soybean meal with only small reductions of gain, although feed efficiency and dressing percentage were decreased. Later studies (Young 1980) showed that with one sample of wheat shorts, containing 18.5% protein, 10.4% acid detergent fiber, and 0.82% lysine, pig gains and feed efficiency were not depressed until the dietary level of shorts exceeded 64.4%. With a second sample, containing 18.6% protein and 16.5% acid detergent fiber, there was a quadratic decline in rate of gain of pigs of comparable size as the level of shorts increased. Young and King (1981) have shown that wheat shorts can supply all of the protein and energy in diets for gestating sows.

Although wheat by-products (offals) are widely used throughout the world and can supply a considerable proportion of the protein and energy in swine diets, the variation in nutritive values reported is of concern. The considerable variation in digestible energy in wheat by-products probably reflects differences in type of wheat used and different degrees of extraction of bran, middlings, and shorts during the milling process (Morgan et al. 1975; Henry 1976; Just et al. 1978; Pals and Ewan 1978; Batterham et al. 1980). Most of the variation is due to fiber content, which has been shown to reduce digestible energy content of feeds.

Maize Bran and Other By-Products

The maize kernel is covered by the hull which contains about 15% crude fiber and has a tip cap (where the kernel attaches to the cob) of similar composition. Just under the hull is the gluten layer which contains about 22% protein. Most of the mass of the kernel (75%) consists of the endosperm, which is composed mostly of starch and has less than 10% protein. The germ or embryo is located at the base of the kernel just above the tip cap and contains about 20% protein and 35% fat. In milling maize for producing maize meal, the germ and hull are removed. *Hominy feed* (hominy meal, hominy grits) is the by-product and consists of the maize bran (hulls), maize germ, and part of the endosperm. It resembles maize meal but is higher in fiber and protein and, unless part of the fat has been removed from the germ, is higher in fat than whole maize or maize meal.

Starch and glucose are obtained from maize by a wet-milling process whereby the maize is first softened in dilute acid followed by milling to free the germs, which are separated by their lower specific gravity (higher fat), resulting in their floating to the top of the liquid. The residue is finely ground and hulls (bran) are separated by filtration. The starch and gluten that remains are separated from each other by centrifugation and filtration. The evaporation of the liquid from which the germ, bran, starch, and gluten have been removed results in a residue called *maize solubles*. This residue includes minerals and some water soluble carbohydrates, protein and vitamins.

Maize-gluten feed is a mixture of gluten meal and bran and may contain maize solubles and part of the germ.

Maize bran feed is too high in fiber and too low in protein quality to be used as a major component of swine feed. *Maize gluten meal* is high in protein (40% or more), but its protein is severely deficient in lysine and tryptophan so it cannot be used as a significant portion of the protein of the diet. Owing to its low mineral and vitamin content and its inferior protein quality, gluten meal must be used sparingly and must be considered chiefly as an energy source for swine. Since gluten feed is merely a mixture of gluten meal and bran whose proportions can be regulated at will by the processor, it is of variable feeding value and its usefulness as an energy source is governed largely by the proportion of bran (fiber percentage).

Hominy feed is an excellent energy source for swine, although its crude fiber content is higher than that of whole maize. It is also higher in fat than whole maize so that it is at least equal to maize in digestible energy.

Rice Bran and Polishings

Rice bran is available in many areas of the world and especially in the tropics and subtropics. The quantity of this by-product available represents about 8% of the paddy rice produced in the world. With a reported world production of 483 million tons of paddy rice (FAO 1988), it is estimated that 38.6 million tons of rice bran were produced.

Pure rice bran is very palatable and well consumed by swine of all ages when used fresh. However, because of its high oil content, it turns rancid on storage and is less palatable to the pig. Rancid rice meal has an unpleasant odor and taste that is offensive to livestock.

Rice bran is of highly variable quality, depending mainly on the quantity of hulls included with the bran. The presence of hulls can be revealed by microscopic analysis if not finely ground; but if they are finely ground and added to bran and polishings, it is more difficult to detect the adulteration. The deleterious effect of hulls is due to the high fiber and silica and the low digestibility of the hulls, which limit dietary energy concentration.

In general, *rice polishings,* which come from the inner part of the grain, do not vary as widely as bran in chemical composition. In most, the protein content is slightly higher and the fiber content much lower than that of bran. Since many rice mills do not separate bran and polishings, and the characteristics of each are so poorly defined that they are difficult to distinguish, what is termed bran is usually a mixture of bran, polishings, and hulls.

Satisfactory gains and efficiency of feed utilization can be obtained when moderate levels of rice bran (30–45%) are used in growing-finishing pig diets. Reduced pig performance can be expected when higher levels are incorporated into the diet. For young growing pigs (5–15 kg), not more than 15–20% rice bran should be included in the diet. Economic consideration may modify these recommendations. Under some feed price conditions, slower and less efficient gains from diets containing high levels of rice bran may be more economical than faster and more efficient grains from high cost diets based on conventional feedstuffs such as maize and soybean meal.

DISTILLERY AND BREWERY BY-PRODUCTS

Distillery By-Products

Production of distilled liquors and alcohol from cereal grain or maize includes grinding, cooking, and the addition of enzymes to hydrolyze the starch to simple sugars and yeast to cause fermentation. After fermentation, the alcohol is removed and the residue is available for livestock feeding. This residue can be fractionated into several components or used as the complete mixture, sometimes called *stillage* or *distillery slop*. The coarser particles are usually removed from the stillage and are called *wet distillers' grains*. After dehydration, this product is sold as *dried distillers' grains*. The water-soluble materials and fine particles remaining are usually dehydrated to form *dried distillers' solubles*. The addition of condensed or dried distillers' solubles to wet distillers' grain produces distillers' grains with solubles which, after they have been dried, are sold as *distillers' dried grains with solubles*. The Association of American Feed Control Officials (AAFCO) has been effective in standardizing the nomenclature, so that each of the above products is further identified by a prefix naming the cereal grain or maize used in its manufacture.

The favorable levels of protein, amino acids, and minerals in distillers' dried grains with solubles would suggest various potential uses in swine rations. Numerous workers have demonstrated the value of distillers' by-products as a protein substitute and as a source of known B vitamins and trace elements. It has also been suggested that these products might be a source of unidentified growth factors but numerous studies have failed to find any evidence of unidentified growth factors in these products.

Work with young starter pigs between 8.0 and 17.9 kg (Wahlstrom and Libal 1980) has demonstrated significant differences in average daily gain among pigs consuming diets containing 0, 10, 20, or 30% distillers' dried grains with solubles. All diets were formulated to contain 19.6% crude protein and 1.05% total lysine. Daily gains decreased in a linear fashion as the level of distillers' dried grains with solubles increased.

Livingston and Livingston (1969) found that when distillers' dried grains with solubles containing 22.8% crude protein and 0.62% lysine were included in the diets of 20–90-kg liveweight pigs at the 0 and 14.7% levels, no difference was found in the growth rate or feed conversion efficiency, but both these measures were significantly poorer when distillers' dried grains with solubles were used at the 25% level. Similar feeding trials (Wahlstrom et al. 1970) indicated that levels of 5 and 10% distillers' dried grains with solubles can be fed in maize-soybean meal diets without adversely affecting rate of gain or feed efficiency of growing-finishing pigs. Increasing the level to 20% reduces digestibility of protein, nitrogen-free extract, and dry matter as well as the gains and the efficiency of feed conversion.

The crude fiber content of distillers' dried grains with solubles is generally higher than that of the original grain. Ewan (1976) reported a metabolizable energy value of 2940 kcal/kg for distillers' dried grains with solubles which is somewhat less than that of maize (3320) and soybean meal (3130). Thus, because of its relatively high fiber content (8.5–9.0%) and the pattern of amino acids, the dietary level of distillers' dried grains with solubles for growing-finishing swine should be limited to 10–15% or a maximum of 20% when supplemental lysine is added. Under most conditions, grains rather than their by-products will be a more economical source of energy.

Distillers' dried grains with solubles have been shown to be a useful feed for gestating swine. Thong et al. (1978) demonstrated that a dietary level of 44.7% distillers' dried grains with solubles could substitute for all of the soybean meal and a portion of the maize without affecting gilt gain or reproductive performance.

Brewery By-Products

The brewery by-product, *dried brewers' grains,* is a bulky, low-energy feed containing 96.7% dry matter, 28.5% crude protein, 6.1% ether extract, 19.1% crude fiber, 3.6% ash, and 39.4% nitrogen-free extract. The protein has been shown to be 72.4% digestible and the digestible energy, metabolizable energy, and metabolizable energy corrected for nitrogen to be 2650, 2500, and 2380 kcal/kg DM, respectively, with only 52.3% of the gross energy being digestible (Kornegay 1973).

Up to 15% brewers' dried grains can be included in maize-soybean meal diets for growing-finishing swine without depressing growth; however, feed required per unit of gain increased progressively with increasing levels of brewers' dried grains. Brewers' dried grains can be used to supply up to 40% of the diet for gestating sows without effecting reproductive performance. It is necessary, however, to increase daily sow intake of diets containing high levels of brewers' dried grains to compensate for lower metabolizable energy content in order to provide adequate energy for gestation weight gains.

ROOTS, TUBERS, FATS, AND OTHER ENERGY SOURCES

One of the constraints on the utilization of roots and tubers for human and animal consumption is the difficulty of processing and storage. Current knowledge on this subject has been reviewed (Plucknett 1979) and should be of special interest to those associated with developing pork production technology in locations where root crops and tubers are an appropriate potential energy source.

Cassava

Swine require large quantities of carbohydrate (energy) and moderate quantities of protein for growth and development. In the leading pork-producing areas, large supplies of grains available at a reasonable price provide this energy source. In other countries, production of cereal grains is insufficient for human consumption needs. Many of these countries, however, have the potential or are already producing large quantities of other feed sources that, if properly fed and supplemented, could support a large and efficient swine industry. One feed source with a great unrealized potential in many tropical areas is *Manihot esculenta* Crantz, more commonly known as cassava, yuca, manioc, tapioca, or mandioca.

Cassava is grown in the lowland tropics throughout the world, generally between latitudes 23° N and 23° S and almost always below 1800 meters altitude within these regions. These areas generally delineate regions of production because of the frost-sensitivity nature of the plant. Cassava has a wide range of adaptability, resistance to drought, tolerance of poor soils, and relative ease of cultivation. As there are varieties adapted to nearly every soil type and degree of annual rainfall, the plant offers an almost unlimited production potential.

The annual production of cassava for 1988 was reported to be 138 million metric tons (FAO 1988). Of the world total approximately 43.4% or 59 million tons are produced in Africa, 33.6% or 45.8 million metric tons in Asia, and 23% or 31.2 million metric tons in Latin America. Although cassava is ninth in terms of world production of all crops and fifth in world production of tropical crops, its importance and potential as a food and feed crop in the tropical regions of the world has not been realized.

The cassava root contains an average of 65% moisture and 35% dry matter. An average of 30.84% of the fresh root is nitrogen-free extract, which is composed principally of starch (80%) and sugar (20%). The starch of cassava contains approximately 20% amylose and 80% amylopectin.

Cassava contains only small quantities of fiber (1.45%) and only a trace of ether extract (0.20%). The whole root contains only 1.5% ash. The levels of calcium (0.12%), phosphorus (0.16%), sodium (0.06%), and magnesium (0.37%) are generally low. However, as is common in other root crops, the level of potassium is relatively high (0.86%).

Most cassava varieties available contain only small quantities of crude protein (nitrogen × 6.25). In fact, the content of the majority of the varieties analyzed does not exceed 3% when expressed on an air-dry basis. Chemical analysis indicate that approximately 40–60% of the total nitrogen is in the form of nonprotein nitrogen. The level of various amino acids such as lysine and tryptophan are promising and constitute more of the protein than those of opaque-2 maize. However, the level of both methionine and cystine are extremely deficient.

The presence of a toxic factor in cassava presents some problems in its utilization as a livestock feed. The toxicity of the roots and the leaves of cassava is due to the presence of hydrocyanic acid (HCN) or prussic acid (Oke 1969). Free HCN does not exist as such in the healthy, intact plant but is formed when normal plant growth is retarded or when the leaves and roots are cut or bruised. The HCN is generally formed by the chemical reaction of two substances, a glucoside and an enzyme that is contained in the plant. Neither the glucoside nor the enzyme is poisonous when considered separately. The two glucosides found in the cassava plant are linamarin and lotaustralin. These glucosides consist of a chemical combination of a sugar with hydrocyanic acid and possibly with another compound, such as a ketone or an aldehyde. Hydrocyanic acid, glucose, and acetone are produced when linamarin is hydrolyzed, and only when the HCN is chemically liberated is it toxic.

Various processing methods can be used to reduce the HCN toxicity of cassava. Some of these methods are more effective than others. Drying in a forced-air oven at moderate temperatures (70–80°C), boiling in water, and sun drying are some of the methods. The method of chipping and sun drying is one that is practical and also effective in reducing both the HCN and glucosides.

Many studies to determine the value of cassava roots as an energy source have been reported and extensive coverage and references are found in *Swine Production and Nutrition* (Pond and Maner 1984). Cassava can be fed either fresh and chopped or as a dried meal.

Growing-finishing pigs fed chopped, raw sweet cassava (low HCN) along with a 42% protein supplement free-choice, gain as rapidly and as efficiently as pigs fed a 16% protein maize-soybean meal control diet. However, there is a tendency for pigs to consume less cassava and more protein supplement than would be considered necessary to provide a balanced 16% protein ration when allowed free-choice selection. In practice, free-choice feeding of chopped cassava and protein supplement is not recommended, good growth and feed efficiency and prevention of overconsumption of supplement is accomplished by mixing the chopped cassava and protein before offering to pigs. It is recommended that the ration of fresh, chopped cassava be mixed daily with a protein supplement containing 35–40% protein in proportions to supply the required proteins, vitamins, and minerals in the final mix.

Extensive studies with dried cassava have been conducted in Colombia by members of the Colombian Agricultural Institute (ICA) and at the International Center for Tropical Agriculture (CIAT). Maner and co-workers (1967) compared graded levels of sun-dried cassava meal up to 68.25%, with and without 10% cane molasses, in diets for growing-finishing pigs and demonstrated that the growth rate was reduced as the level of cassava meal increased in the diet. The addition of 10% molasses improved the performance of all groups, but was not effective in overcoming the small reduction in growth rate caused by the substitution of cassava meal for maize in the diets. It was later shown that the depression observed at the higher levels of cassava substitution, could be completely overcome by the addition of 0.1%–0.2% DL-methionine (Portela and Maner, unpublished data).

J. Mesa and J. H. Maner (unpublished data) have also studied the digestibility of the various nutrient components of cassava meal and found the digestibility of the energy and carbohydrate fractions to be satisfactory and not greatly different from results reported for maize. The digestibility of the dry matter, nitrogen-free extract, and TDN were 93.8%, 98.5%, and 92.5%, respectively.

Cassava meal and fresh cassava have been used to supply energy for gestating and lactating sows (J. H. Maner and J. Buitrago, unpublished data). Both produce satisfactory results for gestating sows but consumption of fresh cassava is low in lactating sows even when the cassava is fed with a quantity of protein supplement calculated to supply a 16% ration.

When cassava meal was fed continually during the entire life cycle, pig performance during the growing-finishing period and reproductive performance during gestation and lactation were satisfactory, but slightly inferior to those of a similar group of gilts fed diets based on maize.

Potatoes

Potatoes (*Solanum tuberosum*) are widely used for livestock feed, and particularly in Europe, they are grown for that specific purpose. Surplus potatoes or small or cut potatoes unfit for human consumption are available in many areas for swine feeding. In addition, increasing quantities of potato waste are becoming available from potato processing industries.

The composition of the potato varies widely according to variety and environmental conditions during growth and storage, and according to processing treatment. The water content is particularly variable, ranging from 63 to 87%, which has a significant effect on nutrient concentration. Expressed on a dry-matter basis, whole potatoes with peel contain 4–6% ash, 10–11% crude protein, 4% crude fiber, 0.2% ether extract, and 82–83% nitrogen-free extract. Approximately 70% of the potato dry matter is starch, which is composed of about 75% amylopectin and 25% amylose.

A concentrated source of starch and energy, whole potato with peel also can make a significant contribution to the supply of both nitrogen and essential amino acids. The nitrogen content of potatoes varies according to variety and location and much of the nitrogen is soluble nonprotein nitrogen composed of approximately equal parts of amide–nitrogen and amino–nitrogen. Although extremely variable, approximately 50% of the nitrogen is in the form of true protein. Potato protein has been shown to have a biological value of 71–75 for the 20-kg pig (Höller 1959), which places it among the highest of the plant proteins. The lysine content is only slightly less than that of soybean meal and almost twice that of maize and barley.

Raw potatoes are unpalatable and support only very slow growth of pigs. McLagan and Nasr (1950) observed that raw potatoes increased the dietary requirements for supplementary protein. With 50% of the dietary dry matter fed as raw potatoes, it was impossible to sustain pigs on a level of intake above that required for maintenance. There is general agreement that raw (uncooked) potato causes a reduction in pig performance and that potatoes are more efficiently utilized by the pig if cooked before feeding.

Although raw potato starch is resistant to α-amylase attack, the digestibility of potato dry matter, organic matter, nitrogen-free extract, and gross energy is consistently high (85–99%) for both raw and cooked potato (Whittemore 1977). Of significance is that most of the cooked potato starch is digested and absorbed anterior to the terminal ileum while only about 30% of raw potato starch is digested and absorbed before reaching this area of the gastrointestinal tract. A large proportion of the raw starch reaches the large intestine, where it accumulates and is degraded by microorganisms, and the end products absorbed in the lower gut. The end products of bacterial digestion may be utilized with an efficiency different from that of the end products of enzymatic digestion in the small intestine. The lower digestible energy values reported for raw potato (3.20 kcal/g DM) compared with those of cooked potato (3.83–4.02 kcal/g DM) support this suggestion.

While the influence of cooking on the apparent digestibility of energy of potatoes appears to be minimal, heat processing has been shown to affect nitrogen digestibility. The nitrogen of raw potato is digested with varying degrees of efficiency by pigs but results are generally poor relative to those obtained with cooked potato. The digestibility of nitrogen is dependent on the degree of cooking.

An alkaloid complex, particularly the glycoside solanine, is a constituent of potatoes and some reports have indicated that the presence of this substance in raw potatoes is responsible for the deleterious effect on consumption and digestibility. However, the raw potato also contains a powerful protease inhibitor that is protein in nature, soluble in water, and heat labile. Whittemore et al. (1975a,b) showed the protease inhibitor activity to be very high in uncooked potatoes but absent in cooked potatoes, and concluded that the presence of this powerful protease inhibitor activity in raw potato adequately explains both the reduction and variability in nitrogen digestibility for uncooked as compared to cooked potatoes.

In practice, potatoes should be heat processed to convert the starch fraction into a form that can be digested by enzymatic action and to denature the protease inhibitor (Livingstone et al. 1980). Cooking can be accomplished by boiling the potatoes in water or by steaming. Various forms of heat drying also contribute to this process. Livingstone et al. (1979) demonstrated that either over- or under-processing with steam can reduce utilization of both nitrogen and energy and suggested that steaming for 20 min at 100°C is adequate to convert starch and to denature the protease inhibitor. Prolonged heating, 1 hr at 100°C, caused rupture of the cell walls with release of the gelatinized starch and resulted in potato protein of low digestibility. However, steaming for 20 min at 70°C did not cause complete gelatinization of the starch and did not effectively denature the protease inhibitor.

The best gains and utilization of both concentrates and potatoes will be obtained if cooked potatoes are fed to appetite along with daily limited quantities of a 19–20% protein concentrate. For economic calculations, 4.7 kg of cooked potatoes are approximately equal to 1 kg of a 16% concentrate. Cooked potatoes should always be allowed to cool to at least a moderate temperature before feeding.

Dried Potato Products

A number of products from the potato are used in livestock feeds. Terminology used to describe these products is sometimes confusing. Potato cossettes, potato meal, potato slices, potato flakes, and potato pulp are all names given to the various products. Waste potatoes that are sliced and then dried give cossettes; when cossettes are ground, potato meal is produced. When potatoes are cooked before drying the resultant product is potato slices; and when they are mashed after cooking and then dried on a roller, potato flakes are produced. Potato pulp is a by-product of the potato starch industry and is the residue from cold water extraction of starch from the pulped, raw potato.

Potato Meal

Dried, raw potato meal included in pig diets in more than moderate amounts may give rise to digestive disturbances and diarrhea. When potato meal makes up more than 30% of the pig's diet, the diet is unpalatable to the young growing pig and scouring normally occurs. Sixty percent potato meal in the diet reduces daily feed intake, suppresses gains, and causes digestive disturbance when fed to pigs between 30 and 105 kg (Seidler et al. 1967; Glaps 1970). Levels of 10–15% potato meal are recommended for the growing pig (20–50 kg) and quantities not to exceed 30% are recommended for the finishing pig (50–100 kg).

Potato Flakes

Potato flakes are prepared by steaming clean, washed potatoes for 30 min in a tank in which the pressure rises 10–15 lb/in.^{2} After they are steam cooked they are mashed and then passed over drying rollers for drying, and finally removed as thin flakes. Flakes have a high digestibility of all components and have a high nutritive value. They may be included in liberal quantities in both growing-finishing diets and diets for older pigs without causing digestive disturbances or diarrhea. The most satisfactory pig performance is obtained with potato flakes if they make up not more than 30–40% of the diet for growing-finishing pigs (Ostrowski and Rys 1969). When potato flakes were used to replace all of the barley in a 16% protein diet for growing-finishing pigs between 30 and 90 kg liveweight, reduced gains, increased feed per kilogram of gain, and decreased carcass quality were recorded even in the presence of adequate protein and supplemental lysine and methionine. With young pigs, Hillyer and Whittemore (1975) showed that 50–58% of cooked potato flakes could replace cereals in diets for suckling or early weaned pigs without affecting intake or performance.

Potato Slices

Potato slices are prepared by passing the slices through a hot-air rotating drier. During the process the potato slices rise to a temperature of about 150°C when they are introduced into the drier and fall to about 80°C at the outlet. The greatest part of the 2-hr processing is accomplished at 80°C. This treatment allows for both cooking and drying. Inadequate cooking of the potato slices could reduce their nutritive value.

Potato slices have been used to replace barley and maize in diets for growing-finishing pigs. A 30% level of substitution during the growing period and a 60% replacement of corn and barley during the finishing period had no effect on gains, feed efficiency, or carcass quality in the studies of Woodman and Evans (1943). Other studies (Zausch and Hölzschul 1960–1961) have shown that potato slices with a high digestibility but with 480 mg of solanine per kilogram could not be used at levels of more than 20% of the diet without causing digestive disturbances, diarrhea, and a reduction in both gains and feed efficiency.

Potato Pulp

Potato pulp is a by-product of the starch industry and is the residue obtained after starch is extracted from the potato with cold water. Only low levels of potato pulp can be used as a grain substitute in growing and finishing swine diets (Friend et al. 1963; Bielinski and Bielinski 1954). When levels of 0, 15, 30, and 45% pulp were used to replace barley in growing diets and oats in finishing diets, linear depressions in both gains and efficiency of feed conversion were recorded. A quantity of as much as 15% dried potato pulp allowed satisfactory gains. Other studies indicated that potato pulp was not as good as potato flakes or steamed potatoes when used to replace either of these ingredients in pig diets.

Sweet Potatoes

Sweet potatoes (*Ipomaea batata L.*) are a high-moisture feed in their fresh state and, except in areas in which cull or surplus potatoes are available at economical prices, are not commonly used as a swine feed by the major swine producers of the world. However, sweet potatoes are traditionally grown for both human and animal food in many tropical and semitropical environments. In Taiwan, it is estimated that 60% of all sweet potatoes produced are fed to swine. In these areas, dry-matter yield per hectare is competitive with many of the other feed sources. The sweet potato in its fresh state contains 68–78% moisture and 22–32% dry matter.

The fresh roots contain 1.35% crude protein (at 32% dry matter) or 4.21% of the dry matter. Other samples have shown higher or lower levels of protein. Although lower in protein than maize, as a dry meal or chip sweet potatoes are a concentrated source of carbohydrates and energy.

Sweet potatoes have been utilized in many forms. They have been fed raw, cooked, and as a silage; they have also been dehydrated to form a meal or chips. When fed wet, either raw, cooked, or as a silage, they are bulky and have a feeding value 25–33% that of maize. Early experience with sweet potatoes seemed to indicate that the sweet potato should be cooked for maximum utilization. However, it has also been stated that no beneficial effect is obtained from cooking (Calder 1960). Cooked sweet potatoes fed as 70% of the dry matter to 25-kg pigs have been shown to have only 80% of the value of maize when the two have been compared as dry matter (Springhall 1969).

The dry matter and energy digestibility of cooked sweet potatoes is slightly higher than that of raw sweet potatoes, but the protein digestibility of the cooked potato is almost double that of the raw potato. Feeding trials with growing-finishing pigs indicate that pigs fed the same level of supplement grow faster and more efficiently when cooked sweet potatoes are substituted for raw potatoes. This increase is probably due to the 10–17% increase in sweet potato consumption observed when cooked potatoes were fed as well as to a small improvement in energy digestibility.

It is common practice in Asia to include all or a portion of the sweet potato vine in pig rations. The vines are fed fresh, as a silage, or as a dehydrated meal. Least-cost formulation in Korea (Lee and Chung 1966) indicated that silage containing 60% roots, 30% vines, and 10% barley bran to control moisture should be fed as only 38% of the dry matter of rations from 50–70 kg of pig weight and up to 68.5% of dry matter from 70 kg to market weight. The silage was not recommended for lighter-weight pigs.

Tseng et al. (1980) fed sweet potato vines along with maize, fresh sweet potatoes, or sweet potato chips in rations for growing-finishing pigs and observed slower growth and reduced feed efficiency when compared to the performance of pigs fed maize–soybean meal diets. Including vines with maize as the energy source had little effect on gain but significantly reduced feed efficiency. In Taiwan studies, Lee and Lee (1979) showed that replacing 20% of the sweet potato chips with dehydrated sweet potato vines caused a 16% reduction in growth rate and an 11% increase in feed requirements. Although commonly used, sweet potato vines have a very low value in diets for growing-finishing pigs and may be more efficiently utilized for breeding animals.

Bananas and Plantains

The banana belongs to the genus *Musa*, comprising 32 or more distinct species and at least 100 subspecies. Of the bananas grown in Central and South America the

Musa sapientum ("Gros Michel") and *M. cavendishii* ("Cavendish") are the two most common, along with the plantain (*M. paradisica*) which is used for cooking. Because of its resistance to Panama (black sigatoka) disease, the "Cavendish" variety has replaced the "Gros Michel" in commercial plantations. New varieties of bananas and plantains resistant to Panama and other diseases are being sought through collection of wild material and through breeding programs. There are approximately 41.9 million metric tons of bananas and over 24 million metric tons of plantains produced annually in the world (FAO 1988).

Although bananas and plantains are grown largely for export and domestic consumption by the human population, a large quantity of this fruit is available for livestock feed. At the packing plants in the areas of commercial banana production, the bananas that are too small or too large, are slightly bruised, have spots or off color, or are not in an optimal stage of maturity for shipping are rejected for export. The "reject" bananas along with smaller quantities of farm-produced bananas and plantains constitute a good source of carbohydrates for pigs during all phases of the production cycle. The fresh whole banana with the peeling contains approximately 80% water, 20% dry matter, 1.0% protein, 1.0% fiber, 0.2% fat, 1.0% ash, and 16.8% nitrogen-free extract. The "crude fiber" of whole bananas consists of 60% lignin, 25% cellulose, and 15% hemicellulose. The ripe pulp itself contains 0.50% lignin, 0.21% cellulose, and 0.12% hemicellulose. Although the most common commercial variety of bananas, "Cavendish," conforms to this generalization of chemical composition, locally grown bananas and plantains vary widely in their chemical composition.

The banana can be utilized fresh or as dried meal. Ecuadorian studies (Clavijo et al. 1970; Oliva et al. 1971) have clearly demonstrated that the pig will daily consume large quantities of bananas if they are allowed to ripen before they are fed. Complementary studies (Clavijo and Maner 1971; Viteri and Maner 1971) have shown that the amount of consumption of ripe and green bananas is associated basically with differences in palatability of the two forms of presentation. Viteri and co-workers (1970) demonstrated that when growing pigs were fed equal daily quantities of either ripe or green bananas along with a controlled quantity of protein supplement, growth rate and efficiency of feed conversion were almost identical in both groups.

The most conspicuous change in the maturation of the banana is the conversion of starch to sugars. During ripening, there is a rapid decline in starch content and a corresponding increase in sugar content. Most of the sugar present is in the form of sucrose, glucose, and fructose, although maltose has been identified to be present in trace amounts. The predominant sugar in bananas is sucrose, some of which is hydrolyzed to glucose and fructose. Glucose accounts for approximately 58% of the total reducing sugars present and fructose for about 42%. Because of the needs for respiration, there is a small but definite decrease in total carbohydrates in all varieties during the ripening process.

The starch and total sugar values for the peel are very small in magnitude when compared to those of the pulp, but follow a pattern of change similar to that of the pulp during the maturation process. Total sugars of the peel continue to increase as the fruit ripens and there is a corresponding decrease in total starch content.

The ripe bananas, which are very palatable and readily consumed by the pig, are fed whole with the peeling. The low protein content and high moisture present in the banana require that a supplemental source of both protein and energy be supplied. Several studies have been conducted to determine the voluntary consumption pattern and performance of growing-finishing pigs fed ripe bananas and protein supplements containing varying levels of crude protein. One such study (Calles et al. 1970), showed that the average daily gain of growing-finishing pigs fed ripe bananas to appetite was significantly improved (770 vs. 660 g) if a 30% protein supplement was fed instead of a 40% supplement. The improvement in gain was assumed to be the effect of the additional energy consumed daily.

Under practical feeding conditions, the recommended level of protein in the supplement will depend upon price relationships between protein and grain or energy sources. Locally available energy sources such as maize, milo, rice bran, and molasses can be efficiently used as an energy source for supplement dilution.

Fresh bananas can be used efficiently for the gestating sow at levels up to 4 or 5 kg supplemented with 600 g of a 40% protein concentrate. The lactating sow requires 5–6 kg of complete, mixed feed per day to meet the requirement for maintenance and milk secretion. If a similar quantity of dry matter were supplied from bananas and supplement, the sow would be required to consume more than 20 kg of bananas daily. This quantity appears to be greater than the sow's physical capacity. Because of the difficulty of adequate consumption and also because of the severe diarrhea that develops when the sow consumes a diet containing 14–15 kg of ripe bananas, ripe bananas are not generally recommended as the major energy source for the lactating sow.

Banana meal prepared from green bananas with the peelings contains approximately 12% moisture, 4% ash, 4% protein, 3% fat, 3% fiber, and 74% nitrogen-free extract. Celleri and co-workers (1971B) used dried green banana meal as a substitute for grain to supply 0, 25, 50, or 75% of the total diet for growing-finishing pigs. The protein of the 16% protein diets was supplied by a combination of fish meal and cottonseed meal. Each increase in level of dried banana meal in the diet caused a small but significant linear reduction in average daily gain, and an increase in the quantity of feed required per kilogram of gain. Similar studies (Oliva et al. 1971) confirm these findings. These workers employed levels of 0, 12, 24, 36, and 48% banana meal to replace maize in the basal ration. When the level of meal was increased, a linear depression in growth rate was reported. This growth depression was associated with increased daily feed intake and reduced feed efficiency. It would appear from these studies that the amount of metabolizable energy is lower in banana meal than in maize.

Dried, green banana meal can supply approximately 50% of the diet of lactating sows. Studies by Ecuadorian workers (Viteri et al. 1971; Celleri et al. 1971A) indicate that the performance of sows and their offspring was not different whether they were fed a control 16% protein diet based on maize, wheat, and fish meal or whether green banana meal was used to substitute for approximately 50% of the maize.

Cane Sugar

Crude and refined sugars, although not commonly used, are an excellent source of energy for swine feeding. Because both products are devoid of protein and contain only very small quantities of vitamins and minerals, their efficient use is determined by the adequacy of the diet. Unlike cane molasses, sugar (sucrose), either refined or crude, is easily incorporated into the diet and does not cause digestive disturbances except in baby pigs whose digestive enzyme system is immature.

Refined sugar was used to replace maize in diets for growing-finishing pigs (Maner et al. 1969). Levels of 0, 15, 30, 45, and 60% were used in diets containing 16% crude protein supplied basically by soybean meal. In this study, each increase in level of sugar up to 45% of the diet was accompanied by an increase in daily gains and a decrease in feed required per kilogram of gain. The performance of the pigs fed 60% sugar was not different from that of those fed 45% sugar diets. Increasing the sugar level in the diet causes an increase in digestible energy and a basic shift in the total amino acid pattern. In the control diet (maize–soya), the protein is supplied by both maize and soybean meal (approximately 50% from maize and 50% from soya). In the high-sugar diets, most of the protein comes from soybean meal.

Other studies with crude sugar (Buitrago et al. 1969, 1970) clearly indicate that this product can completely replace the maize in growing-finishing swine diets. Although the gains and feed conversions were not as outstanding as observed with refined sugars, proportions as large as 60% were well tolerated by the pigs and supported results not different from those observed with the maize–soybean meal control. There was a tendency for improved feed efficiency when sugar composed 15–60% of the diet, in comparison with diets based on only maize.

Only limited information on beet sugar is available; however, it is assumed that beet sugar would have properties and feeding value equal or similar to cane sugar.

Cane Molasses

Molasses, as a major by-product of sugar production, is a satisfacory and economical source of energy for livestock feeding in many tropical and subtropical areas of the world as well as in the United States. An average of 25–50 kg of final molasses are produced for every 100 kg of refined sugar. Both of these quantities result from the processing of one ton of sugar cane. In 1988 over 988 million metric tons of sugar cane and 102 million tons of cane sugar were produced (FAO 1988). Much of the 2.5 to 5.1 million metric tons of molasses calculated to have been produced, was utilized for alcohol production and only limited quantities are used for livestock feeding.

Molasses, whose water content varies between 15 and 25%, contains a large quantity of carbohydrates. The nitrogen-free extract fraction may reach 65% and is composed mainly of water- and alcohol-soluble carbohydrates such as sucrose (32.2%), glucose (8.6%), and fructose (15.5%). The sugars present in the molasses have a high digestibility when fed to pigs of all ages except during the first 21–30 days after birth when the secretion of the enzyme, sucrase is not well developed in the baby pig. Thus, molasses is basically an energy source because of its carbohydrate composition, but because of its high level of both water and ash the gross energy level of the crude product is relatively low (2850 kcal/kg). The level of ash in molasses is high but variable. On the contrary, molasses is a poor source of nitrogen. It contains only about 0.50–0.56% total nitrogen of which only 38–50% is amino acid nitrogen.

Early research indicated that a maximum of 20% final molasses should be used to replace grains in swine diets if satisfactory performance and weight gains were to be obtained. Higher levels of molasses (30–40%) caused a laxative effect on the pig and, owing to the moisture present, tended to dilute the energy concentration of the diet, a fact which is reflected in increased feed per unit of gain. Levels lower than 30% have been used satisfactorily and the "loose-feces" resulting from this amount of dietary molasses has not presented a major problem of either animal health or sanitation. Some workers demonstrated that the tolerance of pigs to molasses increased with age and body weight. Maximum digestible energy is obtained when molasses makes up only 10% of the diet. Levels of more than 20–25% do, however, cause problems of mixing and especially in feeding management, since diets containing more than those amounts are sticky and are difficult to use in automatic feeders.

To overcome the growth and feed efficiency depressions caused by high levels of molasses that reduce dietary energy density, Hawaiian workers (Brooks and Iwanaga 1967; Brooks 1972) have supplemented high-molasses diets with various types of fats. Combinations of molasses, wheat bran, and either 10 or 20% fat have supported pig performance equal to that obtained with molasses-free diets.

Other alternatives that have been proposed include the addition of crude or refined sugar to the final molasses diets and the utilization of high-test molasses. Preston and Willis (1970) demonstrated that increasing the level of sugar in diets containing final molasses improved both gains and efficiency of feed conversion and showed that levels of final molasses as high as 60% of the total diet can be used satisfactorily for growing-finishing pigs if 20% crude or refined sugar is also included in the diet. The replacement of final molasses with high-test molasses improves both rate of gain and efficiency of feed conversion and reduces digestive disturbances.

Cuban investigators have used both high-test molasses and whole molasses to replace all of the maize in swine diets. High-test molasses is a heavy cane syrup from which none of the sugar has been extracted, but rather partially inverted to prevent crystallization. Whole molasses is produced by concentrating cane juice that has not been submitted to the clarification process with sodium hydroxide. High-test molasses differs from final molasses in having a much lower ash content and approximately two-thirds of the total sugars in the form of monosaccharides. The high-test molasses, which has been reported to have a metabolizable energy value of 3750 kcal/kg of dry matter has satisfactorily replaced all of the grain in growing-finishing swine diets.

Molasses has served as an effective laxative agent to prevent constipation of sows and gilts during late gestation and early lactation. Dietary levels of 15–25% molasses increase fecal moisture, increase rate of passage, and eliminate constipation problems often associated with the transfer of gestating animals into the farrowing facility.

Dried Bakery Product

Dried bakery product is produced from reclaimed bakery products that have been blended and processed to provide a product that contains 9.0–9.5% crude protein and 11.0–13.0% crude fat and a maximum of 1.0–1.5% crude fiber, 3.5% ash, and 4.5 kcal/g.

Dried bakery product is an excellent ingredient in starter rations for young pigs. The product is highly palatable, and the high level of fat makes it a good energy source for young pigs. Sewell (1965) replaced 10, 20, or 30% of the maize meal in early weaning starter diets, containing maize, soybean meal, corn sugar (glucose), dried skim milk, animal fat, vitamins, and minerals, and reported that levels of 10 and 20% bakery product slightly increased gains and feed efficiency, but the 30% level of substitution significantly increased gains by 18.6%. Dried bakery product also increased gains when 30% was used to replace corn sugar and maize; however, gains were not improved when an equal quantity replaced corn sugar, dried whey, and maize.

Citrus and Citrus Pulp

In any fresh fruit industry, some fruit falls from the tree or is otherwise unfit for human consumption or for processing. During certain seasons of the year in citrus-producing areas of the world, significant quantities of oranges, grapefruits, and tangerines are available for livestock feeding. Cunha et al. (1950) have shown that tangerines can be used as an energy source for growing-finishing pigs. Pigs normally consume only the pulp of the tangerine and refuse the peel. Some problems of diarrhea or loose feces have been observed when pigs are fed a major portion of their daily ration in the form of reject or fallen tangerines. When offered tangerines, a protein supplement and a mineral mix free-choice, the pigs consumed an average of 12.64 kg of tangerines per day along with 0.97 kg of a protein supplement composed of 50% peanut meal, 25% meat scraps, and 25% dehydrated alfalfa meal. Performance of these pigs was slightly inferior to that in groups of pigs fed either tangerines, maize, and a protein supplement or a diet composed of protein supplement and maize. However, when properly supplemented, the tangerines provided an economical source of dietary energy. Cooked citrus rinds have been shown to be nontoxic to hogs but to have little feeding value.

Baird et al. (1974) fed growing-finishing pigs diets containing 0, 10, 20, and 40% dried citrus pulp. Even though by the addition of prime tallow energy levels were held constant in the diets and monosodium phosphate was added as needed to lower the calcium : phosphorus ratio, daily gains were significantly depressed with increasing levels of citrus pulp. Daily feed consumption also was depressed.

Citrus pulp is not generally recommended for swine diets, but when used in growing-finishing diets, levels not greater than 10% should be used. Baird et al. (1974) stated that citrus pulp can safely form 20% of the swine diet when energy level is corrected in the diet by the addition of tallow.

Garbage

Garbage, also known as swill or kitchen refuse or waste, is a very heterogeneous product that results from food discarded from restaurants, hotels, supermarkets, institutions, military establishments, meat and fish markets, and homes. Because of the variation in origin and content, the chemical composition and nutrient content is also quite variable. Different garbage samples can vary from 8.7–25.6% in dry matter, 14.6–27.3% in crude protein, 14.7–54.5% in ether extract, 4.7–6.4% in gross energy, and 5.2–8.5% in ash. Even with this variable chemical composition, garbage is a significant source of feed for swine production in many areas of the world. It is used particularly in areas close to metropolitan centers It is estimated that in the United States approximately 1% of all pigs slaughtered are fed on garbage. Although still a significant number, this represents a reduction since 1967 when 3% of all pigs slaughtered were fed on garbage.

One of the major problems of feeding garbage to swine is related to sanitation and disease, since garbage can be a vehicle for spreading serious infectious diseases, such as hog cholera, vesicular exanthema, foot and mouth disease and African swine fever, as well as such diseases as erysipelas and brucellosis, tuberculosis, and salmonellosis and trichinosis that not only affect swine but may also infect humans.

To prevent the spread of disease through the feeding of raw garbage, the garbage should be properly processed to destroy completely all of the pathogenic agents that are present. Proper processing is usually accomplished by boiling in an open vat or steam cooker. The mass, in any case, should be thoroughly boiled (100°C) for a period of not less than 30 min in equipment designed to assure that all areas of the garbage reach a boiling temperature. The mass should be allowed to cool for a similar length of time without external cooling. This procedure assures adequate sterilization of the product. Proper handling after cooking also prevents subsequent contamination of the processed material.

If garbage is utilized as the sole source of feed, young pigs grow very slowly and require as much as 230–245 days to reach market weight (Kornegay et al. 1970). The most efficent utilization of cooked garbage can be obtained by supplementing the garbage ration with a high protein concentrate.

Morbidity and mortality is often high among garbage-fed pigs, but proper cooking and the addition of a complete supplement and, in some cases, an antibiotic to the diet appears to reduce both death losses and the level of sickness among the pigs. The greater number of death losses, often encountered in garbage-feeding establishments, is probably related more to poor husbandry, inadequate processing, and unsanitary conditions than to nutritional value of the garbage.

Fats and Oils

Animal fats, including tallow, lard, lamb fat and fish oils, as well as vegetable fats from such plants as coconut, soybean, cottonseed, maize, safflower, sesame and others, are by-products of other food and industrial uses, and are available as such for swine feeding. The distinction between fats and oils is based on physical characteristics. If the product is solid at room temperature, it is generally termed fat; if liquid at room temperature, it is termed oil. The physical properties of fats and oils are determined mainly by their fatty acid composition. Oils are high in polyunsaturated fatty acids, especially linoleic acid, while fats are high in saturated fatty acids and may be nearly devoid of linoleic acid, as is, for example, coconut fat.

The American Association of Feed Control Officials has established categories and definitions of fats and oils used in animal feeds. These include the following:

Animal fats. These are obtained from the tissues of mammals and/or poultry in the commercial processes of rendering or extracting. If the product bears a name descriptive of its kind or origin (tallow, lard, grease), that name must correspond exactly to the content.

Hydrolyzed fat or oil (feed grade). This product is obtained in the fat-processing procedures commonly used in soap making. It must contain not less than 85% total fatty acids, not more than 6% unsaponifiable substances, and not more than 1% insoluble materials. It must be described according to source (hydrolyzed animal fat or hydrolyzed vegetable fat or a mixture).

Fat product (feed grade). Any fat product not meeting definitions for animal or vegetable fat, hydrolyzed fat, or fat ester belongs to this category. Minimum percentage of total fatty acid and maximum percentages of unsaponifiable and insoluble fractions must be specified. The fat product must be adequately tested for safety in animal feeding before being marketed, and its components must be listed.

Vegetable fat or oil. These products are obtained by extracting the oil from seeds or fruits commonly processed for human food. They must contain not less than 90% total fatty acids, not more than 2% unsaponifiable matter, and not more than 1% insoluble matter. The label must specify the source (soybean oil, coconut fat, and so on).

Vegetable oil refinery lipid (feed grade). This is obtained in alkaline refining of vegetable oil for human use. It may contain not more than 17% ash on a dry basis. Since it is the product of alkaline treatment, it must be neutralized with acid before use in animal feeds. This product is often called "soapstock" and consists primarily of the salts of fatty acids, glycerides, and phosphatides. Its feeding value for swine has not been thoroughly studied, but the rate of weight gain may be depressed (Devilat et al. 1971) by replacing the basal diet with approximately 7% soapstock (15% of the calories).

The main nutrient supplied by fats and oils is energy, and their caloric content is approximately 2.25 times that of carbohydrate and protein. The energy concentration of pig diets is commonly increased by inclusion of fats and oils. The increased caloric density of the diet results in less total consumption by the animal. Therefore, if fats and oils are added to the diet the percentage of other nutrients in the diets should be increased to obtain an adequate ratio of calories to other nutrients. The reason is that the pig tends to eat to satisfy energy requirements, and a deficiency of protein or other nutrients can result if the fat content of the diet is increased without otherwise adjusting the diet. The importance of the ratio of calorie to protein in swine feeding was discussed in Chapter 8.

The fatty acid composition of fats and oils from plants and animals is quite variable. The pig tends to deposit fat with a fatty acid composition similar to that of the diet. The higher the fat content of the diet and the longer the duration of feeding of a particular diet, the greater is the effect on depot fat composition. The feeding of whole cooked soybeans, which contain approximately 18% fat with a high degree of unsaturation, produces softer carcasses than are produced with a maize–soybean meal diet. Conversely, feeding hydrogenated coconut fat at a level of only 3% of the diet from weaning to market weight produces a firm carcass as a result of the deposition of highly saturated fat. Addition of corn oil, safflower oil, soybean oil, and rapeseed oil to the diet of growing-finishing pigs resulted in the deposition of unsaturated fatty acids to the depot fat.

The addition of animal fats to swine diets tends to have little effect on depot fat composition. Fish oils, if fed in quantities greater than 5% of the diet, impart a "fishy" flavor to pork. A process of polymerization of menhaden fish oil has been shown to reduce the flavor and odor problem of pork from pigs fed 5% menhaden fish oil, but this oil caused reduced weight gain when it was polymerized.

Following the report of the original research of Seerley et al. (1974), various studies (reviewed by Pettigrew, 1981; Moser and Lewis 1980; and Seerley 1981) have suggested that feeding high fat rations to sows during late pregnancy and lactation can improve piglet survival. In 25 of the 31 comparisons (Moser & Lewis 1980), feeding fat or oil to sows in late pregnancy and/or lactation increased survival rate by an average of 2.6%, which resulted in an increase in litter size at weaning of 0.3 pigs. Individual experiments have in some cases shown greater increases in the survival rate, especially of smaller pigs, owing to the addition of fat to the sow diet during late pregnancy and lactation. Reports from the University of Georgia (Seerley 1981) have shown percentage survival at 21 days of age to be 80.9, 88.3, and 90.2% among pigs in the control, 10% corn oil, and 10% animal fat groups, respectively.

Feeding sows fat (Seerley et al. 1974; Libal et al. 1987) or 1,3-butanediol (Stahly et al. 1985) in late pregnancy has increased the survival of pigs to weaning and, thus, total litter weight at 21 days. However, feeding fat during lactation did not increase survival (Pettigrew 1981; Stahly et al. 1981). Stahly et al. 1981 also reported no improvement in pig weaning weight when sows were fed oleic or linoleic acid as 10% of their lactation diet. More recent data (Seerley 1989) indicates that neither sow weight loss, backfat changes, pig weight, nor survival rate at weaning were affected by the addition of 5% solid fat pellets from gestation day 80 through lactation, nor 10% added solid fat pellets from day 100 of gestation through day 14 of lactation. Survival rates of all pigs including those without added fat to 21 days averaged 90%. Pigs from fat-fed sows had more glycogen per gram of liver, 41% more total liver glycogen, and 16% more serum glucose at birth. These data support the conclusion that when a herd survival rate is 80% or lower, response in survival rate due to feeding sows fat in late gestation is more likely. Because an average of 88.5% of the control pigs in this study survived to weaning, survival of pigs from fat-fed sows would not likely be affected.

It has been postulated that increased survival rate resulting from feeding high fat diets to sows may be due to (1) increased glycogen stores at birth, (2) increased blood glucose at birth, (3) slower disappearance of blood glucose during starvation, (4) slower disappearance of liver glycogen, (5) increased fat content and milk production, and (6) change in the metabolic processes associated with fat utilization.

Almost all researchers who have analyzed milk have reported elevated fat content of milk and colostrum of sows fed added fat. Increase in milk fat content has ranged from 9 to 57% in these studies. Apparently sows metabolize lipids from adipose tissue for milk production, and consequently they lose weight and become thin. Feeding fat prior to farrowing and in early lactation reduces the need to metabolize body fat as the fat for milk synthesis is available in the blood.

There is evidence that adding fat to sow diets during late gestation and lactation can improve pig survival, especially among small pigs and in herds with lower survival rates. If fat is added, a minimum of 7.5% fat should be added to produce a positive response, and there appears to be little advantage of levels greater than 15%. High fat diets should be fed at least 7 days before farrowing, and Seerley (1981) presented data to indicate that a greater response was elicited when rations containing fat were fed during the last trimester of pregnancy. Pettigrew (1979) suggested that at least 1 kg of fat should be fed before the sow farrows. Feeding a diet with 10% added fat for 7 days before farrowing will achieve this goal. Adding fat during both late pregnancy and lactation not only affects prenatal metabolism but also increases the fat content of both colostrum and milk, which appears to be advantageous to the suckling pig.

Fat addition to sow diets is only advantageous if it increases net return for the producer. Strict analysis must be made in each swine operation to determine if it is economical to feed fat to sows.

PROTEIN SOURCES

Next to energy sources, protein constitutes the nutrient class needed by swine in the largest amount. The very young pig requires more than 25% protein in the diet. The protein requirement, as a percentage of the diet, steadily declines to maturity, but even the adult requires 10% or more of protein in the diet for normal life. Since protein sources are generally more expensive than energy souces, the wise choice of feedstuffs to minimize diet costs necessitates inclusion in the diet of the smallest amount of protein compatible with normal productive function (growth, reproduction, or lactation). As indicated earlier in this chapter, most energy sources contain some protein. A wide variety of protein supplements is available. Those most often used in swine diets are discussed here. Emphasis is placed on important special characteristics, deficiencies, and special supplementary qualities for use in complementing or correcting specific amino acid inadequacies of the energy sources in the swine diet.

Processing conditions, including heat and other treatments, often have important effects on the nutritional value of plant and animal protein sources. Aspects of food and feed processing, as related to bioavailability of the proteins and amino acids of a wide variety of feed materials, are important and should be addressed by those associated with swine feed formulation and plant and animal protein supplements.

ANIMAL PROTEINS

Meat Meal and Meat-and-Bone Meal

Meat meal or meat scrap is by definition the finely ground, dried rendered residue from animal tissues, exclusive of hair, hoof, hide, trimmings, blood, and intestinal tract contents. When the dried rendered product contains more than 4.4% phosphorus it is designated as meat-and-bone meal or meat-and-bone scrap.

Although the definition of meat meal specifies the composition of both meat meal and meat-and-bone meal, large variations exist in quality, uniformity, and putrefaction. Meat meals are derived mainly from bones and associated tissues such as tendons and ligaments, gastrointestinal tract, lungs, condemned livers, and some skeletal muscle. However, hoof, hair, hides, horn, wool, and blood are often included. Variations in the proportions of this diverse range of raw materials, with widely different chemical compositions and biological values, undoubtedly account for much of the variation in nutritive values between meals. Quality is greatly affected by the degree of dilution of the muscle protein with tendon and bone. Meat meals may contain 21–61% bone. Bone protein is 83% collagen which contains no tryptophan. Collagen also occurs in skin, connective tissue, cartilage, and tendon, so that 50–65% of the total protein in meat meals may be collagen. Quality is also affected by the amounts of fat, offal, blood, and stomach contents that are allowed into the meal. In addition, keratin protein may occur in meat meals. Keratin is a predominant protein in hoof, horn, and hair and is largely indigestible. Keratin protein contains 3.4–12.2% cystine, while muscle protein contains 1% and collagen protein contains none. Degree of putrefaction of meat meals also affects the palatability, taste, and smell of the diets prepared from these products.

Processing conditions during the manufacture of meat meals can influence their nutritive value. Processing method and temperature are as important as muscle tissue dilution by bone and tendon in determining the final quality of meals. Increasing the final processing temperature or extending the holding period during processing will reduce the essential amino acid content of the meat meal. High-temperature processing causes the destruction of cystine and possibly other sulphur-containing amino acids and reduces the availability of lysine through the binding of its ϵ-amino group.

Use of meat meal and meat-and-bone meal as a protein supplement for swine diets should take into consideration other dietary ingredients, local availability, and price. Least-cost diet formulation will include meat meal in the diet at very low levels in many commercially prepared rations. Many local conditions will justify the use of meat-and-bone meal as a sole protein source. Under these conditions, although growth rate and feed efficiency may be reduced, economic returns may justify their use.

Blood Meal

A considerable amount of blood is available as a by-product of livestock and poultry slaughter. Dried blood meal contains over 80% crude protein and 9% total lysine but is seriously deficient in isoleucine. Its high protein and particularly high lysine content would make it a valuable supplementary protein in pig diets if the protein had a high digestibility and the lysine and other amino acids had a degree of availability equivalent to that of most other high-protein feedstuffs. However, conventional drying methods commonly employed have used a high-temperature (up to 165°C for 10–24 hr) batch-drying process that has resulted in poor availability of certain amino acids, particularly lysine, and a product lacking in palatability for swine. Commercial blood meals produced by vat- or batch-drying have been found to be quite variable in nutritional value. Waibel et al. (1977) reported crude protein digestibility values for vat-dried blood meal to be as low as 55%.

Newer methods of drying blood, which utilize a lower drying temperature, or flash-drying procedures, including both the rotary steam drying process and the ring-drying process, have improved both the palatability and lysine bioavailability of blood meal (Doty 1973). Recent feeding experiments with meals produced from these new processes have shown that levels much higher than the previously recommended levels of 2–3% can be used for pigs without deleterious effects on feed intake or liveweight gain. Wahlstrom and Libal (1977), using maize-based diets, substituted 2 or 4% rotary steam-dried blood meal for an equivalent quantity of soybean meal or an equivalent quantity of soybean meal protein, and were unable to show any significant differences in pig performance. At the 4% level of substitution, lysine supplementation was without effect. In a subsequent trial (Wahlstrom and Libal 1977), levels up to 6% blood meal used to replace an equivalent quantity of soybean protein supported similar feed intake and pig performance; however, a level of 8% significantly reduced feed intake during both the growing and finishing phases and pig gain during the growing phase. King and Campbell (1978) utilized five dietary levels of ring-dried blood meal (0, 3.0, 5.9, 8.9, and 11.8%) to replace soybean meal and meat-and-bone meal in diets for pigs from 21- to 70-kg liveweight. Pigs receiving either 5.9 or 8.9% blood meal grew more efficiently than pigs given either the highest or lowest level of blood meal. A reduction in feed intake at the 11.8% blood meal level retarded both growth and efficiency of feed utilization.

Practical feeding recommendations for dietary levels of blood meal will depend on processing method used to produce the blood meal and the type of diet employed. Conventional vat-dried blood meal can be used to replace 2–4% of the protein supplement or up to 2% of the supplemental protein in grain-based diets. Ring-dried blood meal is an excellent source of lysine and other amino acids and can be used at levels up to 6–8% of the diet. However, at these and higher levels, care must be taken to assure an adequate level of isoleucine.

Fish Meal

One-third of the world catch of more than 50 million metric tons of fish are used for industrial purposes. The most important fish product used in pig feeding is fish meal.

Raw fish is used only in very special circumstances. Fish meals are among the most valuable feed ingredients when they are of high quality, and they are among the most costly and disappointing when they are of poor quality.

Fish meals, basically, are of two types: those made from waste from the human food fishery and those made from whole fish, normally caught specifically for reduction. The second category is by far the more common and is becoming increasingly so. Fish caught for reduction also fall into two general types, those with low oil content, primarily classified as *white fish,* and those with high oil content, which are usually members of the *herring family*. Again, the second type is by far the more common. Meals processed from members of the herring family include *menhaden meals* from the Gulf and Atlantic coasts of the United States;

herring meal from the northeastern United States, Canada, Iceland and Norway; *anchovetta meals* from Peru, Chile and Ecuador; and *pilchard meal* from South Africa.

Fish meal consists essentially of the dried, ground carcasses of fish; most of the oil present may be removed before drying since fish-body oil is also a valuable product. Freshly caught fish of various species may contain from 68 to 85% water, from 15 to 25% protein, and from almost 0 to 25% oil (on a fresh-wet basis). In order to make a fish meal that will be reasonably stable under normal storage conditions, the moisture content must be reduced to around 10%. If the meal contains 15% or more of water, it is possible for molds to grow. Stability requires that the fat content be reduced to about 10% and even a meal with this level of fat may become rancid in storage. With high fat levels, spontaneous heating and even combustion may occur. Fish meal leaving a typical heat dryer would normally be sterile, or nearly so, but it may easily become contaminated with bacteria during succeeding processes or storage.

Since the world's major industrial fisheries in which the fish is reduced to meal are based on high oil fish species, the reduction of these fish to meal requires cooking, usually by a wet reduction with live steam, pressing to remove most of the oil, and the drying and grinding of the pressed cake. Dryer construction and operation materially affect the nutritive quality of fish meal, since overheating and overdrying damage the digestibility of both the protein and the residual oil. Cooling and storage are also important in fish meal quality. Fresh, hot fish meal must be lowered to a temperature of below 49°C immediately after processing since even several hours of high temperature can measurably lower nutritive value.

Fish Protein Concentrate (FPC)

Milk substitutes are needed for use in feeding early-weaned pigs because of the high demand for milk for human use. Probably the most commonly used substitute is soybean protein, but the inferior utilization of soybean protein, in comparison with that of milk protein, by baby pigs suggests a need for alternative protein sources for use in baby pig feeding. Fish protein concentrates (FPC) have been produced by isopropyl alcohol extraction of raw whole fish or steam-dried herring meal or by enzymatic hydrolysis of fresh fish or fish waste. Each of these products contains 80–90% crude protein and favorable concentrations of amino acids.

Early studies utilized a FPC developed by the U.S. Marine Fisheries Service by isopropyl alcohol extraction of raw whole hake (*Urophycis chuss*). Baby pigs were fed liquid diets containing casein, isolated soybean protein, or FPC as the sole protein source. Performance was equal to that obtained with a similar diet containing milk protein (casein) as the sole source of protein and superior to that obtained with isolated soybean protein. Seve et al. (1975) studied the feeding value of a protein concentrate prepared by enzymatic hydrolysis of fresh fish. In baby pigs weaned at 12 days of age, performance with 33% substitution of dried skim milk was equal to that obtained with dried-skim milk alone, but higher levels of substitution reduced both gains and efficiency of feed utilization.

Fish protein concentrate is an excellent source of protein of high biological value for young pigs, but for optimal pig performance it should be limited to supply not more than 33–50% of the total dietary protein.

Fish Silage and Liquefied Fish

The material known as fish silage is a liquid product made from whole or parts of fish, to which no material has been added other than an acid and in which liquefaction of the ground fish mass is carried out by enzymes present in the fish. The principle involved in the manufacture of such silage is that enzymes already present in the fish are spread throughout the fish mass by grinding and mixing and the acidity adjusted so as to favor the rapid action of these enzymes and to inhibit bacterial action. Various acids or mixtures of acids can be used. Mineral acids such as hydrochloric and sulfuric are effective. Even though such acids are relatively inexpensive, they have the disadvantage that the acidity of the fish mass after being lowered to pH 2 in order to control bacterial growth, must be neutralized before it is fed to animals. Formic acid, although more expensive than mineral acid, is an appropriate preservative for fish silage as it reduces the fish mass to pH 4 which is suitable for enzyme activity, is sufficient to restrict bacterial growth, and has the advantage of not requiring neutralization prior to feeding to animals. Since there is no evidence of fermentation during this liquefaction process, the term "fish silage," often used to describe fish preserved in this manner, is a misnomer. True fish silage can be made by the addition of lactic acid bacteria such as *Lactobacillus acidophilus,* a fermenting agent, to the ground fish mass. Such studies have been described by Tibbets et al. (1981). To distinguish these products, fish and fish waste preserved through enzyme action after acid addition is termed *liquefied fish* and fish preserved by bacterial fermentation is called *fish silage*.

Since transporting trash fish and fish waste long distances for processing in fish meal factories is costly, the waste is often dumped into the sea, causing pollution problems and loss of a valuable protein feedstuff. The preparation of products from fish and fish waste by liquefication or ensiling is not widespread. As the production of liquefied fish and fish silage requires relatively low capital expenditures, uses simple technology, and poses no pollution problems, the process is appropriate for fish and fish waste produced from fish processors at ports not possessing a drying plant for fish meal production and particularly appropriate for high quality protein production in developing countries.

Liquefied fish and fish silage are valuable mainly for their content of high quality protein. In feeding trials with growing-finishing pigs, liquefied fish prepared from white fish and fish waste low in oil (cod, haddock, and plaice) have been shown to be acceptable as the sole supplemental protein source; however, liquefied fish high in oil content (up to 40%) such as that produced from herring (*Clupea harengus*) has been shown to be unacceptable as the primary protein supplement to diets for bacon pigs and only can be used to advantage at very low levels. High levels of liquefied herring (above 5.1%) produced pig carcasses with soft yellow fat and fish-tainted meat that is unacceptable for the human market. Fish silage with a low level of lipids can be included successfully at a level up to 6% in pig diets from weaning through lactation without affecting reproduction, growth, or meat taste and acceptability.

Shrimp Meal

Shrimp meal is the dried waste of the shrimp industry, consisting of the heads and shells (hulls) and in some instances whole shrimp. Due to differences in both the species and condition of the raw material as well as processing and effluent recovery techniques, tremendous variability exists between different meals. Although improper storage before processing can reduce the protein level, the proportion of heads and shells significantly influences the protein level and quality. The exoskeleton contains chitin, a nitrogen-containing carbohydrate (N-acetylated glucosamine polysaccharide), that is of little nutritional value. Accurate analysis of protein levels in shrimp meal require correction of the crude protein fraction for the nonprotein nitrogen contributed by the chitin present. Shrimp hulls contain 45–46% crude protein but because of the high content of chitin (54.2%), the corrected protein level is only 22.8%. On the other hand, shrimp heads contain only about 11.1% chitin and have a corrected protein content of 53.5% and have an excellent amino acid profile. In practice, shrimp meal seldom contains only hulls or heads but a variable mixture of both by-products. In usual Kjeldahl determinations, nitrogen values obtained include both protein and chitin nitrogen. It is, therefore, essential to estimate the level of chitin to be able to calculate a corrected protein value.

When the nitrogen present in chitin is used to correct the crude protein content, shrimp meal is a valuable supplemental protein source for swine. Most efficient use is made of shrimp meal when it is combined with other supplemental protein sources such as soybean meal.

Krill Meal

Krill (*Euphausia* sp.) is a phytoplankton feeder that inhabits the waters of the Antarctic. It is a crustacean that resembles a very small shrimp in appearance. Because of their abundance, krill could play an important role in future world protein supplies, both as a human food and as a supplemental protein feedstuff for swine and poultry. Russia, Japan, Germany, Poland, Taiwan, and Chile have been active in research to improve both harvesting and processing techniques. The extremely small size of krill (20–60 mm) necessitates a special netting technology. Because of the green phytoplankton which remains in the krill's digestive tract, their high fat content and the absence of connective tissue, krill exhibit extremely poor keeping qualities and processing must take place within 3 or 4 hr after krill are caught.

Raw krill contain 77–81% moisture, 11–16% crude protein, and 1–4% ether extract. Following processing, krill meal composition compares favorably with herring-type fish meal. It contains a high level of crude protein (52–59%) and is a favorable source of lysine, methionine, and tryptophan. Although a small portion of krill nitrogen is contained in the chitin, protein digestibility coefficients ranging from 71.5 to 77.1% have been reported.

Future use of krill meal as a protein supplement for swine diets will depend on quality, local availability, and price. Good quality krill meal can be used to supply a major portion of the supplemental protein for swine of all ages. Best results will be obtained if dietary levels are limited to 4–6% during the growing period and 2–4% during the finishing period.

King Crab Meal

King crab meal is composed of shells, viscera, and unextracted meat of the crab following processing. It contains 35–40% crude protein, 18% crude fiber, 20% acid detergent fiber, 1.5–3% ether extract, 37% ash, 9–12% calcium, and 1.6–1.8% phosphorus. Although some of the nitrogen is present in a nonprotein nitrogen form bound to chitin, the protein contributed from viscera and unextracted meat is of excellent quality. The crude fiber content is composed mainly of chitin. The molecular structure of chitin is very similar to that of cellulose, differing only in the substitution of an acetylamine group for the hydroxyl group on carbon 2 of the glucose unit. Chitin is not digested by swine and other nonruminant animals. Because of the low level of ether extract and high level of undigestible chitin, the utilizable energy level of crab meal is low. The high ash and specially high level of calcium create an imbalance in swine rations that must be corrected by the addition of phosphorus, such as monobasic sodium phosphate, in order to maintain the preferred 1 : 1–2 : 1 ratio.

Studies at the University of Alaska have evaluated king crab meal as a source of supplemental protein for swine. Husby (1980) has shown that whole king crab meal can replace 50% of the supplemental protein from soybean meal in barley-based diets without reducing rate of growth and with only a small reduction in efficiency of feed utilization.

Dried Milk

Cow milk is, of course, one of the most nearly perfect foods, not only for calves and humans but also for pigs. Normally, it is in too much demand as fluid milk for human consumption to be used as swine feed. Dried whole milk and dried skim milk are also expensive, but because of their high nutritive value and palatability they are often used in early weaning diets or in creep feeds to encourage early consumption of dry feed. The only major difference between dried whole milk and dried skim milk is that in skim milk most of the fat and fat-soluble vitamins have been removed so that all other components are increased proportionately. The only nutrients likely to be deficient in a diet containing a large proportion of dried whole milk would be iron and copper.

Dried whole milk and skim milk can be considered not only as good protein sources, but also as good sources of vitamins and minerals.

Milk Whey

Whey, a by-product from the manufacture of cheese and lactic-acid–precipitated casein from milk, is an abundantly available feed for pigs in certain areas of the world where cheese is produced. Many of these areas have poor or nonexistent fluid milk markets and must rely on cheese for a market for the product. Consumption of cheese products has increased significantly in the last decade, with a proportional increase in the amount of whey produced. This has created serious disposal problems for cheese plants. Many plants dump excess whey, resulting in a loss of a potential feed source and in some cases pollution problems.

The dry matter and moisture content of whey varies according to the processing method used to produce the cheese and the quantity of wash water that is allowed to drain into the whey storage tanks. However, generally, fresh whey contains approximately 93–96% moisture, 4–7% dry matter, 0.7–0.9% crude protein, 5.0% lactose, and 0.7–1.0% ash. To facilitate storage, transportation, and feeding fresh, liquid whey is sometimes condensed or dried. The low concentration of dry matter of whey limits its use as an energy source and the high lactose content reduces its use as a feed for older pigs; these have been regarded as the main reasons for the difficulties of high-whey feeding. The ability of pigs to utilize lactose has been studied in many feeding trials. Satisfactory growth has been reported with 40–57% lactose in baby pig diets. Ekström et al. (1975a) reported that the growing pig tolerated, without symptoms, feeds containing up to 30% lactose. In contrast, other workers have found that levels of more than 15–25% result in reduced food intake, slow growth, and a moderate diarrhea. The main reasons for the slow growth were the poor feed palatability and the diarrhea produced when the diet contained more than 30% lactose.

Hatchery Waste

In very localized areas adjacent to hatcheries, variable quantities of unheated incubator eggs are available. Some of these eggs are wasted.

It would appear either that incubation modifies the egg albumin in a way that makes it readily digestible by the pig or that the pig is unique in its ability to digest raw egg albumin. Whatever the reason, growing-finishing pigs can efficiently utilize raw incubator eggs and dead poultry to satisfy a major portion of their daily feed needs, thereby providing a way to salvage some value from unhatched eggs as well as a convenient method for disposal of the wastes. There is a danger of a biotin deficiency manifested by cracked hooves, dry rough skin, and loss of hair in pigs fed 30% or more of dried, raw egg whites. A protein, avidin, in raw egg white ties up biotin in the gastrointestinal tract and decreases its absorption. Heating destroys the avidin and prevents a biotin deficiency.

Feather Meal

Feather meal with a crude protein content (nitrogen × 6.25) of between 85 and 87% is one of the most protein-rich feedstuffs available for swine feeding. However, because of the keratinous nature of its protein, it is of little, if any, nutritional value because of the lack of digestibility in its natural state. Feather meal is, however, the principal by-product from poultry processing and has, in addition to a very high protein content, an extremely high content of cystine, threonine, and arginine. To be useful for swine feeding it must be hydrolyzed. Steam cooking at elevated pressures (3.2 atm) produces feather meal of the highest digestibility and biological value; acid hydrolysis (HCl at pH 6 for 20 hr) can be employed to produce meals of similar value. Feather meal which has been subjected to steam hydrolyzation has a rather high digestibility (84.4%) with a true availability of single amino acids from 80 to 90%.

In practice, very low levels (3–5%) of properly processed, hydrolyzed feather meal should be used in swine diets if maximum growth and efficiency of feed utilization are desired.

Hog Hair Meal

Significant quantities of hair are available from the annual slaughter of millions of pigs. Unprocessed keratinous proteins, such as hog hair and feathers, are of little or no nutritional value in their natural state because of poor digestibility. However, if properly processed through steam or acid hydrolysis, hydrolyzed hog hair is a potential source of protein and can be used in swine diets.

Hydrolyzed hog hair meal contains 93–95% crude protein which has been shown to be 74.7% digestible. The amino acids in hydrolyzed hog hair meal are not as available as in soybean meal and maize.

Hydrolyzed hog hair meal is of value for swine when utilized at low dietary levels. Maximum value is realized when not more than 2–4% is included in growing-finishing swine diets.

Nonprotein Nitrogen (NPN)

Although some evidence exists to indicate that nitrogen from dietary nonprotein nitrogen compounds is incorporated into amino acids and tissue protein, it has been shown that NPN cannot be used to any great extent and cannot be recommended at the present time as having practical value in swine diets even when limiting essential amino acids are supplied.

Animal and Poultry Waste

As the problem of disposal of animal waste continues to increase, every effort will be made to utilize these residues to the best advantage in animal feeding. Today, it appears that the ruminant animal is best equipped to utilize these fecal wastes and will probably provide the main avenue of useful utilization in the near future (Fontenot et al. 1983). Since the pig makes little use of nonprotein nitrogen which forms a significant proportion of animal and poultry feces, generally animal wastes are of very little value for supplying protein and energy for swine rations. In comparative studies, it has been shown that species from which manure is collected have a significant effect on concentration of digestible nutrients for swine, with poultry manure providing the greatest concentration followed in order by swine manure and cattle manure. Management conditions, diet fed, storage time and conditions, and species from which feces are collected for recycling affect composition and nutritional value. The composition of waste products from cage layers, broiler litter, and cattle manure differs considerably. Therefore the value of poultry litter, poultry manure, cattle manure, swine manure, and oxidation ditch liquor and slurry must be treated separately. This subject is discussed by Pond and Maner (1984).

PLANT PROTEINS

Oilseed Meals

Soybean Meal

Soybean meal (SBM) is one of the most important plant proteins in livestock feeding in the United States as well as in many other countries. The use of soybean meal in swine rations has been under investigation for more than 50 years. Studies have related primarily to the influence of different processing methods on feeding quality of the meal and to the specific nutrient requirements of growing swine when soybean meal is used as a portion or all of the supplemental protein.

The nutritional superiority of heated soybeans over raw soybeans for monogastric animals has been recognized for more than 70 years.

The growth inhibition that results from the consumption of the raw soybean, either full-fat or as a defatted meal, has been attributed to factors such as poor palatability, digestive enzyme inhibition, hemagglutinin, saponins, and decreased availability of amino acids.

The moist heat treatment used by Osborne and Mendel (1917) is still employed to improve the nutritive value of soybeans and soybean meal. A number of other soybean components have been isolated and suggested as anti-nutritional factors. Trypsin inhibitor was one of the first soybean components proposed as an anti-nutritional factor. Other factors include hemagglutinins, saponins, isoflavones, and goitrogens. The role of these anti-nutritional factors in soybeans and other seed legume proteins has been reviewed by Liener (1975).

The physiological response of the growing pig to raw soybean differs from that of the rat and chick in that the pancreas does not enlarge and pancreatic enzyme synthesis is not stimulated in the pig (Yen et al. 1977). The growth failure observed in pigs fed raw soybeans apparently results from reduced feed intake and inhibition of protein digestion.

Properly processed soybean meal contains a protein of excellent quality. Depending on completeness of oil extraction and degree of hull removal, soybean meal contains between 43 and 51% crude protein and has an excellent content of amino acids that are well digested by pigs of all ages except the very young (2–21 days of age) pigs. Soybean protein has a good balance of amino acids essential for the growing pig. Its distribution of amino acids is similar to that of animal protein. It is not severely limiting in any amino acid except methionine. Relative to the requirement of the pig, most amino acids are quantitatively in excess when soybean meal is used to supply all of the dietary protein. However, in practice, soybean meal is seldom used as the only source of protein but is used as a supplemental protein in combination with other feed ingredients, such as maize, sorghum, barley, cassava meal, banana meal, and by-product feeds. The resulting balance of amino acids is of primary importance in practical feeding systems.

A combination of maize and soybean meal provides an excellent diet for pigs of all ages if the proper proportions of maize and soybean meal are used to supply an adequate balance of amino acids.

In studies by Wilson and Leibholz (1981), the substitution of milk protein (skim milk and casein) by soybean meal reduced liveweight gains at each level of soybean meal inclusion. During the period from 7 to 14 days, the weight gains of pigs given the diets in which SBM supplied 50 and 75% of the total dietary protein were 51 and 54%, respectively, of those of the pigs given the all-milk protein diets. By 21–28 days of age, the difference in performance was only 16 and 31%, respectively.

The better performance of baby pigs fed casein is due in part to the higher digestibility of casein as compared to soybean meal. Wilson and Leibholz (1981) demonstrated that each increase in dietary level of soybean meal, at the expense of casein, reduced DM and N digestion and N retention. Mateo and Veum (1980) also reported that the digestibility of dry matter, energy, crude protein, ash, and ether extract were significantly greater for casein than for soybean protein.

Although early weaned neonatal pigs are not able to utilize soybean meal efficiently, weanling pigs (5–15 kg) perform well on practical diets supplemented with soybean meal (Zimmerman 1980; Lewis et al. 1981). However, practical diets based on maize and soybean meal are improved by the addition of lysine, as with older pigs.

All-plant diets based on cereals and soybean meal have provided adequate nutrition and a proper balance of amino acids for satisfactory reproductive performance of sows and gilts.

Coconut Meal

Coconut meal or copra meal, the residual product after extraction of oil from the dried endosperm of the coconut (*Cocos nucifera*) is available in many parts of the world. In large areas of the tropics, coconut meal may be one of the few protein sources available for livestock feeding. Even though the meal contains only moderate levels of crude protein, it is an economically important source of protein in areas in which other sources are not readily available. Hawaii and some of the Caribbean islands are good examples of such areas.

Coconut meal contains between 20 and 26% crude protein, about 10% crude fiber, and depending on processing method and efficiency between 6 and 12% of ether extract. The lysine content of coconut meal on a percent of protein basis is somewhat lower than for soybean meal but the level of total sulfur amino acids is significantly higher.

Different levels of coconut meal have been used to supply a portion of the protein in complete mixed diets for growing-finishing pigs. Generally, each increase in level of coconut meal used to replace soybean meal caused a decrease in feed intake and daily gains and an increase in feed required per unit of gain. Dietary levels up to 20% caused only minor reduction in pig performance; however, 40% dietary coconut meal caused a 39.5 increase in feed required per unit of gain. In a series of three experiments with levels of coconut meal up to 30%, Malynicz (1973) presented data to indicate a reduced performance with increasing levels of coconut meal. More recent studies (Lima et al. 1981), in which growing-finishing pigs were fed levels of 0, 10, 20, and 30% coconut meal as a replacement for soybean meal protein in maize-based diets, have clearly demonstrated a linear depression in both feed intake and daily gains as the level of coconut meal was increased in the diet. Feed required per unit of gain increased linearly with increasing levels of coconut meal. Although it would appear that lysine might be deficient in the diets prepared with expeller-processed (oil extracted by heat and mechanical pressure) coconut meal, supplementing lysine or increasing the protein level by adding more soybean meal had no effect on pig performance or muscle development for diets containing 20 or 40% coconut meal (Cresswell and Brooks 1971).

Because of the low protein and amino acid digestibility, except in unusual circumstances of price relationships and availability of protein supplements, coconut meal should not be included in proportions of more than 15–20% in growing-finishing swine diets if maximum performance is to be expected.

Cottonseed Meal

Cottonseed meal is a protein supplement that is available in many areas of the world for swine feeding. It is commonly used as a portion of the supplemental protein in swine rations in the southeastern and southwestern areas of the United States and in many countries, where it is often the least expensive and most abundant protein supplement available.

Because of reports of poor performance, toxicity, and death of pigs fed moderate to high levels of cottonseed meal, much prejudice has developed against the use of cottonseed meal as a swine feed. Cottonseed meal, when properly used and its limiting factors understood and corrected, is an excellent source of protein for use in swine rations. Excellent pig performance can be obtained when limited amounts of cottonseed meal are fed in combination with other sources of high quality protein.

The limitations of cottonseed meal must be recognized. The seeds of the cotton plant from which cottonseed meal is produced contain a natural pigment, gossypol, that is toxic to pigs. *Gossypol* is a yellow solid, insoluble in water but soluble in most fat solvents.

Free gossypol is that gossypol which is extracted with aqueous acetone, and *total gossypol* is that which is obtained only by acid treatment of the cottonseed product, or by some other treatment harsh enough to release all of the gossypol. "Bound gossypol" is usually estimated as the difference between total and free gossypol.

The bound gossypol is of little significance to the pig, since it is unavailable and passes through the gastrointestinal tract of the pig without being absorbed and is eliminated in the feces. The free gossypol is highly toxic to the pig. Any cottonseed meal containing more than 0.04% free gossypol should be used sparingly in swine diets. It has been determined that a standard 15–16% protein diet should not contain more than 0.01% free gossypol. Percentages higher than this safe level have produced slow growth, gossypol toxicity, and death.

Gossypol toxicity is cumulative. In contrast to an acute toxin that causes a rapid decline in body functions and rapid death, gossypol accumulates in the body and death usually does not occur until after the pig has received the gossypol-containing diet for 4–8 weeks. The exact time of the production of toxicity symptoms and death will depend upon the concentration of gossypol in the ration and the presence of certain mineral elements.

The toxicity of dietary gossypol can be prevented by the addition of iron salts to the diet. The most effective iron salt is ferrous sulfate. It is thought that a strong complex forms between iron and gossypol in the intestinal tract, thereby preventing the absorption of gossypol. The addition of ferrous sulfate on a 1 : 1 weight ratio, iron to free gossypol, to diets containing more than 0.01% free gossypol is effective in reducing gossypol toxicity and improving gains and feed efficiency.

Cottonseed meals manufactured under different processing conditions differ over a wide range with respect to their nutritional value as a protein supplement for pigs. The cottonseed meals with high protein quality and low free gossypol content are the most desirable. Most meals are produced by one of three processing methods; screw press (expeller), prepress-solvent, and direct solvent extraction. The screw press method involves application of heat and mechanical pressure only; the prepress–solvent method involves heat and pressure followed by extraction with a fat solvent; and the direct solvent extraction involves the use of fat solvent only.

An estimation of the nutritive value of a cottonseed meal can be made if the processing procedure used in its production is known. Meals produced by screw press extraction of the oil from the cottonseed are low in free gossypol but have poor quality protein.

Prepress–solvent meals are low in free gossypol and moderate to high in protein quality. The heat generated in the initial prepressing binds much of the free gossypol and the subsequent extraction with solvents removes the remaining oil. During the prepress-solvent extraction, different prepressing methods are employed and, therefore, cottonseed meals of varying protein quality are produced. If the original pressure and temperature during prepressing is excessive, a low quality protein meal will result. If original pressures and temperatures are low, a better quality protein will be produced, but the meal will contain higher levels of free gossypol. Analysis to determine level of free gossypol, available lysine, and nitrogen solubility are employed to measure protein quality and gossypol toxicity levels.

Cotton breeders have developed a variety in which seed is very low in gossypol-containing glands.

Many workers have studied the nutritive value of cottonseed meal and its use as a protein supplement in swine diets and have clearly demonstrated that excellent performance can be obtained when cottonseed meal is properly used in swine rations. Although cottonseed meal is a good source of tryptophan and a fair source of methionine, it is a very poor source of lysine. The lysine content of cottonseed meal is only about two-thirds that of soybean meal.

Perhaps of more importance is the fact that a portion of the lysine in cottonseed meal cannot be digested, absorbed, and utilized by swine. During processing, application of excessive heat may render 10–35% of the lysine unavailable. Tanksley et al. (1981) demonstrated the extreme differences in digestibility of amino acids in cottonseed meals prepared from different processes. Digestibilities of nitrogen and the ten essential amino acids at the end of the small intestine and over the total digestive tract were shown to be higher in laboratory-processed glandless

(GL) cottonseed meal than in commercially-processed direct solvent (DS) and screw press (SP) cottonseed meal. Screw pressed cottonseed meal had higher digestibilities of all the essential amino acids, except lysine, methionine and tryptophan, than did direct solvent cottonseed meal. Lysine was the least digestible, in the lower small intestine, of the essential amino acids in direct solvent (61.7%) and screw press (64.0%); in contrast, it was highly digestible in glandless cottonseed meal (88.5%) which compared favorably with the lysine in soybean meal (87.1%). Lysine availability is, therefore, a major factor in determining the quality of protein in cottonseed meal, and probably no other major protein source varies as much in quality as does cottonseed meal.

It has been demonstrated that one-half of the soybean meal in a maize–soybean meal diet could be replaced by degossypolized cottonseed meal without affecting growth or efficiency of feed utilization of growing-finishing pigs. It was subsequently shown that 25, 50, or 75% of the soybean meal of the diet could be replaced with solvent-processed, degossypolized cottonseed meal without altering pig performance; however, the same levels of expeller cottonseed meal with similar levels of free gossypol but lower levels of soluble nitrogen depressed growth and feed efficiency.

The lysine deficiency of cottonseed meal–cereal grain diets can be overcome by blending the cottonseed meal with other lysine-rich protein feeds such as soybean meal or fish meal or by the addition of crystalline L-lysine. Maize–cottonseed meal diets with 0.1% supplemented L-lysine were superior to unsupplemented rations when fed to growing pigs. In studies by Knabe et al. (1979), increasing levels of dietary lysine (0.60, 0.70, and 0.80%) in sorghum–cottonseed meal rations were associated with a linear increase in gain, feed intake, and efficiency of feed utilization.

Cottonseed meal can be used to supply up to 50% of the supplemental protein in maize– or sorghum–soybean meal diets for gestating and lactating sows. However, small reductions in farrowing and weaning weights may be observed. Maximum performance will depend on the lysine adequacy of the diet.

Linseed Meal

Linseed meal is a protein-rich by-product of drying oil production from linseed or flax (*Linum usitatissimum*). Flax grows best in temperate climates and approximately 60% of the world's production of 2.3 million tons annually is from the United States, Canada, and Argentina (FAO 1980). Solvent-extracted linseed meal constitutes less than 1% of the protein supplements used for livestock feeding in the United States and only an insignificant part is used for swine. It contains approximately 35% protein, 1.7% fat and 9% crude fiber (NRC 1964). It is similar in calcium, phosphorus, and vitamin content to soybean meal but it is seriously limiting in lysine.

In feeding experiments, diets containing up to 25% linseed meal have been fed to growing-finishing pigs without adverse effect. However, because of a serious lysine deficiency and poor amino acid balance, it should always be combined with other lysine-rich protein supplements and should not make up more than one-third of the protein supplement when used with maize or other cereal grains. It tends to have a laxative effect and is sometimes added to sow diets at parturition to prevent constipation.

A vitamin B_6 antagonist has been isolated from linseed meal. Bishara and Walker (1977) have reported that pigs fed diets containing 30% linseed meal grow faster if supplemental pyridoxine is added and suggest that, under some conditions, diets containing linseed meal may be marginally deficient in vitamin B_6.

Peanut Meal

Peanut meal (also called groundnut meal) is the by-product of the extraction of the oil from raw peanuts (*Arachis hypogaea*). The protein content of peanut meal varies from 45 to 55% and the quality is influenced by the quantity of hulls in the meal, and by the type and level of heat it has received during processing.

Early studies with peanut meals indicated that it was of little value for supplementing cereal diets when used as the only source of supplemental protein. Pigs fed cereal-based diets supplemented with peanut meal grow very slowly and have low feed intake and inefficient rates of feed utilization.

The poor nutritive value of the peanut meal protein is not due to reduced digestibility, as diets containing properly processed peanut meal have protein digestion coefficients (approximately 80%) that are similar to those of diets containing soybean meal or fish meal. The poor response of growing pigs fed cereal diets supplemented with peanut meal is due primarily to a lysine deficiency of the mixed diet. Peanut meal is low in methionine and lysine. Performance and nitrogen balance are greatly improved by supplementing the diet with 0.13–0.3 percent L-lysine.

In practice, the amino acid deficiencies of cereal–peanut meal diets are corrected by blending the peanut meal with other sources of protein such as soybean meal, fish meal, and blood meal. One-half of the soybean meal in a 16% protein maize–soybean meal diet can be replaced by peanut meal without lowering performance.

During commercial processing, after crushing, the seed is cooked by means of jacket or live steam, the temperature ranging from 80° to 115°C and the cooking time from 20 to 120 min. If the hydraulic process is used, the temperature may range from 65° to 105°C and the time from 15 to 120 min. Drying temperatures vary from 75° to 115°C. If these or similar conditions are used during processing, little damage is expected and a destruction of the trypsin inhibitor occurs. Variation from a moderate-moist heat would be expected to cause a change in nutritive value of the protein, since dry heat will not bring about a rapid removal of the trypsin inhibitor and, if prolonged, can cause a reduction in lysine availability. A similar but less drastic reduction in biological value of the protein can occur if the peanuts are subjected to moist heat and excessive duration of heating.

Like many other stored feeds, peanuts are subject to the growth of certain fungi which produce aflatoxins. *Aspergillus flavus* occurs rather commonly in peanuts and in peanut meal. The presence of these aflatoxins in diets of pigs causes loss of appetite and reduced growth rate, the severity of reduction depending on the level of aflatoxins present.

Before the significant increase in demand for vegetable oils, raw peanuts were commonly used as a feedstuff for swine and their potential as a livestock feed has been long recognized. The occasional availability of full-fat peanuts, unsuitable for crushing or human consumption, favors their use in swine diets. Peanuts contain 31.2% crude protein and 44.8% oil. Although differences in lysine exist between varieties, differences in total crude protein and oil are greater between grades. Conkerton et al. (1978) found that certain peanut genotypes have more methionine than other similar genotypes. These reports indicate that peanut geneticists may be able to produce cultivars with high levels of lysine and methionine and a more favorable balance of amino acids for swine.

Rapeseed Meal (Canola Meal)

Rapeseed meal is a by-product of the production of vegetable oil from rapeseed (also called canola rape, colza, or raps). In the Cruciferae or mustard family, many species of the genus *Brassica* are known; however, two of these species *B. napus* L. and *B. campestris* L. are most commonly grown. Because of its adaptation to the temperate and subtropical zones, rapeseed is produced as a winter or cool season crop in Europe, Asia, and the Western Hemisphere. Its production in many of these areas is of great importance, since rapeseed and sunflower are the only edible oil crops that can be produced effectively in the colder areas of these regions. Detailed information on rapeseed meal for livestock has been reviewed elsewhere (Clandinin 1981).

Eighty percent of the world's annual production of some 10.6 million metric tons is produced in Asia and Europe. Within the Western Hemisphere, it is an important crop only in Canada and Chile.

Rapeseed contains about 40–50% oil by weight, which may be removed by expeller, solvent extraction, or a combination of both techniques. The oil has a history of industrial use as a special lubricant, but until recently it contained a high level of erucic acid. This is nutritionally undesirable, but plant breeders have been successful in developing low erucic acid rapeseed cultivars. Seed, oil, and meal derived from these cultivars have come to be known as *canola seed, canola oil,* and *canola meal* to distinguish them from high erucic acid cultivars grown extensively in other parts of the world (Clandinin 1981).

Meals produced after removal of the oil may contain 32–44% crude protein with the majority containing 35–37% protein. Average analysis values of 40.5, 1.1, and 0.3% for crude protein, fat, and fiber can be expected from a typical solvent-extracted meal. The amino acid pattern of the rapeseed protein is similar to that of other vegetable protein meals, with rapeseed meal containing more methionine and less lysine than soybean meal. However, the processing method can greatly influence the availability of these amino acids and thus the nutritive value of the meal. The high heat of processing by the expeller method probably is associated with poor amino acid availability; however, a combination expeller–solvent extraction, which is conducted at a lower temperature, produces a higher quality meal.

As a result of research in Canadian and other laboratories on problems associated with the consumption of rapeseed meal, it is clear that the detrimental effects of high levels of conventional rapeseed meal are due to the presence of up to 8% glucosinolates in the meal depending on variety. Glucosinolate breakdown products, isothiocyanates and oxazalidinethione, have been shown to exert a goitrogenic effect on nonruminant animals.

In recent years, perhaps the greatest factor in the increased production and use of rapeseed meal for animal feed has been the development of low (or *zero*) glucosinolate cultivars (Clandinin and Robblee 1978). Two cultivars commonly grown today, "Tower" (in Canada) and "Erglu" (in Europe), and other recently developed varieties, are so-called *double zero* varieties, possessing low levels of both erucic acid and glucosinolates. To distinguish rapeseed meal derived from low erucic acid, low-glucosinolate cultivars of *B. campestris* or *B. napus* from conventional rapeseed meal, the term *canola meal* has been created.

One of the main factors that tends to limit the feeding value of rapeseed meal for pigs is its relatively low digestible energy (DE) or metabolizable energy (ME) values. Rapeseed meal has lower DE and ME values than soybean meal and the high fiber content of rapeseed meal has been suggested as an important reason for the lower energy values. Plant breeders have successfully developed varieties with thinner, less fibrous seed coats. The introduction of yellow seed varieties, such as "Candle," achieve both a 4% reduction of fiber content and a zero level of erucic acid and glucosinolates; such seeds have been termed *triple zero* varieties. To distinguish these new varieties in subsequent discussions in this chapter, the term canola meal will be used for double zero rapeseed meal and low-fiber canola meal for double zero varieties with low fiber content ("Candle").

All swine experiments that have compared canola rapeseed meal ("Tower") with regular rapeseed meal have demonstrated the superiority of the double zero meal as a protein source for pigs. Canola meals are finding increased acceptance as protein sources in swine diets and are being used as a replacement for soybean meal. Some studies have shown that "Tower" canola meal may totally or partially replace soybean meal in starting, growing, and finishing pig diets without adverse effects on performance.

Canola meal ("Tower") contains a substantially higher crude fiber level than soybean meal (13 vs. 6%) and this has been suggested as contributing to the lower performance observed when canola meal is fed as the sole protein supplement in diets of young pigs. The limitations imposed by the high fiber levels of rapeseed meal led to the development by plant breeders of a yellow-coated cultivar ("Candle") that is low in glucosinolate and also contains about 4% less fiber.

Lewis et al. (1978) fed low glucosinolate rapeseed meal (derived from the cultivar "Tower") to replace 50 or 100% of the soybean meal in diets for 72 gilts from 20-kg liveweight through two gestation and lactation periods. The gilts were fed diets containing 17% protein from 20 to 60 kg and diets containing 15% protein from 60 kg until the end of the second lactation. All of the gilts successfully completed one reproductive cycle and 51 animals completed two. Treatments had no significant effect on litter size or weight at birth or weaning, sow weight changes in gestation or lactation, or gestation length in either the first or second parity.

It appears that rapeseed meal produced from the new cultivars with low glucosinolate content (less than 1 mg/g), such as "Tower," "Erglu," and "Candle," can be used to supply all of the supplemental protein for gilts and sows during pregestation, gestation, and lactation without any detrimental effects on reproductive performance.

Rubber Seed and Rubber Seed Meal

The rubber tree (*Hevea brasiliensis*) produces, in addition to latex for rubber, an average of 800–1200 kg of seed per hectare per year. The seeds contain approximately 50% oil and 20% protein. Rubber seeds also contain a cyanogenic glucoside, linamarin, which on decomposition produces hydrocyanic acid, a substance that can cause acute poisoning in pigs and other livestock. Fresh seeds have been shown to produce 2000 mg hydrocyanic acid per kilogram of seed. However, the hydrocyanic acid content diminishes rapidly during the first week of storage and also can be reduced by some physical and chemical treatments. Stosic and Kaykay (1981) reported good performance of growing pigs fed diets containing 40% ground rubber seed that had either been roasted at 350°C for 15 min or soaked in an ash solution for 12 hr. The addition of treated rubber seed to the commercial diet improved both daily gains and efficiency of feed conversion. Palatability of roasted rubber seed was significantly better than that of ash-treated seed.

Rubber seed meal was used by Rajaguru and Ravindran (1979) to replace 10, 20, or 30% of the coconut meal in 16% protein growing pig diets based on wheat, rice polish, meat meal, and coconut meal. Each increase in the level of rubber seed meal depressed pig performance. Replacement levels above 10% significantly reduced both gains and efficiency of feed utilization as compared to those obtained with the control diet. Lysine and methionine appear to be most limiting amino acids in rubber seed protein.

Babatunde et al. (1990) reported that the maximum substitution of rubber seed meal in a cornstarch–soybean meal diet for growing pigs compatible with normal performance was about 30% of the protein in the diet. However, to maintain normal plasma protein and avoid depressed nitrogen retention, the maximum level of rubber seed meal replacement was 10% of the dietary protein.

Rubber seed meal may be used as a partial replacement for higher quality protein sources in swine diets and, if properly supplemented with amino acids, can be used to supply as much as 20% of the total dietary protein for growing-finishing swine. This under-utilized product may be a valuable protein source for swine produced in areas where rubber seed meal is readily available.

Safflower Meal

Safflower (*Carthamus tinctorius*) has long been cultivated as an oilseed crop. Although the seeds contain 36–40% oil, the hull constitutes about 40% of the seed. After the oil is removed, the product that remains contains only 18–22% crude protein and 40% crude fiber. The hulls are about 70% holocellulose, 21% lignin, and 1% ash and greatly limit the energy concentration and utilization of the meal by swine and other classes of livestock. The protein of safflower oil meal is a very poor source of lysine and somewhat deficient in the sulfur-containing amino acids, methionine, and cystine.

Because of its low protein, low lysine, and high fiber, undecorticated safflower meal is of limited value in growing-finishing swine diets. It would appear that it could best be utilized in rations for gestating sows.

Attempts have been made to improve the quality of safflower meal by the physical separation of the hulls from kernels and by the development of thin-hull varieties through plant breeding. Although the complete removal of the hull results in a seed with 60–70% oil and a seed cake of about 60% protein, in practice it is very difficult to effect a complete separation because of the hardness of the seed coat and the extreme softness of the kernel. Screening of undecorticated safflower meal to partially remove the hulls produces a meal with more crude protein (42–46%) and less crude fiber (15–16%).

In studies by Williams and Daniels (1973) and Williams and O'Rourke (1974), decorticated safflower meal containing 46.1% crude protein and 16.2% crude fiber was found to be unsatisfactory as the sole supplementary protein source for sorghum and wheat-based diets for pigs. The growth response to the safflower meal was increased by the addition of synthetic lysine but the performance of the pigs receiving the lysine-enriched safflower meal supplements was always inferior to that of pigs fed either isonitrogenous soybean meal or fish meal supplements in both sorghum and wheat-based diets. Increasing the level of safflower meal at the expense of fish meal resulted in slower rates of growth and reduced efficiency of feed utilization, particularly when fed to pigs less than 44-kg liveweight.

Decorticated safflower meal can supply a limited portion of the supplementary protein in growing-finishing swine diets (5–10%), but when used should always be combined with other "lysine-rich" protein concentrates.

Sesame Meal

Sesame, known botanically as *Sesamum indicum* (also called benne, til, gingerly, simsin, gergelin, and in Latin America ajonjoli), is probably one of the oldest of the cultivated oilseed crops (Cornelius and Raymond 1967). Annual world production of sesame amounts to about 2 million tons with China and India, the chief producers, providing 40% of the total. Sesame also grows in many other countries in the Far and Near East and on the American continents. Sudan is the leading African producer and Mexico the leading American producer. Most of the countries in which sesame is grown consume the bulk of the crop domestically, so that international trade in seed and oil is small in comparison with the quantities produced.

The seeds, which vary from white through black in color, according to variety, are small, weighing usually 2–3.5 g/1000 seeds and are contained in capsules which tend to split open and shed when ripe. Because of this tendency to "pop open," practically all the harvesting is carried out by hand. Nonshattering lines and varieties must be evolved if mechanical harvesting is to be successfully employed to allow the development of sesame as a competitive crop in areas of high labor cost.

The seed has a thin shell which can be separated from the kernel in decorticating machines or by manually soaking and rubbing. Normally however, the seed is milled without decortication to remove the 44–54% oil present in the air-dried seed. The sesame meal produced from the whole seed contains both the kernel and the thin shell.

The chemical analysis of sesame meal varies according to the processing method employed for extracting the oil. Meals prepared by the screw press or hydraulic methods contain higher levels of fat and lower levels of protein than those produced by solvent extraction. Chemically, sesame meals contain 90–95% dry matter, 38–48% crude protein, 0.7–11% fat, 5–7% fiber, 9–13% ash and 25–30% nitrogen-free extract. Sesame meal is a poor source of lysine but an excellent source of both methionine and tryptophan.

When used in swine diets, both the deficiencies and strengths of sesame meal should be taken into consideration. As a protein supplement for diets based on cereal grains it should not be used to supply all of the supplemental protein required but should be blended with other supplemental protein sources such as soybean meal and fish meal, which are good sources of lysine. Sesame meal can replace one-half or all of the soybean meal in a maize–soybean diet containing 4% fish meal, and it can be substituted for soybean meal in quantities as high as 10% of the diet in maize and soybean meal diets. Because of the low level of lysine in both cottonseed meal and sesame meal, blends of these two supplements are inadequate for supplementing cereal-based growing-finishing diets; however, 50–50 blends of these two protein supplements, fed with cereal grains and 5–6% fish meal, produce satisfactory gains and feed utilization.

Sunflower Seed and Meal

In many areas of the world, sunflower seed (*Helianthus annuus*) is the best adapted high-protein crop available. A big advantage of sunflowers is their ability to adapt to a variety of climates and soil compositions, from the heat of Southern United States to the frost of Central Asia. More than 13 million metric tons of the seed are produced each year. The U.S.S.R. and the other formally centrally-planned countries produced more than 6.96 million metric tons or some 53% of the total world crop. Europe produces 2.6 million metric tons and South America 1.7 million metric tons. In the United States, hectarage has been increased steadily since 1968 to more than 2 million hectares in 1990, mostly in the northern great plains region. Production during this time has increased nearly 2 million metric tons. There are two basic types of sunflower: oilseed and confectionary seed. In recent years, production of high-oil varieties of sunflowers has increased substantially.

Sunflower meal is produced from sunflower seed following oil extraction. The composition of the meal will vary according to the composition of the seed and, as with other oil-seed proteins, the nutritive value of the sunflower meal varies with the method of processing. There has been a marked improvement in meal quality during recent years due to advances in oil extraction and meal processing that has markedly lowered the crude fiber and ether extract content of the finished product.

Although new varieties of sunflower seed with lower percentages of hulls are now available, the energy content and protein concentration of sunflower meal vary with the quantity of hulls present. The meals with a high-protein content are produced by removing (decortication) the largest quantity of hulls possible before final processing by screening the meal after extraction. Sunflower meal used to supply all of the supplemental protein in growing-finishing swine diets is a very unsatisfactory source of protein because of its very low lysine content. It produces gains and efficiency of feed utilization inferior to those produced by soybean meal, fish meal, peanut meal, or combinations of these ingredients.

However, sunflower meal diets supplemented with lysine have nutritive value similar to that of soybean meal diets for growing swine (Seerley et al. 1974; Baird 1982). Pigs fed a lysine-supplemented 12% protein corn–sunflower meal diet show a growth response to added threonine and tryptophan (Wahlstrom et al. 1985; Borg et al. 1987). Corn–sunflower meal diets containing 12% protein are capable of promoting growth equal to that obtained with an 18% protein diet if appropriately supplemented with limiting amino acids.

The most efficient utilization of the meal is obtained by combining it with supplements, such as fish meal or soybean meal, which are rich in available lysine, or by supplementation with crystalline lysine.

Grain Legumes

The seeds of grain legumes have many advantages and offer a wide range of germ plasm of wide environmental adaptability with which to meet the protein needs of both the human and animal population. Of the 600 genera with more than 13,000 species that make up the family Leguminosae, only some 20 of the seed-producing species are utilized regularly in appreciable quantities. However, with the exception of soybean and peanuts, which contribute significantly to the total supply of feed protein, the majority are low yielding and contribute only in a limited way to the world's total protein supply (FAO 1981). Because of low yields, many of these crops cannot compete economically with other higher-yielding crops, such as maize, rice, wheat, barley, and cassava. The amount of agronomic research conducted on most of the grain legumes has been limited. Production levels, therefore, remain low because of the inherent characteristics of the varieties grown, the damage from diseases and insect pests, and poor cultural practices. The majority of the food and grain legumes with the exception of soybeans (*Glycine max*), peanuts (*Arachis hypogaea*), beans (*Phaseolus vulgaris.*) and field or garden peas (*Pisum* spp.), which are important in the developed world, are of importance only within the developing countries.

Soybeans

Soybeans (*Glycine max*) are produced in large quantities in both temperate and tropical climates. Annual world production has reached 90 million metric tons, with more than half being produced in the United States where they constitute by far the largest source of protein supplement for animal feeding. Most soybeans are used for oil production with the by-product, soybean meal, also being produced. However, when supply and demand relationships push the price of soybean meal upward in relation to the price of soybeans, the possibility exists for swine producers to utilize full-fat soybeans instead of soybean meal. This practice can be especially economical for soybean producers who are also swine producers. The presence of several inhibitors of protein utilization in raw soybeans requires that they be heated to destroy these inhibitors before the pig can utilize the soybeans efficiently. Several methods are available for applying heat to soybeans, including steaming or autoclaving, infrared gas heating, and extrusion. The development of on-farm soybean cookers has further stimulated interest in on-farm processing of soybeans for swine and has overcome some of the early problems of full-fat soybean feeding.

The unusually favorable amino acid balance of soybean protein, in comparison with most other plant proteins, makes it possible to utilize it as a major source and in some cases the only source of protein in mixed swine diets. Swine, particularly young growing pigs, fed uncooked soybeans gain slowly and inefficiently due to inhibition of the proteolytic enzyme, trypsin. Gestating swine are able to reproduce normally when fed diets containing raw whole soybeans as the only source of supplemental protein. The reason may be that protein required for growth is more critical than for reproduction because of the capacity of the dam to buffer the fetus against nutritional insults. When soybeans are properly cooked either by boiling, steam processing, autoclaving, infrared roasting, or extrusion, they can replace soybean meal in cereal-based diets for growing-finishing pigs with equal or superior results.

The slightly lower feed intake and lower amount of feed required per unit of body weight gain usually encountered when whole soybeans replace soybean meal is related to the greater energy density of the whole soybeans.

The high fat content of soybean and its high proportion of unsaturated fatty acids creates a tendency for soft pork from pigs fed whole cooked soybeans as a major part of the diet. Work at many experiment stations in the United States indicates that whole soybeans can be fed at 22% of a maize-based diet with no adverse effects on pig performance or on consumer acceptance of pork. Although the pork fat from pigs fed whole soybeans was noticeably softer and had a higher proportion of linoleic acid, total fat content of the carcass was not increased.

The decision to use soybean meal or cooked whole soybeans for swine feeding must be based on economic considerations, since the nutritive value of a diet containing soybean meal is about equal to one containing cooked whole soybeans if both are properly supplemented with vitamins and minerals. The cost or inconvenience of providing the power and equipment necessary to heat the bean satisfactorily to approximately 110°C for 3 minutes to destroy inhibitors of protein utilization may be prohibitive in some areas or under some conditions. However, if the facilities are available, cooking offers a valuable means of utilizing home-grown soybeans as a swine feed and, if not available, raw soybeans may be used for gestating sows.

Other Seed Legumes

Dry beans (*Phaseolus vulgaris*), also called snap bean, white kidney bean, navy bean, common bean; *P. aureus* (mung bean); *P. mungo* (urd bean); *P. lunatus* (lima bean); *P. calcaratus* (rice bean); *Cicer arietinum* (chick pea); *Cajanus cajan* or *indicus* (pigeon pea); *Vigna sinensis* or *unguiculata* (cow pea); *Vicia faba* L. (field bean); and *Pisum sativum* or *arvense* (field pea, also called garden pea and arveja) are all used to some extent in swine feeding, but are not discussed here. A description of each is given in Pond and Maner (1984).

Sweet Lupin

Lupin is a legume whose use has been limited largely to green manure and green forage for ruminants because of the presence of high levels of toxic and bitter alkaloids. Genetic selections by plant breeders have resulted in a number of lupin varieties with seeds of low alkaloid content and of improved nutritional value. Although sweet varieties of low toxic alkaloid content were bred and released in 1934 (*Lupinus albus*) and in 1951 (*Lupinus luteus* cv. "Weiko III"), their seeds were not easily distinguishable from those of the bitter varieties, thus detracting from their use as feed. The development of sweet varieties of *Lupinus angustifolius* (Gladstones 1972), which are color-marked with white flowers and seeds, has stimulated considerable interest in Australia and New Zealand in the potential of lupins as an alternative protein source to traditional and increasingly expensive sources of protein for pigs, such as soybean meal, fish meal, and meat meal.

The protein content of lupin seed ranges from 25 to 44% of the dry matter. Based on chemical analysis, only the sulfur–amino acids appear to be limiting. The whole seed contains 5–10% ether extract and about 15% crude fiber. Lupins contain almost no starch. The fatty acid composition of lipids in seeds is remarkably similar to that reported for soybeans. The seed coat, or hull, makes up 20–25% of the weight of the seed and is very low in protein but contains more than 50% fiber. The seed contains no lignins and the seed coats have no detectable phosphorus. Concentrations of trace elements are sufficiently high to make lupins a valuable source of these essential nutrients. *Lupinus albus* has been shown to be a manganese accumulator with levels up to 3750 ppm being reported. At this level of concentration, it is possible that manganese toxicity could develop and the use of *L. albus* should be closely monitored.

Alkaloids are present in all of the better varieties of lupins and appear to be present in varying trace amounts in many of the newer sweet varieties. Among sweet lupin varieties, the alkaloid content has been reported to vary from less than 0.01% to 0.09% which is significantly less than the 0.59% determined for bitter lupin.

It is clear that considerable differences occur between species and varieties of lupin in relation to their suitability as protein sources for pigs. Variation in their alkaloid content is one important determinant of their nutritive value. However, even in cases where the alkaloid content is very low, such as the "Uniwhite," "Uniharvest," and "Unicrop" cultivars, the value of lupin seed meal for pigs may still be limited by factors which suppress voluntary feed consumption (Pearson and Carr 1977).

OTHER PROTEIN SOURCES

Alfalfa

Alfalfa (*Medicago sativa*) is a leafy legume plant that contains large amounts of protein. As a source of crude protein, alfalfa produces 2700 kg/ha as compared to 876 kg for maize and 786 kg for soybeans. Alfalfa hay or dehydrated alfalfa meal contains between 16 and 22% crude protein. Specially prepared alfalfa leaf meal can contain up to 30–50% crude protein. Alfalfa is a good natural source of tryptophan, with a content of 0.45% as compared to 0.65% in soybean meal and only 0.09% in maize. The alfalfa plant is an excellent source of nutrients and dehydrated alfalfa meal is commercially available.

Although the potential protein production from alfalfa is encouraging, attempts to utilize alfalfa meal as a primary source of protein in swine diets and specially during the growing-finishing phase have been discouraging. With few exceptions, published results generally agree that as a percentage of dietary alfalfa increases in swine diets average daily gains decrease and feed-to-gain ratios increase. Although some variations in response have been reported as might be expected on the basis of differences in alfalfa composition, saponin content, methods of processing, and differences in age and condition of animal groups, the level of feed consumption was most commonly cited as the factor responsible for depressed pig performance. Even though the alfalfa diets contained relatively high levels of fiber and lower than normal energy levels, the pigs failed to consume adequate quantities of diet to meet their energy requirements. The consistent absence of an alfalfa-induced increase in dietary feed intake in the studies reported does not agree with the generally observed tendency for feed intake to increase with increased dietary fiber as has been observed with various sources such as purified cellulose, wheat bran, mill feed, cottonseed hulls, rice mill feed, and rice hulls.

Factors other than high fiber and low available energy have been studied to explain why pigs will not voluntarily consume sufficient amounts of a high alfalfa diet to allow satisfactory growth performance. Research results have shown that the saponins in alfalfa play a major role in the growth depression of pigs fed diets containing high levels of alfalfa meal. The mechanism by which alfalfa saponins caused growth depression has not been definitely determined. Possible mechanisms include depressing feed intake, complexing with nutrients to render them unavailable, inhibiting digestive enzyme activity, and inhibiting cellular metabolism.

Although saponins are capable of *in vitro* binding of protein, which could inhibit protein digestion and have also been implicated as trypsin inhibitors, saponin content in the diet (either from high or low saponin alfalfa or from isolated saponin) has no apparent effect on digestibility of crude protein, ether extract, or cell wall constituents.

Method of processing and drying has been indicated as a possible factor limiting the efficient utilization of alfalfa meal by swine. Commercial dehydration of alfalfa involves temperatures in the range of 800°C at the inlet of the drying drum to 125°C at the outlet. High temperature dehydration can cause caramelization and a complexing of the carbohydrates with amino acids (Maillard reaction). Bitterness may be imparted by caramelization.

Alfalfa has shown its most beneficial effect as a feedstuff for swine during the reproductive phase. Since it was shown that factors supplied in alfalfa improve reproductive performance of sows fed in dry lot, it has become a common practice to include 15% or more of alfalfa meal in breeding and gestation rations for swine. Danielson and Noonan (1975) and Pollmann et al. (1979) fed 96.75% sun-cured alfalfa diets through three successive gestation cycles and maintained satisfactory reproductive performance.

Alfalfa in brood sow diets has been reported to increase ovulation rate, litter size at birth, and baby pig survival. A significant reduction in birth weight of pigs farrowed from sows fed diets based primarily on alfalfa appears to be the only negative effect on reproductive performance (Danielson and Noonan 1975; Pollman et al. 1979, 1980).

Although sows fed diets high in alfalfa meal gained less weight during gestation, sow longevity is significantly increased. Results reported by Pollmann et al. (1980) indicate that a higher percentage (89 vs. 68%) of sows fed alfalfa (50% of the diet) completed three reproductive cycles than did similar groups fed the control, maize–soybean meal diet.

Alfalfa protein concentrate and alfalfa juice are products of the alfalfa plant resulting from heat coagulation and a juice-pressing process. Their lower content of fiber makes them attractive as protein and vitamin supplements (Pond and Maner 1984).

Single-Cell Protein

Consciously or not, man has always used microorganisms such as nonpathogenic bacteria, yeast, fungi, and microalgae in his food system. However, little use has, so far, been made of the microorganism-biomass itself as a component of feedstuffs. An exception is the microalgae *Spirulina* used as food in Chad by lakeshore people and in Mexico by the Aztecs living around Lake Texcoco, and yeast from industrial production used in mixed feeds for animals. In recent years, new emphasis has been placed on microorganisms for the purpose of developing methods of food production that would be independent of the traditional soil/climate ecosystem.

Several factors have stimulated the recent research and development interest in single-cell protein. First, the growing demand for human food has created a shortage of protein concentrates such as oilseed and soybean meal, and second, there is concern over environmental pollution resulting from industrial and agricultural waste products such as sulphite waste and animal waste.

Substrates for microbial growth and protein recovery have been evaluated against such criteria as technical suitability, availability and cost. The last is important because substrates, such as molasses and paraffins, may have alternate industrial demands and consequently be competitively priced. On the other hand, the substrate may be an industrial or agricultural by-product or waste material such as sulphite waste liquor, potato starch waste, or animal fecal waste that might even raise disposal problems. At the present time, research and development work is being carried out to recover nutrients from effluents and waste through fermentation technology and to produce single-cell protein from *n*-alkanes, methane, and synthetic methanol and ethanol.

Whatever the substrate may be, the end product is a biomass composed of bacteria, yeast, fungi, or algae. The original term coined for this biomass was *single-cell protein* (SCP), which is somewhat misleading since many of the fungi are not monocellular.

Single-cell protein has been undergoing thorough investigation with regard to nutritional value and safety for human and animal consumption. Such investigation has been centered on SCP grown on such hydrocarbons as *n*-alkanes and methanol. Studies of SCP grown on animal waste have also stressed the effect of accumulated feed additives such as trace elements, antibiotics, and arsenicals. On the other hand, yeast (*Candida utilis*), grown mainly on sulphite waste liquor, molasses, whey, or other carbohydrates, has never been subjected to such scrutiny. Because of differences in chemical and physical characteristics and nutritional value, fungi, yeast, algae, and bacterial protein will be discussed separately.

Fungi

In Finland, a process for the commercial production of a micro-fungus, *Paecilomyces variottii*, has been perfected. The process involves the continuous culture of the fungi on spent sulphite liquor, a waste by-product of the wood pulping industry. In addition to producing a multicellular and filamentous fungus that is readily separated from the broth by simple filtration, the process removes acetic acid and sugars from the spent liquor, hence reducing its biological oxygen demand and pollution characteristics. Following mechanical dewatering and artificial drying, the material, which has been given the name *Pekilo protein,* is cream colored, has very little taste or odor, and is readily accepted by swine as a constituent of their diet.

Pekilo protein has been successfully utilized as a sole protein source to supplement cereal-based diets for early weaned pigs and for growing-finishing pigs. Finnish workers replaced 0, 33, 66, and 100% of the digestible crude protein of skim milk powder with the digestible crude protein of Pekilo in a barley-based diet for growing-finishing pigs. For 20–90-kg liveweight, all levels of Pekilo protein supported pig performance equal to that recorded when skim milk powder was used as a protein supplement.

Others were also unable to show any difference in nutritive value between Pekilo protein and white-fish meal when both were used as the sole protein supplement to cereal-based diets for pigs between 20- and 90-kg liveweight. Pig performance was not different when Pekilo protein replaced white-fish meal at both a standard and low level of protein.

Since no gross or microscopic abnormalities have been observed in pigs fed Pekilo protein, it can be fed as the only supplemental protein source for starter, growing, and finishing pigs. Although no research data are available for reproducing animals, Pekilo should be an excellent supplemental protein source for them.

Bacteria

The intense efforts to make use of microbial fermentation for the production of feeds containing protein have shown the potential of bacterial cells as an alternative to conventional protein sources. The Imperial Chemical Industries of Great Britain have used the bacteria, *Methylophilus methylotrophus* (more recently classified as *Pseudomonas methylotropha*), to produce a bacterial protein of high biological value. The product is produced by continuous fermentation of the bacteria on low concentrations of methanol as the energy substrate and ammonia as the nitrogen source. It contains 72–79% crude protein (N × 6.25) of which approximatey 15% is nucleic acid.

In feeding trials with growing and growing-finishing pigs, bacterial protein has been used as a replacement for white-fish meal in diets based on barley and fine wheat offal. It supported pig performance similar to that of pigs fed white-fish meal as a protein supplement. There were no significant differences in growth rate, feed efficiency, nitrogen retention, or apparent digestibility of nitrogen, energy, or most of the amino acids between diets supplemented with bacterial protein and those supplemented with white-fish meal. The pigs were also shown to be able to dispose of nucleic acid at the levels of intake tested (0.49–1.13% nucleic acid in the diet).

As a protein source for early weaned pigs, this bacterial protein has produced variable results. Whittemore et al. (1976) compared dried bacterial cells with white-fish meal at 10 and 20% levels of inclusion in diets for piglets from 21 to 42 days of age. At both levels of inclusion in mixed cereal–whey diets, pig performance was not different from that of pigs fed white-fish meal. However, in studies by Newport and Keal (1980), who replaced 50, 75, and 100% of a dried skim milk diet with a spray-dried powder containing 30.8% dried bacterial cells, levels of replacement higher than 50% reduced both average daily gains and efficiency of feed utilization of piglets between 2 and 28 days. It also has been used as the only supplemental protein source to cereal-based diets for breeding pigs through six parities (Waterworth et al. 1978). Diets containing 0 or 8% dried bacterial cells, replacing all of the noncereal protein were given to sister pair pigs from 8 weeks of age to weaning of their sixth litter. There were no differences between treatments for growth rate to first service or for the number of gilts reaching sexual maturity. There was no difference between treatments for sow fertility as measured by the period from weaning to effective service. There were no significant dietary effects on the numbers of pigs born alive, average birth weight, or number and weight of pigs at weaning. Twenty-five sows killed for examination revealed no health problems or pathological changes.

These results indicate that, in general, bacterial protein can be used to supply all or a major portion of the supplemental protein in cereal-based diets for both starter and growing-finishing pigs and for reproducing animals during growth, gestation, and lactation. However, additional research is needed to establish satisfactory dietary levels of bacterial protein that can be fed to pigs weaned at a very early age.

Yeast

Bakers' yeast (*Saccharomyces cerevisiae*) and torula yeast (*Candida utilis*) have been used for many years to supply a portion of the supplemental protein in creep and starter diets for early weaned pigs and in some countries as the principal supplemental protein source for swine of all ages. In recent years, increasing attention has been paid to the problem of the world shortage of protein and especially protein of high quality. A search for new sources led to the production of yeast (*Candida lypolytica, Candida boidinii,* and *Pichia aganobii*) on hydrocarbons obtained from oil. A culture of yeast is propagated by blending with petroleum hydrocarbons, a nitrogen source, such as ammonia, and an appropriate mixture of mineral salts. Production of yeast (*Candida utilis*) on sulphite spent liquor also has been increased. Under controlled conditions, the yeast culture multiplies and results in a very efficiently produced protein product which can be washed and dried for animal feed. Although the nutritive content varies somewhat with species and production method, the content of amino acids is quite similar among strains. The methionine–cystine content of all yeasts appears to be a limiting factor.

The use of yeast in animal feeding is subject to certain essential conditions such as: competitive production costs in relation to those of conventional protein sources; an amino acid composition which, together with cereals, can satisfy the needs of animals; and the absence of toxic and carcinogenic effects, both on the animals being fed and on humans who consume the products derived from it. The toxic effect on humans has been fully studied in The Netherlands and U.S.S.R. by administering large quantities of yeast, or products (meat, eggs) from animals fed with yeasts, to thousands of rats and mice over several generations. The results of these tests showed no toxic or carcinogenic effects. In recent years many reports of feeding trials with pigs have been published. Many groups of experimenters have included hydrocarbon yeast in growing-finishing diets in amounts of 5–15%, and in one test up to 30%. On the average, the performance in these trials of pigs fed with yeast was just as good as that of animals which were given soybean meal and/or fish meal.

The inclusion of up to 15% paraffin yeast in piglet feeds also has proved successful. However, with yeast levels of more than 20% in piglet feeds, adjustment of methionine–cystine level is necessary.

Although care should be taken to assure adequate levels of dietary sulphur–amino acids, especially for early weaned piglets, yeast grown on hydrocarbons, sulphite spent liquor, or other products such as molasses and whey can be used at normal levels to supply all or a major portion of the supplementary protein to practical rations for all classes of swine.

Algae

There is keen interest from the theoretical and applied standpoint, in the mass production of algae. Their photosynthetic ability, which eliminates the need for applying a carbon substrate and their efficiency of solar energy utilization give algae particular appeal as a protein source for both human and animal food. Although algae are among the more attractive of the microbial protein sources, having the potential ability to produce approximately 10 times as much protein per unit land area as soybeans, the difficulty of practical production and harvest reduces their immediate potential.

Most researchers have concentrated primarily on the technology of production and nutritional toxicological evaluation. Only a few species of algae have been sufficiently studied from the viewpoint of their growth and production characteristics. Among them are the green algae (*Scenedesmus acutus*) and *Chlorella pyrenidosa*) and the blue–green algae (*Arthrospira platensis* and *Spirulina maxima*).

Considerable industrial effort has been expended on the use of the effluent from methane generation for the production of algae such as *Chlorella* or *Scenedesmus* as a source of single-cell proteins. Yet a major difficulty is the harvesting of the cells because they are very small and require flocculation or centrifugation which are both expensive processes. Blue–green algae (*Arthrospira platensis*) can be successfully grown on swine effluent even after it has been first utilized for methane generation. The major advantage of *Arthrospira* sp. is that, being filamentous, it can be harvested by filtration, and centrifugation or flocculation are not required.

Chemically, the algae contain levels of crude protein similar to or higher than those found in soybean meal and a fiber content similar to that found in barley. The ash and ether extract levels are quite variable. Calcium, phosphorus, silica, magnesium, and potassium are the principal mineral components.

The green algae (*Scenedesmus* and *Chlorella* sp.) have an unpleasant, bitter taste that reduces palatability and limits the quantity that can be used in both human and animal diets. In addition to the problem of acceptability, protein availability is also limited by the durable cell wall of the algae. The cell wall cannot be broken easily by the nonruminant animal and, therefore, protein release and digestibility are reduced. Protein digestibility by swine has been reported to be between 66 and 72%. Because of its bitter taste and protein of low biological value, green algae should not be utilized in quantities greater than about 10% in diets for growing swine. It can be used to replace as much as 75% of the protein supplement in fish meal–barley or meat-and-bone scrap–barley diets without decreasing growth rate or efficiency of feed utilization.

Blue–green algae has a protein content which varies from 50 to 61% (Chung et al. 1978; Devi et al. 1981) and of which nearly 9.9% is nonprotein nitrogen. Chung et al. (1978) have shown the protein of *Arthrospira platensis* to be 76.7 digestible and to have a protein efficiency ratio (PER) of 2.25 and a biological value of 67.7%. Feeding trials with rats and 3-day-old pigs have shown *Arthrospira* to be inferior to casein and soybean meal as a protein source (Yap et al. 1982; Chung et al. 1978). More extensive swine feeding trials are needed before its value as a supplemental protein source can be determined.

Extensive toxicological evaluation of animals fed algae species grown on animal waste and sewage waste water have not shown any evidence of toxicity. Yap et al. (1982) also were unable to show any evidence of toxicity or changes in hemoglobin, hematocrit, serum protein, albumin, or urea when *Arthrospira platensis, Spirulina maxima,* or *Chlorella* sp. were fed for 12 days to 3-day-old pigs.

Seaweed

Seaweed is plentiful in many areas of the world. It offers a cheap source of swine feed if it can be used to supplement the diet. There are many species of seaweed, each undergoing considerable seasonal variation in chemical composition. Seaweed contains relatively low levels of crude protein (2.5–15%) and energy (526 kcal DE/kg DM) and significant quantities of minerals (ash, 15–50%).

The effect of seaweed meal on digestibility of dietary components has been studied by a number of workers. All results agree that the addition of seaweed meal lowers the digestibility of the protein, fiber, organic matter, nitrogen-free extract, and dry matter, and reduces nitrogen retention.

It is obvious from the available data that seaweed is not a good source of either protein or energy and should be used basically as a mineral and especially trace mineral supplement at low dietary levels. Although dietary levels up to 10% have been used successfully, even lower levels have caused severe diarrhea. Maximum value and utilization will probably be realized if no more than 5% is included in the diet.

Synthetic Amino Acids

For many years the feed industry has had pure lysine and methionine available from microbiological production at prices competitive for use in swine diets in limited amounts. However, the cost of producing other amino acids such as tryptophan and threonine, whose concentrations in most plant proteins are below the amounts required by animals, has been prohibitive in the past. Now, gene-splicing techniques are available to produce micoorganisms capable of producing large amounts of these and other amino acids. These breakthroughs in biotechnology (United States Congress, OTA 1984) likely will provide the basis for greater use of synthetic amino acids and a reduction in conventional plant and animal protein use in the future of swine production. In 1982, worldwide sales of amino acids totalled 455,000 metric tons (United States Congress, OTA 1984), one-third of which were used in animal feeds. An annual growth rate of 7–10% is projected for the next few years.

MINERAL AND VITAMIN SOURCES AND NONNUTRITIVE FEED ADDITIVES

Although the largest proportion of the diet is made up of energy and protein, the mineral elements and vitamins are vital to normal growth and reproduction. Most energy and protein sources provide some vitamins and minerals but often it is necessary to supplement the diet with specific mineral and vitamin sources to balance the diet. In recent years, the use of nonnutritive feed additives such as antibiotics has become a common practice as a means of further increasing weight gain and efficiency of feed utilization beyond that obtained with a nutritionally balanced diet. This section is devoted to a description of feedstuffs and substances commonly used to supply vitamins, minerals, and nonnutritive additives so as to provide an optimum diet for all phases of the life cycle of the pig. Physiological functions of minerals and vitamins are covered in Pond and Maner (1984), Church and Pond (1988), and in other nutrition textbooks.

Mineral Sources

Sodium (Na) and Chlorine (Cl)

Almost all feedstuffs are deficient in Na and Cl; therefore, common salt is routinely added to virtually all mixed diets for swine. The level commonly added is 0.4–0.5% of the diet. It has been shown that a level of 0.20–0.25% supplemental NaCl will meet the dietary sodium and chlorine requirements of the growing-finishing pig fed grain–soybean meal diets.

Inadequate dietary levels of NaCl will significantly affect feed and water consumption and result in unthriftiness and slow growth. Long-term salt deficiency may lead to salt craving and death. In the presence of adequate drinking water, swine can tolerate dietary levels of NaCl as high as 2–3% for several days; however, when water is restricted, as little as 1.0% dietary NaCl has produced signs of toxicity which include nervousness, weakness, tremors, incoordination, convulsions, paralysis, coma, and death.

Salt is often used as a convenient carrier for trace elements. Thus, for a nominal cost, essential trace elements such as zinc, copper, iron, and manganese may be incorporated into the salt in amounts appropriate to supply the needed trace elements in the right proportions when the salt is consumed at the accepted dietary level (0.5%). The composition of a typical "trace mineralized salt" available commercially is shown in Table 9.3. It should be emphasized that the mere presence of a trace element in a particular trace mineralized salt preparation does not ensure an adequate intake. For example, in the sample salt preparation shown in Table 9.3, the amount of zinc is only 0.008% or 80 ppm in the salt. When the salt is incorporated into the diet at 0.5%, the amount of zinc contributed by the salt is only 0.4 ppm, a negligible quantity in terms of the pig's requirement of 50 ppm or more. Thus, in the use of micronutrient supplements, special attention must be given to the amount of a particular nutrient provided, for it may be present in only a token amount.

Calcium (Ca)

Calcium compounds are widely distributed in feedstuffs, and Ca is the most abundant cation in the animal body, comprising 1–2% of the total body weight. Approximately 99% of the Ca in the animal body is found in the bone and teeth, with the remaining 1% distributed in the soft tissues.

Almost all feedstuffs contain some Ca; however, the calcium content varies widely and is very low in maize and the cereal grains such as sorghum, barley, wheat, and oats (0.02–0.10%) and in plant protein supplements (0.17–0.66%) such as soybean meal, rapeseed meal, and cottonseed meal that normally supply the major proportion of protein supplement swine diets. Only animal protein sources such as fish meal, meat meal, meat-and-bone meal, and dried milk products contribute significant quantities of Ca. Because most natural feedstuffs are deficient in Ca, specific supplements high in Ca are almost always required. Supplemental sources of Ca used in swine diets include ground calcitic or dolomite limestone or calcium carbonate, oyster shell, high grade calcite, gypsum, and wood ashes. More common supplements are those that contain both Ca and P from sources such as bone meal, dicalcium phosphate, and defluorinated rock phosphate (Table 9.4).

TABLE 9.3 Composition of a Typical "Trace Mineralized Salt"[a] Used at 0.5% of a Mixed Diet for All Ages of Pigs

Mineral[b]	*Percent*	*Amount in complete diet*[c] *when salt is at 0.5%(ppm)*
Sodium chloride (salt)	Not less than 97.00 nor more than 99.00	5,000
Cobalt	Not less than 0.015	0.70
Copper	Not less than 0.023	1.15
Iodine	Not less than 0.007	0.35
Iron	Not less than 0.117	5.4
Manganese	Not less than 0.225	11.25
Sulfur	Not less than 0.040	2.0
Zinc	Not less than 0.008	0.40

[a] Guaranteed analysis.

[b] The ingredients supplying the above minerals are: salt, cobalt carbonate, copper oxide, calcium iodate, iron carbonate, manganese oxide, sodium sulfate, zinc oxide; yellow prussate of soda added as an anti-caking agent.

[c] The amount of zinc, and in some cases other trace elements as well, contributed by salt at 0.5% of the diet is negligible in terms of the pig's needs in the above preparation.

TABLE 9.4 Calcium and Phosphorus Content of Common Mineral Supplements (in Percent)

Supplement	*Calcium*	*Phosphorus*
Bone meal, raw	21	9–10
Bone meal, steamed	24–29	12–14
Bone black, spent	27	12
Dicalcium phosphate	23–26	18–21
Defluorinated rock phosphate[a]	31–34	13–17
Raw-rock phosphate[b]	24–29	13–15
Soft phosphate[a]	18	9
Curaçao phosphate[a]	35	15
Oyster shell	38	—
Limestone	38	—
Calcite, high grade	34	—
Gypsum	22	—
Dolomite limestone	22	—
Wood ashes	21	—

[a] Usually less than 0.2% F.
[b] 2.0–4.0% F.

Source: Sievert 1967.

Dietary calcium toxicity *per se* is not considered a problem with swine, as the homeostatic mechanism of the animal tends to protect against the absorption of excessive quantities of the element. Calcium excesses are important because of the interrelationship of the element with other nutrients and especially with P and zinc (Zn). Large excesses of Ca can upset the desired calcium-to-phosphorus ratio, reduce gains and feed efficiency, and induce zinc deficiency (parakeratosis).

Phosphorus (P)

Although a total phosphorus requirement of between 0.40 to 0.70% has been established for swine, many dietary factors may influence these requirements. The amount absorbed is dependent on source, intestinal pH, lactose intake, and dietary levels of Ca, P, Vitamin D, iron (Fe), aluminum (Al), manganese (Mn), potassium (K), magnesium (Mg), and fat. Signs of phosphorus deficiency include poor growth, rickets, and osteomalacia and are closely associated with those for Ca and vitamin D.

Phosphorus is present in variable amounts in almost all feedstuffs used by swine. Unfortunately, the principal form of P in most cereal grains and plant protein sources is phytate (mixed salt of *myo*-inositol hexaphosphoric acid or phytic acid), which can be utilized by swine only after the compound has been hydrolyzed to yield P. Intestinal phosphatase or phytases are ineffective in hydrolyzing phytate in swine diets and only phytases of plant or microbial origin are effective in hydrolyzing significant amounts of phytate. Although available phosphorus values for plant sources are usually less than the content of nonphytate phosphorus, the higher availability values for wheat and barley compared to other grains are probably related to the amount of phytase present in the grains. Phytate may undergo some hydrolysis during certain types of feed or ingredient treatment or processing such as pelleting, fermentation, and acid treatment; however, the amount released is small in relation to the animal needs. Estimates of the available phosphorus in some commonly used feedstuffs are shown in Table 9.5.

Most swine diets require supplemental sources of P in addition to that present in the common feedstuffs. Unlike phytate phosphorus, most of the inorganic phosphorus sources are quite high in available P (Table 9.6). The supplements most often used in swine diets, dicalcium phosphate, defluorinated rock phosphate, and steamed bone meal, are of similar bioavailability. Phosphorus in meat-and-bone meal is as available as that in dicalcium phosphate and the same is assumed to be true for fish meal, tankage, dried milk, and other animal products.

Supplemental phosphates are not considered to be highly toxic. Single oral doses can be tolerated with little if any effects. On the other hand, long-term consumption of dietary levels two to three times the requirement level will cause severe problems due to induced changes in calcium metabolism. The effect of high phosphorus levels is very dependent on the calcium level of the diet, as well as other factors such as vitamin D. A relative excess of P in relation to Ca can result in increased bone resorption in young and adult animals and reduced growth in young swine. When tolerance is expressed in terms of a multiple of the requirement level, P has one of the lowest tolerance factors of any mineral element.

Zinc (Zn)

Zinc absorption is affected by dietary level and chemical form used in the diet. Zinc in swine feed is supplied predominantly as zinc oxide; however, other inorganic salts of Zn such as the carbonate, sulfate, chloride, nitrate, and acetate are well absorbed. Miller et al. (1981) indicated that for the young growing pig, Zn from metallic dust was about 30% more available than from zinc oxide powder. Zinc

TABLE 9.5 Biological Availability and Total and Available Phosphorus in Some Commonly Used Feedstuffs

Feedstuff	*Biological availability of P, %*	*Total P, %*	*Available P, %*
Maize	12	0.25	0.03
Grain sorghum	12	0.25	0.03
Oats	23	0.35	0.08
Barley	35	0.35	0.12
Wheat	48	0.35	0.17
Wheat bran	29	1.15	0.33
Wheat middlings	34	0.90	0.31
Rice bran	18	1.50	0.27
Soybean meal, dehulled	22	0.65	0.14
Cottonseed meal	0	1.00	0

Source: Adapted from Cromwell 1980.

TABLE 9.6 Bioavailability of Phosphorus in Some Inorganic and Animal Product Sources for Swine[a]

Source of P	*Availability of P, %*
Monosodium phosphate	100
Monocalcium phosphate	100
Dicalcium phosphate	98
Meat-and-bone meal[b]	98
Steamed bone meal	95
Diammonium phosphate	95
Defluorinated phosphate	92
Curaçao Island phosphate	69
Soft rock phosphate	34

Source: [a] adapted from Cromwell 1980. [b] Cromwell et al. 1976.

from certain ores such as sphalerite and franklinite are largely unabsorbed. The percentage of Zn in several supplements is shown in Table 9.7. Some feedstuffs high in Zn are listed in Table 9.8.

Iron (Fe)

The amount of Fe that must be absorbed from feed in order to maintain normal body iron levels is determined by the amount excreted or lost from hemorrhage and the needs for maintenance, growth, and reproduction. Iron is most important as a supplement during the suckling period. Although the intestine of the newborn pig has fully functional absorption capacity for Fe, suckling pigs are particularly prone to iron deficiency because (1) they have limited iron stores at birth; (2) their growth rate is phenomenal during the first few weeks of life; and (3) their main source of nutrients, sow's milk, contains a very limited concentration of Fe (about 1 mg/liter). The feeding of high levels of many different iron compounds to gestating and lactating sows has not effectively increased the iron-content of milk. Thus, unless a supplemental source of Fe is provided during the first few days after birth, anemia

TABLE 9.7 Zinc Content of Zinc Supplements and Amounts of Each Needed to Provide Zinc at 100 ppm (0.01%)

Zinc supplement	*Zinc content*	*Amount needed to provide 100 ppm Zn (ppm)*
Zinc carbonate	54	185
Zinc oxide	80	125
Zinc sulfate	40	250
Zinc metal powder	100	100

TABLE 9.8 Good Sources of Zinc for Swine

Source	*Zinc content, ppm—dry basis*
Nonfat dried milk	41
Blood meal	306
Brewers' dried grains	98
Dried whey, roller process	27
Defluorinated rock phosphate	38
Calcium caseinate, food grade	57
Coastal bermuda grass, dried	32[a]
Fish meal (menhaden)	147
Meal meal	103
Rapeseed meal (solvent)	66
Sesame meal (expeller)	100
Soybean meal (solvent)	27[b]
Sow milk	25[c]
Wheat bran	95
Wheat shorts	106

[a] Content varies with location where grown.

[b] The bioavailability may be low. Parakeratosis is often observed in pigs fed a maize-soybean meal diet unsupplemented with inorganic zinc salts.

[c] Pond et al. 1965.

Source: National Research Council 1988; Miller and Miller 1963.

develops. Injectable forms of Fe (iron dextran, iron dextrin) can be given intramuscularly at 2 or 3 days of age to supply 150–200 mg. Because of the higher biological availability of injectable Fe (over 90% of injectable iron is utilized), a much smaller quantity is required to maintain normal hemoglobin than when dietary Fe is the sole source. If one assumes a metabolic requirement of 15 mg of Fe per day, the total need for the first 10 days of life can be met by a single injection of 150 mg. Rapid growth and the pig's need to meet the total energy requirement by consumption of dry feed, beginning at about the third week of postnatal life, mean that attention must be given to the amount and biological availability of feed iron.

The postweaning dietary iron requirement is about 80 ppm and diminishes in later growth and maturity, since there is a smaller increase in blood volume with increasing body weight. Feed ingredients in a balanced diet usually supply enough Fe to meet postweaning requirements. Some feedstuffs high in Fe are listed in Table 9.9. Only a small part of the Fe consumed by the pig is absorbed. Efficiency of iron absorption from the diet can be influenced by many factors such as iron status of the animal, size of the dose, adequacy of major nutrients, presence and excess of other minerals, and biological availability of the iron source. Bioavailability values of Fe from natural feedstuffs fed to swine are not available. Relative biological value (RBV) of Fe in a few feedstuffs has been determined for the rat and chick. For maize, blood meal, oat flour, isolated soybean protein, and fish protein concentrate, RBV has been shown to be 73, 35, 21, 97, and 33–59%, respectively. Natural iron sources for growing pigs should come from animal products rather than forages because of the high fiber content of forages.

TABLE 9.9 Good Sources of Iron and Manganese for Swine

Feedstuffs	*Iron as-fed (ppm)*	*Manganese as-fed (ppm)*
Alfalfa meal	310	2.8
Blood meal	3000	Low
Bermuda grass, fresh	400	36.7
Bluegrass hay, all analyses	200	83.9
Bone meal	800	30.4
Brewers' dried grains	250	38
Clover hay, all analyses	800	73.5
Coconut meal	190	55.4
Cowpeas hay, all analyses	800	43.9
Fish meal, menhaden	440	33
Limestone	3800	279.6
Maize dried distillers' solubles	600	73.5
Molasses, cane	200	42.2
Rice bran	190	138
Vetch hay, all analyses	400	53.7

Source: National Research Council 1964, 1985.

Many chemical substances are being used as sources of oral iron. The availability of Fe varies from source to source. Ferrous sulfate heptahydrate ($FeSO_4 \cdot 7H_2O$) is the standard against which other iron sources usually are evaluated. It is highly available to animals. Ferrous sulfate monohydrate appears to be equally available and has better handling and mixing properties. Often one hears that the ferrous compounds are much more available than ferric sources of Fe. Although this is frequently true, as is the case when ferric oxide and ferrous sulfate are compared, not all ferric compounds have a low availability. The relative biological values for rats, chicks, and pigs of different iron compounds are listed in Table 9.10. Many of the ferric compounds are readily available to animals. The differences in availability of the various iron compounds are related to solubility in acid; absorption of Fe occurs in the upper duodenum where the influence of stomach acid secretion is greatest. Thus iron salts that are soluble in acid are more available for absorption than those of lower solubility. Such differences among iron sources in biological availability are large enough to be of practical importance in diets for growing-finishing swine.

Many compounds appear to form complexes with Fe in the gastrointestinal tract and to aid absorption by keeping Fe in solution during transit through the upper part of the small intestine. Some amino acids, simple sugars, and organic acids as well as ascorbic acid have been shown to increase iron absorption. Other dietary factors such as copper and phosphorus also influence iron absorption and toxicity; acute iron toxicity may occur in vitamin E–selenium-depleted baby pigs given Fe parenterally.

Copper (Cu)

Copper is absorbed from the upper gastrointestinal tract and especially from acid media in the stomach. Efficiency of copper absorption depends on age, the form of

TABLE 9.10 Relative Biological Value of Iron from Various Sources Using Ferrous Sulfate as a Standard

Iron source	*Chick–rat, %*	*Pig, %*
$FeSO_4 \cdot 7H_2O$	100	100
$FeSO_4$, anhydrous	100	100
$FeSO_4$, feed grade	100	100
$FeSO_4 \cdot H_2O$	—	100
$FeSO_4 \cdot 2H_2O$	—	100
$Fe(NH_4)_2(SO_4)_2 \cdot 6H_2O$	99	—
$FeCO_3$	2–6	0,14,27,60
$FeCl_2 \cdot 4H_2O$	98	—
$FeNH_4$ citrate	107	100
Fe choline citrate	102	—
Fe–Cu–Co choline citrate	—	140
Fe–citrate	73	100
Fe–glycerophosphate	93	—
$Fe_2(SO_4)_3$	83	—
$FeCl_3 \cdot 6H_2O$	44	—
NaFe pyrophosphate	5–26	33
Fe_2O_3	4	12
Reduced iron	18,37,63	33
Iron powder	—	70
Ferric poly phosphate	—	97
di Na EDTA	—	90

Source: Adapted from Miller et al. 1981; Zimmerman 1980; Fritz et al. 1970.

copper consumed, the copper status of the animal, and the presence of other dietary compounds such as phytates, calcium, zinc, iron, lead, silver, molybdenum, and ascorbic acid, which tend to chelate copper. Although in practice, copper sulfate is the most commonly used form of supplemental copper for swine, other soluble copper compounds such as copper carbonate, copper–methionine, and copper chloride can be utilized as well as copper sulphate in supplying Cu to swine diets. The Cu from all these compounds is more readily absorbed than that supplied by copper oxide, although this compound is commonly used as a copper supplement.

Iodine (I)

Iodine is widely distributed in animal feeds but in very small amounts. The iodine content of plant products varies widely depending on species of plant and the iodine content of the soil where grown. Samples of oilseed meals have been shown to contain 0.11–0.2 ppm I. Products of animal origin, other than fish meal, do not contain significant levels of I. To assure adequate dietary iodine, small quantities of supplemental iodine often are added to the diet as a constituent of the salt (NaCl) supplement. Iodine sources permitted by the Food and Drug Administration as feed additives include calcium iodate, calcium iodobehenate, cuprous iodide, 3,5-diiodosalicylic acid, ethylenediamine–dihydriodide (EDDI), pentocalcium orthoperiodate, potassium iodate, potassium iodide, sodium iodate, sodium iodide, and thymal iodide. Generally, iodine from these compounds is highly available to swine; however, loss of I can occur from some of these compounds when exposed to heat, moisture, or light and when improperly protected when mixed with other minerals. Calcium and potassium iodate and pentacalcium orthoperiodate are more stable in salt and feed mixtures than sodium or potassium iodides. If salt contains 0.007 ppm I, the requirement can be met by incorporating 0.2% salt in the diet.

Selenium (Se)

Although the quantitative dietary requirement for Se has not been established precisely for all classes of swine nor for all dietary and environmental circumstances, it is generally agreed that the selenium requirement is between 0.1 and 0.3 mg/kg of diet, and dietary adequacy is dependent on the bioavailability of the Se in the feed and upon concentrations of a number of other components, especially vitamin E. Because of their shared antiperoxidant role, the selenium requirement is inversely related to the vitamin E concentration in the diet. However, although a high dietary level of vitamin E will reduce the level of Se required, it will not completely eliminate the need for dietary selenium. Conversely, if natural diets are very low in vitamin E (0.3 IU/kg), selenium levels as high as 1.3 ppm may not prevent deficiency lesions in all pigs. Excess dietary levels of arsenic (As), cadmium (Cd), Cu, Zn, Fe, and oxidized unsaturated fat, as well as low dietary protein, moldy grain, infections, cold damp environments, and physical stress have been shown to interact with Se and vitamin E metabolism and to influence the development and severity of deficiency lesions.

There is a great geographical variability in selenium content of soils as reflected in crop content and in the natural incidence of selenium deficiency in swine in various parts of the world (Table 9.11). Supplemental selenium can be adequately

TABLE 9.11 Some Good Sources of Selenium of Canadian Origin for Swine

Feedstuffs	*Selenium content (ppm)*	
	Mean	*Range*
Linseed meal	1.05	0.65–1.51
Rapeseed meal	0.98	0.46–1.90
Brewers' grains	0.94	0.81–1.07
Malt sprouts	0.60	0.29–0.97
Herring meal	2.84	2.21–3.36
White-fish meal	1.83	1.65–2.18
Fish meal	1.96	1.25–2.94
Feather meal	0.72	0.60–0.88
Blood meal	0.69	0.51–1.16
Corn gluten meal	0.31	0.20–0.57
Distillers' grains	0.35	0.16–0.56

[a] The concentration of selenium in plant materials varies with the geographic location and growing conditions.

Source: Arthur 1971.

supplied by sodium selenite, sodium selenate, selenomethionine, or seleniferous maize. Because of the complexities of selenium nutrition and metabolism, bioavailability of various compounds has not been determined for swine. Bioavailability of selenium for poultry in some feedstuffs is presented in Table 9.12.

Of practical concern in some locations is selenium toxicity in areas where soil levels and, in turn, crop levels of selenium are excessive. Although a level of 5 mg Se per kilogram of diet supplied by sodium selenite, sodium selenate, selenomethionine, or seleniferous maize does not produce toxicity, levels from 7.5 to 10.0 mg/kg may produce toxicity.

Cobalt (Co)

No dietary requirement for Co has been established. Cobalt is found in variable quantities in plants and most feeds supply adequate cobalt for B_{12} synthesis by microorganisms in the gastrointestinal tract. Supplemental sources for ruminants include cobalt acetate, carbonate, chloride, oxide, and sulfate. Bioavailability of different cobalt compounds has not been determined for swine.

Adding Co as cobalt chloride to the diet at levels up to 200 ppm did not result in toxicosis in pigs fed a diet adequate in Fe. The addition of 400 or 600 ppm Co caused anorexia, stiff-leggedness, humpback, incoordination, and extreme muscular tremors. Decreased serum iron and anemia may occur in the presence of excess cobalt.

Manganese (Mn)

Manganese concentration varies greatly among plant species but it is widely distributed in feedstuffs (Table 9.9). Therefore, diets containing commonly used feed ingredients should contain more than adequate quantities of Mn. Manganese is

TABLE 9.12 Biological Availability for Poultry of Selenium in Feedstuffs

Feedstuffs	*Biological availability (%)*
Alfalfa meal	100
Brewers' grains	89
Brewers' yeast	81
Maize	83
Cottonseed meal	78
Herring meal	31
Menhaden meal	35
Meat-and-bone meal	36
Poultry by-product meal	33
Soybean meal	64
Tuna meal	33
Wheat	100

Source: Scott and Cantor 1972.

absorbed through the intestinal tract; however, there is evidence that absorption is decreased by feeding isolated soybean protein or excess levels of Ca, P, and Fe.

Manganous oxide and manganous sulfate are the two most commonly used forms in swine feeding; however, other sources such as manganese acetate, carbonate, citrate, chloride, gluconate, orthophosphate, and dibasic phosphate can be used. Data with chicks showed equal bioavailability of Mn from sulfate, chloride, carbonate and oxide.

The toxic level of Mn is many times the requirement.

Magnesium (Mg)

Magnesium is one of the major mineral elements recognized as essential for swine. However, because it is abundant in commonly used feedstuffs and supplementation of practical diets is not normally required, its significance in swine nutrition is often overlooked. Excellent reviews on magnesium functions, use, availability, and toxicity have been published (Sell 1979; Sell and Fontenot 1980; National Research Council 1980).

Since the levels of Mg present in common feed ingredients are quite high (Table 9.13), information about magnesium in swine nutrition is meager.

The adequacy of the Mg in common ingredients for meeting nutritional needs depends primarily on the degree of availability of this element for absorption by the animals. Since deficiencies are not commonly seen in swine fed natural ingredients, little bioavailability data are available for swine. From balance studies with young pigs, Bartley et al. (1961) obtained Mg retention values ranging from 16 to 35% for magnesium carbonate and magnesium sulfate. Miller (1971) reported that growing-finishing pigs retained 10–40% of the Mg ingested when fed maize–soybean meal diets. Related data of true and apparent Mg bioavailability of some organic and inorganic sources obtained with rats and chicks are tabulated in Tables 9.14 and

TABLE 9.13 Magnesium Content of Selected Feedstuffs

Feedstuff	*Mg, %*	*Feedstuff*	*Mg, %*
Barley	0.14	Alfalfa meal, dehyd. 17%	0.26
Maize	0.12	Blood meal, spray or ring dried	0.22
Oats	0.16	Cottonseed meal, solvent	0.40
Sorghum	0.20	Feather meal	0.20
Wheat	0.17	Fish meal, menhaden	0.16
Brewers' dried grains	0.16	Meat-and-bone meal, 50%	1.12
Maize gluten feed	0.29	Peanut meal, solvent	0.40
Maize distillers' solubles, dehyd.	0.64	Rapeseed meal, solvent	0.51
Molasses, cane	0.35	Safflower meal, solvent	0.37
Oat groats (dehulled)	0.09	Sesame meal, expeller	0.86
Rice bran	0.95	Soybean meal, solvent	0.27
Rice polishings	0.65	Sunflower meal	0.75
Wheat bran	0.52	Skim milk, dried	0.11
Wheat middlings	0.29	Whey, dried	0.13

Source: National Research Council 1988.

TABLE 9.14 Apparent Absorption of Magnesium in Various Chemical Compounds by Rats

Form of magnesium	*Apparent absorption, %*[b]		*Bioavailability*[b]
$MgCO_3$	68.3	64.9	75.1
$MgCl_2$	—	61.0	73.3
MgO	—	58.0	73.0
$MgPO_4$	66.5	54.1	74.2
$MgSO_4$	—	53.3	73.0
$MgSiO_3$	—	54.2	—
Mg phytate, acidic	62.4	—	—
Mg phytate, basic	68.1	—	—
Mg lactate	—	—	75.1
Mg citrate	—	—	73.8
Mg acetate	—	—	76.0

[a] First column data from Meintzer and Steenbock 1955; second column from Cook 1973.
[b] Data from Ranhotra et al. 1976.
Source: Adapted from Sell and Fontenot 1980.

9.15. Although comparative data are not available, it is assumed that swine are able to utilize Mg at rates similar to those of rats and chicks.

Potassium (K)

The potassium content of most feedstuffs for swine is considered to be adequate to meet metabolic requirements. Supplemental potassium apparently has a sparing effect on the dietary lysine requirement. Further work is needed to clarify the conditions of this potassium–lysine interaction in promoting weight gain in young pigs, but it seems clear that the balance of dietary electrolytes affects lysine utilization in pigs and that potassium supplementation of diets marginal in lysine

TABLE 9.15 True Absorption of Magnesium in Magnesium Sulfate and Selected Feedstuffs as Determined in Chickens

Magnesium source	*True absorption, %*
$MgSO_4 \cdot 7H_2O$	57.4
Maize	55.7
Wheat	56.8
Oats	82.7
Barley	54.5
Rice	60.3
Soybean meal	62.0
Dried skim milk	42.5

Source: Guenter and Sell 1974.

may be economical under certain conditions. Supplemental potassium levels of 0.55–1.44% (as potassium bicarbonate) improved weight gain of pigs fed diets containing 0.43% lysine, but not when higher lysine was present (Austic et al. 1982). Optimum ratios of potassium to lysine in the diet are unknown and warrant further study.

Vitamin Sources

Many feedstuffs commonly used as energy and protein sources for swine contain appreciable amounts of fat-soluble vitamins (A, D, E, and K) and water-soluble vitamins (thiamin, riboflavin, pantothenic acid, vitamin B_6, vitamin B_{12}, niacin, choline, biotin, and folacin), but many supply one or more vitamins in amounts inadequate to meet metabolic needs. Inclusion in the diet of a variety of feedstuffs of plant and animal origin made it possible to avoid serious vitamin deficiencies in earlier times before the identities and metabolic functions of the vitamins were known and, indeed, before the term vitamin was even coined. In general, animal products, including milk, liver, and muscle, and green forages are rich sources of most of the vitamins while seeds (such as maize and cereal grains) and roots and tubers (such as potatoes, and cassava) are poor to fair sources. In practice, the following vitamins are most likely to be in limited supply in natural diets: riboflavin, choline, pantothenic acid, niacin, vitamin B_{12}, vitamin A, vitamin E, and vitamin D. Rich sourcs of these are listed in Tables 9.16–9.20. Vitamin K and biotin are synthesized in adequate quantities by intestinal microflora to overcome dietary deficiencies under most conditions. Recently, field reports have indicated the possibility of clinical signs of vitamin K and biotin deficiency among growing pigs raised on slatted floors. Under such housing conditions, coprophagy would be minimized and the ingestion of vitamins synthesized in the lower intestinal tract could be decreased. Most commercial swine feeds now contain supplemental vitamin K and biotin in addition to those routinely added. Some evidence indicates a beneficial effect of vitamin C addition to diets of weaned pigs under certain conditions (Yen and Pond 1981). Still unknown are (1) the mode of action under which vitamin C synthesis becomes insufficient to meet the needs of the animal and (2) the exact conditions under which a positive growth response to supplemental vitamin C can be expected. As husbandry conditions change, nutritionists need to keep alert to the need for corresponding changes in vitamin fortification of diets to meet new conditions.

The increase in confinement rearing of swine, in indoor environment on concrete floors or slats, has been accompanied by a decrease in the utilization of pastures and forages. While such crops cannot be considered important sources of energy, protein, and major minerals (calcium, phosphorus, sodium, chlorine) for swine, they most certainly have provided, in the past, a vital source of fat-soluble and water-soluble vitamins as well as of trace elements. The move to confinement, in addition to removing an important source of nutrients associated with forage consumption, has deprived the pig of rooting in the soil, which is a rich source of trace elements (discussed earlier in this chapter) and has often deprived the pig of sunlight, important in vitamin D_3 synthesis by conversion from 7-dehydrocholesterol present in the pig's skin and other tissues.

Through technology it is now possible to add all of the vitamins economically to the diets of swine in crystalline or highly purified form. Thus, a typical diet for a growing-finishing pig in the United States contains maize (as the energy source), soybean meal (as the protein source), trace mineralized salt, dicalcium phosphate (or another calcium-phosphorus supplement), and a commercially prepared vitamin supplement added at 0.1–0.5% of the diet in a soybean meal carrier. However, in many situations it may be more economical to provide vitamins from natural products as components of the energy or protein fraction of the diet. Fortunately, the B vitamins, riboflavin, pantothenic acid, niacin, and choline, tend to occur together in abundant quantities in specific plant and animal products, so that selecting a source high in riboflavin, for example, generally results in a substantial supply of the others as well. Table 9.16 illustrates this tendency. The only exception among the feedstuffs listed is dried milk whey, which is a good source of riboflavin, pantothenic acid, and niacin but a rather poor source of choline. These four vitamins are found in abundant amounts in forages, distillers' and fermentation by-products, fish and animal muscle, and some of the oilseed meals. Niacin, although present in abundance in many of the cereal grains, is almost completely biologically unavailable from these grains as well as from maize, so that these seeds must be considered as zero as sources of niacin.

Vitamin B_{12}, on the other hand, is strictly a product of microbial origin and is found in large quantities in fish, milk, and meat products but not in plants. This phenomenon is of historical interest in that vitamin B_{12}, until its discovery and isolation in 1948, was called APF (animal protein factor). It occurs in animal

TABLE 9.16 Good Sources of Riboflavin, Pantothenic Acid, Niacin, and Choline for Swine (mg/kg)

Feedstuffs	*Riboflavin*	*Pantothenic acid*	*Niacin*	*Choline*
Alfalfa meal, dehydrated	13.4	27.1	19.1	1,016
Clover hay, red, all analyses	15.6	9.9	37.2	—
Cottonseed meal	5.3	10.3	33.9	2,776
Dried fermentation solubles	171.4	202.8	215.6	3,113
Dried milk whey	29.9	47.7	11.2	Low
Dried whole milk	19.6	22.7	8.4	—
Dried citrus pulp	2.4	13.0	21.6	845
Dried fish solubles	7.7	44.9	231.1	5,223
Fish meal, herring	9.0	11.4	88.9	4,004
Fish meal, menhaden	4.8	8.8	55.9	3,080
Linseed meal	3.5	17.8	35.6	1,863
Maize distillers' dried grains	3.1	5.9	42.2	1,859
Meat meal	5.3	4.8	56.9	1,955
Milo distillers' dried solubles	14.7	26.4	140.8	2,222
Peanut meal	5.3	48.2	169.0	1,683
Soybean meal	3.3	14.5	26.8	2,743
Wheat germ meal, defatted	5.1	12.3	51.3	3,375
Yeast, brewers' dried	35.0	109.8	447.5	3,885
Yeast, torula dried	44.4	82.9	500.3	2,911

Source: National Research Council 1964, 1968; Morrison 1956.

products in smaller quantities (a few micrograms per kilogram) than other vitamins (Table 9.17) but is also required in smaller quantities by swine.

Among the fat soluble vitamins, A and E are abundant in green forages (Tables 9.18 and 9.19). Vitamin A occurs mainly as its precursor, carotene, and vitamin E is present as a mixture of tocopherols, the most biologically active of which is alpha-tocopherol. Yellow maize is a good source of carotene, but the cereal grains and most roots and tubers (except carrots, sweet potatoes, and yams) are low or devoid of carotene and vitamin A. Because vitamin E is contained in high concentrations in the germ fraction of seeds, whole seeds are fair sources while germ-rich portions are excellent sources. Fish meal and liver are also good sources. Until recent years the tocopherol content of yellow maize was considered adequate, but frequent reports of vitamin E and selenium deficiency in swine have necessitated reevaluation and to avoid a deficiency it is recommended that alphatocopherol be added to maize–soybean meal diets for swine.

The metabolic requirement for vitamin D can be met in two ways: by ultraviolet irradiation of the skin sterols of the pig to produce cholecalciferol (vitamin D_3) or by ingestion of plants that have undergone ultraviolet irradiation of plant sterols to produce ergocalciferol (vitamin D_2). In the first method, no dietary requirement exists, but when sunlight is withheld, a dietary source is required. Table 9.20 lists some rich sources of vitamin D. Sun-cured forages, irradiated yeast, and cod-liver oil meal are the best sources of vitamin D. As the trend toward life-cycle confinement in the production of swine continues, more emphasis must be placed on meeting the vitamin D needs for normal skeletal development.

Nonnutritive Feed Additives

The use of nonnutritive additives in the feed of pigs has become common in feed preparations because they increase growth rate, increase efficiency of feed utilization, and reduce mortality and morbidity from clinical and subclinical infections.

Antibiotics were among the first nonnutritive additives to be routinely used in

TABLE 9.17 Good Sources of Vitamin B_{12} for Swine

Source	*Vitamin B_{12} (μg/kg)*
Blood meal[a]	440
Condensed fish solubles[b]	660
Crab meal[b]	418
Dried skim milk[c]	20–55
Dried whey[c]	22
Fish meal[a]	172–218
Meat-and-bone scrap[c]	110–125
Poultry by-product[b]	264

Source: [a] National Research Council 1988; [b] Sievert 1967; [c] Becker et al. 1963.

TABLE 9.18 Good Sources of Carotene[a] and/or Vitamin A[b] for Swine

Source	Vitamin A (IU/kg)	Carotene (mg/kg)
Alfalfa hay, barn cured	63,200	37.9
Alfalfa leaf meal	104,200	62.4
Clover hay	99,000	59.4
Guinea grass, fresh	105,500	63.3
Lespedeza hay	71,300	42.8
Maize, yellow dent	6,800	4.1
Soybean hay	53,000	31.8

[a] 0.6 μg β-carotene = 1 IU vitamin A activity; 1.0 mg β-carotene = 1667 IU vitamin A activity.

[b] 1 IU vitamin A = 1 U.S.P. unit vitamin A = 0.300 μg crystalline vitamin A alcohol = 0.344 μg vitamin A acetate = 0.550 μg vitamin A palmitate.

Source: National Research Council 1964.

swine diets. Antibiotics have been proven effective in improving pig performance and/or effective in control of specific and nonspecific diseases.

Low, subtherapeutic levels of antibiotics are fed primarily to improve growth rate and feed efficiency as well as to control infections. The magnitude of the response to antibiotics varies with stage of life cycle or stage of production and the environmental conditions to which the animals are exposed. The response is greater in young animals than in more mature animals. It is also greater during

TABLE 9.19 Good Sources of Vitamin E for Swine

Feedstuffs	α-tocopherol (mg/kg)
Alfalfa leaf meal[a]	383
Alfalfa meal, dried[b]	125
Orchard grass, dried[a]	222
Bluegrass, Kentucky, dried[a]	345
Fish meal[a]	209
Maize, yellow[a]	24
Maize distillers' solubles, dried[b]	55
Soybean seed[a]	36.5
Soybean hay[a]	27
Oats[a]	36.5
Rice bran[b]	60
Wheat[a]	34
Wheat germ oil[a]	1897

Source: [a] Morrison 1956. [b] National Research Council 1988.

TABLE 9.20 Good Sources of Vitamin D for Swine

Source	*Vitamin D activity (IU/kg)*
Alfalfa leaves from sun-cured hay	1,043
Alfalfa hay, sun-cured	1,800
Alfalfa hay, barn-cured	500
Clover hay, field-cured	1,910
Corn distillers' dried grains with solubles	550
Cod liver oil meal	39,930
Cocoa shells, sun-dried	3,434
Timothy hay, sun-cured	1,577
Activated animal sterol, 2000 IU/g	1,995,400
Activated 7-dehydro cholesterol, 2000 IU/g	1,995,400
D-activated plant sterol, 9000 IU/g	8,976,000
Irradiated yeast, 9000 IU/g	8,976,000

Source: National Research Council 1964; Morrison 1956.

TABLE 9.21 Pig Response to Antimicrobials in Feed of Growing-Finishing Pigs (16 Kilograms to Market Weight)

	Number			*Wt, kg*[a]		*Avg daily gain, g*			*Feed/gain*		
Antimicrobial	*Exp.*	*Rep*	*Pig*	*I*	*F*	–	+	*% imp.*	–	+	*% imp.*
CTC-P-S[b]	7	26	356	25	92	738	787	6.6	3.16	3.11	1.8
Bacitracin MD	7	38	334	26	98	724	742	1.2	3.01	3.00	.4
Bacitracin, Zn	1	4	55	31	100	695	695	0	3.36	3.42	−1.8
Chlortetracycline	17	85	1,063	27	94	694	717	3.3	3.23	3.23	0
CTC:P:ST	2	8	56	21	94	790	845	7.0	3.02	2.97	1.8
Copper sulfate	4	48	108	21	90	704	724	2.9	3.07	2.96	3.7
Bambermycin	30	146	1,632	37	94	669	687	2.7	3.48	3.36	3.4
Lincomycin	7	22	292	38	99	714	750	5.1	3.49	3.43	1.7
Nosiheptide	3	17	74	11	92	623	670	7.5	3.35	3.25	3.0
Salinomycin	9	41	282	17	94	735	779	6.0	3.05	2.94	3.7
Tiamulin	9	36	206	14	72	610	665	8.9	2.97	2.86	3.8
Tylosin	45	205	2,144	30	90	665	689	3.6	3.37	3.26	3.1
Virginiamycin	23	146	872	24	93	726	745	2.5	3.13	3.09	1.4
Sum		822	7,474								
Weighted avg				27	92	690	715	3.6	3.22	3.14	2.4

[a] Initial (I) and final (F) weights.
[b] Chlortetracycline:penicillin:sulfamethazine (2:1:2) or chlortetracycline:penicillin:sulfathiazol (2:2:2).

Source: Zimmerman 1986.

TABLE 9.22 Percentage Improvement in Performance of Pigs Fed Antimicrobials for Specific Years

		Improvement, %	
Years	*Periods*[a]	*Daily gain*	*Feed/ gain*
1950 to 1977[b]	Starter	16.1	6.9
	Grower-Finisher	4.0	2.1
1978 to 1985	Starter	15.0	6.5
	Grower-Finisher	3.6	2.4

[a] Starter period from about 8 to 26 kg and grower-finisher period from 27 to 92 kg body weight.

[b] Hays 1979.

Source: Zimmerman 1986.

critical stages of production such as weaning, breeding, and farrowing. Environmental stresses such as inadequate nutrition, crowding, moving and mixing of animals, and poor sanitation also contribute to increased response to antibiotics (Hays 1978; Zimmerman 1986).

Examples of the improvement in weight gains and feed conversion to be expected from the use of antibiotics in pig feeding are shown in Tables 9.21 and 9.22. The response of the poorly performing pigs to antibiotics is much greater than the response of those performing at a higher rate (Table 9.23). The degree of improvement now is similar to that in the 1950s when antibiotics were first being used in swine feeding (Table 9.22). Feeding high levels of antibiotics to female pigs at breeding time increased conception rate by 8.7% (Tables 9.24 and 9.25) and average number of pigs born alive by 0.5 pig per litter (Table 9.25).

TABLE 9.23 Effect of Antibiotics in the Feed on Weight Gain and Efficiency of Feed Utilization of Growing Pigs—A Summary

		Improvement over controls[a]	
Daily gain of controls (kg)	*No. of comparisons*	*Weight gain (%)*	*Feed/gain ratio (%)*
0.09–0.18	2	22.0	8.2
0.18–0.27	3	27.0	4.5
0.27–0.36	4	20.4	5.6
0.36–0.45	7	16.1	11.1
0.45–0.54	9	12.3	6.4
0.54–0.63	9	9.4	1.9
0.63–0.72	20	5.6	4.7
0.72 and up	7	3.8	1.8
	Total 61	Average 10.7	Average 5.1

[a] Antibiotics combination of penicillin and streptomycin.

Source: Hays 1969.

TABLE 9.24 Effect of Antimicrobials Fed During Breeding Period on Farrowing Rate[a]

Drug	No. sows	Drug, g/d	Farrowing rate, %	
			Control	Treated
Chlortetracycline	179	1.0	62	79
Chlortetracycline	198	0.5	74	86
CTC-P-S[b]	96	1.25	81	96
CTC-P-S	126	1.25	53	56
CTC-P-S	79	1.25	30	36
Tylosin	192	0.6	75	77
Tylosin	143	0.5	81	84
Tylosin-S[c]	197	0.5	70	83
Furazolidone	87	0.5	63	93
Chlortetracycline	249	1.0	67	75
Chlortetracycline	239	1.0	71	72
CTC-P-S	184	1.0	61	70
Chlortetracycline	101	1.0	90	94
Sum	2,070			
Weighted avg.			68	77

[a] In general, antimicrobials were fed for 1 wk prebreeding and 1 to 2 wk postbreeding. Farrowing rate is equal to the percentage of sows allotted that were bred on first estrus.

[b] Chlortetracycline:penicillin:sulfamethazine (2:1:2).

[c] Tylosin;sulfamethazine (1:1).

Source: Zimmerman 1986.

TABLE 9.25 The Effect of Antibiotics at Breeding Time on Farrowing Rate and Litter Size

No. of sows	Farrowing rate (%)		Live pigs/litter	
	Control	Treated	Control	Treated
377	68.5	82.9[a]	9.8	10.1[a]
59	—	—	7.1	9.7[b]
96	87.5	95.8[c]	9.0	10.3[c]
182	60.9	70.0[c]	9.8	10.0[c]
249	66.9	75.4[a]	9.9	10.2[a]
192	93.8	91.6[d]	10.9	11.3[d]
Weighted av	73.0	81.7	9.8	10.3

[a] Chlortetracycline, 0.5–1.0 g/sow/day.

[b] Chlortetracycline, 0.5–1.0 g/sow/day, 0.54 g/sow/day.

[c] Chlortetracycline, sulfamethazine, and penicillin at 0.5, 0.5, and 0.25 g/sow/day, respectively.

[d] Tylosin phosphate, 0.6 g/sow/day.

Source: From Hays 1978.

The use of antibiotics in animal feed for a period of about four decades has not created a public health problem (CAST 1981). This conclusion was reached by a panel of scientists appointed to prepare a document for the U.S. Congress and the Food and Drug Administration (FDA) from a comprehensive review and analysis of the benefits and potential hazards of the use of antibiotics in animal feeds and more recently by other experts on the subject (American Society of Animal Science, 1986).

The FDA conducts a comprehensive program in a conscientious effort to guard against problems with approved drugs. Furthermore, before new products are approved for use they must be proven beyond reasonable doubt to be both *safe* and *effective* in promoting growth or in meeting whatever claims have been made for them. Each year a feed additive compendium is published that provides information on identities, levels, combinations, and conditions under which all approved drugs can be legally used in animal feeds.

REFERENCES

Common maize

Church, D. C., and Pond, W. G. 1988. Basic Animal Nutrition and Feeding, 3rd ed. Wiley, New York.

FAO. 1988. Food and Agricultural Organization Production Yearbook, vol. 42.

Mertz, E. T. 1963. Corn proteins–A chemical and nutritional perspective. *In* Proc. 18th Annu. Hybrid Corn Industry-Research Conf., December 12 and 13, 1963, Publ. *18*, pp. 7–12. Washington, D.C., American Seed Trade Assoc.

Morrison, F. B. 1956. Feeds and Feeding, 22nd ed. Morrison, Ithaca, N.Y.

National Research Council. 1983. Underutilized Resources as Animal Feedstuffs, pp. 1–253. National Research Council, Natl. Acad. of Sci. Natl. Academy Press, Washington, D.C.

Osborne, T. B., and Mendel, L. B. 1914. J. Biol. Chem. *18*, 1.

Patience, J. F., and Thacker, P. A. 1989. Swine Nutrition Guide. Prairie View Swine Center, University of Saskatchewan, Saskatoon, Canada.

Quality protein mutant maize

Baker, D. H., Becker, D. E., Jensen, A. H., and Harmon, B. G. 1970. J. Anim. Sci. *30*, 364.

Beeson, W. M., Pickett, R. A., Mertz, E. T., Cromwell, G. L., and Nelson, O. E. 1966. *In* Proc. Distillers Feed Research Council. *21*, 70.

Gallo, J. T., Maner, J. H., and Corzo, M. 1968. J. Anim. Sci. *28*, 1152 (Abstr.).

Mahan, D. C., Becker, D. E., and Jensen, A. H. 1971a. J. Anim. Sci. *32*, 470.

Mahan, D. C., Becker, D. E., and Jensen, A. H. 1971b. J. Anim. Sci. *32*, 476.

Maner, J. H., Gallo, J. T., Obando, H., and Grant, J. 1969. Am. Soc. Agron. Detroit, Michigan (Abstr.).

Mertz, E. T. 1963. *In* Proc. 18th Annu. Hybrid Corn Industry-Research Conf. *18*, 7–12.

Pickett, R. A. 1966. *In* Proc. High-Lysine Corn Conf., pp. 19–22. Corn Industry-Research Foundation, Inc., Washington, D.C.

Other maize mutants

Pond, W. G., and Maner, J. H. 1984. Swine Production and Nutrition. AVI, Westport, Conn.

Grain sorghum (milo)

Hulse, J. H., Laing, E. M., and Pearson, O. E. 1980. Sorghum and the Millets: Their Composition and Nutritive Value. Academic Press, New York.

McGinty, D. D. 1969. Proc. 6th Bienn. Int. Grain Sorghum Res. Util. Conf., p. 20.

Pond, W. G., and Maner, J. H. 1984. Swine Production and Nutrition. AVI, Westport, Conn.

Barley

Bhatty, R. S., Christison, G. I., and Rossnagel, B. G. 1979. Can. J. Anim. Sci. *59,* 585.

Dinusson, W. E., Nystuen, P. A., and Bolin, D. W. 1956. J. Anim. Sci. *15,* 1256.

FAO. 1988. Food and Agricultural Organization Production Yearbook, vol. 42.

Fuller, M. F., and Livingstone, R. M. 1979. Br. J. Nutr. *41,* 321.

Fuller, M. F., and Mennie, I. 1979. Br. J. Nutr. *41,* 333.

Gill, D. R., Oldfield, J. E., and England, D. C. 1966. J. Anim. Sci. *25,* 34.

Graham, H., Fadel, J. G., Newman, C. W., and Newman, R. K. 1989. J. Anim. Sci. *67,* 1293.

Joseph, W. E. 1924. Mont. Agric. Exp. Stn. Bull. *169,* 47.

Mitchell, K. G., Bell, J. B., and Sosulski, F. W. 1976. Can. J. Anim. Sci. *56,* 505.

National Research Council. 1988. Nutrient Requirements of Swine. Natl. Acad. Sci. Washington, D.C.

Newman, C. W., Thomas, O. O., and Eslick, R. F. 1968. J. Anim. Sci. *27,* 981.

Newman, C. W., and Eslick, R. F. 1970. Proc. West. Sec. Am. Soc. Anim. Sci. *21,* 111.

Newman, C. W., Eslick, R. F., Pepper, J. W., and El-Negoumy, A. M. 1980. Nutr. Rep. Int. *22*(6), 833.

Oats

Christison, G. I., and Bell, J. M. 1980. Can. J. Anim. Sci. *60,* 465.

Jensen, A. H., Becker, D. E., and Terrill, S. W. 1959a. J. Anim. Sci. *18,* 701.

Jensen, A. H., Becker, D. E., and Terrill, S. W. 1959b. J. Anim. Sci. *18,* 1356.

Katz, R. S., Baker, D. H., Sasse, C. E., Jensen, A. H., and Harmon, B. G. 1973. J. Anim. Sci. *37,* 1165.

Meade, R. J., Dukelow, W. R., and Grant, R. S. 1966. J. Anim. Sci. *25,* 58.

Owen, J. B., and Ridgman, W. J. 1967. Anim. Prod. *9,* 107.

Robbins, G. S., Pomeranz, Y., and Briggle, L. W. 1971. J. Agric. Food Chem. *19,* 536.

Wahlstrom, R. C., Reiner, L. J., and Libal, G. W. 1977. J. Anim. Sci. *45,* 948.

Rye

Bowland, J. P. 1966. Univ. Alberta 45th Annu. Feeder's Day Rep. *51,* 1. (Cited by Friend and MacIntyre 1969.)

Esnoala, M. A., Rosa, J. G., and Romero, O. 1976. Agric. Tec. *26,* 18.

Hale, O. M., Johnson, J. C. Jr., and Southwell, B. L. 1967. Univ. Georgia Agric. Exp. Stn. Res. Bull. *14,* 17. September, 1967.

Hölzschul, W., and Trautmann, H. 1962. Deutsch. Landwirtsch. *13,* 24. (Cited in Nutr. Abstr. Rev. *32,* 989.)

Wieringa, G. W. 1967. On the occurrence of growth inhibiting substances in rye. Thesis, Landboumhogesch., Wageningen, p. 68. (Cited in Nutr. Abstr. Rev. *38,* 91, No. 581.)

Wheat

Allee, G. L. 1977. Kansas State Univ. Swine Day Rep. 1977, p. 11.

Beames, R. M. 1963. Queensl. J. Agric. Sci. *20,* 231.

Braude, R., Townsend, J., Harrington, G., and Rowell, J. G. 1961. J. Agric. Sci. Cambridge *57,* 257.

Cole, D. J. A., Clent, E. G., and Luscombe, J. R. 1969. Anim. Prod. *11,* 325.

Erickson, J. P., Miller, E. R., Hill, G. M., Block, J. R., Behiak, D. M., and Ku, P. K. 1980. J. Anim. Sci. *51,* 1065.

Hale, O. M., Morey, D. D., and McCormick, M. C. 1979. Ga. Res. Rep. *311,* 1–14.

Ivan, M., and Farrell, D. J. 1975. Anim. Prod. *20,* 77.

Ivan, M., Farrell, D. J., and Edey, T. N. 1975a. Anim. Prod. *20,* 267.

Ivan, M., Farrell, D. J., and Edey, T. N. 1975b. Anim. Prod. *20,* 277.

Jensen, A. H., Baker, D. H., Becker, D. E., and Harmon, B. G. 1969. J. Anim. Sci. *29,* 16.

Lawrence, T. L. J. 1967. J. Agric. Sci. Camb. *69,* 271.

Lennon, A. M., Clawson, A. J., and Alsmeyer, W. L. 1972. J. Anim. Sci. *35,* 778.

Wu, J. F., and Ewan, R. C. 1979. J. Anim. Sci. *49,* 1470.

Triticale

Allee, G. L., and Hines, R. H. 1972. J. Anim. Sci. *35,* 1101a (Abstr.).

Bowland, J. P. 1968. *In* 47th Annu. Feeders' Day Rep., Dep. Anim. Sci. Univ. Alberta, Edmonton, Alberta, Canada.

Erickson, J. P., Miller, E. R., Elliott, F. C., Ku, R. K., and Miller, D. E. 1979. J. Anim. Sci. *48,* 547.

Harrold, R. L., Dinusson, W. E., Haugse, C. N., and Buchannan, M. L. 1971. N. D. Agric. Exp. Stn. Farm Res. *28,* 34.

Larter, E. N. 1974. *In* Triticale: First Man-made Cereal. (C. C. Tsen, Editor), pp. 35–52. Am. Assoc. Cereal Chem. St. Paul, Minn.

Leighty, C. E. 1915. J. Am. Soc. Agron. *7,* 209.

Longernecker, T. C., Langford, L., and Stoggs, D. 1969. High Plains Res. Found. Rep. *331.* Plainsview, Tex.

Martynov, V., and Shtan'ko, I. 1979. Svuinovodstvo *1,* 14.

Noland, P. R., Sharp, R. N., and McGhee, R. W. 1972. Arkansas Farm Res. *21* 7.

Stothers, S. C. 1970. Res. Bull. A. S. *70–1,* p. 23. Dep. of Anim. Sci., Univ. of Manitoba, Winnipeg.

Stothers, S. C., and Shebeski, L. H. 1965. 14th Rep. of Livestock Res., p. 19. Dep. of Anim. Sci., Univ. of Manitoba, Winnipeg.

Villegas, E. 1973. Proc. Wheat, Triticale, and Barley Seminar.: (ed., R. G. Anderson). CIMMYT, January 22–26, 1973. El Batan, Mexico.

Wilson, R. F. 1972. Ohio Agric. Res. Dev. Center Swine Res. Rep. *61–5,* 25. Wooster, Oh.

Rice

Bray, C. I. 1943. La. Agric. Exp. Stn. Circ. *33.*

FAO. 1988. Food and Agricultural Organization Production Yearbook, vol. 42.

Kik, M. C. 1956. J. Agric. Food Chem. *4,* 170.

Noland, P. R. 1960. Arkansas Farm. Res. *9,* 30–38.

Buckwheat

Anderson, D. M., and Bowland, J. P. 1979. 58th Annu. Feeders' Day Rep., p. 12. Dep. of Anim. Sci., Univ. Alberta, Edmonton.

Wheat by-products

Batterham, E. S., Lewis, C. E., Lowe, R. F., and McMillan, C. J. 1980. Anim. Prod. *31,* 259.

Beames, R. M., and Natoli, W. J. 1969. Aust. J. Exp. Agric. Anim. Husb. *9,* 594.

Forsythe, W. A. 1979. Mich. Res. Rep. *386,* 71. September, 1979.

Henry, Y. M. 1976. Proc. 1st Int. Symp. Feed Composition, Anim. Nutr. Req. Computerization Diets. pp. 270–281. Utah State Univ., Logan.

Just, A., Jørgensen, H., and Fernández, J. 1978. Natl. Inst. Anim. Sci. Copenhagen, Denmark.

Morgan, D. J., Cole, D. J. A., and Lewis, D. 1975. J. Agric. Sci. Cambridge *84,* 7.

Nehring, K., Klippel, W., Uhlemann, H., and Hoffmann, B. 1970. Archiv Tierernährung *20,* 69.

Patience, J. F. 1977a. J. Anim. Sci. *45,* 1294.

Patience, J. F. 1977b. J. Anim. Sci. *45,* 1302.

Pals, D. A., and Ewan, R. C. 1978. J. Anim. Sci. *46,* 402.

Saunders, R. M., Walker, M. A., Jr., and Kohler, G. O. 1969. J. Poult. Sci. *48,* 1497.

Saunders, R. M., Connor, M. A., Kohler, G. O., and Balylock, L. B. 1974. J. Anim. Sci. *38,* 1272.

Young, L. G. 1980. J. Anim. Sci. *51,* 1113.

Young, L. G., and King, G. L. 1981. J. Anim. Sci. *52,* 551.

Rice bran and polishings

FAO. 1988. Food and Agricultural Organization Production Yearbook, vol. 42.

Distillery by-products

Ewan, R. C. 1976. Proc. 31st Distillers' Feed Conf., p. 16.

Thong, L. Q., Jensen, A. H., Harmon, B. G., and Cornelius, S. G. 1978. J. Anim. Sci. *46,* 674.

Wahlstrom, R. C., and Libal, G. W. 1980. S. D. Swine Day Res. Rep. November 20, 1980. S. D. State Univ., Brookings.

Wahlstrom, R. C., German, C. S., and Libal, G. W. 1970. J Anim. Sci. *30*, 532.

Brewery by-products

Kornegay, E. T. 1973. J. Anim. Sci. *37*, 479.

Roots, tubers, fats, and other energy sources

Plucknett, D. L. 1979. Small-Scale Processing and Storage of Tropical Root Crops. Westview Press, Boulder, Colo. 461 pp.

Cassava

Buitrago, J., and I. Jiménez. 1968. Unpublished data. ICA, Cali, Colombia.

Couch, J. F. 1932. USDA Leaflet 88.

FAO. 1988. Food and Agricultural Organization Production Yearbook, vol. 42.

Maner, J. H., and Jiménez, I. 1967. Dia de Campo sobre Porcinos. April 1967. ICA, Palmira, Colombia.

Oke, O. L. 1969. World Rev. Nutr. Diet. *11*, 170.

Pond, W. G., and Maner, J. H. 1984. Swine Production and Nutrition. AVI, Westport, Conn.

Potatoes and dried potato products

Bielinski, K., and Bielinski, K. 1954. Rocz. Nauk. Roln. *68*, 297. (Cited in Nutr. Abstr. Rev. *25*, 841, 1955.)

Friend, D. W., Cunningham, H. M., and Nicholson, J. W. G. 1963. Can. J. Anim. Sci. *43*, 241.

Glaps, J. 1970. Rocz. Nauk. Roln. *92*, 247. (Cited in Nutr. Abstr. Rev. *41*, 709.)

Hillyer, G. M., and Whittemore, C. T. 1975. J. Sci. Food Agric. *26*, 215.

Höller, H. 1959. Arch. Tierernähr. *9*, 236.

Livingstone, R. M., Baird, B. A., Alkinson, T. and Crofts, R. M. J. 1980. J. Sci. Food Agric. *31*(7), 695.

McLagan, J. R., and Nasr, H. 1950. Scot. J. Agric. *30*, 143.

Ostrowski, H., and Rys, R. 1969. Rocz. Nauk. Roln. *92*, 71. (Cited in Nutr. Abstr. Rev. *40*, 1469.)

Seidler, S., Warchol, P., and Weckowicz, E. 1967. Zesz. Nauk. WSR. Szczecinie *26*, 35. (Cited in Nutr. Abstr. Rev. *39*, 274.)

Whittemore, C. T. 1977. Anim. Feed Sci. Tech. *2*, 171.

Whittemore, C. T., Moffat, I. W., and Taylor, A. G. 1975a. J. Sci. Food Agric. *26*, 1567.

Whittemore, C. T., Taylor, A. G., Moffat, I. W., and Scott, A. 1975b. J. Sci. Food Agric. *26*, 255.

Woodman, H. E., and Evans, R. E. 1943. J. Agric. Sci. *33*, 1.

Zausch, M., and Hölzschul, W. 1960–1961. Jahrb. Arbeitsgem. Fiueitturungsberat. *3*, 173. (Cited in Nutr. Abstr. Rev. *32*, 1385.)

Sweet potatoes

Calder, A. 1960. Rhod. Agric. J. *57,* 293.

Lee, J. H., and Chung, C. H. 1966. Res. Bull Chinju Agric. College *5,* 29–38.

Lee, P. K., and Lee, M. S. 1979. J. Taiwan Livestock Res. *12,* 49.

Springhall, J. A. 1969. Papua New Guinea Agric. J. *21,* 76.

Tseng, C. M., Tsai, A. H., and Chan, M. P. 1980. J. Agric. Assoc. China. *112,* 54.

Bananas and plantains

Calles, A., Clavijo, H., Hervas, E., and Maner, J. H. 1970. J. Anim. Sci. *31,* 197 (Abstr.).

Celleri, H., Oliva, F., and Maner, J. H. 1971A. Reunión Latinamericana de Producción Anim. April 26–30, 1971, p. 91, Bogota, Colombia.

Celleri, H., Oliva, F., and Maner, J. H. 1971B. Unpublished data. INIAP, Quito, Ecuador.

Clavijo, H., and Maner, J. H. 1971. III Reunión Latinamericana de Producción Anim. April 26–30, 1971, p. 93, Bogota, Colombiá.

Clavijo, H., Calles, A., Oliva, F., and Maner, J. H. 1970. Unpublished data. INIAP. Santo Domingo, Ecuador.

FAO. 1988. Food and Agricultural Organization Production Yearbook, vol. 42.

Oliva, F., Alvarado, F., Celleri, H., and Maner, J. H. 1971. Unpublished data. INIAP. Santo Domingo, Ecuador.

Oliva, F., Viteri, J., Calles, A., and Maner, J. H. 1971. INIAP. April, 1971. Quito, Ecuador.

Viteri, J., Oliva, F., and Maner, J. H. 1971. Unpublished data. INIAP. Quito, Ecuador.

Viteri, J., Alvarado, F., and Maner, J. H. 1970. Unpublished data. INIAP. Quito Ecuador.

Cane sugar

Buitrago, J. 1972. Reproduction of gilts and long-term effects on progeny: Effect of energy intake during gestation. Ph.D. Thesis. Cornell Univ. Ithaca, New York.

Buitrago, J., García, E., Maner, J. H., and Gallo, J. T. 1969. ICA Mimeo, Palmira, Colombia.

Buitrago, J., García, E., Corzo, M. A., Maner, J. H., and Gallo, J. T. 1970. Rev. ICA *5,* 419.

Maner, J. H., Obando, H., Portela, R., and Gallo, J. 1969. J. Anim. Sci. *29,* 139 (Abstr.).

Cane molasses

Brooks, C. C. 1972. J. Anim. Sci. *34,* 217.

Brooks, C. C., and Iwanaga, I. I. 1967. J. Anim. Sci. *26,* 741.

FAO. 1988. Food and Agricultural Organization Production Yearbook, vol. 42.

Maner, J. H., Gallo, J., Corzo, M., and Buitrago, J. 1969a. J. Anim. Sci. *29,* 139 (Abstr.).

Maner, J. H., Obando, H., Portela, R., and Gallo, J., 1969b. J. Anim. Sci. *29,* 139 (Abstr.).

Preston, T. R., and Willis, M. B. 1970. Feedstuffs *42*(13), 20.

Dried bakery product

Sewell, R. F. 1965. Georgia Agric. Res. *7,* 9.

Citrus and citrus pulp

Baird, D. M., Allison, J. R., and Heaton, E. K. 1974. J. Anim. Sci. *38,* 545.

Cunha, J. T., Lamar, G. A., Shauver, C. B., Pearson, A. M., Folks, S. J., and Glasscock, R. S. 1950. Fla. Agric. Exp. Stn. Cir. *S11*, 1–7.

Hollis, G. R., West, R. L., Carpenter, J. W., and Palmer, A. Z. 1969. Fla. Agric. Exp. Stn. Mimeo Ser. No. SUS 70–3.

Kirk, W. G., and Crown, R. M. 1947. Fla. Agric. Exp. Stn. Bull. 428.

Garbage

Kornegay, E. T., Vander Noot, G. W., Barth, K. M., Graber, G., MacGrath, W. S., Gilbreath, R. L., and Bielk, F. J. 1970. Rutgers Exp. Stn. Bull. 829, 31 pp.

Fats and oils

Devilat, J. B., Recabarren, I., and Skoknic, A. 1971. III Reunión Latinoamericana de Producción Animal, p. 83 (Abstr.).

Lawrence, N. J., Maxwell, C. V., and Belcher, C. G. 1980. Okla. Stn. Univ. Anim. Sci. Res. Rep. *MP-107*, p. 159. April, 1980.

Libal, G. H., Hoppe, M. K., and Wahlstrom, R. C. South Dakota Swine Day Rep. 1987–1988. p. 34.

Moser, B. D., and Lewis, A. J. 1980. Feedstuffs *49* (15), 20.

Pettigrew, J. E. 1979. 71st Annu. Meeting Amer. Soc. Anim. Sci. July 1979, Univ. of Arizona, Tucson.

Pettigrew, J. E. 1981. J. Animal Sci. *53*, 107.

Seerley, R. W. 1981. Feedstuffs *53*(10), 34.

Seerley, R. W., Pace, T. A., Foley, C. W., and Scarth, R. D. 1974. J. Anim. Sci. *38*, 64.

Seerley, R. W. 1989. J. Animal Sci. *67*, 1989.

Stahly, T. S., Cromwell, G. L., and Simpson, W. S. 1981. J. Animal Sci. *51*, 352.

Stahly, T. S., Cromwell, G. L., and Moneque, H. J. 1985. J. Animal Sci. *61*, 1485.

Animal Protein

Doty, D. M. 1973. *In* Alternate Sources of Protein for Animal Production. Natl. Acad. Sci., Washington, D.C.

Fontenot, J. P., Smith, L. W., and Sutton, A. L. 1983. J. Anim. Sci. *57* (Suppl. 2), 221.

Frank, G. R., Jensen, A. H., and Day, D. L. 1980. J. Anim. Sci. *50*, 1031.

Husby, F. M. 1980. Agroborealis *12*, 4–8.

King, R. H., and Campbell, R. G. 1978. Anim. Feed Sci. Technol. *3*, 191.

Seve, B., Aumaitre, A., and Tord, P. 1975. Ann. Zootech. *24*, 21.

Tibbetts, G. W., Seerley, R. W., McCampbell, H. C., and Vezey, S. A. 1981. J. Anim. Sci. *52*, 93.

Wahlstrom, R. C., and Libal, G. W. 1977. J. Anim. Sci. *44*, 778.

Waibel, P. E., Cuperlovic, M., Hurrell, R. F. and Carpenter, K. J. 1977. J. Agric. Food Chem. *25*(1), 171.

Plant Proteins

Soybean meal

Khayambashi, H., and Lyman, R. L. 1966. J. Nutr. *89,* 455.

Lewis, A. J., Peo, E. R., Jr., Moser, B. D., and Crenshaw, T. D. 1981. J. Anim. Sci. *51,* 361.

Liener, I. E. 1975. *In* Protein Nutritional Quality of Foods and Feeds (ed., M. Friedman), vol. 1, part 2, p. 523. Marcel Dekker, New York.

Mateo, J. P., and Veum, T. L. 1980. J. Anim. Sci. *50,* 869.

Osborne, T. B., and Mendel, L. B. 1917. J. Biol. Chem. *32,* 369.

Robison, W. L. 1930. Ohio Agric. Exp. Stn. Bull 452.

Shrewsbury, C. L., Vestal, C. M., and Hauge, S. M. 1932. J. Agric. Res. *44,* 267.

Wilson, R. H., and Leibholz, J. 1981. Brit. J. Nutr. *45,* 301, 321, 337, 347, 351.

Yen, J. T., Jensen, A. H., and Simon, J. 1977. J. Nutr. *107,* 156.

Zimmerman, D. R. 1980. J. Anim. Sci. *51*(Suppl. 1), 78.

Coconut meal

Creswell, D. C., and Brooks, C. C. 1971. J. Anim. Sci. *33,* 370.

Lima, A. M., Maner, J. H., and Simas, R. 1981. CEPED Annu. Rep., December, 1981.

Malynicz, G. L. 1973. Papua New Guinea Agric. J. *24,* 142.

Cottonseed meal

Knabe, D. A., Tanksley, T. D., and Hesby, J. H. 1979. J. Anim. Sci. *49,*134.

Tanksley, T. D., Jr., Knabe, D. A., Purser, K., Zebrowska, T., and Corley, J. R. 1981. J. Anim. Sci. *52,* 769.

Linseed meal

Bishara, H. N., and Walker, H. F. 1977. Brit. J. Nutr. *37,* 321.

FAO. 1980. FAO Production Yearbook 1980.

Peanut meal

Conkerton, E. J., Blanchet, E. D., Ory, R. L., and Hammons, R. O. 1978. Peanut Sci. *5,* 75.

Rapeseed meal (canola meal)

Clandinin, D. R. 1981. Canola Meal for Livestock and Poultry. Canola Council of Can. Publ. *59,* Winnipeg, Manitoba.

Clandinin, D. R., and Robblee, A. R. 1978. Proc. 5th Int. Rapeseed Conf., pp. 204–212. Malmo, Sweden.

Lewis, A. J., Aherne, F. X., and Hardin, R. T. 1978. Can. J. Anim. Sci. *58,* 203.

Rubber seed and rubber seed meal

Babatunde, G. M., Pond, W. G., and Peo, E. R., Jr. 1990. J. Anim. Sci. *68,* 392.

Rajaguru, A. S. B., and Ravindran, V. 1979. J. Natl. Sci. Council, Sri Lanka *7,* 101.

Stosic, D. D., and Kaykay, J. M. 1981. World Anim. Rev. *39,* 29.

Safflower meal

Williams, K. C., and Daniels, L. J. 1973. Aust. J. Exp. Agric. Anim. Husb. *13*, 48.

Williams, K. C., and O'Rourke, P. K. 1974. Aust. J. Exp. Agric. Anim. Husb. *14*, 12.

Sesame meal

Cornelius, J. A., and Raymond, W. D. 1967. Trop. Sci. *9*, 75–89.

Sunflower seed and meal

Baird, D. M. 1982. Proc. Georgia Nutr. Conf., Atlanta, Ga., pp. 74–83.

Borg, B. S., Libal, G. W., and Wahlstrom, R. C. 1987. J. Anim. Sci. *64*, 1070.

Seerley, R. W., Burdick, D., Russom, W. C., Lowrey, R. S., McCampbell, H. C., and Amos, H. E. 1974. J. Anim. Sci. *38*, 947.

Wahlstrom, R. C., Libal, G. W., and Thaler, R. C. 1985. J. Anim. Sci. *60*, 720.

Grain legumes and other protein sources

Abu-Shakra, S., Mirza, S., and Tannons, R. 1970. J. Sci. Food Agric. *21*, 9.

Aykroyd, W. R., and Doughty, J. 1964. Legumes in Human Nutrition. FAO Publ. *19*.

Bell, J. M., and Wilson, A. G. 1970. Can. J. Anim. Sci. *50*, 15.

FAO. 1981. Food and Agricultural Production Yearbook.

Chung, P., Pond, W. G., Kingsbury, J. M., Walker, E. F., Jr., and Krook, L. 1978. J. Anim. Sci. *47*, 319.

Danielson, D. M., and Noonan, J. J. 1975. J. Anim. Sci. *41*, 94.

Devi, M. A., Subbulakshmi, G., Devi, K. M., and Venkataraman, L. V. 1981. J. Agric. Food Chem. *29*, 522.

Gaye, A., and Moal, J. 1972. J. Rech. Porcine France, pp. 169–173. Institut Technique du Porc, Paris.

Gladstones, J. S. 1972. Bull. Dep. Agric., Western Aust. Perth. *3834*, 1.

Newport, M. J., and Keal, H. D. 1980. Brit. J. Nutr. *44*, 161.

Pirie, N. W. 1977. World Anim. Rev. *22*, 11.

Pollmann, D. S., Danielson, D. M., and Peo, E. R., Jr. 1979. J. Anim. Sci. *48*, 1385.

Pollmann, D. S., Danielson, D. M., Crenshaw, M. A., and Peo, E. R., Jr. 1980. J. Anim. Sci. *51*, 294.

Pond, W. G., and Maner, J. H. 1984. Swine Production and Nutrition. AVI, Westport, Conn.

United States Congress. 1984. Commercial Biotechnology–An International Analysis. Office of Technology Assessment, U.S. Congress, Washington, D.C., pp. 195–198.

Yap, T. N., Wu, J. F., Pond, W. G., and Krook, L. 1982. Nutr. Rep. Int. *25*, 543.

Waterworth, D. G., Cross, M. E., and Smith, K. L. 1978. Anim. Prod. *26*, 397 (Abstr. 92).

Whittemore, C. T., Moffot, I. W., and Taylor, A. G. 1976. J. Sci. Food Agric. *27*, 1163.

Mineral and vitamin sources and nonnutritive feed additives

American Society of Animal Science. 1986. Public Health Implications of the Use of Antibiotics in Animal Agriculture. J. Anim. Sci. 62 (Suppl. 1), 1–106.

Arthur, D. 1971. Can. J. Anim. Sci. *51,* 71.

Austic, R. E., Madubuike, F. N., Boyd, R. D., Klasing, K. C., and Riley, W. W., Jr. 1982. Proc. Cornell Nutr. Conf. Cornell Univ., pp. 47–64. Ithaca, N. Y.

Bartley, J. C., Reber, E. F., Yusken, J. W., and Norton, H. W. 1961. J. Anim. Sci. *20,* 137.

Becker, D. E., Jensen, A. H., and Harmon, B. G. 1963. Ill. Agric. Exp. Stn. Circ. *866.*

Cast. 1981. Antibiotics in Animal Feeds. Council for Agric. Sci. and Tech. Rep. No. *88,* March, 1981.

Church, D. C., and Pond, W. G. 1988. Basic Animal Nutrition and Feeding, 3rd ed. John Wiley & Sons Inc., New York.

Cook, D. A. 1973. J. Nutr. *103,* 1365.

Cromwell, G. L. 1980. Feedstuffs *53*(10), 38, March 3, 1980.

Cromwell, G. L., Hays, V. W., Overfield, J. R., and Krug, J. L. 1976. J. Anim. Sci. *42,* 1350.

Fritz, J. C., Pla, G. W., Roberts, T., Boehne, J. W., and Hove, E. L. 1970. J. Agric. Food Chem. *18,* 647.

Hays, V. W. 1979. Drugs in Livestock Feed, vol. 1. Tech. Rep. pp. 29–36. OTA, Congress of the U.S., Washington, D.C.

Guenter, W., and Sell, J. L. 1974. J. Nutr. *104,* 1446.

Hays, V. W. 1969. Proc. Symp. Drug Use. Natl. Acad. Sci. Publ. *1679.* Washington, D.C.

Hays, V. W. 1978. *In* Nutrition and Drug Interrelations. (John N. Hathcock and Julius Coon, Editors), pp. 545–576, Academic Press, N. Y.

Meintzer, R. B., and Steenbock, H. 1955. J. Nutr. *56,* 285.

Miller, E. R. 1971. Res. Rep. 148. Mich. State Univ. Exp. Stn., pp. 59–62, East Lansing.

Miller, W. J., and Miller, J. K. 1963. J. Dairy Sci. *46,* 581.

Miller, E. R., Ku, P. K., Hitchcock, J. P., and Magee, W. T. 1981. J. Anim. Sci. *52,* 312.

Morrison, F. B. 1956. Feeds and Feeding, 22nd ed. Morrison, Ithaca, N. Y.

National Research Council. 1964. Joint U.S.–Canadian Tables of Feed Composition. Natl. Acad. of Sci. Publ. *1232.* Washington, D.C.

National Research Council. 1968. Nutrient Requirements of Swine. Natl. Acad. of Sci. Publ. *1599.* Washington, D.C.

National Research Council. 1980. Mineral Tolerance of Domestic Animals. Natl. Acad. of Sci., Washington, D.C.

National Research Council. 1988. Nutrient Requirements of Swine, 9th ed. Natl. Acad. of Sci., Washington, D.C.

Pond, W. G., and Maner, J. H. 1984. Swine Production and Nutrition. AVI, Westport, Conn.

Pond, W. G., Veum, T. L., and Lazar, V. A. 1965. J. Anim. Sci. *24,* 668.

Ranhotra, G. S., Loewe, R. J., and Puyat, L. V. 1976. Cereal Chem. *53,* 770.

Scott, M. L., and Cantor, A. H. 1972. Proc. Cornell Nutr. Conf., p. 66.

Sell, J. L. 1979. The 2nd Int. Minerals Conf. Proc., p. 33–55. January 17, 1979. St. Petersburg Beach, Fl.

Sell, J. L., and Fontenot, J. P. 1980. *In* Magnesium in Animal Nutrition, Natl. Feed Ingredients Assoc., 1980.

Sievert, C. W. 1967. Feed ingredients analyses. Feed Bag Red Book. Editorial Service Co., Milwaukee, Wis.

Underwood, E. J. 1977. Trace Elements in Human and Animal Nutrition, 4th ed., Academic Press, N.Y.

Yen, J. T., and Pond, W. G. 1981. J. Anim. Sci. *53,* 1292.

Zimmerman, D. R. 1980. Iron in Swine Nutrition. *In* NFIA Literature Review on Iron in Animal and Poultry Nutrition (eds., H. Russell Conrad, Dean R. Zimmerman, and G. F. Combs, Jr.). Natl. Feed Ingredients Assoc., West Des Moines, Iowa.

Zimmerman, D. R. 1986. Role of subtherapeutic levels of antimicrobials on pig production. J. Anim. Sci. 62 (Suppl. 3), 6–17.

10

Production Systems: Design and Management

Following the review in Chapters 5 through 9 of the understanding of the biological mechanisms as developed from decades of research concerning swine, we return to the systems view presented in Chapter 3. We will consider the principles of how the modern pork production manager needs to organize the production system. Particular attention will be given to the interactions and interrelationships among the components of the system. These interactions mandate numerous compromises and trade-offs as the manager attempts to ensure that economic efficiency is achieved. That is, for a profitable enterprise, the value of the output products must exceed the costs of the input components by a reasonable amount. Profit is necessary in order for the operation to continue. The task is challenging due to fluctuations in prices paid and received and due to the risks from unplanned occurrences.

DESIGNING A PORK PRODUCTION SYSTEM

In this section, the numerous choices and decisions a pork producer must make as he/she designs his/her production system for economic efficiency will be reviewed in relation to inputs and outputs of the system. To review the characteristics of the inputs that influence the performance of the system, the producer has opportunity to control the performance of the system by choosing:

1. the location which, in turn, determines the expected climatic conditions over the year
2. the configuration of the facilities, including number of buildings of each type,

size, spacing, pen arrangement, pigs/pen, feed delivery system and manure handling system

3. the environmental control capabilities of each facility through ventilation, insulation, floor type, supplemental heat and cooling, etc.
4. the crossbreeding system, including both breeds and performance potentials of the specific animals brought into the system
5. the pathogen and parasite control program to reduce the disruptive effects of disease
6. the feeding program, including specific diets for different stages in the life cycle and whether the daily ration is voluntary or restricted to specified amounts
7. the scheduling of animal movement between facilities, of ration changes, of mating, of weaning and of marketing, all according to either the age, date, or weight these specific changes are planned to occur

The other major category of choice concerns the marketing of the output by choices of market and manner of price determination plus negotiation of price. Of course, the first three choices are only open when starting a new facility. However, few of these several decisions can be made independent of other decisions and choices in other categories. These interdependencies of decisions are due to the interactions and feedbacks stressed in Spedding's (1979) definition of a system. Such interactions are prevalent in swine production systems and major groupings of interactions will be described below.

But first, an idealized production environment and system to maximize animal performance will be described to serve as a reference point for assessing more practical production systems. This ideal production system can serve as a goal, but development toward that goal will be limited by practical considerations with most of these limitations dictated by reducing the costs of the inputs and components of the system to be less than the income produced from the products of the system.

Idealized Production System

The ideal production environment would be completely free of all disease-producing pathogens and managed by isolation and traffic control to eliminate the risk of introducing any of these pathogens. The environmental control of the buildings would be such that the internal conditions are dust- and odor-free with the ambient temperature always in the thermo-neutral range for all the pigs in that facility. This range means the temperature would never be cold enough to require the animal to utilize feed energy to keep warm except for the normal heat generated by digestion and metabolism. Nor would the ambient temperature ever be great enough to cause the animals to reduce feed intake to reduce their heat load from digestion and metabolism. Similarly, the temperatures would be such as to not reduce reproductive performance or lactation. The parent stock would have genetic capabilities for producing large litters of healthy pigs, would lactate heavily so that the pigs can be weaned young and at heavy weights, would rebreed quickly for the next litter, and would continue producing large, healthy litters for several parities. The progeny would exhibit fast and lean growth. The dust-free rations would be controlled in composition and amount to fully satisfy each animal's needs for maximum reproduction, lactation, and lean growth and maintenance of body tissues and normal activity. The daily ration would not be allowed to exceed those needs, so as to not produce wasteful fat in excess of a minimum required for body functioning. The sizes and pen arrangements of the facilities would be coordinated with the needs of the maximized performance so that all the buildings are fully utilized all the time with minimal vacant periods between cycles of pigs. Manure would be removed frequently to eliminate odors and gases and handled so as to fully utilize its value, probably as fertilizer, not pollute ground water nor create odors offensive to neighbors. Each market animal would be marketed at the point of weight and fatness where the income from wholesale cut yield would exceed the feed costs and the prorated labor and facilities costs by a maximum amount (the point of diminishing net returns).

The primary flaw in the utopian system described in the previous paragraph is that the facilities to provide such temperature and disease control might be so prohibitively expensive that the value of output could never exceed the costs of production. A second flaw is that the variation in animal performance from animal to animal, even in the maximum performing crossbreeding system, is going to be great enough that control of daily feed rations would need to be precise to the individual animal and changed frequently for that animal. To make this more difficult, the performance potential of individual animals will never be known precisely. Similarly, this variability in performance would disrupt the possibility for full facility utilization.

These noted flaws lead to the necessity of compromise. Most of the compromises of the ideal with practicality lead to choices of those portions of the ideal for maximum performance that can be accomplished on a cost-effective basis. Among these are the compromises of nutritional, environmental, and disease control being precise only to larger groups of animals, specific buildings, or rooms, or, at most, to specific pens in the building realizing the variability of performance of individual animals. We will now review the practical choices open to the producer, with special attention to the interactions that make decisions interdependent.

Location and Environmental Control

The first choice has to be the location of the enterprise. This choice usually involves availability of land, feed ingredients, and market; but, with that decision, comes a range of climate and weather conditions. As mentioned previously, more tropical climates require less emphasis on retaining body heat or supplemental heat with insulation but require greater emphasis on dissipating excess body heat in the warmer portion of the day and the year. For colder climates, the emphasis is more on retaining body heat or providing supplementary heat and is less on high ventilation rates to dissipate heat. But the range of conditions over the year is great enough for most pork-production areas in the United States that there will be some concern for keeping baby pigs warm during the winter and for keeping large animals cool during midday during the summer. Therefore, flexibility to compensate for daily and annual cycles plus fluctuations is usually necessary.

The manner of modifying environment varies from tropical to temperate climates. In warm climates, many units provide houses with curtains or panels that restrict air movement during winter, but that may open to allow natural ventilation during hot summer months. For units in cooler climates, more insulation is needed. Buildings tend to be of sturdier construction, with windowless walls and thick insulation in ceilings and walls. Without windows, the buildings are ventilated by forced air with electric fans. With automated electric ventilation, more direct control of temperatures and air movement is possible than in the open houses that might be appropriate in tropical environments. Thermostats can control the fans so that all fans will be running during the heat of a summer day down to just enough fans during cold nights to keep the internal conditions from becoming stale. Thermostats may also control curtains in natural ventilation facilities.

Breeding System and Disease Control

It has already been stated that the breeding stock and crossbreeding system provides the potential for performance of the system and most producers will seek at least their male replacement animals from specialists in seedstock improvement. However, these introduced animals are potential sources of infection for new pathogens into the production herd. Thus, the producer must face a risk of pathogen introduction as he attempts to secure genetic improvement from a seedstock supplier. For some pathogens, the effect of infection on pathogen-free stock is great because of the absence of controlling antibodies. Breeders have become increasingly concerned about their pathogen status and disease-control programs. Many producers prefer to obtain all replacement breeding stock from one source over time so that the pathogen status is relatively consistent. Fortunately, there are tools of vaccination and eradication, along with laboratory tests for presence of pathogens, so that improved genetics and pathogen control can be developed simultaneously. Most breeders now accept this dual responsibility of genetic improvement and pathogen control. For each specific pathogen, choices have to be made whether the best route of control is by vaccination to eliminate detrimental effects of pathogens when present, by securing and maintaining seedstock free of other specific pathogens, or by sanitation measures to keep the level of infection low enough that vulnerable animals, such as young pigs, do not become infected at levels high enough to lead to the spread of the disease. Many breeding herds are established by (and additions limited to) SPF (specific-pathogen free) procedures, whereby baby pigs are taken from the mother by surgery and reared in laboratory conditions. Eliminating contact between the mother and her pigs breaks the infection cycle for many pathogens.

Of course, breeding stock is not the only source of outside infection to the producers herd. Many pathogens can be carried into a facility by rodents, birds, people, and even feed. Some pathogens are airborne or transmitted with dust. Traffic control even to the point of requiring employees to shower and change clothes before contact with animals is becoming commonplace (Fig. 10.1). Isolation to prevent reinfection is more difficult in production areas with heavy swine populations, but is possible.

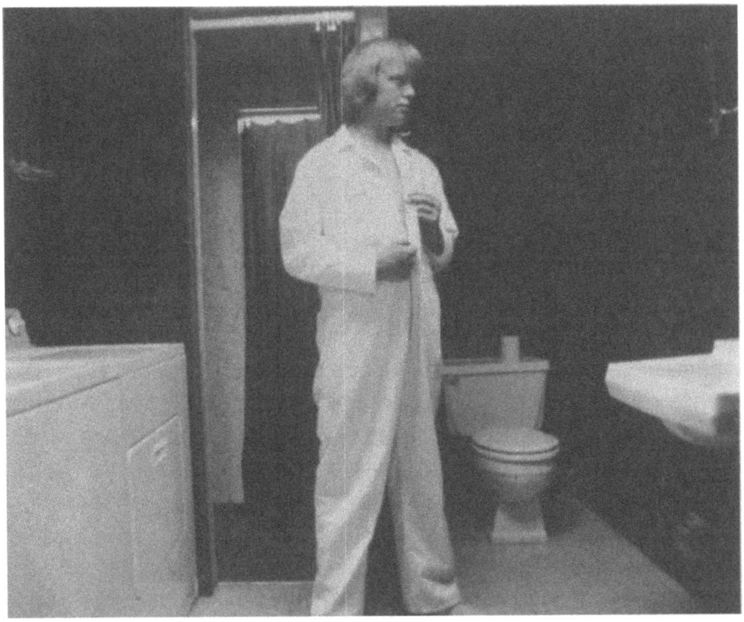

FIGURE 10.1 Shower and clothes changes are sometimes required for modern production systems minimizing risk of pathogen introduction. Photo: Sands Livestock Systems, Inc., Columbus, Nebraska.

Genotype and Nutrition Interactions

No matter how high the potential for growth, reproduction, or lactation performance of the genetic stock, actual performance cannot exceed that supported by the nutritional intake. Conversely, no matter how high the nutritional value of the feed intake, the actual performance will not exceed the genetically influenced performance capabilities of the animals. For this reason, ration formulation needs to be directed towards meeting differing "minimum requirements" for efficient production or reproduction. The challenge is that the production potentials and thus, minimum requirements, differ for alternative breeds or crosses. The optimum ration that meets, but does not wastefully exceed, requirements requires precise knowledge of genetic differences in performance. For full economic efficiency, ration differences (both protein and energy intake) need to be synchronized with differences in performance potential due to genetic makeup, sex, and stage in life-cycle, including gestation and lactation requirements. Intake of minerals and vitamins must also be adequate to support the performance potential.

Genotype, Nutrition, and Carcass Composition

Voluntary feed intake of swine when feed is readily available will often be greater than that needed to support lean growth, especially if the ration has a high percent of well balanced protein plus requirements for minerals and vitamins. Voluntary intake of energy tends to be related to the body protein mass of the individual.

Balanced protein intake will be partially used to maintain that body protein mass—that is, replace protein losses in the continuous degradation and resynthesis processes for body tissues (see Chapter 5). The remainder of digested balanced protein intake is available for the accretion of new protein tissues up to the potential for protein accretion for that animal at that stage of development. Any feed protein intake in excess of that level may be utilized as energy along with the carbohydrate and lipid content of the intake. These energy sources are utilized first for supporting these body processes. Most of the excess after these needs are met will be converted to lipids and deposited as fat. Hence, on voluntary intake, most growing animals and even breeding animals, open or early in gestation, will fatten. Since this is often undesirable, methods of controlling intake to predicted needs offer possibilities for modifying performance of the animals and, thus, the system.

Two types of products are being researched and developed to enhance lean growth potential—pork somatotropins and beta-adrenergic-agonists (see Chapter 11). Although not currently approved for use, these products may become available in the near future. Their use does not seem to lead to an increase in absolute feed intake and voluntary intake relative to body protein mass is often less. Thus, there is less excess energy from the consumed ration to lead to fat deposition. The result is, therefore, not only greater lean amounts, but less fat amounts in the carcass at a specific age; or, at a specific market weight, carcass lean and the lean cut yield are greater portions of the total carcass. However, as with genetic differences, the balanced protein intake has to be adequate to support the increased lean growth potential. As a result, the ration content needs to be synchronized with enhanced performance potential from causes in addition to genetics.

Controlling energy intake to control the low-valued and inefficiently produced fat content of the carcass is obviously a possible method of modifying animal performance. United States producers have not used this approach, but British and European producers have, probably because of greater price incentives for reducing fatness. Limiting intake for gestating gilts is routinely done. There are two costs of such practices—increased labor or equipment for feed handling and record-keeping to accomplish this control, and an increase in time to market. Thus, there is an increased labor and facilities cost per animal to reach a specified market weight. In the past, compensation for leaner carcasses did not seem sufficient in the U.S. market to justify such practices. However, this inadequacy of marketing is being corrected and computerization can facilitate the necessary record-keeping.

An added difficulty to practical control of nutrient intake is the variability of lean growth potential among individuals within a herd. A ration appropriate in content and amount for an average animal may provide excess energy—leading to excess fatness—for a lower-than-average lean growth potential animal, but restrict the lean growth of an animal with higher-than-average potential. To achieve uniformly low levels of fatness, flexibility in levels of feeding may be necessary. Such systems might be costly but should be considered for cost-effectiveness.

Past efforts at selection for lean growth have been confounded by the increased frequency of the related symptoms called PSS (porcine stress syndrome) and PSE (pale soft exudative) pork. This condition has been found to have a genetic basis, the manifestation of which is exacerbated by stressful handling during marketing. A review of this problem is given in PIH 26 (Appendix 5a).

Environmental Control and Nutrition

Whenever the environmental control is not complete enough to maintain the thermo-neutral range of temperature, compensating adjustments in the system may have to be in the form of diet and ration changes. When conditions become too cold, greater energy intake becomes necessary if growth is not to be restricted. When conditions become too hot, less voluntary energy intake will occur, so that adjustments in protein, mineral, and vitamin content is needed to not restrict lean growth. These interactions lead to the desirability of ration adjustments in response to the actual ambient temperature conditions. Adjustments this precise have not often been done in the past, but need to be considered more in the future in conjunction with closer relationships between the producer and the feed manufacturer. In many modern systems, the producer manufactures his prepared rations from purchased or home-grown ingredients. Under such arrangements, there is considerable potential for timely adjustments in diet.

Configuration, Performance, and Scheduling

The scheduling of animal movement between facilities needs to be synchronized with the capacities of the buildings so as to achieve the best facility utilization. But the optimum combination depends on the performance characteristics of the animals due to genetics, feeding, and other factors. The reproductive rate of pigs weaned per sow per year influences the needed relative sizes of the nursery-growing-finishing complex to the breeding-gestation-farrowing complex. Reduction of time from weaning to market reduces the needs for nursery-growing-finishing space. Early weaning and early rebreeding reduces the needs for breeding and farrowing space. Figure 10.2 and Table 10.1 show an example schematic diagram and schedule for utilization of facilities with three groups of sows and their progeny cycling through four buildings. The plan allows empty time periods in farrowing for cleanup and sanitation, but not for the other buildings unless they are split into smaller buildings or rooms for the separate groups. All-in–all-out patterns are becoming more prevalent for all buildings in a complex and greater physical separation and traffic control between the breeding herd, the nursery, and the growing-finishing facility. The depicted schedule is for breeding of each group every six months. Many modern producers prefer to wean early and rebreed more frequently, primarily to achieve greater facility utilization and utilization of breeding females. An average of two and one-half litters per sow per year is possible.

The manner of scheduling of movement between facilities deserves attention. Movement can be based on date, age, or weight. Movement of all animals from a specific building on a specific date to a specific new building reduces to one day each the periods for filling and emptying the houses, making greater utilization possible. But this might not provide the best environmental conditions and diets for some pigs. Younger animals might not be ready for moving to cooler temperatures or different rations. Alternately, movement on age would decrease building utilization but provide better environment for animal performance. More desirable for animal performance might be movement by weight, but this would increase the

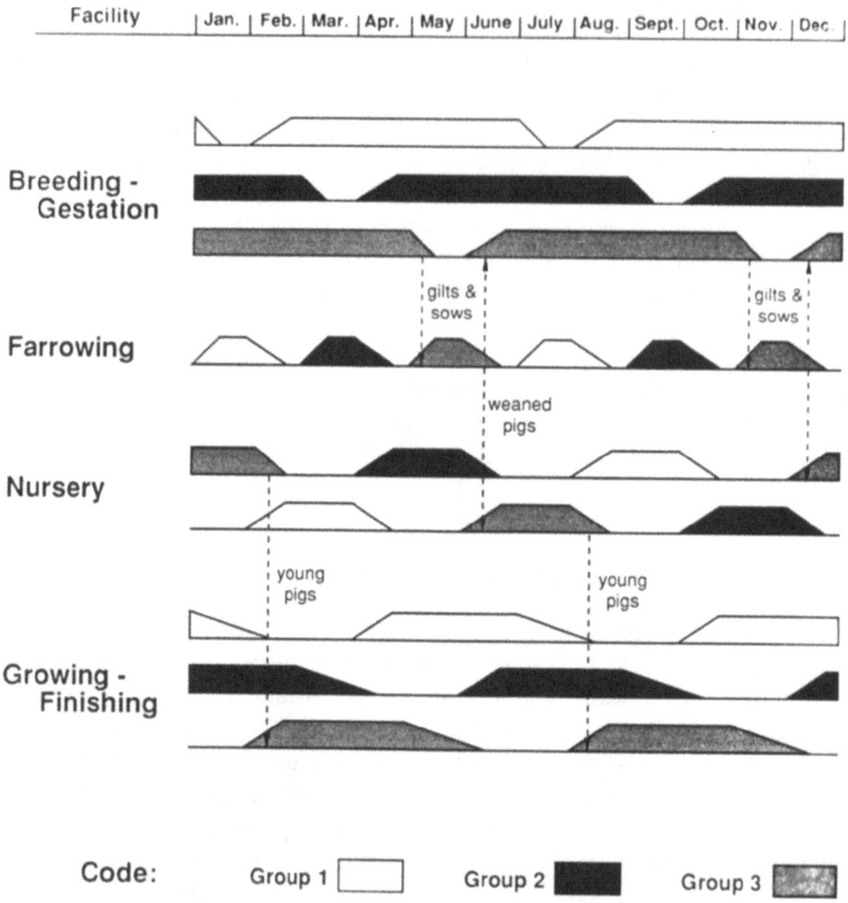

FIGURE 10.2 Schematic diagram representing time schedule for facility utilization with three groups of sows occupying farrowing facility in rotation, with empty time for cleanup and sanitation between groups of sows. Animal flow between facilities indicated for group three by dashed arrows. Figure adapted from **Pork Industry Handbook,** PIH 15, with permission of authors, D. H. Bache and J. Foster, Purdue University.

time periods for filling and emptying the house due to the variability in growth rates. In practice, most systems will establish a plan striving to move by weight, but when the scheduled time to empty the house arrives, slow growing animals will be moved even if they are light in weight.

The Pork Industry Handbook includes a series of publications prepared by Extension Service workers and pork producers in several states covering many aspects of pork production. The table of contents for this series is present in Appendix Table 5a. These include recommended production systems coordinating configurations, performance and scheduling. Elaboration on many points of this chapter can be found in PIH-13, PIH-14, PIH-15, PIH-16, PIH-17, PIH-48, PIH-49, PIH-75, PIH-83, PIH-88, and PIH-115. The examples in these publications are oriented towards representative performance standards and might need to be modified for the performance standards pertinent to a specific crossbred and for other departures in performance.

TABLE 10.1 Schedule Showing Major Activities for the Three Sow Group Facility Diagrammed in Figure 10.2.

	Management calendar for a 150-sow operation	
Stage	*Days from breeding*	*Management practices*
Pre-breeding	−30	Co-mingle gilts with sows or provide for fence-line contact; also provide new boars fence-line contact with sows.
	−28 to −14	Vaccinate sows for lepto; also vaccinate for erysipelas, if a problem.
	−14 to −7	Increase feed intake for gilts to 6-7 pounds per gilt per day for 1 to 2 weeks prior to breeding.
	−2	Spray for lice and mange.
Gestation	0	Breed, rotating boars daily.
	21 to 24	Remove boars.
	35 to 60	Pregnancy check, and sell any open females.
	90	Vaccinate sows for erysipelas, if a problem.
	105	Worm sows with a broad-spectrum wormer.
		Clean the farrowing house.
	108 to 110	Wash sows with warm water and soap, spray for lice and mange, and move to farrowing house; isolate farrowing house from all visitors.
		Start hand-feeding 3 to 4 pounds per sow per day of a farrowing ration containing 20% wheat bran (or other bulky ingredients) and continue until 2 to 3 days after farrowing.
	112	Prepare auxiliary heat for pigs.
	Days after farrowing	
Farrowing-nursery	0 to 1	Dip navel cord in mild iodine solution.
		Clip needle teeth.
		Cut off tails.
		Ear notch gilts from good litters of 9 or more pigs.
		Complete the farrowing records, including date, number pigs farrowed, etc.
		Observe sows for MMA (mastitis, metritis, agalactia).
	1 to 3	Transfer pigs to equalize litters.
		Give iron shots.
	3 to 14	Castrate males.
	7 to 10	Provide creep feed for pigs.
	14 to 21	Give second iron shot, if needed.
	28 to 42	Wean pigs, group by size and move to nursery.
Growing-finishing	42	Worm pigs with a broad-spectrum wormer, and spray for lice and mange.
	56	Vaccinate pigs for erysipelas, if a problem.
	72	Worm pigs with a wormer other than the one used at 42 days.
	150	Select replacement gilts from those ear-notched at birth.
	150 to 200	Observe withdrawal regulations for feed additives, and market the hogs at 215 to 235 pounds.

Source: Pork Industry Handbook, PIH 15, with permission of the authors, D. H. Bache and J. Foster, Purdue University.

Configuration, Scheduling, and Disease Control

As suggested earlier, effective minimization of disease risk entails all-in–all-out movement of animals between facilities and requires cleanup and disinfection to control pathogens not controllable by vaccination or by eradication from breeding stock. This includes several bacteria leading to digestive disorders especially affecting young pigs. Since down time for a facility means lower average utilization and, thus, increased costs per animal, these costs have to be offset for by less mortality and better performance. Some of these pathogens can be transmitted through the air or on dust so that distance between facilities, especially for different age groups, and separate labor are often desirable.

Performance, Carcass, and Marketing

Concern for excessive fatness in pork products is increasing. As outlined above, combinations of genetics, growth enhancers, nutritional programs, including controlled intake, offer a variety of tactics to control fatness. But all of these approaches increase costs—more expensive breeding stock, higher protein rations, costs of administering growth enhancers, more elaborate feed delivery, reduced facility utilization, etc. These approaches will obviously reduce profit if the resulting market animals are only sold on a live animal basis with no premiums for reduced fat. But even with payment-for-value programs with premiums for lower fatness, close assessment of the relative benefits and costs for several alternative combinations of approaches is mandatory.

Most purchasing programs reflect preference for an optimum carcass weight depending on that processor's preferred line of products. This leads the producer to marketing at a preferred live weight. However, full consideration of his total production system might lead the producer to marketing some animals heavier or lighter than that target. Many processors desire heavier but lean animals. Considering feed utilization and indicated fatness, some slow-growing, fat animals may become inefficient before they achieve that target weight. Conversely, some fast-growing, lean animals can continue to be efficient up to weights in excess of that weight, even if they are discounted for being heavier than the target. Marketing decisions can be more complex than marketing animals that are close to a target weight on a certain day of each week.

This survey of interactions, interrelationships, and interdependencies in the design of swine production systems may bring concern to the reader about the complexity of pork production systems. If so, the authors' purpose has been partially achieved. Hopefully, this warning about the complexity of what is involved justifies the full comprehension of the chapters of this book where the biological basis for pork production has been elaborated. If this systems view has motivated the reader to study diligently these principles, along with further specific recommendations from the Pork Industry Handbook publications, to prepare for opportunities to design and manage modern pork production systems with all these interdependencies, the authors' purpose has been realized.

HUSBANDRY PRACTICES

The management of swine for profitable production is perhaps as varied as are the environments in which swine are raised, and yet there are many factors over which the successful swine manager has control and which must be considered if the enterprise is to be profitable. These factors include the applications of economic, agronomic, biologic, and engineering principles to management practices. Specific management practices related to production and marketing are described here. Although local conditions will have a bearing on specific decisions, the principles that apply to temperate environments also apply to tropical environments and therefore the same general guidelines can be used for all environmental conditions. A book by Curtis (1981), a report by the National Research Council (1981), and the series of Fact Sheets on Management in the Pork Industry Handbook listed in Appendix Table 5a deal exhaustively with environmental management and address the principles of environmental effects on the biological responses of animals under a broad array of environmental conditions. Most of the data and examples presented here were obtained under conditions in the United States, but inferences can be drawn and applied to all areas of the world.

Suckling Pigs

The baby pig begins to explore his environment within a few minutes after birth and soon finds his way to a nipple and begins to nurse. Because of his curiosity, it is essential that the pen environment be clean to minimize chances of exposure to disease and parasites. In addition, the following tasks should be performed.

1. The umbilical cord should be tied off to prevent loss of blood, and it should be cut so that 3–5 cm remain and this portion dipped into a solution of iodine. This section will soon dry up and drop off, leaving a clean, noninfected navel. The umbilical cord is a crucial area for the entrance of infection. A common malady, "navel ill," which can cause lameness or death, may result from failure to disinfect this cord at birth.
2. The "needle" teeth should be clipped. Baby pigs have four pairs of sharp teeth (see dental pattern, Fig. 5.6), two on each jaw called needle or "wolf" teeth. They are of no practical value to the pigs themselves and they may irritate the sow's udder during nursing or cause injury to other pigs when they fight or play among themselves. Therefore, it is advisable to clip these teeth shortly after birth. Side-cutting pliers are suitable for this purpose. It is important to avoid loosening the base of the tooth or leaving jagged edges or causing injury to the gums.
3. Ear-notching should be done during the first day to establish identity. This is a simple and effective method of identifying individual pigs and litters and is essential for good record keeping. Tools especially designed for the purpose should be used and are available commercially. Any systematic notching scheme is suitable, but the one recommended by most swine breed associations in the United States has been found effective and is shown in Fig. 10.3.

FIGURE 10.3 Suggested ear-notching system for pigs. The right ear is used for the litter mark and all pigs in the same litter must have the same ear notches in this ear. The right ear is on the pig's own right. The left ear is used for notches to show the individual pig number in the litter. Each pig will have different notches in this ear. One can mark up to 181 litters with this system. Litter number and notches for that number are shown in each square. From Yorkshire Journal, August 1971.

4. Iron should be administered to prevent anemia. The newborn pig has only a limited reserve of iron for hemoglobin synthesis owing to inefficient placental transfer of iron to the fetus. The newborn pig contains only about 50 mg of iron, stored mainly in the liver, and the daily need is about 5 or 10 mg. Since sow milk is also very low in iron, the suckling pig must have supplemental iron during the first few days to prevent anemia. Pigs depend mainly on milk for nutrients until they begin to nibble on sow feed or creep feed at 2–3 weeks of age. Thus, the amount of supplemental iron should be 150–200 mg to meet the needs until iron-containing dry feed is eaten in significant amounts at about 3 weeks. In the past, this iron need was normally met by the pig rooting in the soil and otherwise picking up iron from the environment. When pigs are kept indoors in concrete-floor pens, this source of iron is eliminated. Anemia can be prevented and cured by supplying iron either orally or by injection. Commercial sources of iron in the form of pills or pastes, to be administered to the pigs individually, are also available. Also, iron-rich sweetened feeds or pellets may be provided on the pen floor. A more dependable method is the intramuscular injection of iron-dextran compounds. These compounds are available commercially and a single injection given at a few days of age provides protection until dry feed is consumed. A concentration of 100–150 mg of iron/cc is customary, so that a 1- or 2-cc injection is adequate. Iron toxicity can be produced by exceeding the recommended dosage level, but there is about a five-fold safety factor.

5. Cutting the tails is advisable whenever pigs are to be raised in total confinement. This is conveniently done at the time the needle teeth are clipped. The tail should be cut to as short a stub as possible, with the use of side-cutting pliers or other blunt pinchers to crush the bone and reduce bleeding.
6. Creep feed should be supplied during the second week to encourage dry feed consumption. Baby pigs begin to develop an appetite for dry feed at about this time as their feed requirement begins to exceed the amount supplied by the sow milk. Palatability is an important factor in creep diets and for this reason sweeteners, such as sugar, and components that improve texture, such as lard, are common constituents.

   The creep area may be partitioned from the other feeders so that only the young pigs have access to it, or it may be a separate feeder built or positioned so that only the young pigs can eat from it.
7. Castration of male pigs can be done at any age but a good time is at 1–3 weeks of age when the problem of restraining the animal is minimum and the stress on the animal is not great. Procedures are described in PIH-18 (Appendix Table 5a). Cryptorchids cannot be castrated without surgery into the body cavity. To avoid loss of identity in pigs with only one testis descended, it is preferable not to remove the descended one until both can be removed.
8. The final task to be performed on the suckling pig is weaning. With the availability of highly fortified diets for young pigs, it is possible to remove them from the sow at birth and feed sow milk substitutes. However, although this is energetically the most efficient method of producing pork, the technology of devising diets and sanitation practices that can compete with sow milk and the nursing intervals and similar factors that innate maternal-offspring relationship provides has not yet been developed. There are now experimental as well as commercially available baby pig brooder units that utilize this concept, but the industry as a whole is not ready for it yet, although certain producers with superior managerial ability and technical knowledge may find it a profitable approach.

It is customary to wean pigs at 4 to 6 weeks of age, although some producers are routinely weaning pigs at an earlier age. Some advantages of early weaning are these: sows require less feed than during lactation; sows can be rebred earlier if desired; there is a possibility of heavier, more uniform pigs at 6 weeks of age. However, it requires superior management after weaning (the sow can no longer cover up poor management with a constant balanced and sterile milk supply), and the highly fortified diets required to replace sow milk are expensive.

If pigs are weaned at 5 weeks or younger, they should weigh a minimum of 9.5 kg at 5 weeks, 6.8 kg at 4 weeks, 5.4 kg at 3 weeks, 4.1 kg at 2 weeks, and 2.4 kg at 1 week of age.

For other general information on husbandry practices associated with suckling pigs, growing pigs, sows during pregnancy, farrowing and lactation, and boars, the series of Fact Sheets published in the Pork Industry Handbook (Appendix Table 5a) should be consulted.

Orphan Pigs

If a sow dies after farrowing, or if udder trouble, lactation failure, or litters larger than the sow is able to raise occur, the result will be "orphan pigs." Two possibilities exist for the successful raising of such pigs: adoption by a foster sow, or the use of either cow milk or a sow milk replacer. In either solution, it is important that the newborn pig receive colostrum as the first feed. The phenomenon of "gut closure," in which the absorption of large molecules such as the gamma globulins of colostrum ceases following ingestion of feed, dictates that colostrum be the first food to ensure immune globulin transfer to the blood. Cow colostrum apparently provides some protection and, though not as satisfactory as sow colostrum, is greatly superior to no colostrum at all. Injectable gamma globulin preparations are also available commercially and should be used for good survival rates in colostrum-deprived pigs.

If another sow has farrowed within a short time interval before or after birth of the orphaned litter, the pigs may be transferred to her. This transfer must be made within a few days after farrowing because those glands that are not used will soon cease their milk production. To ensure acceptance of new pigs, the sow should be separated from her own litter while the new pigs are combined, and a disinfectant or other odorous material should be sprinkled on all the pigs to disguise odors.

Pigs raised on cow or synthetic milk have a strong suckling instinct and will suckle the ears and navels of other pigs when kept together, causing commotion and unrest that drains their strength. For this reason, they should be housed individually during the first 2 weeks. Feeding should be as frequent as feasible to simulate normal nursing pattern. Automatic feeding devices are available commercially that allow time- and volume-controlled dispensing at 90-minute intervals or less.

The newborn pig can survive for about 24 hours without food if kept warm, but the first feeding should be given during the first 12 hours if possible. It is possible to train baby pigs to drink from a shallow bowl by pushing their snout into the milk. After one or two feedings, they learn to consume milk quickly from a bowl, and bowl-feeding is less time-consuming than bottle feeding.

Scouring (diarrhea) is a major problem in artificially raised pigs. It is easily induced by overfeeding. Therefore, the amount of milk fed at each feeding should be reduced by one-half for one or two feedings if diarrhea appears. It is not necessary to warm refrigerated milk before feeding except as it may encourage consumption in very weak pigs for the first few feedings.

Disease Control Programs

Many infectious, metabolic, and nutritional diseases affect swine. Some of the infectious diseases are of special significance because the organisms are pathogenic to humans. Examples are brucellosis, leptospirosis, erysipelas, and tuberculosis. No attempt is made here to describe the etiology, pathogenesis, treatment, or prevention of specific diseases. The practice of veterinary medicine has gradually evolved from a major focus on treatment of diseases in individual animals to emphasis on preventive herd health maintenance in which the veterinarian and swine producer work together in applying modern principles of disease prevention and control.

Those interested in detailed coverage of specific diseases, including their prevention and treatment, have available several excellent resources. The book *Diseases of Swine* (Leman 1986) provides thorough coverage of current knowledge of all aspects of infectious and metabolic swine diseases. Other sources of information with a more applied approach are found in Chapter 10, "Infectious and Noninfectious Diseases," of *Swine Production and Nutrition* (Pond and Maner 1984) and in the series of Fact Sheets covering herd health in the Pork Industry Handbook (Appendix Table 5a). The concepts of sanitation, isolation, and specific pathogen free (SPF) pig production as components of commercial pork production systems are considered in several places elsewhere in this book.

Animal Behavior

The impact of innate behavioral patterns of the pig on the design and application of appropriate husbandry practices has been addressed by others (Hafez et al. 1962; Houpt and Wolski 1982; Curtis 1983). General behavioral traits of the pig were described in Chapter 1. Pork producers have always been concerned about providing humane treatment to their animals, and their husbandry practices and facility designs for proper care and environment continue to reflect these concerns. The first-class treatment of pigs aimed at complementing and enhancing their natural behavior rather than thwarting it as an objective in tune with development of efficient production systems.

MANAGING A PORK PRODUCTION SYSTEM

Even after a production system is designed to achieve the planned purpose of economically efficient pork production, major tasks remain in the management of that system. These tasks will be discussed in four categories:

1. Execution of plan
2. Adjustments for unplanned occurrences
3. Problem identification
4. Deciding among alternative solutions

Execution of Plan

If the plan involves new facilities, the first task would require seeing that these buildings are constructed with the planned capabilities. Financing the start-up costs for the system will nearly always be required, including consideration of the costs for the capital. Then comes purchase of breeding stock with the desired performance capabilities. Prior to delivery, competent labor should be secured with appropriate negotiation of wages. If the facilities do not include the capabilities for feed mixing, prepared feeds must be purchased and feed deliveries scheduled. If feed is to be mixed as a part of the production system, feed ingredients in the appropriate quantities and delivery schedules should be obtained. Then comes implementation of the planned schedule for matings, weaning, movement between buildings, ration changes, feed delivery, manure disposal, ventilation control, and marketing as the animals reproduce, grow, and develop as planned.

Adjustments for Unplanned Occurrences

Everything may not occur precisely as planned. Prices of feedstuffs and other supplies may change. Market prices may depart from what was expected. The weather certainly will depart from any plan. Animal performance may not be what was expected and certainly will not be uniform. Scheduling problems will occur if animals have not reached the target weight for movement to the next building when it is time to empty the building to prepare for the next group. Maybe not enough females conceived to fill the farrowing house. Or, maybe more than expected conceived and there are not enough vacant farrowing stalls to receive all the females ready to farrow. Unplanned diseases may occur if isolation and control is not as stringent as intended. New products may become available that would improve animal performance, but are expensive. Managing a pork production system can be full either of exciting challenges or of troublesome problems, depending on the perspective of the manager.

A key to successful management is problem identification. Especially important is the identification of small problems before they become large problems.

Problem Identification

As pork production has evolved to require an increasingly businesslike viewpoint and as production has become more intensive, capabilities for problem identification have become more necessary, but have improved.

In the past, problem identification came largely from the experience and intuition of the manager and focused largely on identifying animals not performing as expected. Record keeping (Fig. 10.4) has assumed a major role in modern pork production. Primary are financial records to assure that the business is suitably profitable. However, financial records may not always provide the reason for negative profitability even if they help identify that it is occurring. With a large number of animals in an enterprise, especially in the breeding herd, records become a necessary support for scheduling. For example, records on when breeding females were last bred become mandatory to scheduling which bred sows should be moved into the farrowing facility each week. These sorts of records easily extend to records evaluating reproductive performance (conception rate, litter size, rebreeding interval, etc.) for the herd as a whole, specific farrowing seasons, and individual boars and sows. When these records are complemented by records from the nursery, growing, and finishing facilities, the manager is better able to locate sources of inefficiency in his system. Knowing in which building, in which farrowing group, in which trait, in which month, etc., was performance less than expected is the first step towards identifying a problem specifically enough to search for alternative solutions.

The trend towards computerization has supported the need for more records. Some pork production enterprises include microcomputers and appropriate software packages to aid the managerial staffs in scheduling, monitoring performance, and financial accounting for the operation. Other production units utilize data processing services from computer service bureaus to support their efforts. PigTales and PigChamp are two examples of computer services to provide summaries of records to aid the manager in managing the details of the operation and identi-

FIGURE 10.4 Records have become a necessity for successful management of swine production systems; many computer packages are available to assist in this important task. Photo: Sands Livestock Systems, Inc., Columbus, Nebraska.

fying weak points in the functioning of the system. Such computerized procedures are described as Management Information Systems (MIS). As the name implies, a good MIS provides the manager with information about the performance of the system. Often these summaries present critical measures of aspects of the system, such as average conception rate, number of services, litter size, survival, growth rate, feed conversion, facility utilization, costs per animal, costs per pound of pork produced, etc., and compare these performance measures to performance goals. Performance goals may be specified by the manager, such as the performance expected as a basis for the system design, or the goals may come from the performance of other production systems. For service bureaus providing data processing service to numerous producers, the performance goals might be the recent performance of the top third of producers utilizing the service. Those measures which depart most from the goals probably indicate the aspects of the system that are most in need of improvement and, thus, require more of the manager's attention to find alternatives. Tables 10.2, 10.3, and 10.4 provide performance levels and goals as developed by the National Swine Improvement Federation. Following are formulas for some of the herd performance statistics (from Guidelines for Uniform Swine Improvement Programs, 1987, National Swine Improvement Federation and from PIH-100).

1. Breeding phase

Mated female-to-service boar ratio =

$$\frac{\text{average mated female inventory}}{\text{average service boar inventory}}$$

TABLE 10.2 Breeding and Farrowing Production Levels for Determining Production Profile and Goals.

		Performance Level				
Performance measure	*Unit*	*Excellent*	*Average*	*Poor*	*Attainable goal*	*Your farm*
Breeding Performance						
Pregnancy rate—gilts	%	Over 80	70–80	Under 70	90	____
Pregnancy rate—sows	%	Over 90	80–90	Under 80	95	____
Farrowing rate—gilts	%	Over 80	70–85	Under 70	85	____
Farrowing rate—sows	%	Over 90	80–90	Under 80	95	____
Farrowing Performance						
Pigs born alive						
gilts	no.	Over 10.5	9.5–10.5	Under 9.5	11.0	____
sows	no.	Over 11.5	10.5–11.5	Under 10.5	12.0	____
Stillborn/litter	no.	Under 0.8	0.8–1.5	Over 1.5	0.5	____
Avg. birth weight	lb.	Over 3.5	3.0–3.5	Under 3.0	3.5	____
Sow mortality	%	Under 2	2–5	Over 5	Under 2	____

Sources: Guidelines for Uniform Swine Improvement Programs, National Swine Improvement Federation and PIH-100 Performance Guidelines for the Swine Operation.

TABLE 10.3 Weaning Production Levels for Determining Production Profile and Goals.

Performance measure	*Unit*	*Performance Level*			*Attainable goal*	*Your farm*
		Excellent	*Average*	*Poor*		
Pigs weaned/litter						
gilts	no.	Over 9.5	7.6–9.5	Under 7.6	10.0	____
sows	no.	Over 10.5	8.5–10.5	Under 8.5	11.0	____
Weaning percentage	%	Over 90	80–90	Under 80	90.0	____
Pigs weaned/female/ year	no.	Over 20	16–20	Under 16	22	____
Litters/sow/year	no.	Over 2.0	1.6–2.0	Under 1.6	2.2	____
Litter weaning wt.						
3 weeks	lb.	Over 12	9–12	Under 9	14	____
4 weeks	lb.	Over 16	11–16	Under 11	18	____
Farrow-to-feeder (feed conversion)	lb. feed/ lb. gain	Under 3.8	3.8–4.3	Over 4.3	3.6	____
Birth to weaning mortality	%	Under 10	10–20	Over 20	Under 10	____
Weaning to 50 pounds mortality	%	Under 2	2–4	Over 4	Under 2	____

Sources: Guidelines for Uniform Swine Improvement Programs, National Swine Improvement Federation and PIH-100 Performance Guidelines for the Swine Operation.

Pregnancy rate (%) =

$$\frac{\text{number of females checked positive for pregnancy within 40 days post mating}}{\text{number of mated females within a contemporary group}} \times 100$$

Farrowing rate (%) =

$$\frac{\text{number of females that farrow within a contemporary group}}{\text{total number of mated females for this contemporary group}} \times 100$$

Replacement rate (%) =

$$\frac{\text{number of added mated females}}{\text{average mated female inventory}} \times 100$$

Mated female death loss (%) =

$$\frac{\text{number mated females that die per year}}{\text{average mated female inventory}} \times 100$$

TABLE 10.4 Finishing Production Levels for Determining Production Profile and Goals.

		Performance Level				
Performance measure	*Unit*	*Excellent*	*Average*	*Poor*	*Attainable goal*	*Your farm*
Pigs weaned/litter						
gilts	no.	Over 9.5	7.6–9.5	Under 7.6	10.0	____
sows	no.	Over 10.5	8.5–10.5	Under 8.5	11.0	____
Whole-herd avg. daily gain						
birth-to-market	lb.	Over 1.25	1.0–1.25	Under 1.0	1.38	____
40 lb.-to-market	lb.	Over 1.5	1.3–1.5	Under 1.3	1.7	____
Age at 230 lb.	days	Under 182	182–227	Over 227	165	____
Feed conversion for the herd						
farrow-to-finish	lb. feed/ lb. gain	Under 3.3	3.3–3.8	Over 3.8	3.1	____
40 lb.-to-market	lb. feed/ lb. gain	Under 3.0	3.0–3.5	Over 3.5	2.8	____
Loin muscle area at 230 lb.	sq. in.	Over 5.5	4.7–5.4	Under 4.7	6.0	____
Fat depth, 10th rib, at 230 lb.	in.	Under 0.9	0.9–1.3	Over 1.3	0.8	____
Avg. backfat at 230 lb. or last rib fat thickness	in.	Under 1.1	1.1–1.4	Over 1.4	1.0	____
Percent lean	%	Over 56	52–56	Under 52	57	____
Hogs discounted at market	%	Under 2	2–5	Over 5	1.5	____
Growing-finishing mortality	%	Under 2	2–4	Over 4	Under 2	____
Hogs raised/female/ year	no.	Over 19.0	15.5–19.0	Under 15.5	21	____

Sources: Guidelines for Uniform Swine Improvement Programs, National Swine Improvement Federation and PIH-100 Performance Guidelines for the Swine Operation.

2. Farrowing Phase
Average number of live pigs born per litter =

$$\frac{\text{live pigs born}}{\text{number of litters farrowed}}$$

Pigs born live per female per year (or per time period) =

$$\frac{\text{pigs born live per year (or per time period)}}{\text{average total female inventory}}$$

Pigs born live per mated female per year (or per time period) =

$$\frac{\text{pigs born live per year (or per time period)}}{\text{average mated female inventory}}$$

Stillborn (%) =

$$\frac{\text{number of stillborn pigs}}{\text{total pigs born}} \times 100$$

Mummified pigs (%) =

$$\frac{\text{number of mummified pigs}}{\text{total pigs born}} \times 100$$

Pre-weaning death loss (%) =

$$\frac{\text{pre-weaning death loss}}{\text{live pigs born}} \times 100$$

Average pig birth weight =

$$\frac{\text{litter birth weight}}{\text{total number of live pigs weighed}}$$

3. Weaning Phase
Pigs weaned per female per year (or per time period) =

$$\frac{\text{number of pigs weaned per year (or per time period)}}{\text{average total female inventory}}$$

Pigs weaned per mated female per year (or per time period) =

$$\frac{\text{number of pigs weaned per year (or per time period)}}{\text{average mated female inventory}}$$

Pigs weaned per litter farrowed =

$$\frac{\text{number of pigs weaned}}{\text{number of litters farrowed}}$$

Average pigs produced per litter =

$$\frac{\text{pigs produced per year (or per time period)}}{\text{number of farrowings per year (or per time)}}$$

Pigs weaned per crate per year (or per time period) =

$$\frac{\text{number of pigs weaned/year (or time period)}}{\text{average number of farrowing crates}}$$

Weaning percentage (i.e., % live pigs weaned) =

$$\frac{\text{number of pigs weaned}}{\text{number of live pigs born}} \times 100$$

Herd weaning age = average number of days from farrowing to weaning for the herd (synonymous—"average weaning age")

Average weaning weight =

$$\frac{\text{total weight of weaned pigs}}{\text{number of pigs weaned}}$$

Litters per female per year (or per time period) =

$$\frac{\text{total number farrowings per year (or per time period)}}{\text{average total female inventory}}$$

Litters per mated female per year (or per time period) =

$$\frac{\text{total number farrowings per year (or per time period)}}{\text{average mated female inventory}}$$

4. Growing-Finishing Phase

Average daily gain =

$$\frac{\text{total ending weight} - \text{total beginning weight} - \text{total beginning weight of dead}}{\text{number of pigs at the end} \times \text{total number of days}}$$

Days to 230 pounds = actual age +

$$\frac{(230 - \text{actual weight}) \times (\text{actual age} - 38)}{(\text{actual weight})}$$

Backfat at 230 pounds = actual backfat +

$$\frac{(230 - \text{actual weight}) \times (\text{actual backfat})}{(\text{actual weight} - 25)}$$

Average market weight =

$$\frac{\text{total weight all market hogs}}{\text{number market hogs}}$$

Marketable pork produced per year (or per time period) = weight of ending inventory + sales + home consumption, minus beginning inventory, minus purchases (usually expressed in lbs. or cwt.)

Pigs produced per year (or per time period) = ending inventory + sales + home consumption, minus beginning inventory, minus purchases

Whole herd feed efficiency for total pork produced =

$$\frac{\text{total pounds of feed consumed per year (or per time period)}}{\text{total pork produced per year (or per time period)}}$$

Whole herd feed efficiency for marketable pork produced =

$$\frac{\text{total pounds of feed consumed per year (or per time period)}}{\text{marketable pork produced per year (or per time period)}}$$

Post-weaning death loss (%) =

$$\frac{\text{post-weaning death loss}}{\text{pigs weaned}} \times 100$$

Deciding Among Alternative Solutions

After problem identification, several alternative avenues for modifying the system may occur to the alert manager. Because of the same interactions, interrelationships, and interdependencies discussed in the section "Designing a Pork Production System," decisions on managing and, particularly, problem-resolution may be just as complex. Even the experienced manager may have difficulty envisioning all the consequences of changes to the system.

What is needed to support the manager's decision making, either in designing or managing a system, is the ability to predict the consequences of an alternative course of action. If all consequences of all alternatives could be accurately predicted, the manager would only have to choose that alternative where the expected result is closest to his purpose or goal. That goal is probably summarized as being most profitable.

Decades of research on the numerous biological details of swine reproduction and growth has advanced knowledge to the point that it seems possible to capture that knowledge in the form of a set of computer programs that could simulate numerous alternative courses of action in pork production and predict the likely performance of the system. Such computer programs are termed *computer simulation models*. Together with management information systems, these programs can provide potential support to the manager's decision making. Thus, these computerized systems can be termed *integrated decision support systems* (IDSS). Research toward such procedures for guiding pork production has been done by Whittemore (1981, 1983, 1986) in the United Kingdom; Moughan and Smith (1984, 1987) in New Zealand; Black et al. (1986) in Australia; Tess, Bennett, and Dickerson (1983) in the United States; and Pomar, Harris, and Minvielle (1990a, 1990b) in the United States and Canada. The procedures developed as of this writing may not completely fulfill all the needs of today's modern manager. However, the potential and likelihood of IDSS seems great enough that the authors of this book have provided some review of the perspective and approaches. Expectations about the likely use of such procedures are included in the next chapter.

REFERENCES

Black, J. L., Campbell, R. G., Williams, I. H., James, K. J. and Davies, G. T. 1986. Res. and Dev. in Agric. *3,* 121.

Curtis, S. E. 1983. Environmental Management. Iowa State University Press, Ames.

Hafez, E. S. E., and Signoret, J. P. 1969. *In* The Behavior of Domestic Animals, 2nd ed., ed., E. S. E. Hafez. Williams & Wilkins, Baltimore Md., pp. 349–390.

Houpt, K. A., and Wolski, T. R. 1982. Domestic Animal Behavior of Veterinarians and Animal Scientists. Iowa State University Press, Ames, 356 pp.

Leman, A. D. 1986. Disease of Swine. Iowa State University Press, Ames, 930 pp.

Moughan, P. J., and Smith, W. C. 1984. N.Z. J. Agric. Res. *27,* 501.

Moughan, P. J., Smith, W. C., and Pearson, G. 1987. N.Z. J. Agric. Res. *30,* 481.

National Swine Improvement Federation. 1987. Guidelines for Uniform Swine Improvement Programs, ed. C. Christians. 93 pp.

Pomar, C., Harris, D. L., and Minvielle, F. 1991a. J. Anim. Sci. (In press)

Pomar, C., Harris, D. L., and Minvielle, F. 1991b. J. Anim. Sci. (In press)

Pond, W. G. and Maner, J. H. 1984. Swine Production and Nutrition. AVI, Westport, Conn., ch. 10, pp. 156–200.

Spedding, C. R. W. 1979. An Introduction to Agricultural Systems. Appl. Sci. Publ., London.

Tess, M. W., Bennett, G. L., and Dickerson, G. E. 1983. J. Anim. Sci. *63,* 615.

Whittemore, C. T. 1981. *In* Computers in Animal Production, eds., G. M. Hillyer, C. T. Whittemore, and R. G. Gunn, Occ. Publ. no. 5, p. 47. Br. Soc. Anim. Prod.

Whittemore, C. T. 1983. Agric. Sys. *11,* 159.

Whittemore, C. T. 1986. J. Anim. Sci. *63,* 615.

11

The Future of Pork Production

The world pig population continues to increase faster than the human population, which is now about 5.5 billion and expected to reach 7 to 8 billion by the year 2020. One often hears the assertion that because pigs compete directly with humans for some of the available food supplies, the importance of pork in the food supply must decline, both on moral and economic grounds, as the population increases. Science and technology are creating changes—in the pig, as well as in production systems—that may assure continued growth of swine numbers and pork consumption throughout the world in the foreseeable future.

In this chapter, we look to the future of pork production and attempt to predict how the industry and production practices will change in the next few decades. Part of these predictions will be based on our interpretation of recent trends and prophesying that these trends will continue. These predictions seem safe. More risky is the prediction of whether many of the areas being researched by scientists in the USDA, in the land-grant universities, and in the support industries will successfully lead to practical procedures that will be adopted by producers. When predictions seem too uncertain, we will review possible approaches that are being researched.

INDUSTRY TRENDS

In a paper presented at the 1989 World Pork Expo, Dr. Andrew J. Thulin (1989) of Michigan State University stated:

> Agriculture in general and the swine industry in particular continue to evolve and change. One of the few things that we know with certainty is that tomorrow's food

industry will not be the same as today's. Change is inevitable. We need to think about the future in order to position ourselves to anticipate, direct, and benefit from these changes in the world around us.

Thulin's projections of the continuing trends were listed as "Twenty steps into the future of the U.S. swine industry." They are:

1. Swine production industry will be increasingly demand-driven, not supply-driven.
2. Increased size of units. The 500-sow unit will be increasingly common. The commercial farrow-to-finish units could conceivably be 10,000–12,000, depending on the estimate of reproductive efficiency.
3. These units can be family owned and organized. However, because of the capitalization and managerial expertise required, only the top 10–20% of the current operations will prosper going into the twenty-first century.
4. Contractors will increase their share of the swine industry. Those contractors who are successful will share risks and have long-term commitments to the industry.
5. Vertical integration will increase their share of the swine industry but not with the rapidity or extent of the broiler and turkey industry, due to the productive complexities of the species.
6. Involvement of the feed industry in owning swine operations and breeding stock will increase, as the need to maintain feed volume, control risks, and earn profits rises.
7. There will be fewer feed manufacturing firms, with increased purchasing of only premixes by the larger swine operations. The larger, more efficient swine production operations will purchase more individual ingredients on a bulk basis and manufacture diets on the farm.
8. Swine production profit margins are expected to shrink, necessitating increased efficiency for production marketing and financial management activities.
9. New production technology employed will include use of somatotropin and beta-agonists, pending their legal approval, in an attempt to increase carcass leanness and growth performance.
10. Marketing technology employed will include techniques to manage risk of adverse price movements. These techniques might employ increased use of currently available forward-pricing techniques as options, futures markets, or direct shipment and price contracting with the packer.
11. Financial management technology employed will include techniques to manage risk of adverse interest rate movements. These techniques might employ increased use of the options, futures market, or direct contracting with outside traditional agriculture capital sources.
12. Computer technology will be increasingly used. Examples of use could range from early weaning monitoring systems employed in livestock housing to

decision support systems that include models to assist in determining strategy and tactics with regard to production, financial, and marketing decisions.

13. Greater emphasis will be placed on increasing sow productivity. There will be fewer purebred producers and a concomitant increase in the market share for replacement stock supplied by commercial seedstock companies. The seedstock will have to be merchandised on the basis of deliverable specifications.
14. Less emphasis and activity placed on production of own feedstuffs. The land base will be more concerned with adequate space for manure disposal.
15. Vigorous competition in the food marketplace will require increasing response to changing consumer demands. Hogs must become more uniform to meet the tighter specifications of the processing, packing, retailing, and food-service industries.
16. Consumers will be increasingly active in expressing their desires for a healthy, nutritious, tasty, and safe pork product that satisfies their many life-style desires.
17. Legal restrictions on swine production activities will increase. Pork producer organizations will need to become increasingly active in representing interests that are good for the pork industry as a whole.
18. It is likely that permits to produce may be required in the more densely populated areas as environmental concerns escalate.
19. The animal rights groups will become increasingly sophisticated and active. Animal welfare problems and the application of possible legal provisions to animal husbandry will remain an area of dispute for many years as the problems involved intensify.
20. Legal restrictions and requirements will probably be imposed on some facility designs and management practices.

ANIMAL PERFORMANCE

For many years to come, animal performance will improve for most of the relevant traits of economic performance to pork production schemes. That is easy to predict. Specifically *how* that improvement will occur is more difficult. The improvement will probably come from a combination of breeding, nutrition, housing, and disease-control approaches.

Major improvements will occur in lean growth rate. There is considerable variation between and within breeds for the genetic factors involved. Breeders now have the tools to achieve improvement and the motivation to use them. These tools can yield steady improvement within breeds over time. The production industry will probably move to those terminal crosses that are superior in the traits involved in rapid lean growth.

However, the research on porcine somatotropin (PST) or beta-adrenergic-agonists (ractopamine and others) show promise of leading to more dramatic increases in lean growth. It has already been demonstrated that they do so (Beerman 1989). Remaining research is directed to how to practically administer the PST to the animals. Daily injections will not be practical; slow-release implantation methods are being researched. The other problem is public acceptance and the verification of an absence of adverse influences upon the pork consumer of the products of these hormonally and chemically modified animals. The consuming public has become quite wary of chemicals and hormones. Regulatory agencies of the government will be very cautious about approving such products.

Another approach being researched is termed *transgenics*. This refers to the incorporation of genetic material controlling the synthesis of specific proteins, such as the growth hormone gene, of other species into the genome of pigs. This has successfully been done, but the resulting animals exhibited numerous adverse effects of the incorporation (Pursel et al. 1989). Earlier efforts in the laboratory mouse showed more positive results. The difficulty likely lies in the interaction with other genes controlling the timing and tissues of the manifestation of the effects of that gene. Research on these approaches will continue to be emphasized, and they may become practical, but their practicality does not seem likely in the short-term.

However the changes in lean growth rate occur, they will require changes in feeding programs. A concurrent objective with increased lean growth is a reduction in fatness. With the same feed intake, fat reduction is an inevitable consequence of increased protein synthesis. To what degree either more genetic selection against appetite or limit feeding will be necessary to control excess fatness is not altogether clear. There might be less need at current market weights but the trend to slaughter at heavier weights seems likely to maintain concern for fatness control. Conversely, control of fatness by either genetics or nutrition makes practical the cost-efficient production of lean pork at heavier slaughter weights.

The trend to more efficient reproductive performance is likely to continue. For F_1 parent females, the industry will move to one or two best breed crosses showing best maternal performance, and the breeds going into those will be further improved by genetic selection. Feeding and management systems will be synchronized to the performance capabilities of specific F_1 sows so that they can efficiently produce several parities without reproductive or lactation failure.

An approach to improving reproduction in swine is stimulated by the introduction of Chinese breeds of pigs. These pigs, evolved and bred separately from most other breeds, have a quite different set of characteristics. They mature early and produce and lactate to larger litters than U.S. and European breeds. They are decidedly inferior in lean growth rate and tend to become exceedingly fat. At present, there are considerably diverse opinions about the likelihood of genetically separating the desirable characteristics from the undesirable characters in new crosses or in new composite breeds, followed by selection in appropriate directions.

Improvement of animal performance through disease control is likely to receive continued or increased emphasis in the future. Control and treatment by medication is costly. Where practical, vaccination will be used, but that also can be costly. Newer approaches to vaccine production, such as vaccines produced by recombinant DNA (gene splicing) technology are already here (for example, vaccines for foot and mouth disease [aftosa] and for pseudorabies [Aujezky's disease]). Such developments may significantly reduce costs of infectious disease control. Most purebred herds in the future will be derived (through surgery to produce specific pathogen-free pigs by caesarian section or by hysterectomy, or through heavy medication after early weaning) to be free of major pathogens and parasites. After establishment, that status will be rigidly controlled by isolation and traffic control. These strict measures will also be used in many production herds. As new facilities are built, there will be more separation of age groups and stringent cleanup between groups.

Housing and equipment will continue to be improved to provide more effective temperature and humidity control along with labor savings. In addition, dust and odor control will be improved by appropriate techniques. All of these changes will be manifested in improved animal performance.

The use of recombinant DNA technology in pork production probably will include the cost-competitive production of individual amino acids whose incorporation into the diet at the appropriate levels and combinations will enhance efficiency of animal growth and save significant amounts of protein supplement. Recombinant DNA techniques may also find application in production of hormones, digestive enzymes, and growth factors to be administered in the feed or by effective delivery systems (implants or similar devices). Porcine growth hormone (somatotropin) is effective in improving animal lean growth and development of delivery systems is currently under active investigation. As the use of nutrient repartitioning agents, such as beta-adrenergic agonists and other compounds with a similar effect on lean tissue growth, becomes more common, adjustments will be made in diet formulation to accommodate changes in nutrient requirements associated with the altered body composition induced by the agents.

Females and castrated males eventually may be housed separately during the growing period and fed diets of different protein content to more nearly meet their respective nutrient requirements. As technology for suppressing secondary sex characteristics of intact boars is refined, and as pork purveyor attitudes change, the use of boars rather than castrates for pork production will become more commonplace in the United States, in line with the already accepted practice in many other countries. If economic incentives justify the practice of producing market animals of one sex, the technology of sex selection by appropriate sorting of male and female gametes or by other procedures may be refined.

The development and use of nonconventional and newly identified sources of energy and protein from plants and animals for feeding swine will continue, and specific crops such as grain amaranth, several seed legumes, and forages may find greater use. Blue-green algae, yeasts, and bacterial cells grown in culture may be genetically engineered to produce feed resources with specific desired nutritional composition. As the knowledge of intestinal microbial populations and large intestine physiology in the pig expands, greater utilization of feedstuffs high in lignocellulose and other fermentable substrates may be forthcoming. The identification of new microbes with high cellulose-splitting ability that thrive and compete well in the milieu of the large intestine lumen will enhance the prospects for greater utilization of fibrous feeds. Cloning of such organisms and their introduction into the large intestine environments might create the opportunity for greatly expanded feeding strategies in areas where more conventional high concentrate feedstuffs are scarce.

ANIMAL WELFARE

In the future, producers will continue to be motivated to enhance animal welfare. To begin this discussion we present a definition (Runnels 1990):

> ANIMAL WELFARE is the concept of using animals for human ends (food, biomedical research, companionship, entertainment, etc.) while minimizing pain, stress, suffering and deprivation and enhancing their well-being during their lifetimes.

Another term, *animal rights,* often has been used interchangeably with *animal welfare,* but the two terms represent two distinctly different concepts and should not be confused. Again, paraphrasing Runnels (1990):

> ANIMAL RIGHTS is the concept that animals have rights that are equivalent to or exceed those of humans. The supporters of this concept espouse the view that humans should not use animals for the benefit of anything other than the benefit of the individual animal.
>
> Such a notion, carried to its logical conclusion, declares that the use of animals for food, fiber, biomedical research, recreation, education, and companionship is unacceptable.

A small but vocal group has been effective in gaining support for this second concept and has resorted to violence and illegal acts to impose its views on the rest of society. Clearly, animal welfare is a concept supported by all reasonable pork producers and others associated with animal agriculture. It is this concept on which our discussion in this section focuses. Most livestock producers are concerned about the welfare of their animals. Nearly all the trends in pork production, from the mudlot to the modern controlled environment facility, has improved the "well-being" of the pigs involved. Cleanliness, nutrition, temperature control, and protection from parasites and diseases have undoubtedly reduced "pain, stress, suffering, and deprivation." No one who has observed, and understands, feels that the "well-being" of the pig was better when he went to the sun-baked mudhole to wallow because he was too hot and that the mudhole was full of algae and bacterial growth. Nor was its "well-being" good when its water trough was frozen and it had to find its feed in a trough covered with snow. The "well-being" of the baby pig, crushed when the sow lay on it in an open pen, wasn't very good either. Most changes in husbandry practices have improved the well-being of pigs, and the pigs have responded to these improvements with faster growth, greater survival, and more efficient reproduction. These are the same attributes that contribute to profitability.

On the other hand, there have probably been some uncaring farm employees who, without appropriate respect for the animals, have inflicted pain, suffering, neglect, and mismanagement on the animals. There have been enough abuses that the welfare of animals has become a public concern. This concern was manifested in Britain and Europe before the United States, but is increasing here, as well. The use of animals for some types of biomedical research has received the most criticism.

Practices that have received the most attention in animal agriculture are those in which animals have only limited space and restricted movement. Particular attention has been given to the practice of restraining veal calves into small stalls, the placement of several laying hens into a restricted size cage, the restraint of sows in farrowing crates, and the restriction of gestating females to individual stalls. The purpose of farrowing crates is to prevent the sow from lying on, and crushing, her baby pigs. Certainly, not being crushed improves the welfare of those pigs. The restriction to individual stalls during gestation is for the purpose of providing more control of nutrition to meet the sow's precise needs.

Nevertheless, the question arises whether more opportunity to move about and exercise would not improve the well-being of the animals. How stressful is restricted movement? Human interpretations would suggest that being caged or restrained is stressful. Is it to the pig? Attempts to study this question objectively leads to the question of: What is stressful to a pig and how do you measure the stress?

The consequence of this public concern is that pork producers should continue to search for practices that contribute to the well-being of their stock. They need to search for alternative ways of controlling feed and protecting baby pigs that will neither be stressful nor be perceived as stressful. Legislation has been passed in some countries restricting animal agriculture practices. It seems preferable that the producers' ethical attitudes and effective public communications resolve these public concerns before legal controls are imposed.

COMPUTERIZED DECISION SUPPORT

During recent decades, many industries concerned with manufacturing products from raw materials and other resources, supported by facilities and equipment and operated by labor and management, have evolved to use computerization to make the labor and management more effective and the processes more efficient. Increasingly, computers and computer software extend the capabilities of management and labor. One tool used to aid the development of plans in many manufacturing industries is simulation modeling of production system alternatives. These simulations support the decision making termed *strategic planning*. Computerization is also involved in day-to-day execution of plans and controlled modification of those plans as appropriate responses to specific occurrences not predicted in the plan. The latter use of computers to aid decision making is termed *tactical control*.

In the future, computers and computer software are expected to be used by the pork industry to plan and control the processes involved in production. Computers have been used for many years in the feed-manufacturing-allied industry for formulating diets that satisfy specified nutrient requirements for minimal costs from ingredients available for specified current prices (linear programing, least-cost diet formulation). In addition, computerized controls of the mixing of these rations are becoming more commonplace. In Chapter 6, STAGES is described as a computerized decision-support system for the breeding segment of the swine industry as it develops the breeding stock with specified characteristics for input into the production segment of the industry. As presented in Chapter 10, computerized management information systems are currently being used by many producers.

Computers will support the decision-making efforts of a farm manager as he or she *plans* the combinations of breeding stock, mating systems, rations, housing, equipment, and scheduling needed to produce pork efficiently in response to fluctuating markets. Computers will assist this same manager as he or she implements those plans but needs to respond to the fluctuations of climate and season, shifts in performance due to extraneous influences, genetic variation, changes in the price of feed ingredients, fluctuations in market prices, etc. Up to now, the manager of a swine production facility has had to make these decisions based—to a large degree—upon personal observation, experience, and intuition.

The opportunity to use computers in decision support for swine farm managers requires the ability to simulate and predict accurately the consequences of the many alternatives available to the manager at each decision. If such predictions are reliable, the decision maker will merely choose the alternative scenario in which the simulated performance comes closest to his or her objective. The computer will have to make such predictions more accurately than the manager's intuitive expectations based on experience and training.

Decades of research into the biological processes of swine growth and reproduction seems adequate to provide a sound basis for developing simulation models that can be more precise and accurate and, thereby, extend the manager's decision-making abilities. We briefly project here our conceptualization of how, at some time in the next few decades, a swine farm manager will use computers and computer-based simulation models to guide him or her in making decisions to implement improved systems. Thus, we are anticipating how the computers might impact and improve pork production, both in strategic planning and tactical control.

Strategic Planning

Innovative, progressive managers will implement this planning process either on their own on-the-farm microcomputers or on microcomputers in the offices of a swine management consultant (either a consultant who advises the manager for a fee, a public-supported swine extension specialist, or a service manager for a feedmill, a breeding stock supplier, or a pork processing plant with whom the producer has a contract). A local microcomputer (in the farm manager's office or in the office of his management consultant) may or may not do the "number-crunching" calculations for the simulations. That might be done at a larger centralized computer, with the local computer transmitting instructions over the telephone for new alternatives to be evaluated and compared. The larger centralized computer might execute the simulations that predict numerous details of the biology, physical movement, and economic implications of these alternatives for the producer's farm for a time period into the future. Results will then be transmitted back to the local microcomputer for display to the manager.

However, as a preferable alternative to tables and tables of numbers summarizing simulation results, most of the results will be presented in a graphical form on the microcomputer display screen. A special form of graphics termed "animation" will likely be used to present the results of the expected farm operation with 1 second of time on the display representing 1 day in the operation of the farm; thus, a prediction of activities and performance for 1 year into the future of the farm can be observed in about 6 minutes. However, this animated view may be interrupted, zoomed in to a part of the view, restarted at any point in simulated time, and replayed at a different speed so that the viewer can dynamically study all parts of the simulated system. These changes can be done in a manner in which both the details of the system and the interrelationships between these details can be carefully scrutinized. However, other summaries will be displayed in static bar charts or graphs of feed consumption and composition relative to time, growth, and body composition relative to time and facility inventory and utilization charts.

Even though the calculations might be done on a centralized computer, the simulations will have to be tailored to each manager's specific housing configuration and characteristics (especially capacities, floor space, air space, and manure-handling capabilities), his expected climate and seasonal cycles of weather conditions (especially temperature and humidity), and perhaps, the specific breeding stock (including expected performance characteristics) he now has. Many of these characteristics interact with other factors that will be strategically changed to improve the production system. These possible changes include nutritional characteristics of the feeds, management schedules for ration changes, movement between facilities, intended age, weight and/or fatness at marketing, temperature, humidity, and air movement targets for ventilation system, and procedures for obtaining replacement breeding stock. In some cases, the alternatives to be simulated and evaluated will include changes in breeding stock or crossbreeding system and modifications of, or additions to, current housing facilities.

A simplistic view would be that the producer-manager will only be interested in finding the system that maximizes profitability, as reflected in an economic function, such as "discounted net returns for a fixed time period per unit of initial investment." However, he or she will need to comprehend the impact of differences in more fundamental performance measures in order to intelligently continue the search through the multitude of alternatives. This calls for presentation of additional relevant descriptions of the system performance and biological details in graphic and animated form as described above.

Because of the specificity of the facilities of each producer and because climatic and seasonal factors differ widely, there will not be just one "best" pork production system; thus, the need will be to do the simulations specific for each unique condition. On the other hand, when all combinations of rations, breeding stock, crossbreeding systems, scheduling of weaning, mating, ration changes, moving between buildings, etc., are considered, the number of alternatives possible will make it difficult to simulate all possible sets of alternatives for each and every producer. Strategies will be needed to search logically through the many alternatives.

Tactical Control

Even if simulation modeling leads to the design of a near optimum production system for the producer's resources, actual conditions will not always be the same as simulated. The weather may be warmer or cooler than expected. This weather characteristic will influence the internal operating temperature of each house, which in turn, will influence the feed consumption of individuals in that house. If severe, the temperature will affect nutrient utilization in that house. In particular, energy utilized for maintaining a pig's body temperature, when cold conditions prevail in the facility, will reduce the energy from the intake that will be deposited as fat. Due to genetic variation, pigs may consume feed, grow, and fatten differently than predicted for the specific cross. Due to variation in nutritional characteristics of feed ingredients, animal performance may differ from that predicted. However, all of these possibilities can be monitored during the execution of the production program. Elements of weather, housing, feed, and animal performance will be routinely monitored as a basis for tactical production control. Presence of pathogens can be included in the monitoring. The deviations from what is predicted will be used to adjust the model to give better predictions. The simulation model will be used to adjust the diet mixture or amount to give performance closer to that desired. For example, if some pigs were getting too fat, as indicated by ultrasonically monitoring backfat level, adjustments would be made in the diet mix and daily amount to limit further fatness to what is economically allowable in the current production and market situation. Much of the data for this monitoring will be automatically recorded on electronic equipment and transferred to an on-the-farm computer as a part of the management information system (see Chapter 10).

At what level will the adjustment of rations be made—for the building, for the pen, or for the individual pig? The most common system for growing and finishing pigs in the United States today is full-feeding of a single diet for each building. Feed delivery systems are primarily designed for this approach because it is the simplest to automate. However, feed delivery systems are feasible that would deliver different rations to each pen. For these tactical control decisions to be computerized, an on-the-farm microcomputer will be necessary. Electronically controlled feeders can control ration allotment to individual animals. Such feeders are being used in Europe and in experimental situations to limit-feed individual gestating sows. There is no doubt that such systems could increase efficiency of lean growth by reducing fatness, and this is increasingly being demanded in the market place. The question is whether the potential improvement in market value will more than compensate for computerized feed delivery systems to control feeding to individual animals or, at least, to specific pens. It is expected that, if the potential usage is great enough, manufacturers will develop practical feed-delivery equipment with such capabilities for widespread use in the pork industry.

If daily control of feed mix and amount to each pen were practiced, the potential for performance control would be increased further if the pigs within that pen were uniform for nutrient requirements. This will be accomplished through monitoring of the growth and fatness tendencies of pigs during the growing period by periodic weighing and ultrasonic probing of backfat, and by recording feed intake. This monitoring will allow sorting into new pen arrangements in the finishing house so that diet mix and amount will lead to maximum lean growth but will control fatness. The feeding level will be chosen which provides an optimum compromise between market penalties for excess fatness and cost penalties for extending the time to market. Pertinent production factors will be tactically controlled as responses to monitored performance, weather, and related conditions.

Similarly, controlled feeding will be done for female animals in the breeding herd according to monitored weights, fatness, and history of performance. Controlled feeding is currently common for gestating gilts and sows, but is usually based simply upon stage in gestation and visual condition observations. Relative to what the manager and caretaker can now do, there is a potential for considerable refinement in limiting intake. But this will require that the process be computerized with the computers controlling the feeding.

The above discussion may seem conjectural, but it represents the research goals of one of the authors as he, in collaboration with other scientists researching swine at the Roman L. Hruska U.S. Meat Animal Research Center in Clay Center, Nebraska, attempts to package the results of decades of swine research into a form that will support the decision-making activities of progressive, innovative pork farm managers. The primary limitation to accomplishing these goals is in describing complex physiological processes in the precise language of mathematics. The computer technology and computer software to accomplish this is not a limitation. For now, pork farm managers are not all comfortable using computers, but that will change as computer training becomes an essential part of public schools, as computer software becomes more "user friendly," and as computer service bureaus provide computer capabilities to clients. The use of information and knowledge can enable the technology of computers and computer software to make this and other agricultural manufacturing processes more efficient and more effective.

REFERENCES

Beerman, D. H. 1989. *In* Animal Growth Regulation, eds., D. R. Campion, G. J. Hausman, and R. J. Martin. Plenum Press, N.Y. 377 pgs.

Pursel, V. G., Pinkert, C. A., Miller, K. F., Bolt, D. J., Campbell, R. G., Palmiter, R. D., Brinster, R. L., and Hammer, R. E. 1989. Science *244*, 1281.

Runnels, L. J. 1990. Yorkshire Journal *51*(1), 106.

Thulin, A. J. 1989. Pigs *5*(3), 33.

APPENDIX 1

Swine Industry Directory

NATIONAL ORGANIZATIONS

American Association of Swine Practitioners
15921 Fleur Drive
Des Moines, IA 50309
Thomas A. Neuzil, DVM, Executive Secretary

American Feed Industry Association, Inc.
1501 Wilson Blvd., Suite 1100
Arlington, VA 22209
Oakley M. Roy, President

Organizational Purpose

The national trade association representing feed manufacturers and suppliers to that industry with the mission of helping to establish a favorable business climate for the industry.

American Meat Institute
P. O. Box 3556
Washington, DC 20007
J. Patrick Boyle, President

Organizational Purpose

To create a regulatory, legislative, and consumer environment in which the meat packing and processing industry can operate efficiently and profitably. The institute will provide programs and services to improve the industry's economic environment and to expand the industry's knowledge base through education and research.

American Registry of Professional Animal Scientists
Business Office: 309 W. Clark Street
Champaign, IL 61820
Phone: (217) 356-3182

Monthly Publication: Professional Animal Scientist

Organizational Purpose

(1) To certify qualified members, (2) to strengthen Animal Science among the professions, and (3) to promote the profession of Animal Science.

American Society of Animal Science
Business Office: 309 W. Clark Street
Champaign, IL 61820
Phone: (217) 356-3182

Robert B. Zimbelman, Executive Vice President
c/o FASEB
9650 Rockville Pike
Bethesda, MD 20814
Phone: (301) 571-1875

Monthly Publication: Journal of Animal Science

Organizational Purpose

To foster communication and collaboration among all individuals and entities associated with animal science research, education, industry, or governance to serve human needs.

Midwest Plan Service
122 Davidson Hall
Iowa State University
Ames, IA 50011
Glenn A. Church II, Manager

Organizational Purpose

MWPS is to produce high quality materials involving scientific consensus on complex engineering concepts to assist member universities in meeting their educational objectives. Our customers include farmers, ag business people, vo-ag instructors, bookstores, and others interested in improving their environments.

National Association of Swine Records
P. O. Box 2417
West Lafayette, IN 47906
Phone: (317) 463-3593
Darrell D. Anderson, President

Organizational Purpose

To promote the development and use of purebred seedstock in the United States and the world.

National Live Stock and Meat Board
444 North Michigan Avenue
Chicago, IL 60611
Tim Downey, Vice President, Pork & Deli/Prepared Meats

Organizational Purpose

To conduct demand building programs for pork.

National Pork Board
P. O. Box 9114
Des Moines, IA 50306
Phone: (515) 223-2620
Michael Simpson, Executive Vice President

Organizational Purpose

To carry out effective and coordinated programs for promotion, research and consumer information designed to (1) strengthen the position of the pork industry in the marketplace, and (2) maintain, develop, and expand markets for pork and pork products.

National Pork Producers Council
P. O. Box 10383
Des Moines, IA 50306
Phone: (515) 223-2600
Russ Sanders, Exec. Vice President

Organizational Purpose

The Council's mission is to influence and effect, when possible, the internal and external factors relating to pork production on the farm, marketing the live hog, and processing and merchandising the product to the consumer in ethical ways in order to enhance the producer's opportunity for profit.

National SPF Swine Accrediting Agency, Inc.
P. O. Box 280
Conrad, IA 50621
Wayne Whipps, Executive Secretary

Organizational Purpose

Swine health accrediting and monitoring with stringent health regulations.

National Swine Improvement Federation
Department of Animal Science
University of Minnesota
St. Paul, MN 55108
C. Christians, Executive Secretary

Organizational Purpose

The purpose of this federation of organizations, businesses, and individuals interested or involved in performance evaluation of swine is to bring about uniformity of procedures, development of programs, cooperation among interested entities, education of its members and the ultimate consumers of performance evaluation methods, and to build confidence of the swine industry in the principles and potentials of performance testing.

U.S. Animal Health Association
P. O. Box 28176
6924 Lakeside Avenue–Suite 205
Richmond, VA 23228
Ella R. Blanton, Executive Director

Organizational Purpose

The study of national and international animal diseases and problems with diseases, with input from all interested segments providing for use of all of the information available regarding specific problems. Goal: The control and eradication of animal diseases.

SWINE INDUSTRY PERIODICALS

Hog Farm Management
Miller Publishing Company
P. O. Box 2400
12400 Whitewater Dr.
Minnetonka, MN 55343
Phone: (612) 931-0211
Editor: Steve Marbery

Hogs Today
Farm Journal Inc.
230 West Washington Square
Philadelphia, PA 19105
Editor: John Russell

National Hog Farmer
Division, Intertec Publishing Corp.
7900 International Drive
Minneapolis, MN 55425
Phone: (612) 851-4660
Editor: Bill Fleming

PIGS
Misset International
P. O. Box 4, 7000BA
Doetinchem, the Netherlands
Editor: Wiebe van der Sluis

Pig Farming
Wharfedale Road
Ipswich IPI 4LG
England
Editor: Bryan Kelly

Pork '90
Livestock Division
Vance Publishing Corporation
7950 College Blvd.
Shawnee Mission, KS 66201
Editor: Bill Newham

PUREBRED SWINE BREED ASSOCIATIONS

Berkshire

American Berkshire Association (ABA)
P. O. Box 2436
West Lafayette, IN 47906
Phone: (317) 497-3618
Jack Wall, Director of Breed Activities

Monthly Publication: Purebred Picture (with ALA and PCRA)

Description: Black with 6 white points (feet, tail, and face); any or all of which may be missing; white splashes on the body are acceptable.

Recommended Uses: As a terminal sire to produce fast growing, heavy muscled and sound progeny; replacement gilts, if used, will equal or surpass other dark breeds.

Chester White

Chester White Swine Record Association (CWSRA)
1803 W. Detweiller Drive
Peoria, IL 61615
Phone: (309) 691-0151
Daniel Parrish, Executive Secretary

Monthly Publication: Chester White Journal

Description: White with dropped ears and possessing extra muscle, width of chest, leanness, soundness and heaviness of bone.

Recommended Uses: Maternally to produce F_1 or other crossbred gilts to increase litter size born alive, litter size and weight at 56 days, conception rate, and total index.

Duroc

United Duroc Swine Registry (UDSR)
1803 W. Detweiller Drive
Peoria, IL 61615
Phone: (309) 691-8094
Gary E. Huffington, Executive Secretary

Monthly Publication: Duroc News

Description: Dark red in total body color, drooping ears.

Recommended Uses: Use in rotational cross, use in terminal cross as a sire breed, or use in a maternal cross—normally at 25% level. Breed is specifically used for growth, durability and efficiency.

Hampshire

Hampshire Swine Registry (HSR)
6748 N. Frostwood Pkwy.
Peoria, IL 61615
Phone: (309) 692-1571
Rick Maloney, Executive Secretary

Monthly Publication: Hampshire Herdsman

Description: Black with white belt

Recommended Uses: Primarily used as terminal sires to utilize advantages for leanness, muscle and efficiency of lean gain; also used in production of F_1 gilts for extra pigs and milk production.

Landrace

American Landrace Association (ALA)
P. O. Box 2340
West Lafayette, IN 47906
Phone: (317) 497-3718
Donald E. Verhoff, Executive Director

Monthly Publication: Purebred Picture (with ABA and PCRA)

Description: All white hair with limited isolated spots of skin pigmentation allowed, long profile, bell-shaped ear extending straight forward but not in front of eyes.

Recommended Uses: In terminal or rotational systems as a source of sow productivity including pigs born alive, birth weights, 21-day litter weights, extended parities, farrowing interval, underline quality and quantity.

Poland China

Poland China Record Association (PCRA)
P. O. Box 2537
West Lafayette, IN 47906
Phone: (317) 497-3818
Todd Boicken, Executive Secretary

Monthly Publication: Purebred Picture (with ABA and ALA)

Description: Black with six white points (feet, switch, and nose) and a drooped ear.

Recommended Uses: To produce boars for rotational crossing.

Spotted Swine

National Spotted Swine Record, Inc. (SSR)
P. O. Box 2807
West Lafayette, IN 47906
Phone: (317) 497-3918
Rex Hoppes, Executive Secretary

Monthly Publication: Spotted News

Description: Black and White "Spotted" markings with small drooping ears; extremely durable, production oriented, high cutability swine.

Recommended Uses: Totally universal. Prepotent. Aggressive males are known for high conception rates and producing large vigorous pigs in terminal or rotational crossing systems. Females in F_1 or other crossbred crosses with white breeds to add milk production, high fertility, durability, and reduced management.

Yorkshire

American Yorkshire Club (AYC)
P. O. Box 2417
West Lafayette, IN 47906
Phone: (317) 463-3593
Darrell D. Anderson, Executive Secretary

Monthly Publication: Yorkshire Journal

Description: White, large-framed, long-bodied with erect ears; adequate in muscle, quite lean, structurally correct with excellent underlines.

Recommended Uses: To produce F_1 gilts in combination with several other breeds; popular in rotational crossbreeding; certain bloodlines may be used as terminal sires. Primary role is on the maternal side for large litters and excellent milking.

FIGURE A1.A **U.S. Breeds of Swine: Upper left, Berkshire boar; upper right, Chester White gilt; lower left, Duroc (boar); lower right, Hampshire boar.**

TABLE A1.1 1989 Breed Data for Membership, Registration, Transfers, AI Certificates, and Participation in Performance Testing Programs

	Breed abbreviations[a]							
	ABA	*CWSRA*	*UDSR*	*HSR*	*ALA*	*PCRA*	*SSR*	*AYC*
Membership and Registrations								
Senior	420*	872	1,336	1,300*	156	264	722	1,100
Junior	260*	256	4,352	525*	71	192	367	400
No. litters registered	1,940	6,375	20,175	19,326	3,849	2,100	8,408	23,103
No. animals transferred	1,711	9,340	9,787	17,380	4,506	1,692	7,902	20,004
No. AI certificates	33	175	2,167	510			***	750
Central Tests								
No. boars	180	**	1,400*	1,050	163	110*	***	1,500*
STAGES								
Maternal								
No. herds	42	***	142	71	84	40	***	599
No. litters	416	***	4,967	4,419	2,642	636	***	16,665
Growth								
No. herds	4	***	31	14	3	0	***	20
No. pigs	154	***	1,900	2,241	1,463	0	***	5,845*
Other On-Farm Programs								
Maternal								
No. herds	60*	**	**	30*	**	**	**	75*
No. litters	**	**	**	3,000*	**	**	**	10,000*
Growth								
No. herds	60*	**	**	50*	**	**	**	50*
No. pigs	**	**	**	8,000*	**	**	**	10,000*

[a] See directory for abbreviations
* Approximate data
** Unknown or not available
*** Confidential data
Data supplied by Executive Officers of breeds with membership in National Association of Swine Records and National Swine Improvement Federation.

FIGURE A1.B U.S. Breeds of Swine: Upper left, Landrace gilt; upper right, Poland China boar; lower center, Yorkshire boar.

FIGURE A2 Representative Genetic Products of Corporate Breeding Programs: Upper left, DEKALB DK31 gilt with DK77 boar; upper right, McLean County MxD boar; lower left, P.I.C. L26 boar; lower right, Farmers Hybrid Boar Power R1 boar.

CORPORATE BREEDING PROGRAMS

(Members of the National Swine Improvement Federation, with programs not included in National Association of Swine Records member breeds)

Babcock Swine Herds, Inc. (BSH)
P. O. Box 759
Rochester, MN 55903
Jim McPeak–President
Dr. Jim Schneider–Swine Geneticist

Primary Genetic Products

All genetic input including Babcock Parent Boars and, alternatively, Babcock Parent Gilts or Babcock Grandparent Boars and Babcock Grandparent Gilts necessary for rota-terminal crossbreeding systems. Genetic input includes live animals, semen (fresh and frozen) and/or embryos. Domestic customers may utilize paternal parent males, maternal parent females and/or separate maternal grandparent lines. International customers may utilize separate paternal grandparent lines as well as separate maternal grandparent lines. P.I.G.S. Program is an on-the-farm performance program. It is available to customers using GP lines. Results are all FAXed to the Babcock Research Center where PD's are calculated.

Genetic Evaluation Procedure

Reduced animal model (RAM) used to calculate predicted differences (PDs) on all nucleus animals. Selection index procedures are used for nucleus animals and for almost all of the genetic products sold. Traits and indexes incorporated differ between genetic lines and products, but in all cases maximizing efficiency of lean growth remains the primary objective.

DEKALB Swine Breeders, Inc. (DSB)
3100 Sycamore Road
DeKalb, IL 60115
Phone: (815) 758-2409
Roy Poage–President
Dr. David Fox–Director of Swine Genetic Research
Dr. Sam Buttram, Research Geneticist

Primary Genetic Products

Specific Cross(R) Program

Hybrid Boars		*Hybrid Gilts*	
DK77 or DK88	bred to	DK30	Bred and developed for confinement or pasture production systems to produce terminal cross market animals.
DK77 or DK88	bred to	DK31 or DK33	Bred and developed for confinement production systems to produce terminal cross market animals.

Custom Genetics(R) Program—Grandparent boars and gilts are used for production of replacement hybrid gilts to be used in *Specific Cross(R) Program*.

Hybrid Boar Rotation Program—Hybrid boars: DK77, DK88, DK61, DK63. Used in a rotational crossbreeding system.

Genetic Evaluation Procedure

A selection index combining growth rate, feed efficiency, and backfat has been used as a selection criteria. Animals are also closely evaluated for feet and leg soundness. BLUP evaluations are now being made on lean gain per day.

Farmers Hybrid Companies, Inc. (FHC)
P. O. Box 4528
Des Moines, IA 50306
Dr. Olin N. Andrews, Jr., President and CEO
Dr. Hal I. Sellers, Director, Research

Primary Genetic Products

Boar Power (R) R-1 boars—A combination of Elite(R) Lines 4 and 5. Used principally in rotational systems. Characteristics include rapid growth and high feed efficiency and high milk production.

Boar Power(R) R-2 boars—A combination of Elite(R) Lines 2 and 9. Used principally in rotational systems. Characteristics include durability, high libido and lean yield.

Boar Power(R) M-1 boars—A combination of Elite(R) Lines 4 and 7. Used principally in rotational, rota-terminal and terminal systems. Strengths include litter size, prominent nipples and feed efficiency.

Boar Power(R) M-2 boars—A combination of Elite(R) Lines 2 and 8. Used principally in rotational, rota-terminal and terminal systems. Strengths include litter size, milk production and leanness.

Boar Power(R) CX boars—A combination of Elite(R) Lines 5 and 9. Used principally in terminal and rota-terminal systems. Strengths include libido, feed efficiency and lean production.

Golden(R) Gilts—Sisters to the R-1, R-2, M-1 and M-2 boars and recommended for use in the same breeding systems as the boars.

Genetic Evaluation Procedure

Traits: Average daily gain, probe backfat, feed intake, total litter size farrowed (feed intake is measured on gilt pigs only).

Procedure: Mixed model procedure (BLUP) within Elite lines. In nucleus herds, top 20% of gilts and top 5% of boars selected as replacements.

Genetic Improvement Service (GIS)
P. O. Box 447
Burlington, IN 46915
Phone: (317) 566-3391
Craig Martin, General Manager
Stanley R. Williams, Director of Research and Development

Primary Genetic Products

Blackline Plus—Terminal Boar, ¾ Hampshire and ¼ Duroc

V-line Gilt—Gilt for terminal cross production, ½ Yorkshire ¼ Landrace ¼ Duroc. Adaptable to a wide range of facilities and management conditions.

E-L Gilt—Gilt for terminal cross production, ½ Landrace ¼ Yorkshire ¼ Duroc. Recommended for crated, tethered or groups less than 10 in gestation.

Rotational Boars—Hampshire, Yorkshire, Duroc, Landrace, W-Line (Y × D F_1), G-Line (L × D F_1), M-Line (Y × L F_1), F-Line (Synthetic line predominantly Hampshire and Yorkshire)

Genetic Evaluation Procedure

Within-herd EBV program developed by Dr. Allan Schinckel, Purdue University.

Terminal Index—Used on Hampshire, Duroc, Hamp × Duroc F_1, Blackline Plus Terminal Sires and F-Line Synthetic to optimize lean growth. Includes backfat, loineye, and days of age adjusted to 230 lb and litter size.

Maternal Index—Used on Landrace, Yorkshire, W-Line, G-Line and M-Line. Includes litter size and litter weight as well as terminal traits.

Real time ultrasound used to get accurate live animal backfat and loineye measurements.

Purebred—All boars, barrows, and gilts tested for growth and fatness; complete sow productivity.

F_1—All boars tested and a limited amount of gilts tested for growth and fatness; complete sow productivity.

Other crossbreds—All boars tested for growth and fatness; limited sow productivity.

McLean County Hog Service, Inc. (MCHS)
Rt. 2, Box 96
LeRoy, IL 61752
Robert Builta, President

Primary Genetic Products

M × $D^{(R)}$ line—Inbred Montana and purebred Duroc. Development for the standpoint of more hybrid vigor in the offspring, aggressive breeder. Crosses well for terminal or rotational crossing program.

Y × $L^{(R)}$ line—Inbred Lacombe and purebred Yorkshire. Noted for large litters and adapted well for confinement. Line works well for terminal or rotational crossing programs with other hybrid lines.

Sp × 1 line—Spotted Poland and Inbred Minnesota #1. This line is noted for its feed efficiency and rate of grain, but carries mothering and milking ability. Can be used as a first cross boar or used in a rotational program.

H × LW line—Hampshire and Large White. Not only noted for good carcass, but good feed efficiency and extremely large litters. The H × LW complements all other lines and can be used as terminal cross as well as in rotational cross programs.

Genetic Evaluation Procedure

Every boar sold is individually performance tested for rate of gain and backfat. Minimum standards must be met for all animals for litter size and mothering ability. Selection process is monitored every thirty days for all production herds and all test information is recorded monthly.

Pig Improvement Co., Inc. (PIC)
P. O. Box 348
Franklin, KY 42135-0348
Phone: (502) 586-9224
K. W. Woolley, President
Phillip J. David, Ph.D., Director, USA Technical & Development Dept.
Eldon R. Wilson, Ph.D., Research Manager

Primary Genetic Products

Camborough 15—Crossbred gilt designed for use in an indoor farrowing system. Recommended for production of slaughter pigs only.

Camborough Blue—Crossbred gilt designed for use in an outdoor farrowing system. Recommended for production of slaughter pigs only.

HY Boar—Crossbred boar produced by crossing synthetic lines which incorporate Continental European muscular breeds. Recommended for use in slaughter pig production only. Especially designed for markets where high lean content is important.

L26 Boar—Crossbred boar which incorporates lines selected for efficient production of lean meat. Recommended for production of slaughter pigs only.

Genetic Evaluation Procedure

Traits measured include average daily gain, ultrasonic measures of external fat, average daily feed intake on individual boars, number of pigs born and weaned per litter. Have used selection index. Now implementing the use of an animal model, multiple-trait BLUP procedure.

TABLE A1.2 Summary of 1989 Sales and Performance Testing for Corporate Breeding Programs not Included in Breed Organizations

				Performance test	
Corporation[a]	*No. boars sold for breeding*	*No. gilts sold for breeding*	*Central-boar tests*	*Growth (pigs)*	*Maternal (litters)*
BSH	***	***	0	19,554	3,713
DSB	***	***	0	***	***
FHC	***	***	0	35,733	1,630+
GIS	***	***	0	22,000*	9,600*
MCHS	***	***	0	***	***
PIC	***	***	0	51,235	16,900*

[a] See directory for abbreviations
* Approximate data
*** Confidential data
\+ Data for nucleus herds only
Data provided by Corporate Geneticist.

SWINE ARTIFICIAL INSEMINATION ORGANIZATIONS

These are members of the National Swine Improvement Federation who publish semen catalogs:

Birchwood Genetics, Inc. (BGI)
465 Stephens Road
West Manchester, OH 45382
David L. Flory, President
Breeds of boar: Duroc, Hampshire, Landrace, Yorkshire

Cole Swine Farms (CSF)
17006 SR37
Mt. Blanchard, OH 45867
Lynn A. Cole, President
Breeds of boar: Duroc, Hampshire, Landrace, Yorkshire, Herdbuilder (L × HD)

International Boar Semen (IBS)
P. O. Box 496
Eldora, IA 50627
Carol Pettyjohn, C.E.O.
Breeds of boar: Berkshire, Chester White, Duroc, Hampshire, Landrace, Poland China, Spotted, Yorkshire

Lone Willow Genetics (LWG)
Rt. 2, Box 49
Roanoke, IL 61561
Phone: (309) 923-7234
Bruce Leman, General Manager
Breeds of boar: Duroc, Hampshire, Yorkshire

Stoney Creek Farms (SCF)
Rt. 2, Box 262
Farmland, IN 47340
Dr. Dale A. Hendrickson, Owner
Breeds of Boar: Chester White, Duroc, Hampshire, Landrace, Spotted, Yorkshire

Swine Genetics International (SGI)
Rt. 1, Box 3
Cambridge, IA 50046
Dr. Harold H. Hodson, Jr., Executive Director
Breeds of Boar: Berkshire, Chester White, Duroc, Hampshire, Landrace, Large White, Poland China, Spotted. Yorkshire

CENTRAL AND ON-FARM SWINE TESTING ORGANIZATIONS

From the following states and Canada:

Alabama —Sand Mountain Swine Evaluation Station
c/o Jerry Davis
Rt. 3, Box 421
Crossville, AL 35962
(205) 659-4392

—Southeastern Swine Evaluation Station
c/o Robert Barker
P. O. Box 773
Florence, AL 35631
(205) 764-7074

TABLE A1.3 Summary of Semen Sales, Boars Used, Performance and Progeny Test of Boars Used for 1989 in National Swine Improvement Federation Member Organizations Who Publish Semen Catalog

	AI organization[a]					
	BGI	*CSF*	*IBS*	*LWG*	*SCF*	*SGI*
No. boars producing semen	38	20	96	11	51	98 (Fresh)[b] 137 (Frozen)
No. units semen sold						
Fresh	14,835+	2,000*	60,000*	2,678	12,000+	36,208 (98 boars)
Frozen	0	0	10,000*	0	0	8,554 (137 boars)
			No. Boars Tested			
Central						
Performance	8	0	73	0	9	20 of 98, 36 of 137
Progeny	16	0	40*	1	24	40–50%*
On-Farm						
Performance	27	20	10–12*	11	17	72 of 98, 92 of 137
Progeny	38	20	**	11	38	90%*

[a] See directory for abbreviations
[b] 55 to 60 at a time
* Approximate data
** Data unknown or not available
*** Confidential data
\+ Includes use on owner's farm
Data supplied by executive officer of organization.

Arkansas —Arkansas Testing Station
c/o Gary Murphy
University of Arkansas
Reef Sub Station
P. O. Box 671
Newport, AK 72112

California —California Swine Testing Station
c/o Robert Busch
Plumas 200
Chico State University
Chico, CA 95929-0315
(916) 895-4551 or 895-6025

Florida —Florida Swine Evaluation Unit
c/o Dan Bembry
Rt. 2, Box 2181
IFAS
Live Oak, FL 32060
(904) 362-3067

Georgia —Georgia Swine Testing Station
c/o Scott Utley
Box 1209, Rural Development Center
Tifton, GA 31793
(912) 386-3407

Illinois —Illinois Swine Test Program
c/o Dr. Lief H. Thompson
1301 W. Gregory Dr.
University of Illinois
Urbana, IL 61801
(217) 333-7349

—WIU Swine Evaluation Center
c/o Bruce A. Engnell
Rt. 5 Tower Rd.
Macomb, IL 61455
(309) 298-2613

Indiana —Indiana Boar Test Station
c/o Wayne Yake
Purdue University
Dept. of Animal Sciences
Lilly Hall
West Lafayette, IN 47907
(317) 583-2311 or 494-8570

Iowa —Iowa Swine Producers Association
c/o Andy McVay
Box A, ISU Station
Ames, IA 50010
(515) 292-1888

	—Farmland Livestock Services c/o Kevin Shaffer Lisbon, IA 52253 (319) 455-2396
	—Iowa Swine Testing Station c/o Andy McVay Rt. 4, Box 271 Ames, IA 50010 (515) 292-1888
	—Northeast Iowa Swine Improvement Assn. c/o Tim Klaustermeier Rt. 2 New Hampton, IA 50659 (515) 394-2246
	—Purebred Genetics Swine Test Station c/o Dwayne Naab Rt. 1, Box 31C Ida Grove, IA 51445 (712) 364-2725
Louisiana	—Louisiana Testing Station c/o John E. Pontif Dean Lee Agricultural Center LSU-A, Rt. 2 Alexandria, LA 71303
Maryland	—Maryland Swine Test Station c/o Ricky Bauer 711 Andover Rd. Linthicum, MD 21090 (301) 850-4045
Michigan	—Michigan Swine Breeders Association c/o Dr. Andy Thulin Animal Science Department Michigan State University East Lansing, MI 48824
	—Michigan Swine Evaluation Station c/o Paul Matzat Michigan State University 205 Anthony Hall East Lansing, MI 48824
Minnesota	—Cenex/Land O'Lakes Ag. Services c/o Clair Fralick P. O. Box 64089 St. Paul, MN 55164-0089
	—Minnesota Pork Producers Assoc. Rt. 1, Box 134A New Ulm, MN 56073

—Minnesota Swine Testing Station
c/o Avery Wills
Rt. 1, Box 134A
New Ulm, MN 56073
(507) 354-6513

Mississippi

—Mississippi Swine Testing Station
c/o Howard Miller
Animal Science Department
Mississippi State University
State College, MS 39762

Missouri

—Farmland Livestock Services
c/o Dallas Reeve, Coordinator of Livestock Services
10380 N. Executive Hill Blvd.
Kansas City, MO 64190
(800) 821-8000

—NEMO Swine Test Station
c/o Butch Howard
Rt. 1, Box 8
Canton, MO 63435
(314) 288-5558

—Northwest Missouri Swine Test Station
c/o Dale Buhman
Rt. 1
Maryville, MO 64468
(816) 582-4280

—SEMO Swine Testing Station
c/o Dr. Wm. E. Ellis
Southeast MO State University
109 A Magill Hall
Cape Girardeau, MO 63701
(314) 651-2106

—UMC Boar Test Station
c/o Brad Belew
University of Missouri
Animal Science Dept.
S-111 An. Sci. Center
Columbia, MO 65211

Montana

—C.H. Perry Swine Test Station
c/o Wayne F. Gipp
Montana State University
219 Linfield Hall
Bozeman, MT 59717-0003
(406) 586-3414

Nebraska

—Nebraska SPF Swine Accrediting Agency
c/o Dr. Thomas E. Socha
1840 North 48th Street
Lincoln, NE 68504

	—SENEK Swine Test Station c/o John McKeever Rt. 1 Wymore, NE 68466 (402) 674-3311
New York	—New York Test Station c/o Dr. Tro V. Bui Dept. Animal Science 250 Morrison Hall Cornell University Ithaca, NY 14853 (607) 255-4505
North Carolina	—North Carolina Swine Evaluation Station c/o Ron Hughes 1508 Hwy. 70 West Clayton, NC 27520 (919) 553-7696
Ohio	—Ohio Pork Improvement Association c/o Dr. Gene Isler Animal Science Dept. Ohio State University Columbus, OH 43210
	—Ohio Swine Evaluation Station c/o Rod Limes 2029 Fyffe Rd. Columbus, OH 43210 University Ph. (614) 451-3750 Station Ph. (614) 292-2098
Oklahoma	—Oklahoma Swine Evaluation Station c/o Dr. Bill Luce Oklahoma State University 201 Animal Science Stillwater, OK 74078 (405) 744-6058
Pennsylvania	—Meat Animal Evaluation Center c/o Glenn Eberly 651 Fox Hollow Rd. State College, PA 16803 (814) 238-2527
South Carolina	—South Carolina Swine Evaluation Center c/o Dr. Bill Brown Clemson University Box 280 Elgin, SC 29045 (803) 788-2162
South Dakota	—South Dakota Swine Testing Station c/o Lawrence (Pete) Eckhoff Rt. 3, Box 4 Brookings, SD 57006 (605) 693-3632

Tennessee	—University of Tennessee Swine Testing Program c/o Glenn Conatser 103 McCord Hall Knoxville, TN 37901
	—Middle Tennessee Swine Test Station Location: Spring Hill, TN c/o Dr. Richard Simms Box 110019 Nashville, TN 37222-0019 (615) 832-8341
	—West Tennessee Swine Test Station c/o James Tracy 605 Airways Blvd. Jackson, TN 38301 (901) 422-1583
Texas	—Texas Swine Test Center c/o Archie Abrameit The Luling Foundation Drawer 31 Luling, TX 78648 (512) 875-2438
Virginia	—Virginia Purebred Swine Breeders Association c/o William M. Stanley Rt. 1, Box 2306 Ashland, VA 23005
	—Virginia Swine Evaluation Station c/o Phillip K. Taylor Extension Animal Science-Swine Box 7219 Suffolk, VA 23437-2019 (804) 539-2064
Wisconsin	—Wisconsin Swine Genetic Center (Cenex/Land O'Lakes Ag. Services) c/o Ron Schmaling Rt.1 Belleville, WI 53508 (608) 424-3022
Ontario (Canada)	—Agriculture Canada c/o Randy Nelson Animal Production Ottawa, Ontario Canada K1A 0C5
	—Quality Swine Co-op Testing Program c/o Jim F. Hunter Box 53 Shedden, Ontario Canada N0L 2E0

APPENDIX 2

Conversion Tables

TEMPERATURE CONVERSIONS

To convert a temperature, in either Celsius (Centigrade) or Fahrenheit, to the other scale, find that temperature in the center column, and then find the equivalent temperature in the other scale either in the Celsius column to the left or in the Fahrenheit column to the right. For example, if a given temperature is 40°F, the equivalent temperature on the Celsius scale will be 4.44°C (shown in the lefthand column); if the given temperature is 40°C, the corresponding Fahrenheit reading will be 104°F (shown in the righthand column).

On the Celsius scale the temperature of melting ice is 0° and that of boiling water is 100° at normal atmospheric pressure. On the Fahrenheit scale, the equivalent temperatures are 32° and 212° respectively. The formula for converting Fahrenheit to Celsius is C = 5/9(F − 32), and the formula for converting Celsius to Fahrenheit is F = (9/5C) + 32.

Temperature Conversions

C	*F or C reading*	*F*	*C*	*F or C reading*	*F*
−73.33	−100	−148.0	−45.56	−50	−58.0
−70.56	−95	−139.0	−42.78	−45	−49.0
−67.78	−90	−130.0	−40.0	−40	−40.0
−65.00	−85	−121.0	−37.23	−35	−31.0
−62.22	−80	−112.0	−34.44	−30	−22.0
−59.45	−75	−103.0	−31.67	−25	−13.0
−56.67	−70	−94.0	−28.89	−20	−4.0
−53.89	−65	−85.0	−26.12	−15	5.0
−51.11	−60	−76.0	−23.33	−10	14.0
−48.34	−55	−67.0	−20.56	−5	23.0

(continued)

Temperature Conversions *(continued)*

C	*F or C reading*	*F*	*C*	*F or C reading*	*F*
−17.8	0	32.0	8.89	48	118.4
−17.2	1	33.8	9.44	49	120.2
−16.7	2	35.6	10.0	50	122.0
−16.1	3	37.4	10.6	51	123.8
−15.6	4	39.2	11.1	52	125.6
−15.0	5	41.0	11.7	53	127.4
−14.4	6	42.8	12.2	54	129.2
−13.9	7	44.6	12.8	55	131.0
−13.3	8	46.4	13.3	56	132.8
−12.8	9	48.2	13.9	57	134.6
−12.2	10	50.0	14.4	58	136.4
−11.7	11	51.8	15.0	59	138.2
−11.1	12	53.6	15.6	60	140.0
−10.6	13	55.4	16.1	61	141.8
−10.0	14	57.2	16.7	62	143.6
−9.44	15	59.0	17.2	63	145.4
−8.89	16	60.8	17.8	64	147.2
−8.33	17	62.6	18.3	65	149.0
−7.78	18	64.4	18.9	66	150.8
−7.22	19	66.2	19.4	67	152.6
−6.67	20	68.0	20.0	68	154.4
−6.11	21	69.8	20.6	69	156.2
−5.56	22	71.6	21.1	70	158.0
−5.00	23	73.4	21.7	71	159.8
−4.44	24	75.2	22.2	72	161.6
−3.89	25	77.0	22.8	73	163.4
−3.33	26	78.8	23.3	74	165.2
−2.78	27	80.6	23.9	75	167.0
−2.22	28	82.4	24.4	76	168.8
−1.67	29	84.2	25.0	77	170.6
−1.11	30	86.0	25.6	78	172.4
−0.56	31	87.8	26.1	79	174.2
0	32	89.6	26.7	80	176.0
0.56	33	91.4	27.2	81	177.8
1.11	34	93.2	27.8	82	179.6
1.67	35	95.0	28.3	83	181.4
2.22	36	96.8	28.9	84	183.2
2.78	37	98.6	29.4	85	185.0
3.33	38	100.4	30.0	86	186.8
3.89	39	102.2	30.6	87	188.6
4.44	40	104.0	31.1	88	190.4
5.00	41	105.8	31.7	89	192.2
5.56	42	107.6	32.2	90	194.0
6.11	43	109.4	32.8	91	195.8
6.67	44	111.2	33.3	92	197.6
7.22	45	113.0	33.9	93	199.4
7.78	46	114.8	34.4	94	201.2
8.33	47	116.6	35.0	95	203.0

(continued)

Temperature Conversions *(continued)*

C	*F or C reading*	*F*	*C*	*F or C reading*	*F*
35.6	96	204.8	127	280	536
36.1	97	206.6	132	290	554
36.7	98	208.4	138	300	572
37.2	99	210.2	143	310	590
37.8	100	212.0	149	320	608
43	110	230	154	330	626
49	120	248	160	340	644
54	130	266	177	350	662
60	140	284	182	360	680
66	150	302	188	370	698
71	160	320	193	380	716
77	170	338	199	390	734
82	180	356	204	400	752
88	190	374	210	410	770
93	200	392	216	420	788
99	210	410	221	430	806
100	212	414	227	440	824
104	220	428	232	450	842
110	230	446	238	460	860
116	240	464	243	470	878
121	250	482	249	480	896
166	260	500	254	490	914
171	270	518	260	500	932

Area Conversions

Metric		
1 square centimeter	=	0.155 sq inch
	=	100 sq millimeters
1 square meter	=	1,550 sq inches
	=	10,764 sq feet
	=	1,196 sq yards
	=	10,000 sq centimeters
1 square kilometer	=	0.3861 sq mile
	=	1,000 sq meters
1 hectare	=	2.471 acres
		10,000 sq meters
Imperial		
1 square inch	=	6.452 sq centimeters
	=	1/144 sq foot
	=	1/1296 sq yard

(continued)

Area Conversions (continued)		
1 square foot	=	929.088 sq centimeters
	=	0.0929 sq meter
1 square yard	=	8,361.3 sq centimeters
	=	0.8361 sq meter
	=	1,296 sq inches
	=	9 sq feet
1 square mile	=	2.59 sq kilometers
	=	640 acres
1 acre	=	0.4047 hectare
	=	43,560 sq feet
	=	4,840 sq yards
	=	4,046.87 sq meters
1 kilometer	=	3,281 feet
	=	1,094 yards
	=	0.621 mile
	=	1,000 meters
Imperial		
1 inch	=	25.4 millimeters
	=	2.54 centimeters
1 foot	=	30.48 centimeters
	=	0.3048 meter
	=	12 inches
1 yard	=	0.9144 meter
	=	91.44 centimeters
	=	3 feet
1 mile	=	1,609,347 meters
	=	1,609 kilometers
	=	5,280 feet
	=	1,760 yards

Yield Conversions		
Metric		
1 kilogram per hectare	=	0.89 pound per acre
1 cubic meter per hectare	=	14.2916 cubic feet per acre
Imperial		
1 pound per acre	=	1.121 kilograms per hectare
1 ton (2,000 lb) per acre	=	2,242 metric tons per hectare
1 cubic foot per acre	=	0.0699 cubic meter per hectare
1 bushel (60 lb) per acre	=	67.26 kilograms per hectare

Volume Conversions

Metric		
1 milliliter	=	1 cubic centimeter (cc)
1 liter	=	1.057 U.S. quarts liquid
	=	0.9081 quart, dry
	=	0.2642 U.S. gallon
	=	0.221 Imperial gallon
	=	1,000 milliliters or cc
	=	0.0353 cubic foot
	=	61.02 cubic inches
	=	0.001 cubic meter
1 cubic meter	=	61,023.38 cubic inches
	=	35,314 cubic feet
	=	1,308 cubic yards
	=	264.17 U.S. gallons
	=	1,000 liters
	=	28.38 U.S. bushels
	=	1,000,000 cu. centimeters
	=	1,000,000,000 cu. millimeters
Imperial		
1 fluid ounce	=	1/128 gallon
	=	29.57 cubic centimeters
	=	29.562 milliliters
	=	1.805 cubic inches
	=	0.0625 U.S. pint (liquid)
1 U.S. quart liquid	=	946.3 milliliters
	=	57.75 cubic inches
	=	32 fluid ounces
	=	4 cups
	=	1/4 gallon
	=	2 U.S. pints (liquid)
	=	0.946 liter
1 quart dry	=	1.1012 liters
	=	67.20 cubic inches
	=	2 pints (dry)
	=	0.125 peck
	=	1/32 bushel
1 cubic inch	=	16.387 cubic centimeters
1 cubic foot	=	28,317 cubic centimeters
	=	0.0283 cubic meter
	=	28.316 liters
	=	7.481 U.S. gallons
	=	1,728 cubic inches
1 U.S. gallon	=	16 cups
	=	3.785 liters

(*continued*)

Area Conversions (continued)		
	=	231 cubic inches
	=	4 U.S. quarts liquid
	=	8 U.S. pints liquid
	=	8.3453 pounds of water
	=	128 fluid ounces
	=	0.8327 British Imperial gallon
1 British Imperial gallon	=	4.546 liters
	=	1.201 U.S. gallons
	=	277.42 cubic inches
1 U.S. bushel	=	35.24 liters
	=	2,150.42 cubic inches
	=	1.2444 cubic feet
	=	0.03524 cubic meter
	=	2 pecks
	=	32 quarts (dry)
	=	64 pints (dry)

APPENDIX 3

USDA Standards for Slaughter and Feeder Pigs

The USDA standards include classes of slaughter and feeder pigs. Definition of the respective classes are as follows:

Barrow—A barrow is a male swine castrated when young and before development of the secondary physical characteristics of a boar. *Gilt*—A gilt is a young female swine that has not produced young and has not reached an advanced stage of pregnancy. *Sow*—A sow is a mature female swine that shows evidence of having reproduced or has reached an advanced stage of pregnancy. *Stag*—A stag is a male swine castrated after development or beginning of development of the secondary physical characteristic of a boar. Typical stags are somewhat coarse and lack balance—the head and shoulders are more fully developed than the hindquarter parts; bones and joints are large, the skin is thick and rough, and the hair is coarse. *Boar*—A boar is an uncastrated male swine.

USDA grades of slaughter barrows and gilts are predicted on the same two general considerations that provide the basis for the grades of barrow and gilt carcasses: (1) quality of the lean and (2) expected combined carcass yields of the four lean cuts (ham, loin, picnic shoulder, and Boston butt). In the determination of quality, the amount and distribution of external fat, firmness of fat and firmness of lean are used as quality-indicating factors. Barrows and gilts with characteristics which indicate that the carcass lean will have an unacceptable level of quality and/or the belly will be of unacceptable thickness are graded U.S. Utility. Also graded U.S. Utility—regardless of their development of other quality-indicating characteristics—are slaughter barrows and gilts with indications that their carcasses will be soft and oily.

The USDA system of grading the second general consideration, yield, is as follows.

Four grades—U.S. No. 1, U.S. No. 2, U.S. No. 3, and U.S. No. 4—are provided for barrows and gilts with characteristics which indicate that the carcass lean will have an acceptable level of quality. These grades are based entirely on the expected combined carcass yields of the four lean cuts.

Average back fat thickness in relation to carcass length or liveweight is used as a guide to expected yields of the four lean cuts in these standards. In grading barrows and gilts, these factors normally are appraised visually; liveweight is the only one that can be readily determined in the live animal. The average back fat thickness appraisal includes consideration of the distribution of fat on other parts of the animal in addition to those points on the back where it is measured on the carcass.

Barrows and gilts will produce carcasses that vary in yields of four lean cuts because of variations in their degree of fatness (expressed as average thickness of back fat), and in their degree of muscling (thickness of muscling in relation to skeletal size). Since many slaughter swine have a normal development of muscling for their degree of fatness, in determining their grade the average thickness of back fat and carcass length or liveweight are the only factors considered.

The degree of muscling specified for each of the four grades decreases progressively from the U.S. No. 1 grade through the U.S. No. 4 grade. This reflects the fact that among barrows and gilts of the same weight, the fatter animals normally have a lesser degree of muscling. For purposes of these standards six degrees of muscling are recognized: very thick, moderately thick, slightly thick, slightly thin, thin, and very thin. These degrees are intended to cover the entire range of muscling present among slaughter swine. The degrees specified as typical for barrows and gilts at the minimum of the U.S. No. 1, U.S. No. 2, U.S. No. 3, and U.S. No. 4 grades are, respectively, thick, moderately thick, slightly thin, and thin.

For animals having a development of muscling which is different from that normally associated with their degree of fatness, the average back fat thickness-carcass length or average back fat thickness-liveweight relationships for the various grades are different. Consideration is given such unusual developments of muscling as follows: In each grade, superior muscling is permitted to compensate for greater fatness at the rate of one full degree of superior muscling for a 0.1-in. increase in average back fat. Except for the U.S. No. 1 grade, the reverse type of compensation is also permitted and at the same rate; 0.1-in. less fat compensates for full degree of inferior muscling. In the U.S. No. 1 grade, this type of compensation is limited to one full degree of inferior muscling; barrows and gilts which have less than moderately thick muscling but which otherwise qualify for the U.S. No. 1 grade are graded U.S. No. 2. In no case, however, may variations from normal muscling alter the final grade more than one full grade.

The following specifications for official U.S. standards for grades of slaughter barrows and gilts are taken directly from the USDA bulletin on the subject (USDA, 1970, Grades for pork carcasses. Marketing Bull. No. 19, USDA Consumer and Marketing Service, U.S. Govt. Printing Office, Washington, D.C.)

U.S. No. 1—Barrows and gilts in this grade will produce carcasses with acceptable lean quality and acceptable belly thickness and a high percentage of lean cuts.

Swine near the borderline between the U.S. No. 1 and U.S. No. 2 grades are thickly muscled in the hams, loins, and shoulders. The width through the ham is nearly equal to the width through the shoulders and both are wider than the back. The back is slightly full, is well rounded, and blends smoothly into the sides. The sides are moderately long and smooth. The rear flank is slightly full and its depth is less than the depth of the fore flank. The jowls are firm and slightly thick and full.

A development of muscling superior to that specified as minimum for the U.S. No. 1 grade may compensate for a development of fatness which is greater than that indicated as maximum for the U.S. No. 1 grade at the rate of one full degree of muscling for 0.1-in. greater thickness of back fat. For example, a barrow or gilt with very thick muscling may have 0.1-in. more back fat than that indicated as maximum for this grade and remain eligible for the U.S. No. 1 grade. The reverse type of compensation is also permitted at the same rate—except that in no case may a barrow or gilt be graded U.S. No. 1 with less than moderately thick muscling.

U.S. No.2—Slaughter barrows and gilts in this grade will produce carcasses with acceptable lean quality, acceptable belly thickness, and a slightly high percentage of lean cuts. Swine near the borderline between the U.S. No. 2 and U.S. No. 3 grades are moderately thickly muscled in the hams, loins, and shoulders. The width through the shoulders is slightly more than the width through the hams. The back is moderately full and is slightly rounded but with a slight break into the sides. The sides are slightly short and smooth. The rear flank is moderately full and its depth is slightly less than the depth of the fore flank. The jowls are moderately thick and full.

A development of muscling superior to that specified as minimum for the U.S. No. 3 grade may compensate for a development of fatness which is greater than that indicated as maximum for the U.S. No. 3 grade at the rate of one full degree of muscling for 0.1-in. greater thickness of back fat. For example, a barrow or gilt with moderately thick muscling may have 0.1-in. more back fat than that indicated as maximum for this grade and remain eligible for the U.S. No. 3 grade. The reverse type of compensation is also permitted at the same rate. For example, a barrow or gilt with thin muscling may have 0.1-in. less back fat than that indicated as maximum for this grade and remain eligible for the U.S. No. 2 grade.

U.S. No. 3—Slaughter barrows and gilts in this grade will produce carcasses with acceptable lean quality and acceptable belly thickness and a slightly low percentage of four lean cuts. Swine near the borderline between the U.S. No. 3 and U.S. No. 4 grades are slightly thinly muscled in the hams, loins, and shoulders. The width through the shoulders is definitely greater than the width through the hams. The back is full and nearly flat with a pronounced break into the sides. The sides are short and smooth. The rear flank is full and its depth is equal to the depth of the fore flank. The jowls are thick and full.

A development of muscling superior to that specified as minimum for the U.S. No. 3 grade may compensate for a development of fatness which is greater than that indicated as maximum for the U.S. No. 3 grade at the rate of one full degree of muscling for 0.1-in. greater thickness of back fat. For example, a barrow or gilt with moderately thick muscling may have 0.1-in. more back fat than that indicated as maximum for this grade and remain eligible for the U.S. No. 3 grade. The reverse type of compensation is also permitted at the same rate.

U.S. No. 4—Barrows and gilts in this grade will produce carcasses with acceptable lean quality and acceptable belly thickness. However, they are fatter and less muscular and will have a lower carcass yield of the four lean cuts than those in the U.S. No. 3 grade.

U.S. Utility—Barrows and gilts typical of this grade will have a thin covering of fat. The sides are wrinkled and the flanks are shallow and thin. Barrows and gilts in this grade will produce carcasses with unacceptable lean quality and/or unacceptable belly thickness.

APPENDIX 4

Glossaries of Terms Used in Pork Production, Feed Processing, and Breeding and Genetics

ANIMAL TERMS

barrow. Male castrated before sexual maturity.

boar. Male swine of any age.

breeding stock. Female or male young or mature swine kept for breeding to produce piglets.

feeder pig. Young swine after weaning and before reaching slaughter weight.

Source: Many of the terms defined under Physiological Terms were taken directly from the glossary on pp. 499–510 of "Principles of Nutrition" by E. D. Wilson, K. H. Fisher, and M. E. Fuqua, 3rd ed., 1975, John Wiley and Sons, New York, and from the glossary on pp. 387–396 of "Basic Animal Nutrition" by D. C. Church and W. G. Pond, 2nd ed., 1982, John Wiley and Sons, New York. Terms defined under International Feed Nomenclature were taken from Appendix B of "Names for Feed Processes and Their Use in Evaluating the Nutrient Content of Feeds" by L. E. Harris and E. W. Crampton, *in* "Effect of Processing on the Nutritional Value of Feeds" (1973), National Research Council, National Academy of Sciences, Washington, D.C. Terms defined under Feed Manufacturing and Feed Ingredient Terms were taken from the list of Official Feed Terms, Association of American Feed Control Officials, State Capitol Building, Charleston, West Virginia. Terms defined under Animal Terms and Husbandry Terms are those in common use in the swine industry. The breeding, genetics and production systems terms are taken from the Guidelines for Uniform Swine Improvement Programs (edited by C. Christians), published by the National Swine Improvement Federation (1987).

finishing pig. Young swine generally weighing more than 70 kg but not yet heavy enough for slaughter.

gilt. Female swine of any age to second pregnancy.

growing pig. Young swine after weaning, generally weighing less than 70 kg.

hog. Swine of either sex, generally referring to immature gilts, barrows or boars.

market pig. *See* Slaughter pig.

mummified fetus. Dead partially resorbed fetus, characterized by shrunken, brownish appearance, expelled at the time of birth of normal live littermates.

mummy. *See* Mummified fetus.

pig. In United States, refers to young swine of either sex; in Europe, refers to all ages and either sex and is used in preference to swine.

piglet. Baby pig.

runt. Piglet or growing pig considerably smaller than littermates or contemporaries and often unhealthy in appearance.

shoat. Young swine of either sex, generally weighing less than 50 kg.

slaughter pig. Young swine ready for slaughter, usually weighing 90–120 kg.

sow. Female swine having produced one or more litters.

stag. Male castrated after reaching sexual maturity.

suckling pig. Young swine before weaning.

swine. Broad term applying to all animals belonging to *Sus suis*.

weanling pig. Young swine at or shortly after weaning.

HUSBANDRY TERMS

back fat probe. Measure the depth of back fat over the shoulder, loin, and rump on market pigs or breeding stock using a scalpel and metal rule or an ultrasonic instrument.

castrate. Remove the gonads (testes in males, ovaries in females).

creep. An area near the lactating sow accessible only to the suckling pigs.

creep feed. Baby pig feed provided in the creep area beginning at about 2 weeks.

dock. Remove the tail by cutting or pinching.

farrow. To give birth to a litter of pigs.

farrowing crate. Enclosure for the sow before farrowing and during early lactation to minimize accidental newborn pig deaths and injuries caused by overlaying by the dam.

hand breeding. *See* Hand-mating.

hand mating. Supervised mating of selected males and females for purposes of recording pedigree.

pen breeding. *See* Hand breeding.

pen mating. Mating of groups of breeding females by one or more boars penned with the females through one or more estrous cycles.

wean. Remove the piglets from their dam.

PHYSIOLOGICAL TERMS

acetylcholine. An acetic acid ester of choline; normally present in many parts of the body; functions in the transmission of nerve impulses.

active transport. The movement of substances (particularly electrolyte ions) across cell membranes, usually against a concentration gradient. Unlike diffusion or osmosis, active transport requires the expenditure of metabolic energy.

adenine. One of four nitrogenous bases found in DNA.

adenosine triphosphatase (ATPase). An enzyme in muscle that catalyzes the hydrolysis of the terminal phosphate group of adenine triphosphate.

adenosine triphosphate (ATP). A compound that consists of one molecule each of adenine (a purine) and ribose (a 5-carbon atom sugar) and three molecules of phosphoric acid; it is required for energy transfer and the phosphorylation of compounds.

ADF. Acid detergent fiber, the fraction of a feedstuff analyzed by the Van Soest scheme of detergent analysis used to divide carbohydrates in plant constituents into those highly available and poorly available to animals; ADF content of a feed reflects the amounts of carbohydrates not solubilized by acid detergent.

adrenaline. *See* Epinephrine.

albumin. Group of globular proteins; a major constituent of blood serum protein.

amine. A chemical compound formed from ammonia (NH_3) by replacement of one or more of the hydrogen atoms with hydrocarbon groups ($—CH_3$ or $—C_2H_5$ and others).

amino acids. Simplest organic structure of which proteins are formed; many different amino acids occur in nature of which about 10 (for most animals) are required in the diet; all have the common property of containing a carboxyl group and an amino group on the adjacent C atom.

amino group. A chemical structure, $—NH_2$; a constituent of all amino acids, attached to the C atom adjacent to the carboxyl group.

anemia. A deficiency in the blood, either in quality or quantity.

hyperchromic anemia. A decrease in hemoglobin that is proportionately much less than the decrease in the number of erythrocytes.

hypochromic anemia. A decrease in hemoglobin that is proportionately much greater than the decrease in the number of erythrocytes.

macrocytic anemia. A condition in which the erythrocytes are much larger than normal.

megaloblastic anemia. A condition in which there are megaloblasts in the bone marrow.

microcytic anemia. An anemia in which the erythrocytes are smaller than normal.

nutritional anemia. Hypochromic and microcytic anemia due to insufficient iron.

pernicious anemia. Macrocytic and hypochromic anemia due to lack of secretion of the intrinsic factor by the gastric mucous membrane (this factor is essential for the absorption of vitamin B_{12}).

antagonist. A substance that exerts a nullifying or opposing action to another substance.

antibody. A protein produced by the body in response to the presence of a foreign agent (antigen). Antibodies are part of the body's natural defense against invasion by foreign substances.

antigen. Any substance not normally present in the body that produces an immune reaction upon introduction into the body.

antioxidant. A substance that inhibits the oxidation of other compounds.

antivitamin. A substance that interferes with the synthesis or metabolism of a vitamin.

anus. Distal opening of the gastrointestinal tract through which undigested feed residues are excreted as feces.

apoferritin. A protein in the mucosal cells of the small intestine; together with iron it forms the compound ferritin.

appetite. A desire for food or water; generally a long-term phenomenon, in contrast to short-term satiety.

arachidonic acid. A 20-carbon-atom fatty acid with four double bonds; in the body it is synthesized from the essential fatty acid linoleic acid.

ascorbic acid (vitamin C). Water soluble vitamin required in oxidation-reduction reactions (electron transport) and for normal tyrosine oxidation and collagen metabolism and in hydroxylation reactions (i.e., hydroxyproline from proline).

ash. The residue remaining after complete combustion at 500° to 600°C of a feed or animal tissue or excreta during proximate analysis.

basal metabolic rate. The basal metabolism expressed as kilocalories per unit of body size [square meter of body surface; weight to the three-fourths power ($W^{3/4}$)].

basal metabolism. Energy expenditure of the body at rest, under comfortable environmental conditions, and in the postaborsorptive state (12 hr after the ingestion of food).

bile. Secretion from the liver, containing metabolites such as cholesterol and important bile acid conjugates which aid in emulsification of fats for digestion and absorption.

bile duct. Tube carrying bile from liver to duodenal lumen.

biopsy. The removal and examination of tissue or other material from the living body, usually for diagnosis.

biotin. Water soluble vitamin required as a constituent of several enzymes related to energy and amino acid metabolism.

BMR. Basal metabolic rate.

calcitonin (thyrocalcitonin). A polypeptide secreted by the thyroid glands that lowers the level of calcium and phosphorus in the plasma.

calcium (Ca). Mineral required as a structural component of the skeleton and for controlling excitability of nerve and muscle.

calorimeter. The equipment used to measure the heat generated in a system. In nutrition it is an instrument for measuring the amount of heat produced by a food on oxidation or by an indiviual.

carboxyl group. A chemical structure $-\mathrm{C}\genfrac{}{}{0pt}{}{=\mathrm{O}}{-\mathrm{OH}}$; a constituent of all amino acids and the characteristic group of all organic acids.

carboxypeptidase. A proteolytic digestive enzyme secreted by the pancreas; hydrolyzes peptides to amino acids.

cartilage. A connective tissue characterized by nonvascularity (without blood vessels) and firm texture; consists of cells (chondrocytes), interstitial substance (matrix), and a ground substance (chondromucoid).

cassava. A tropical plant of the spurge family with edible starchy roots.

catalyst. A substance that either increases or decreases the rate of a particular chemical reaction without itself being consumed or permanently altered; enzymes are special chemical catalysts of biological origin.

cecum. A blind pouch just distal to the small intestine and containing a large population of anerobic bacteria that ferment complex carbohydrates; appendix in humans corresponds loosely with a rudimentary cecum.

cellulose. Polymer of glucose in linkage resistant to hydrolysis by digestive enzymes.

ceruplasmin. A copper-containing protein (an alphaglobulin) in blood plasma.

chlorine (Cl). Mineral required in regulation of extracellular osmotic pressure and in maintaining acid-base balance.

cholecystokinin. Hormone secreted by duodenal cells and perhaps by certain brain cells; release stimulated by fat in lumen of duodenum and perhaps by concentrations of metabolites in blood circulating in brain; causes gall bladder contraction and may exert short term effect on food intake.

cholic acid. A family of steroids comprising the bile acids; they are derived from cholesterol.

choline. Water soluble vitamin required as a structural component of tissues (i.e., lecithin, sphingomyelin) and of acetylcholine and as a donor of labile methyl groups; there is no evidence for a role as a cofactor in enzymatic reactions.

chromium (Cr). Trace mineral required as trivalent ion for normal glucose metabolism.

chylomicron. A particle of emulsified fat present in lymph; they are especially numerous after a meal high in fat content.

chyme. A semiliquid material produced by the action of gastric juice on ingested food; it is discharged from the stomach into the duodenum.

chymotrypsin. A proteolytic digestive enzyme secreted by the pancreas.

cirrhosis of the liver. A progressive destruction of the liver cells and an abnormal increase of connective tissue.

citric acid cycle. The major series of pathways through which carbohydrates are oxidized to produce energy in animal cells.

cobalt (Co). Trace mineral required as a constituent of vitamin B_{12}.

coenzyme. An organic molecule that is required for the activation of an apoenzyme to an enzyme. The vitamin coenzymes are niacin, pyridoxine, thiamin, riboflavin, pantothenic acid, and folic acid.

coenzyme A. A complex molecule containing pantothenic acid; it is required for fatty acid oxidation and synthesis and for the synthesis of cholesterol and phospholipids; it combines with acetate to form acetyl coenzyme A (active acetate), which in turn combines with oxaloacetate to form citrate and enter the tricarboxylic acid cycle.

collagen. Group of fibrous proteins; resistant to digestive enzymes, prevalent in muscle from aged animals; become gelatin on acid or alkaline hydrolysis.

colon. Large intestine, distal to cecum and small intestine; contains a large population of anaerobic bacteria that ferment complex carbohydrates which may represent a significant energy contribution in some animals.

convulsion. An involuntary spasm or contraction of muscles.

copper (Cu). Trace mineral required for normal red blood cell formation, apparently by allowing normal iron metabolism, and as a component of several oxidases involved in normal wool and hair growth.

cornea. The clear, transparent membane covering of the anterior part of the eye.

corpus luteum. A yellow mass formed in the ovary at the site of the ruptured ovarian follicle; if impregnation takes place, the corpus luteum grows and persists for several months; if impregnation does not take place, the corpus luteum shrinks and degenerates.

creatine. A nonprotein nitrogenous substance in muscle; it combines with phosphate to form phosphocreatine, which serves as a storage form of high energy phosphate required for muscle contraction.

creatinine. A nitrogenous compound that is formed as a metabolic end product of creatine; creatinine is produced in the muscle, passes into the blood, and is excreted in the urine.

crude fibre. The insoluble carbohydrates remaining in a feed after boiling in acid and alkali during proximate analysis; this fraction represents the poorly digested part of a feed by animals.

crude protein. The content of nitrogen in a feed or animal tissue or excreta, multiplied by a factor (usually 6.25, since most proteins contain about 16% N) to provide an estimate of protein content; both nonprotein N (amino acids, amines, ammonia, etc.) and true protein may be present.

cyanocobalamin (B_{12}). Water soluble vitamin required as a coenzyme in several enzymes including isomerases, dehydrases and enzymes involved in methionine biosynthesis.

cystine. A sulfur-containing, nonessential amino acid that occurs notably in keratin and insulin; in the diet, it exerts a sparing effect on methionine.

cytochromes. Hemoproteins that contain iron in their structure; their principal biological function is electron and/or hydrogen transport.

cytosine. One of four nitrogenous bases found in DNA.

deamination. Removal of the amino group from a compound.

degradation. Conversion of a chemical compound to one that is less complex.

dehydrogenase. Any of a class of enzymes found in plant and animal tissues that catalyze oxidation by the transference of hydrogen ions to hydrogen acceptors.

deoxyribonucleic acid (DNA). A large complex molecule in the cell nucleus that carries the code of genetic information; it is composed of the (nitrogen-containing) bases: adenine, thymine, cytosine, and guanine, each of which is attached to a pentose (deoxyribose) and to a phosphate.

diet. A regulated selection or mixture of feedstuffs or foodstuffs provided on a continuous or prescribed schedule; a balanced diet supplies all nutrients needed for normal health and productive functions.

digestibility, apparent. That percentage of a feed nutrient that is apparently digested and absorbed from the gastrointestinal tract as indicated by intake minus fecal output of the nutrient; differs from true digestibility in that feces contain endogenously derived as well as feed-derived nutrients so that true digestibility must be estimated indirectly by feeding a diet devoid of the nutrient of concern.

digestibility, true. *See* Digestibility, apparent.

disaccharide. Dimer or simple sugars; sucrose (common cane or beet sugar) yields glucose and fructose.

diuresis. Excretion of urine; commonly denotes production of unusually large volumes of urine.

diuretic. A substance that promotes the excretion of urine; diuretic drugs are used chiefly to rid the body of excess fluid.

dry matter. That portion of a feed or tissue remaining after water is removed, usually expressed as a percent.

duodenum. Upper part of small intestine into which empty the secretions of the exocrine pancreas and liver bile.

edema. An abnormal accumulation of fluid in the intercellular spaces of the body.

elastin. Group of fibrous proteins; similar to collagen but not converted to gelatin; a major constituent of tendons and cartilage.

electrolyte. Any substance that, when in solution, dissociates into charged particles (ions) capable of conducting an electrical current.

emaciation. Excessive leanness; a wasted condition of the body.

endemic. A disease of low morbidity that persists over a long period of time in a certain region.

endocrine. Pertains to internal secretions; endocrine glands are those that produce one or more internal secretions (hormones) that enter directly into the blood and affect metabolic processes.

endogenous. Produced within or caused by factors within the organism.

enterocrinin. Hormone secreted by jejunal cells; release perhaps stimulated by food digestion products; stimulates secretion of intestinal fluids and enzymes.

enterogastrone. Hormone secreted by duodenal cells; release stimulated by fat and fatty acids and bile in lumen of duodenum; inhibits gastric secretion and motility.

enzyme. A protein formed in plant or animal cells that acts as an organic catalyst in initiating or speeding up specific chemical reactions.

epinephrine. Adrenal medulla hormone that stimulates autonomic nerve action; increases heart rate, perspiration.

ergosterol. A sterol found chiefly in plant tissues; on exposure to ultraviolet irradiation, it becomes vitamin D.

erythrocyte. Red blood cell.

erythropoiesis. Formation of erythrocytes.

estrogens. Ovarian steroid hormones responsible for promoting estrus and the development and maintenance of secondary sex characteristics; estriol $C_{18}H_{24}O_3$ and estrone $C_{18}H_{22}O_2$ are two important estrogens.

ether extract. The fraction of a feed or animal tissue that is soluble in a fat solvent such as ethyl ether and removed by extraction during proximate analysis.

exogenous. Originating outside or caused by factors outside the body.

fatty acids. Organic acids, composed of carbon, hydrogen, and oxygen, which combine with glycerol to form fat.

feces. Excretory product of undigested feed residues, plus endogenously produced digestive secretions, sloughed cells of the intestinal lining and metabolites reexcreted into the intestinal lumen in the bile from the liver.

feed. Food for animals.

feedstuff. Any substance suitable for animal feed; several feedstuffs are normally combined to provide a balanced diet.

ferritin. Form in which iron is stored in the body; it is an iron-protein complex made up of iron and the protein apoferritin.

fistula. An abnormal passage from an abscess, cavity, or hollow organ to the surface or from one abscess, cavity, or organ to another.

fluorine (F). Trace mineral required for protection against dental caries.

fluorosis (dental). A mottled discoloration of the enamel of the teeth due to chronic ingestion of excessive amounts of fluorine.

folacin (folic acid). Water soluble vitamin required as a cofactor in a variety of metabolic reactions involving incorporation of single C units into large molecules.

folic acid. *See* folacin.

food. Any material, usually of plant or animal origin, containing essential nutrients.

foodstuff. Any substance suitable for food; several foodstuffs are normally combined to provide a balanced diet.

fortification. The addition of one or more nutrients to a food in amounts so that the total amount will be larger than that contained in any natural (unprocessed) food of its class, for example, the fortification of fruit juices with vitamin C; the FDA has not established standards for fortification.

galactosemia. The accumulation of galactose in the blood due to a genetic lack of the enzyme galactose-1-phosphate uridyl transferase, which is necessary for the conversion of galactose to glucose; it is characterized by vomiting and diarrhea, abdominal distension, enlargement of the liver, and mental retardation.

gastrin. Hormone secreted by intestinal cells near pyloris; release stimulated by distension and movement of stomach; stimulates acid secretion by gastric glands in stomach.

globulin. Group of globular proteins; alpha, beta and gamma globulins are major constituents of blood serum protein.

glossitis. Inflammation of the tongue.

glucagon. A compound secreted by the alpha cells of the islets of Langerhans that is hyperglycemic (produces a rise in the blood glucose concentration), glycogenolytic (hydrolyzes glycogen to glucose), and stimulates gluconeogenesis.

gluconeogenesis (glyconeogenesis). The formation of glucose from noncarbohydrate sources, chiefly certain amino acids and the glycerol portion of the fat molecule.

glucose tolerance test. A test indicating the efficiency of the body in its use of glucose; changes are noted in the concentration of glucose in the blood at determined intervals after ingestion of a standard amount of sugar.

glutelins. Group of globular proteins.

glyceride. A compound (ester) formed by the combination of glycerol and fatty acids and the loss of water (the ester linkage); according to the number of ester linkages, the compound is a mono-, di-, or triglyceride.

glycerol (glycerin). The 3-carbon atom alcohol derived from the hydrolysis of fat.

glycogenesis. The formation of glycogen from glucose.

glycolysis. The conversion of glucose to lactic acid in various tissues, notably muscle; since molecular oxygen is not consumed in the process, it is referred to as "anaerobic glycolysis."

goitrin. The antithyroid, or goitrogenic, compound obtained from turnips and the seeds of cruciferous plants.

gram (g). A metric unit of weight that is equivalent to about 1/28 of an ounce (1 oz = 28.4 g), 1/1000 of a kilogram, or 1000 milligrams.

growth hormone. *See* Somatrotropin.

guanine. One of four nitrogenous bases found in DNA.

hematocrit. The volume percentage of erythrocytes in whole blood; it is determined by centrifuging a blood sample to separate the cellular elements from the plasma; the results of the test indicate the ratio of cell volume to plasma volume and are expressed as cubic millimeters of packed cells per 100 ml of blood.

hematopoiesis. The production of various types of blood cells and blood platelets.

heme. The nonprotein, insoluble pigment portion of the hemoglobin molecule; the prosthetic group of the hemoglobin molecule.

hemodilution. An increase in the fluid content of the blood, resulting in the diminution of the proportion of formed elements (red and white blood cells and platelets).

histones. Group of globular proteins.

homeostasis. The tendency of the body to maintain uniformity or stability in its internal environment or fluid matrix.

homeothermal (homothermal). Warm-blooded animals.

hormone. A secretion produced in the body (chiefly by the endocrine glands) that is carried in the bloodstream to other parts of the body; each hormone has a specific effect on cells, tissues, and organs.

hunger. Desire for food; the antithesis of satiety.

hydrogenation. The addition of hydrogen to a compound, especially to an unsaturated fat or fatty acid; adding hydrogen at the double bond will solidify soft fats or oils.

hydrolysis. A chemical process whereby a compound is broken down into simpler units with the uptake of water.

hypoglycemia. A condition characterized by a lower than normal level of glucose in the blood.

hypomagnesemia. An abnormally low magnesium content of the blood plasma.

hypothalamus. A portion of the brain, lying beneath the thalamus at the base of the cerebrum, and forming the floor and part of the walls of the third ventricle; it contains centers for temperature regulation, appetite control, and others.

ileum. Lower part of small intestine, distal to the jejunem and proximal to the large intestine; absorption of the most nutrients occurs in jejunem and ileum.

immunoglobulin (Ig). Serum globulin having antibody activity; most of the antibody activity appears to be in the gamma fraction of globulin.

inorganic. Compounds that do not contain carbon.

insulin. A hormone secreted by the beta cells of the islets of Langerhans of the pancreas; it facilitates glucose oxidation, as well as the synthesis of glycogen, fat, and protein.

interstitial. Relates to spaces in any structure, such as tissue.

intestinal peptidase. A group of poly-, tri-, and dipeptide digestive enzymes secreted by epithial cells in the small intestine.

intolerance. An allergy or sensitivity, for example, to certain foods or medicines.

intrinsic factor. A transferase enzyme (mucoprotein) secreted by the mucosal cells of the stomach which is required for the absorption of vitamin B_{12} through the intestinal wall; a lack of this factor will produce pernicious anemia.

iodine (I). Trace mineral required as a constituent of thyroxine.

iron (Fe). Trace mineral required for hemoglobin and myoglobin formation as a constituent of several heme enzymes and of several nonheme metalloenzymes.

isomer. A compound having the same percentage composition and molecular weight as another compound but differing in chemical or physical properties; for example, L (levorotary) amino acids are biologically active but D (dextrorotary) amino acids generally are not (*see also* Racemic mixture).

isotope. One of two or more atoms, the nuclei of which have the same number of protons but different numbers of neutrons; for example, ^{14}N and ^{15}N are both stable isotopes of nitrogen.

jejunem. Middle segment of small intestine, distal to the duodenum and proximal to the ileum; absorption of most nutrients occurs in jejunem and ileum.

joule (J). A unit of energy that is equal to 0.0002 kcal.

kcal. Kilocalorie.

keratin. Group of fibrous proteins; resistant to digestive enzyme; major constituent of skin, hooves, horns.

Krebs cycle. *See* Citric acid cycle.

ketogenic. That which is conducive to the formation of ketone bodies, such as high-fat, low-carbohydrate diet.

α-ketoglutaric acid. A compound that is common to the metabolic pathways of carbohydrates, fats, and certain amino acids.

ketone. A chemical compound that contains the carbonyl group (CO); ketone bodies include betahydroxybutyric acid, acetoacetic acid, and acetone.

kilocalorie (kcal). The quantity of heat required to raise the temperature of 1 kg of water 1°C (or, more precisely, from 15° to 16°C).

kilogram (kg). A metric unit of weight that is equivalent to 2.2 pounds, or 1000 g.

lactic acid. A compound formed in the body during anaerobic glycolysis; it is also produced in milk by the bacterial fermentation of lactose.

lecithin. The traditional term for phosphatidyl choline.

leucine. An essential amino acid.

ligament. A band or sheet of fibrous tissue connecting two or more bones, cartilages, or other structures, or serving as a support for muscles.

lignin. Biologically unavailable mixture of polymers of phenolic acids; it is a major structural component of mature plants and of trees; its presence affects the bioavailability of cellulose and hemicellulose in plants.

linoleic acid. The essential fatty acid; it is unsaturated; and occurs widely in plant glycerides.

lipids. Substances that are diverse in chemical nature but soluble in fat solvents (such as ethanol, ether, chloroform, benzene); lipids include fats and oils; phospholipids, glycolipids, and lipoproteins; fatty acids, alcohols [glycerol, sterols (including vitamin D and vitamin A), and carotenoids].

liver. A large internal organ of the body, which lies in the upper right section of the abdomen directly beneath the diaphragm. Among other things, it produces bile, glycogen, antibodies; interconverts proteins, carbohydrates, and fats; stores iron, copper, vitamins A and D; and detoxifies harmful substances.

lymph. A fluid that circulates within the lymphatic vessels and is eventually added to the venous blood circulation; it arises from tissue fluid and from intestinal absorption of fatty acids; it is colorless, odorless, slightly alkaline, and slightly opalescent.

lymphatic system. All of the vessels and structures that carry lymph from the tissues to the blood.

lysine. An essential amino acid.

malnutrition. An overall term for poor nourishment; it may be due to an inadequate diet or to some defect in metabolism that prevents the body from using the nutrients properly.

magnesium (Mg). Mineral required as a constituent of bone, for oxidative phosphorylation of mitochondria of cells, and for activation of numerous energy-related enzymes.

manganese (Mn). Trace mineral required for formation of chondroitin sulfate and as a component of metalloenzymes and as a cofactor in several enzymes.

megaloblast. A large, nucleated, embryonic type of cell; it is found in the blood in cases of pernicious anemia, vitamin B_{12} deficiency, and folacin deficiency.

metabolic pool. The total amount of a specific substance in the body that is in a state of active turnover, such as the amino acid pool or vitamin D pool; subtractions from and additions to the pool are constantly being made.

metabolic size. The body weight raised to the three-fourths power ($W^{3/4}$).

metabolite. Any compound produced during metabolism.

metalloenzyme. An enzyme containing a metal (ion) as an integral part of its active structure.

methionine. A sulfur-containing essential amino acid.

microgram (μg). A metric unit of weight that is equivalent to one-millionth of a gram or 1/1000 milligram.

micelle. An aggregate of molecules of lipids and bile acids formed in the lumen of the gastrointestinal tract during preparation of dietary lipids for absorption.

milligram (mg). A metric unit of weight that is equivalent to 1/1000 gram or 1000 micrograms.

milliliter (ml). A metric unit of liquid measure that is equivalent to 1/1000 liter.

millimeter (mm). A metric unit of length that is equivalent to 1/10 centimeter.

mitochondria. The largest organelles in the cytoplasm; the cell mitochondria contain important enzymatic, oxidative, and respiratory systems.

molecular weight. The weight of a molecule of a chemical compound as compared with the weight of an atom of hydrogen; it is equal to the sum of the weights of the constituent atoms.

molybdenum (Mo). Trace mineral required as a constituent of the metalloenzyme, xanthine oxidase.

monogastric. Simple stomach; the term often used for simple stomached (nonruminant) animals; technically a misnomer, since ruminants have only one stomach with four compartments.

monosaccharide. Simple sugar of 5 (pentoses) or 6 (hexoses) C atoms; glucose is a prime example.

mucin. Secretions containing mucopolysaccharides.

mucopolysaccharide. A complex of protein and polysaccharides.

mucoprotein. A complex of protein and oligosaccharides (yield on hydrolysis: two to ten monosaccharides).

NDF. Neutral detergent fiber, the fraction of a feedstuff analyzed by the Van Soest scheme of detergent analysis, containing mostly cell wall constituents of low biological availability.

necrosis. Death of a cell or group of cells due to irreversible damage.

nephritis. Inflammation of the glomeruli of the kidney causing impairment of the filtering process so that blood and albumin are excreted.

neuritis. Inflammation of the peripheral nerves (which link the brain and spinal cord with the muscles, skin, organs, and other parts of the body).

NFE (nitrogen-free extract). Consists primarily of readily available carbohydrates such as sugars and starches, but may contain hemicellulose and lignin; NFE is calculated by difference during proximate analysis by subtracting all measured components from 100.

niacin (nicotinic acid). Water soluble vitamin required as a constituent of coenzymes, nicotinamide adenine dinucleotide (NAD), and nicotinimide adenine dinucleotide phosphate (NADP); (another form of the vitamin is nicotinamide).

nicotinamide adenine dinucleotide (NAD) [formerly called diphosphopyridine nucleotide (DNP)]. Attached as a prosthetic group to a protein, it serves as a respiratory enzyme (part of an oxidative-reduction system, converting substrates to CO_2 and H_2O and the transfer of electrons removed to oxygen).

nicotinamide adenine dinucleotide phosphate (NADP) [formerly triphosphopyridine nucleotide (TPN)]. A coenzyme that participates in oxidation-reduction processes.

nicotinic acid. *See* Niacin.

norepinephrine. Hormone secreted by sympathetic nerve endings; it is a vasoconstrictor.

nucleic acid. Complex, high molecular weight molecules that contain phosphate, ribose, and four bases—adenine, guanine, cytosine, and thymine; DNA and RNA, which are nucleic acids, are responsible for inherited characteristics.

nucleoprotein. A complex of protein and nucleic acid.

nucleotide. A combination of purine and pyrimidine bases, 5-carbon atom sugar, and phosphoric acid; a hydrolytic product of nucleic acid; a constituent of the coenzymes NAD and NADP.

nucleus. Typically the largest structure within cells; it contains DNA, RNA, and usually a distinct body—the nucleolus.

nutrient. A chemical substance that nourishes; such as protein, carbohydrate, mineral or vitamin.

obesity. The accumulation of body fat beyond the amount needed for good health.

oleic acid. An 18-carbon fatty acid that contains one double bond; it is found in animal and vegetable fat.

ornithine cycle. *See* Urea cycle.

osmosis. The passage of a solvent through a membrane separating two solutions from the solution of lesser concentration to that of greater concentration.

osteomalacia. Adult rickets; a softening of the bones caused by a vitamin D deficiency in adults.

osteoporosis. A reduction in the normal quantity of bone.

oxidation. The increase of positive charges on an atom or loss of negative charges; oxidation is one of the changes that take place when fats become rancid.

oxygen (O). A gaseous element that combines with other elements to form oxides.

palmitic acid. A saturated fatty acid with 16-carbon atoms; it is common in fats and oils.

pancreas. Organ located in the abdominal cavity whose functions are both endocrine and exocrine; that is, its specialized cells produce the hormones insulin and glucagon for control of energy and protein metabolism and other cells produce several digestive enzymes for food digestion.

pancreatic duct. Tube carrying pancreatic secretions, including lipases, amylase, chymotrypsin and trypsin, from pancreas to duodenal lumen.

pancreozymin. Hormone secreted by duodenal cells; release stimulated by acid and nutrients in lumen of duodenum; stimulates pancreatic secretion of enzymes.

pantothenic acid. Water soluble vitamin required as a constituent of coenzyme A, needed for acetylation of numerous compounds in energy metabolism.

parturition. The birth of offspring.

phosphorus (P). Mineral required as a structural component of the skeleton, as a component of phospholipids which are important in lipid metabolism and cell membrane structure and as a component of ATP, creatine phosphate, and other essential compounds in energy metabolism.

phosphocreatine. A creatine–phosphoric acid compound that occurs in muscle; the energy source in muscle contraction.

phospholipids. Fat-like substances consisting of glycerol, two fatty acids, a phosphate group, and a nitrogen-containing compound, such as choline (found in the phospholipid lechithin).

photosynthesis. Process by which chlorophyll-containing cells in green plants convert sunlight to chemical energy and synthesize glucose from carbon dioxide and water and release oxygen.

phytic acid (inositolhexaphosphoric acid). A phosphorus-containing organic acid; phytin is the mixed salt of phytic acid with calcium and magnesium.

placenta. Vascular specialized tissue associated with the uterus during pregnancy and through which the fetus is nourished.

plasma. The fluid portion of the blood in which corpuscles are suspended; serum is plasma from which the fibrinogen has been removed.

polyneuritis. An inflammation encompassing many peripheral nerves.

polyunsaturated fatty acids. Fatty acids containing two or more double bonds, such as linoleic, linolenic, and arachidonic acids.

polyuria. Excessive excretion of urine.

potassium (K). Mineral required for maintenance of acid–base balance, for activation of several enzymes, for normal tissue protein synthesis, for integrity of heart and kidney muscle, and for a normal electrocardiogram; it is located mostly within cells.

progesterone. Ovarian steroid hormone produced by corpus luteum; required for maintenance of pregnancy.

prolamins. A group of globular proteins.

protamines. Group of globular proteins.

protein-bound iodine (PBI). The binding of almost all thyroxine in the blood to protein (PBI); the measure of the amount of PBI is useful as an indicator of the quantity of circulating iodine.

proximate analysis. A combination of analytic procedures used to quantify the protein, lipid, dry matter, ash and carbohydrate (nitrogen-free extract) content of feed, animal tissues, or excreta.

PTH (parathyroid hormone). Hormone secreted by the parathyroid gland; maintains blood plasma calcium homeostasis by promoting calcium resorption from bone in hypocalcemia.

purine. The parent substance of the purine bases; adenine and guanine are the major purine bases of nucleic acids; other important purines are xanthine and uric acid.

phyloris. Muscular sphincter separating stomach from duodenum, which controls rate of movement of ingesta out of stomach.

pyridoxine (B_6). Water soluble vitamin required as a coenzyme in a large array of enzymes associated with protein and nitrogen metabolism (other forms of the vitamin are pyridoxamine and pyridoxal).

pyrimidine. The parent substance of several nitrogenous compounds found in nucleic acids—uracil, thymine, and cytosine.

pyruvic acid. A keto acid of 3-carbon atoms; it is formed of carbohydrate in aerobic metabolism; pyruvate is the salt or ester of pyruvic acid.

radioactive. The emission of particles during the disintegration of the nuclei of radioactive elements; the emissions include alpha particles, beta particles, and gamma rays.

radioisotope. A radioactive form of an element; the nucleus of a stable atom when charged by bombarding particles (in a nuclear reactor) becomes radioactive and is called "labeled" or "tagged."

reduction. The gain of one or more electrons by an ion or compound; for example, ferric iron (Fe^{3+}), which is the common form that is found in food, is reduced to ferrous iron (Fe^{2+}) in the acid medium of the stomach.

racemic mixture. Pertaining to a chemical compound containing equal amounts of dextrorotary (D) and levorotary (L) isomers so that it does not rotate the plane of polarized light.

ration. A fixed portion of feed, usually expressed as the amount of a diet allowed daily.

rectum. Distal portion of gastrointestinal tract, joined proximally to the colon and opening to the exterior via the anus.

riboflavin (B_2). Water soluble vitamin required as a constituent of coenzymes, flavin adenine dinucleotide (FAD), and flavin mononucleotide (FMN).

ribonucleic acid (RNA). A nucleic acid found in all living cells; on hydrolysis it yields adenine, guanine, cytosine, uracil, ribose, and phosphoric acid; it takes part in protein synthesis.

ribosome. A ribonucleic acid-containing particle in the cytoplasm of a cell; the site of protein synthesis.

satiety. The condition of being fully satisfied with food; the antithesis of hunger.

secretion. Hormone secreted by duodenal cells; release stimulated by acid and nutrients in lumen of duodenum; stimulates pancreatic secretion of water and electrolytes.

sedimentation. The setting out of a precipitate.

erythrocyte sedimentation rate (ESR). The rate at which erythrocytes settle out of unclotted blood; inflammatory processes cause an aggregation (clumping together) of the red blood cells which makes them heavier and more likely to fall.

selenium (Se). Trace mineral required as a constituent of the enzyme glutathione peroxidase which destroys peroxides arising from tissue lipid oxidation.

serum. The clear portion of any fluid from an animal's body that remains after the solid elements have been separated out; blood serum is the clear, straw-colored liquid that remains after blood has clotted; the plasma from which fibrinogen has been removed.

silicon (Si). Trace mineral required for initiation of the mineralization process in bones.

sodium (Na). Mineral required for maintenance of acid–base balance, transfer of nerve impulses and as an extracellular component of an energy dependent sodium "pump"; it is located mostly in extracellular fluids.

somatotropin. Anterior pituitary hormone active in promoting protein and mineral accretion in animals and stimulating animal growth.

sphingomyelin. A group of phospholipids found in the brain, spinal cord, and kidney; on hydrolysis it yields phosphoric acid, choline, sphingosine, and a fatty acid.

spleen. A large organ situated under the ribs in the upper left side of the abdomen; it functions in the normal destruction of old red blood cells.

starch. Polymer of glucose readily hydrolyzed by digestive enzymes.

stearic acid. A saturated fatty acid composed of 18-carbon atoms.

sterol. An alcohol of high molecular weight, such as cholesterol and ergosterol.

stillbirth. Birth of a dead fetus.

stomach. The first portion of the gastrointestinal tract on which chemical action occurs on ingested feeds in most animal species; in humans, pigs, and many other species it is a single compartment (simple stomach); in ruminants such as cattle, sheep, goats, deer it contains several compartments.

sulfur (S). Mineral required as a constituent of several organic metabolites, including methionine, cystine, thiamin, biotin, coenzyme A and mucopolysaccharides.

taste. To distinguish flavors between or among feed or water components.

testosterone. Testicular steroid hormone; stimulates sperm production and controls secondary sex characteristics and accessory reproductive organs; increases muscle protein accretion.

tetany. A condition characterized by sharp bending (flexion) of the wrist and ankle joints, muscle twitchings, cramps, and convulsions; inadequate calcium in the blood causes irritability of the nerves and muscles so that they respond to a stimulus with greater sensitivity and force than normally.

tetraiodothyronine. Thyroid hormone active in regulation of basal metabolic rate (BMR); chemical formula is $C_{15}H_{11}I_4NO_4$.

thiamin (B_1). Water soluble vitamin required as a constituent of the enzymes cocarboxylase and lipothiamide.

threonine. An essential amino acid.

thymine. One of the four nitrogenous bases in nucleic acid.

thymus gland. A two-lobed ductless gland located behind the upper part of the sternum and extending into the neck; it is fairly large in young animals but usually shrinks in adulthood; it is structured like a lymph node and contains lymphatic follicles; it may play a role in immune reactions.

thyroxine. An iodine-containing hormone that is produced by the thyroid gland; it is a derivative of the amino acid tyrosine and has the chemical name tetraiodothyronine.

transferrin (siderophilin). An iron-binding protein in the blood that transports iron.

tricarboxylic acid cycle (Krebs cycle and citric acid cycle). A series of biochemical reactions by which carbon chains of sugars, fatty acids, and amino acids are metabolized to yield carbon dioxide, water, and energy.

triglyceride (fat). An ester composed of glycerol and three fatty acids.

tristearin. A triglyceride of stearic acid.

tryptophan. An essential amino acid; a precursor of niacin.

trypsin. A proteolytic digestive enzyme secreted by the pancreas; its precursor trypsonogen is activated to trypsin by enterokinase.

tyrosine. A nonessential amino acid; it spares the essential amino acid phenylalanine.

undernutrition. A condition resulting from insufficient food.

urea. The chief end product of protein metabolism in mammals and one of the chief nitrogenous constituents in the urine.

urea cycle. Major pathway of nitrogen excretion in mammals.

uric acid. A nitrogenous end product of purine metabolism; it is present in the blood and excreted in the urine.

valine. An essential amino acid.

vasoconstrictor. An agent (motor nerve or chemical compound) that acts to decrease the caliber of blood vessels.

vitamin A. Fat soluble vitamin required for normal night vision, normal epithelial cells, and normal bone growth and remodeling.

vitamin B_1. *See* Thiamin.

vitamin B_2. *See* Riboflavin.

vitamin B_6. *See* Pyridoxine.

vitamin B_{12}. *See* Cyanocobalamin.

vitamin D. Fat soluble vitamin required for normal absorption of calcium from the intestinal tract as a precursor of 1,25 dihydroxycholecaliferol which acts in the control of mobilization of calcium from bone and stimulation of calcium absorption from the gastrointestinal tract.

vitamin E. Fat soluble vitamin required as an antioxidant in preventing a wide variety of diseases relating to maintaining integrity of cellular membranes.

vitamin K. Fat soluble vitamin required for normal blood clotting, specifically for synthesis of prothrombin in liver.

xanthurenic acid. A metabolite of tryptophan; it is found in normal urine, but appears in increased amounts in cases of vitamin B_6 deficiency.

xerosis. Abnormal dryness, as of the conjunctiva (xerophthalmia) or skin (xeroderma).

zinc (Zn). Trace mineral required as a constituent of an array of enzyme systems in several body tissues, as an activator of several metalloenzymes and in binding reactants to the active site of enzymes.

INTERNATIONAL FEED NOMENCLATURE

acid-hydrolyzed. *See* Hydrolyzed.

acidified. Treated with an acid to provide a pH lower than pH 7. Sometimes a precipitate forms that may be removed by mechanical means (e.g., curd from whey).

air-ashed. Reduced by combustion in air to a mineral residue.

alcohol-extracted. Treated with alcohol to remove all alcohol-soluble substances.

ammoniated. Combined or impregnated with ammonia or an ammonium compound.

artificially dried. Moisture removed by other than natural means. *See* Fan air-dried with heat.

as fed. *See* Dry-matter content of feed samples.

aspirated. Removal of light materials from heavier materials by use of air. Refers to chaff, dust, or other light materials.

barn-cured. Dried with forced ventilation in an enclosure. Refers to forage. *See* Fan air-dried without heat (international term).

bitterness-extracted. Treated to remove bitter taste.

blended. Mingled or combined. Refers to ingredients of a mixed feed; does not imply uniformity of dispersion.

blocked. Compressed into a large, solid mass.

bolted. Separated from parent material by means of a bolting cloth. Refers to two ingredients (e.g., bran separated from flour).

calcined. Heated to high temperature in the presence of air.

canned. Processed, packaged, sealed, and sterilized in cans or similar containers.

centrifuged. Separated by a force moving away from a center.

chipped. Cut or broken into fragments or cut into small, thin slices.

chopped. Reduced in particle size by cutting.

cleaned. Subjected to any process (e.g., scalping, screening, aspiration, or magnetic separation) by which unwanted material is removed.

clipped. Refers to removal of ends of whole grain.

close-planted. Planted with less than normal distance between rows.

coagulated. Curdled, clotted, or congealed, usually by the action of a coagulant.

coarse-bolted. Separated from parent material by means of a coarsely woven bolting cloth.

coarse-sifted. Separated according to particle size by passage through coarsely woven wire sieves.

condensed. Reduced in volume by removal of moisture.

conditioned. Brought to predetermined moisture characteristics or temperature (or both) prior to further processing. *See also* Tempered (international term).

cooked. Heated in the presence of moisture to alter chemical or physical characteristics (or both) or to sterilize. *See also* Pressure-cooked.

cracked. Reduced in size by a combined breaking and crushing action. Refers to particles of grain.

crimped. Rolled with corrugated rollers. The grain to which this term refers may be tempered or conditioned before it is crimped, and may be cooled afterward.

crumbled. Broken with corrugated rollers. Refers to pellets.

crushed. *See* Rolled.

cultured. Produced in a culture. Refers to biological material.

cured. Prepared for keeping or use (e.g., by drying, smoking, or salting or by using a chemical preservative).

D-activated. Activated with vitamin D (e.g., by ultraviolet light). Refers to plant or animal sterol fractions.

debittered. Having had bitter substances removed.

defluorinated. Having had fluorine partially removed.

degermed. Having had the embryos wholly or partially separated from the starch endosperms. Refers to seeds.

dehulled. Having had the hulls removed.

dehydrated. Having had most of the moisture removed by heat.

deribbed. Having had the primary veins removed. Refers to leaves.

digested. Subjected to prolonged heat and moisture, or to chemicals or enzymes, with a resultant change or decomposition of the physical or chemical nature.

dressed. Made uniform in texture by breaking or screening lumps or by applying liquid(s). Refers to feed.

dried. *See* Dehydrated (international term).

dry. *See* Dry-matter content of feed samples.

dry-matter content of feed samples. The dry matter content of feed samples and other materials as expressed on three dry-matter bases: as fed, partially dry, and dry. Definitions of these terms follow:

as fed. As fed refers to the feed as it is consumed by the animal; the term "as collected" is used for materials which are not usually fed to the animal, i.e., urine, feces, etc. If the analyses on a sample are affected by partial drying, the analyses are made on the "as fed" or "as collected" sample. Similar terms: air dry, i.e., hay; as received; fresh; green; wet.

partially dry. Partially dry refers to a sample of "as-fed" or "as-collected" material that has been dried in an oven (usually with forced air) at a temperature usually at 60°C or freeze-dried and has been equilibrated with the air; the sample after these processes would usually contain more than 88% dry matter (12% moisture); some materials are prepared in this way so they may be sampled, chemically analyzed, and stored. This analysis is referred to as "partial dry matter % of 'as-fed' or 'as-collected' sample." The partially dry sample must be analyzed for dry matter (determined in an oven at 105°C) to correct subsequent chemical analyses of the sample to a "dry" basis. This analysis is referred to as "dry matter % of partial dry sample." Similar term: air dry (sometimes air dry is used for "as fed"; *see* As fed).

dry. Dry refers to a sample of material that has been dried at 105°C until all the moisture has been removed. Similar terms: 100% dry matter; moisture free. If dry matter (in an oven at 105°C) is determined on an "as-fed" sample it is referred to as "dry matter of as-fed sample." If dry matter is determined on a partial dry sample it is referred to as "dry matter of partial dry sample." It is recommended that analyses be reported on the "dry" basis (100% dry matter or moisture free), and in addition the "as fed-dry matter" should be reported.

dry-milled. Milled by tempering with a small amount of water or steam to facilitate separation into component parts. Refers to kernels of grain.

dry-rendered. Having undergone (1) cooking in open steam-jacketed vessels until the water has evaporated and (2) removal of fat by draining and pressing. Refers to residues of animal tissues.

ensiled. Preserved by ensiling, a process in which finely cut parts of plants, packed in an air-tight chamber (e.g., a silo), undergo an acid fermentation that retards spoilage.

etiolated. Grown in reduced light. Refers to plants.

eviscerated. Subjected to removal of all organs. Refers to the great cavity of an animal's body.

expanded. Increased in volume as a result of abrupt reduction in pressure. Refers to a feed or feed mixture that is extruded after being subjected to moisture, pressure, and temperature to gelatinize the starchy part.

expeller extracted. *See* Mechanically extracted.

extracted, mechanical. *See* Mechanically extracted.

extracted, solvent. *See* Solvent-extracted.

extruded. Pushed through orifices of a die under pressure. Refers to feed.

extruded with steam. Pushed through orifices of a die under pressure and after preconditioning with steam. Refers to feed.

extruded without steam. Pushed through orifices of a die under pressure and in the absence of steam. Refers to feed.

fan air-dried. Dried with a device producing a current of air.

fan air-dried with heat. Dried with a device producing a current of heated air. Syn: artificially dried.

fermented. Acted upon by yeasts, filamentous fungi, or bacteria in a controlled aerobic or anaerobic process. Refers to products (e.g., grains and molasses) used in the manufacture of alcohols, acids, vitamins of the B-complex group, and antibiotics.

finely ground. Reduced to very small particles by impact, shearing, or attrition.

finely screened. Separated according to particle size by passage through a finely woven meshed material.

finely sifted. Separated according to particle size by passage through a finely woven meshed material.

flaked. (1) Prepared by a method involving the use of high heat, tempering, and rollers set close together. (2) Cut into flat pieces (e.g., potato flakes). Syn: steam flaked.

fused. Blended by melting.

gelatinized. Ruptured by a combination of moisture, heat, and pressure. Refers to starch granules of a feed.

ground. Reduced in particle size by impact shearing or attrition.

heat- and acid-precipitated. Separated from a suspension or solution by action of heat and acid.

heat-hydrolyzed. *See* Hydrolyzed.

heat-processed. Prepared by a method involving the use of elevated temperatures, with or without pressure.

heat-processed flaked. *See* Flaked.

heat-rendered. Melted, extracted, or clarified by heating. (Water and fat are usually removed.)

homogenized. Broken down into evenly distributed globules small enough to remain as an emulsion for long periods. Refers to particles of fat.

hydraulically extracted. *See* Mechanically extracted.

hydrolyzed. Subjected to hydrolysis, a process by which complex molecules (e.g., those in proteins) are split into simpler units by chemical reaction with water molecules. (The reaction may be produced by an enzyme, catalyst, or acid or by heat and pressure.)

iodinated. Treated with iodine.

irradiated. Treated, prepared, or altered by exposure to radiant energy.

kibbled. Cracked or crushed. Refers to baked dough or to extruded feed that was cooked before or during the extrusion process.

lactase hydrolyzed. *See* Hydrolyzed.

leached. Affected by the action of percolating water or other liquid.

magnetic separation. Removal of ferrous material of magnets (e.g., removal of iron objects from mixed feed).

malted. Converted into malt or treated with malt or malt extract.

maltase processed. Treated with the enzyme maltase.

mechanically extracted. Extracted by heat and mechanical pressure. Refers to removal of fat or oil from seeds. Syn: expeller-extracted, hydraulically extracted, old process.

mechanically extracted caked. Extracted from seeds by heat and mechanical pressure in such a way that the remaining product (e.g., cottonseed meal) is caked. Refers to fat or oil.

micronized. Heated by gas-fired infrared generators to 150°C (300°F). Microwaves are emitted in the heating process. Refers to grain.

mixed. Two or more materials combined by agitation to a specific degree of dispersion.

molded. Overgrown or otherwise acted upon by fungi.

new process. *See* Solvent-extracted.

old process. *See* Mechanically extracted.

partially dry. *See* Dry-matter content of feed samples.

partially extracted. Partially removed from a feed by a chemical or mechanical process. Refers to fat or oil.

pearled. Reduced by machine brushing to smaller, smooth particles. Refers to dehulled grains (e.g., pearled barley).

pelleted. Made into pellets.

polished. Smoothed by a mechanical process. Refers to grain (e.g., polished rice).

popped. Heated with accompanying expansion until explosion or eruption occurs. Refers to grain [e.g., popped corn (maize)].

precipitated. Separated from suspension or solution as a result of a chemical or physical change.

premixed. Mixed with a diluent or carrier (or both) preliminary to final mixing with other ingredients.

prepressed solvent-extracted. Removed from materials partly by heat and mechanical pressure and (later) partly by organic solvents. Refers to fat or oil.

pressed. Compacted or molded by pressure, or extracted under pressure. Refers to fat, oil, or juice.

pressure-cooked. Heated either wet or dry and under pressure greater than atmospheric pressure.

pulverized. *See* Ground.

retort-charred. Partly burned in a closed retort. Refers to bone black.

roasted. Cooked in an oven by dry heat. For grains, cooked or heated with dry heat with an exit temperature of 150°C (300°F). Usually accompanied by expansion and toasted appearance.

rolled. Compressed between rollers. Rolling may entail tempering or conditioning. *See* Steam-rolled.

scalped. Removed by screening. Refers to large particles.

scoured. Cleansed by impact or friction. Refers to removal of the beard from the wheat kernel.

screened. Separated into different sizes by being passed over or through screens.

separation. Classification of particles by size, shape, or density.

separation, magnetic. *See* Magnetic separation.

shredded. Cut into long, thin pieces.

sifted. Separated into different sizes by being passed through wire or nylon sieves.

sizing. *See* Screened.

skimmed. Removed by settling, flotation, or centrifuging. Refers to removal of the lighter part of a liquid from the heavier part (e.g., removal of cream from milk).

solvent extracted. Removed from materials (e.g., soybean seeds) by organic solvents. Refers to fat or oil. Syn: new process.

solvent-extracted caked. Removed from materials by organic solvents and pressed to make a cake. Refers to fat or oil.

spent. Exhausted of absorbing properties (e.g., spent bone black).

stabilized. Made more resistant to chemical change by an added substance.

stack-ensiled. Ensiled while in a pile above ground.

steamed. Treated with steam, as in steam cooking. Syn: steam cooked, steam rendered, tanked.

steam-rolled. Preconditioned in steam under pressure for a short period and then compressed between rollers.

steep-extracted. Soaked in water or other liquid to remove soluble materials. Refers to grain (e.g., corn that is being wet milled).

Steffen process. A process for treating beet molasses to recover additional sugar through precipitation of calcium sucrate.

sulfur-fertilized. Supplied with a form of elemental sulfur as a nutrient. Refers to plants.

sun cured. Dried by exposure to the sun.

sun-cured brown. Partially dried by exposure to the direct rays of the sun, then put in a stack or bale, where heat from microbial action causes browning.

sun cured on riders. Stacked on tripods made of poles, then sun-cured. Refers to forage.

tanked. *See* Steamed.

tempered. Brought to predetermined moisture characteristics or temperature (or both) before further processing. Syn: conditioned.

threshed. Separated from straw by impaction and subsequent screening. Refers to grain, peas, or beans.

toasted. Browned, dried, or parched by exposure to a wood fire or to gas or electric heat.

toxicity extracted. Treated to remove a poisonous substance.

vacuum dehydrated. Dehydrated under vacuum. *See* Dehydrated.

wafered. Agglomerated by compressing into a form that usually measures more in diameter or cross section than in length. Refers to fibrous feeds (e.g., wafered alfalfa hay).

water extracted. Removed with water. Refers to a product from which soluble substances have been removed.

weathered. Exposed to air, sunlight, and precipitation.

wet milled. Steeped in water, which may contain sulfur dioxide, to facilitate separation of the parts. Refers to kernels of corn.

whole pressed. Pressed to remove oil. Refers to seeds with hulls (e.g., cotton seeds).

FEED MANUFACTURING TERMS

additive. An ingredient or combination of ingredients added to the basic feed mix or parts thereof to fulfill a specific need. Usually used in micro-quantities and requires careful handling and mixing.

(*Note:* A *food additive* is defined by federal law as any substance which becomes a component of or affects the characteristics of a feed or food if such substance is not generally recognized among experts qualified by scientific training and experience to evaluate its safety as having been adequately shown through scientific procedures to be safe under the conditions of its intended use. Excepted are substances having "prior sanction" and pesticide chemicals under certain conditions. See Public Law 65-929 for details.)

antibiotics. A class of drug. They are usually synthesized by a living microorganism and in proper concentration inhibit the growth of other microorganisms.

artifically dried. (Process.) Moisture having been removed by other than natural means.

aspirated, aspirating. Having removed chaff, dust, or other light materials by use of air.

balanced. A term that may be applied to a diet, ration, or feed having all known required nutrients in proper amount and proportion based upon recommendations of recognized authorities in the field of animal nutrition, such as the National Research Council, for a given set of physiological animal requirements. The species for which it is intended and the functions such as maintenance or maintenance plus production (growth, fetus, fat, milk, eggs, wool, feathers or work) shall be specified.

blending. (Process.) To mingle or combine two or more ingredients of feed. It does not imply a uniformity of dispersion.

blocked, blocking. (Process.) Having agglomerated individual ingredients or mixtures into a large mass.

blocks. (Physical form.) Agglomerated feed compressed into a solid mass cohesive enough to hold its form and weighing over 2 pounds, and generally weighing 30–50 pounds.

brand name. Any word, name, symbol or device or any combination thereof identifying the commercial feed of a distributor and distinguishing it from that of others.

bricks. (Physical form.) Agglomerated feed, other than pellets, compressed into a solid mass cohesive enough to hold its form and weighing less than two pounds. (*See also* Blocks).

by-product. (Part.) Secondary products produced in addition to the principal product.

cake. (Physical form.) The mass resulting from the pressing of seeds, meat, or fish in order to remove oils, fats, or other liquids.

carcass residue, mammals. (Part.) Residues from animal tissues including bones and excluding hair, hoofs, horns and contents of the digestive tract.

carriers. An edible material to which ingredients are added to facilitate uniform incorporation of the latter into feeds. The active particles are absorbed, impregnated, or coated into or onto the edible material in such a way as to physically carry the active ingredient.

chaff. (Part.) Glumes, husks, or other seed coverings together with other plant parts separated from seed in threshing or processing.

cleaned, cleaning. (Process.) Removal of material by such methods as scalping, aspirating, magnetic separation, or by any other method.

cleanings. (Part.) Chaff, weed seeds, dust, and other foreign matter removed from cereal grains.

clipped, clipping. (Process.) Removal of the ends of whole grain.

commercial feed. The term commercial feed (as defined in the Uniform Feed Bill) means all materials which are distributed for use as feed or for mixing in feed, for animals other than man except: (1) Option A—Unmixed seed, whole or processed, made directly from the entire seed. Option B—Unmixed or unprocessed whole seeds. (2) Hay, straw, stover, silage, cobs and hulls (a) when unground, and (b) when unmixed with other materials. (3) Individual chemical compounds when not mixed with other materials.

complete feed. A nutritionally adequate feed for animals other than man; by specific formula is compounded to be fed as the sole ration and is capable of maintaining life and/or promoting production without any additional substance being consumed except water.

concentrate. A feed used with another to improve the nutritive balance of the total and intended to be further diluted and mixed to produce a supplement or a complete feed.

condensed, condensing. (Process.) Reduced to denser form by removal of moisture.

conditioned, conditioning. (Process.) Having achieved predetermined moisture characteristics and/or temperature of ingredients or a mixture of ingredients prior to further processing.

cooked, cooking. (Process.) Heated in the presence of moisture to alter chemical and/or physical characteristics or to sterilize.

cooled, cooling. (Process.) Temperature reduced by air movement, usually accompanied by a simultaneous drying action.

cracked, cracking. (Process.) Particle size reduced by a combined breaking and crushing action.

cracklings. (Part.) Residue after removal of fat from adipose tissue or skin of animals by dry heat.

crumbled, crumbling. (Process.) Pellets reduced to granular form.

crumbles. (Physical form.) Pelleted feed reduced to granular form.

cubes. (Physical form.) *See* Pellets.

cubes, range. (Physical form.) *See* Pellets and Range cubes.

degermed. (Process.) Having had the embryo of seeds wholly or partially separated from the starch endosperm.

dehulled, dehulling. (Process.) Having removed the outer covering from grains or other seeds.

dehydrated, dehydrating. (Process.) Having been freed of moisture by thermal means.

diet. Feed ingredient or mixture of ingredients including water, which is consumed by animals.

diluent. (Physical form.) An edible substance used to mix with and reduce the concentration of nutrients and/or additives to make them more acceptable to animals, safer to use, and more capable to being mixed uniformly in a feed. (It may also be a carrier.)

dressed, dressing. (Process.) Made uniform in texture by breaking or screening of lumps from feed and/or the application of liquid(s).

dried, drying. (Process.) Materials from which water or other liquid has been removed.

drug. (As defined by FDA as applied to feed.) A substance (a) intended for use in the diagnosis, cure, mitigation, treatment, or prevention of disease in man or other animals or a substance (b) other than food intended to affect the structure or any function of the body of man or other animals.

dry-milled. (Process.) Tempered with a small amount of water or steam to facilitate the separation of the various component parts of the kernel in the absence of any significant amount of free water.

dust. (Part.) Fine, dry pulverized particles of matter usually resulting from the cleaning or grinding of grain.

emulsifier. A material capable of causing fat or oils to remain in liquid suspension.

ensiled. (Process.) Aerial parts of plants that have been preserved by ensiling. Normally the original material is finely cut and blown into an air-tight chamber such as a silo, where it is pressed to exclude air and where it undergoes an acid fermentation that retards spoilage.

evaporated, evaporating. (Process.) Reduced to a denser form; concentrated as by evaporation or distillation.

expanded, expanding. (Process.) Subjected to moisture, pressure, and temperature to gelatinize the starch portion. When extruded, its volume is increased, due to abrupt reduction in pressure.

extracted, mechanical. (Process.) Having removed fat or oil from materials by heat and mechanical pressure. Similar terms: expeller extracted, hydraulic extracted, "old process."

extracted, solvent. (Process.) Having removed fat or oil from materials by organic solvents. Similar term: "new process."

extruded. (Process.) A process by which feed has been pressed, pushed, or protruded through orifices under pressure.

fat. (Part.) A substance composed chiefly of triglycerides of fatty acids, and solid or plastic at room temperature.

fatty acids. (Part.) Aliphatic monobasic acids containing only the elements of carbon, hydrogen, and oxygen.

feed(s). Edible material(s) that are consumed by animals and contribute energy and/or nutrients to the animal's diet. (Usually refers to animals rather than man.)

feed additive concentrate. (Part.) (As defined by FDA.) An article that is intended to be further diluted to produce a complete feed or a feed additive supplement and is not suitable for offering as a supplement or for offering free choice without dilution. It contains, among other things, one or more additives in amounts in a suitable feed base such that from 100 to 1000 pounds of concentrate must be diluted to produce 1 ton of a complete feed. A feed additive concentrate is unsafe if fed free choice or as a supplement because of danger to the health of the animal or because of the production of residues in the edible products from food producing animals in excess of the safe levels established.

feed additive premix. (As defined by FDA.) An article that must be diluted for safe use in feed additive concentrate, a feed additive supplement or a complete feed. It contains, among other things, one or more additives in high concentration in a suitable feed base such that up to 100 pounds must be diluted to produce 1 ton of complete feed. A feed additive premix contains additives at levels for which safety to the animal has not been demonstrated and/or which may result when fed undiluted in residues in the edible products from food producing animals in excess of the safe levels established.

feed additive supplement. (As defined by FDA.) An article for the diet of an animal which contains one or more food additives and is intended to be the following: (1) Further diluted and mixed to produce a complete feed; or (2) Fed undiluted as a supplement to other feeds; or (3) Offered free choice with other parts of the rations separated available.

(*Note:* A feed additive supplement is safe for the animal and will not produce unsafe residues in the edible products from food producing animals if fed according to directions.)

feed grade. Suitable for animal consumption.

feed mixture. *See* Formula feed.

feedstuff. *See* Feed(s).

fermentation aid. A substance added to assist in providing proper conditions that result in action by yeasts, molds, or bacteria in a controlled aerobic or anaerobic process used for the manufacture of certain products.

fermented, fermenting. (Process.) Acted upon by yeasts, molds, or bacteria in a controlled aerobic or anaerobic process in the manufacture of such products as alcohols, acids, vitamins of the B-complex group, or antibiotics.

fines. (Physical form.) Any material which will pass through a screen whose openings are immediately smaller than the specified minimum crumble size or pellet diameter.

flakes. (Physical form.) An ingredient rolled or cut into flat pieces with or without prior steam conditioning.

flaked, flaking. (Process.) *See* Rolled.

flour. (Part.) Soft, finely ground and bolted meal obtained from the milling of cereal grains, other seeds, or products. It consists essentially of the starch and gluten of the endosperm.

food(s). When used in reference to animals, is synonymous with feed(s). *See* Feed(s).

formula feed. Two or more ingredients proportioned, mixed, and processed according to specifications.

free choice. A feeding system by which animals are given unlimited access to the separate components or groups of components constituting the diet.

gelatinized, gelatinizing. (Process.) Having had the starch granules completely ruptured by a combination of moisture, heat, and pressure, and in some instances by mechanical shear.

germ. (Part.) The embryo found in seeds and frequently separated from the bran and starch endosperm during the milling.

gluten. (Part.) The tough, viscid nitrogen substance remaining when the flour or wheat or other grain is washed to remove the starch.

gossypol. (Part.) A phenolic pigment in cottonseed that is toxic to some animals.

grain. (Part.) Seed from cereal plants.

GRAS. Acronym for the phrase "generally recognized as safe." A substance that is generally recognized as safe by experts qualified to evaluate the safety of the substance for its intended use.

grease. Animal fats with a titer below 40°C.

grits. (Part.) Coarsely ground grain from which the bran and germ have been removed, usually screened to uniform particle size.

groats. (Part.) Grain from which the hulls have been removed.

ground, grinding. (Process.) Reduced in particle size by impact, shearing, or attrition.

heat-processed, heat-processing. (Process.) Subjected to a method of preparation involving the use of elevated temperatures with or without pressure.

hulls. (Part.) Outer covering of grain or other seed.

hydrolized, hydrolizing. (Process.) Complex molecules having been split to simpler units by chemical reaction with water, usually by catalysis.

ingredient, feed ingredient. A component part or constituent of any combination or mixture making up a commercial feed.

kibbled, kibbling. (Process.) Cracked or crushed baked dough, or extruded feed that has been cooked prior to or during the extrusion process.

lard. (Part.) Rendered fat of swine.

lecithin. (Part.) A specific phospholipid. The principal constituent of crude phosphatides derived from oil-bearing seeds.

malt. (Part.) Sprouted and steamed whole grain from which the radicle has been removed.

malted, malting. (Process.) Converted into malt or treated with malt or malt extract.

mash. (Physical form.) A mixture of ingredients in meal form. Similar term: mash feed.

meal. (Physical form.) An ingredient which has been ground or otherwise reduced in particle size.

medicated feed. Any feed that contains drug ingredients intended or represented for the cure, mitigation, treatment, or prevention of diseases of animals other than man or that contains drug ingredients intended to affect the structure or function of the body of animals other than man. Antibiotics included in a feed at growth promotion and/or feed efficiency levels are drug additives and feeds containing such antibiotics are included in the definition of medicated feed.

microingredients. Vitamins, minerals, antibiotics, drugs, and other materials normally required in small amounts and measured in milligrams, micrograms, or parts per million (ppm).

middlings. (Part.) A by-product of flour milling comprising several grades of granular particles containing different proportions of endosperm, bran, and germ, each of which contains different levels of crude fiber.

mill by-product. (Part.) A secondary product obtained in addition to the principal product in milling practice.

mill dust. (Part.) Fine feed particles of undetermined origin resulting from handling and processing feed and feed ingredients.

mill run. (Part.) The state in which a material comes from the mill, ungraded and usually uninspected.

mineralize, mineralized. (Process.) To supply, impregnate, or add inorganic mineral compounds to a feed ingredient or mixture.

mixing. (Process.) To combine by agitation two or more materials to a specific degree of dispersion.

molasses. (Part.) The thick, viscous by-product resulting from refined sugar production or the concentrated, partially dehydrated juices from fruits.

pearled, pearling. (Process.) Dehulled grains reduced by machine brushing into smaller smooth particles.

pellets. (Physical form.) Agglomerated feed formed by compacting and forcing through die openings by a mechanical process. Similar terms: pelleted feed, hard pellet.

pellets, soft. (Physical form.) Pellets containing sufficient liquid to require immediate dusting and cooling. Similar term: high molasses pellets.

pelleted, pelleting. (Process.) Having agglomerated feed by compaction and forced through die openings.

pomace. (Part.) Pulp from fruit. *See* Pulp.

premix. A uniform mixture of one or more microingredients with diluent and/or carrier. Premixes are used to facilitate uniform dispersion of the microingredients in a larger mix.

premixing. (Process.) The preliminary mixing of ingredients with diluents and/or carriers.

product. (Part.) A substance produced from one or more other substances as a result of chemical or physical change.

protein. (Part.) Any of a large class of naturally occurring complex combinations of amino acids.

pulp. (Part.) The solid residue remaining after extraction of juices from fruits or stems. Similar terms: bagasse and pomace.

pulverized, pulverizing. (Process.) *See* Ground, grinding.

range cake. (Physical form.) *See* Cake.

range cubes. (Physical form.) Large pellets designed to be fed on the ground. Similar term: range wafer.

ration. The amount of the total feed that is provided to one animal over a 24-hr period.

rolled, rolling. (Process.) Having changed the shape and/or size of particles by compressing between rollers. It may entail tempering or conditioning.

scalped, scalping. (Process.) Having removed larger material by screening.

scratch. (Physical form.) Whole, cracked, or coarsely cut grain. Similar terms: scratch grain, scratch feed.

screened, screening. (Process.) Having separated various sized particles by passing over and/or through screens.

self-fed. A feeding system where animals have continuous free access to some or all component(s) of a ration, either individually or as mixtures.

separating. (Process.) Classification of particle by size, shape, and/or density.

separating, magnetic. (Process.) Removing ferrous material by magnetic attraction.

shorts. (Part.) Fine particles of bran, germ, flour or offal from the tail of the mill from commercial flour milling.

sizing. (Process.) *See* Screened, screening.

skimmed. (Process.) Material from which floating solid material has been removed. It is also applied to milk from which fat has been removed by centrifuging.

soluble. Liquid containing dissolved substances obtained from processing animal or plant materials. It may contain some fine suspended solids.

solvent extracted. (Process.) A product from which oil has been removed by solvents.

spray dehydrated. (Process.) Material that has been dried by spraying on the surface of a heated drum. It is recovered by scraping from the drum.

starch. (Part.) A white, granular polymer of plant origin. The principal part of seed endosperm.

steamed, steaming. (Process.) Having treated ingredients with steam to alter physical and/or chemical properties. Similar terms: steam cooked, steam rendered, tanked.

stillage. (Part.) The mash from fermentation of grains after removal of alcohol by distillation.

supplement. A feed used with another to improve the nutritive balance or performance of the total and intended to be (1) fed undiluted as a supplement to other feeds; or (2) offered free choice with other parts of the ration separately available; or (3) further diluted and mixed to produce a complete feed.

tallow. (Part.) Animal fats with titer above 40°C.

tankage. (Part.) *See* Carcass residue.

tempered, tempering. (Process.) *See* Conditioned, conditioning.

titer. A property of fat determined by the solidification point of the fatty acids liberated by hydrolysis.

toasted. (Process.) Browned, dried, or parched by exposure to a fire, or to gas, or electric heat.

trace minerals. Mineral nutrients required by animals in microamounts only (measured in milligrams per pound or smaller units).

vitamins. Organic compounds that function as parts of enzyme systems essential for the transmission of energy and the regulation of metabolisms of the body.

wafered, wafering. (Process.) Having agglomerated a feed of a fibrous nature by compressing into a form usually having a diameter or cross-section measurement greater than its length.

wafers. (Physical form.) A form of agglomerated feed based on fibrous ingredients in which the finished form usually has a diameter or cross-section measurement greater than its length.

whey. (Part.) The watery part of milk separated from the curd.

whey solids. (Part.) The solids of whey (proteins, fats, lactose, ash, and lactic acid).

FEED INGREDIENT TERMS

alfalfa meal or pellets, dehydrated. The aerial portion of the alfalfa plant, reasonably free of other crop plants, weeds, and mold, which has been finely ground and dried by thermal means.

alfalfa meal or pellets, sun-cured, or ground alfalfa hay. The aerial portion of the alfalfa plant, reasonably free of other crop plants, weeds, and mold, which has been sun cured and finely ground. If it is chopped instead of finely ground, it must be designated as "sun-cured chopped alfalfa" or "chopped alfalfa hay."

animal fat. Obtained from the tissues of mammals and/or poultry in the commercial processes of rendering or extracting. It consists predominantly of glyceride esters of fatty acids and contains no additions of free fatty acids or other materials obtained from fats. It must contain not less than 90% total fatty acids, not more than 2.5% unsaponifiable matter, and not more than 1% insoluble matter. If the product bears a name descriptive of its kind or origin—i.e., tallow, lard, grease—it must correspond thereto. If antioxidant(s) is(are) used, the common name(s) must be indicated, followed by the word "preservative(s)."

blood meal, conventional cooker dried. Produced from clean, fresh animal blood, exclusive of all extraneous material such as hair, stomach belchings, and urine except in such traces as might occur unavoidably in good manufacturing processes. The product usually has a dark blacklike color and is rather insoluble in water.

blood meal, flash dried. Produced from clean, fresh animal blood, exclusive of all extraneous material such as hair, stomach belchings, and urine except in such traces as might occur unavoidably in good manufacturing processes. A large portion of the moisture (water) is usually removed by a mechanical dewatering process or by condensing by cooking to a semisolid state. The semisolid blood mass is then tranferred to a rapid drying facility where the more tightly bound water is rapidly removed. The minimum biological activity of lysine shall be 80%.

brewers' dried grains. The dried extracted residue of barley malt alone or in mixture with other cereal grain or grain products resulting from the manufacture of wort or beer and may contain pulverized dried spent hops in an amount not to exceed 3%, evenly distributed.

coconut meal. The ground residue that remains after removal of most of the oil from dried meat of coconuts, by mechanical or solvent extraction process. May also be called "copra meal."

corn gluten feed. That part of the commercial shelled corn that remains after extraction of the larger portion of the starch, gluten, and germ by the processes employed in the wet milling manufacture of corn starch or syrup. It may or may not contain fermented corn extractives and/or corn germ meal.

corn gluten meal. The dried residue from corn after the removal of the larger part of the starch and germ, and the separation of the bran by the process employed in the wet milling manufacture of corn starch or syrup, or by the enzymatic treatment of the endosperm. It may contain fermented corn extractives and/or corn germ meal.

cottonseed hulls. Consists primarily of the outer covering of the cottonseed.

cottonseed meal. Cottonseed meal, mechanical or solvent extracted, is the product obtained by finely grinding the cake that remains after removal of most of the oil from cottonseed by a mechanical or solvent extraction process. It must contain not less than 36% crude protein. Low gossypol cottonseed meal is a meal in which the gossypol is not more than 0.4% free gossypol.

distillers' dried grains. Obtained after removal of ethyl alcohol by distillation from the yeast fermentation of a grain or a grain mixture by separating the resultant coarse grain fraction of the whole stillage and drying it by methods employed in the grain distilling industry. The predominating grain shall be declared as the first word in the name.

distillers' dried solubles. Obtained after the removal of ethyl alcohol by distillation from the yeast fermentation of a grain or a grain mixture by condensing the thin stillage fraction and drying it by methods employed in the grain distilling industry. The predominating grain shall be declared as the first word in the name.

distillers' dried grains with solubles. The product obtained after the removal of ethyl alcohol by distillation from the yeast fermentation of a grain or a grain mixture by condensing and drying at least three-fourths of the solids of the resultant whole stillage by methods employed in the grain distilling industry. The predominating grain shall be declared as the first word in the name.

dried bakery product. Mixture of bread, cookies, cake, crackers, flours and doughs that has been mechanically separated from nonedible material, artifically dried, and ground. If the product contains more than 3.5% salt, the maximum percentage of salt must be part of the name.

dried beet pulp, plain. The dried residue from sugar beets that has been cleaned and freed from crowns, leaves and sand, and that has been extracted in the process of manufacturing sugar.

dried citrus meal. The finer particles obtained by screening dried citrus pulp.

dried citrus pulp. The ground peel, residue of the inside portions, and occasional cull fruits of the citrus family that have been dried, producing a coarse, flaky product. It may contain dried citrus meal or pellets and whole citrus seeds. If calcium oxide or calcium hydroxide is added as an aid in processing, the maximum percentage present, expressed as calcium, must be shown. If it bears a name descriptive of its kind or origin, it must correspond thereto.

dried skimmed milk, feed grade. The residue obtained by drying defatted milk. It contains 8% maximum moisture. The words "feed grade" are not required when listing as an ingredient in manufactured feed.

dried whey. The product obtained by removing water from whey. It contains not less than 11% protein nor less than 61% lactose.

dried whey product. The residue obtained by drying whey from which a portion of the lactose has been removed. The minimum percentage lactose must be prominently declared on the label.

fat, hydrolyzed, or oil, feed grade. Obtained in the fat processing procedures commonly used in edible fat processing or soap making. It consists predominantly of fatty acids and must contain not less than 85% total fatty acids, not more than 6% unsaponifiable matter, and not more than 1% insoluble matter. Its source must be stated in the product name—i.e., hydrolyzed animal fat, hydrolyzed vegetable oil, hydrolyzed animal and vegetable fat. If antioxidant(s) is(are) used, the common name(s) must be indicated, followed by the word "preservative(s)."

feeding oat meal. Obtained in the manufacture of rolled oat groats or rolled oats and consists of broken oat groats, oat groat chips, and floury portions of the oat groats, with only such quantity of finely ground oat hulls as is unavoidable in the usual process of commercial milling. It must not contain more than 4% crude fiber.

fish meal. Clean, dried, ground tissues of undecomposed whole fish or fish cuttings, either or both, with or without the extraction of part of the oil. If it contains more than 3% salt, the amount of salt must constitute a part of the brand name, provided that in no case must the salt content of this product exceed 7%.

fish solubles, condensed. Obtained by condensing the stickwater. It must contain not less than 30% crude protein.

fish solubles, dried. Obtained by dehydrating the stickwater. It must contain not less than 60% crude protein.

guar meal. Obtained from whole guar beans after removal of most of the endosperm. If the product is heat treated, it may be designated as heat treated or toasted.

hemicellulose extract. A by-product of the manufacture of pressed wood. It is the concentrated soluble material obtained from the treatment of wood at elevated temperature and pressure without use of acids, alkalis, or salts. It contains pentose and hextose sugars and has a total carbohydrate content of not less than 55%.

hominy feed. A mixture of corn bran, corn germ, and part of the starchy portion of either white or yellow corn kernels or mixture thereof, as produced in the manufacture of pearl hominy, hominy grits, or table meal, and must contain not less than 4% crude fat. If prefixed by the words white or yellow, the product must correspond thereto.

linseed meal. Linseed meal, mechanical extracted, is the product obtained by grinding the cake or chips which remain after removal of most of the oil from flaxseed by a mechanical extraction process. It must contain no more than 10% fiber.

Linseed meal, solvent extracted, is the product obtained by grinding the flakes that remain after removal of most of the oil from flaxseed by a solvent extraction process. It must contain no more than 10% fiber.

malt sprouts. Obtained from malted barley by the removal of the sprouts that may include some of the malt hulls, other parts of malt, and foreign material unavoidably present. It must contain not less than 24% crude protein. The term malt sprouts when applied to a corresponding portion of other malted cereals must be used in qualified form, i.e., rye malt sprouts, wheat malt sprouts, etc.

meat meal. The dry-rendered product from mammal tissues, exclusive of hair, hoof, horn and hide trimmings, manure, and stomach contents, except in such amounts as may occur unavoidably in good factory practice. It shall not contain added blood meal. It shall contain not more than 14% pepsin indigestible residue. Not more than 11% of the crude protein in the product shall be pepsin indigestible. It must be designated according to its protein content. If the product bears a name descriptive of its kind, composition, or origin, it must correspond thereto.

meat-and-bone meal. The dry-rendered product from mammal tissues, exclusive of hair, hoof, horn, hide trimmings, manure and stomach contents, except in such amounts as may occur unavoidably in good factory practice. It shall not contain added blood meal. It shall contain more than 4.4% phosphorus and not more than 14% pepsin indigestible residue. Not more than 11% of the crude protein shall be pepsin indigestible. It must be designated according to its protein content. If the product bears a name description of its kind, composition, or origin it must correspond thereto.

meat meal tankage. The dry product from mammal tissues, exclusive of hair, hoof, horn, hide trimmings, manure, and stomach contents, except in such amounts as may occur unavoidably in good factory practice, especially prepared for feeding purposes by tankage under live steam or dry rendering. It shall contain not more than 14% pepsin indigestible residue. Not more than 11% of the crude protein in the product shall be pepsin indigestible. It must be designated according to its protein content. If the product bears a name descriptive of its kind, composition, or origin it must correspond thereto.

molasses, beet. A by-product of the manufacture of sucrose from sugar beets. It must contain not less than 48% total sugars, expressed as invert and its density determined by double dilution must not be less than 79.5° Brix.

molasses, starch. A by-product of the manufacture of dextrose from starch derived from corn or grain sorghums in which the starch is hydrolized by use of enzymes and/or acid. It must contain not less than 43% reducing sugars expressed as dextrose and not less than 50% total sugars expressed as dextrose. It shall contain not less than 73% total solids.

peanut meal. A ground product of the shelled peanuts, composed principally of the kernels, with such portion of the hulls, or fiber, and oil, as may be left in the ordinary course of manufacture by a mechanical or solvent extraction process. It must contain not more than 7% crude fiber.

poultry by-product meal. Consists of the ground, dry-rendered or wet-rendered clean parts of the carcass of slaughtered poultry, such as heads, feet, undeveloped eggs and intestines, exclusive of feathers, except in such trace amounts as might occur unavoidably in good factory practice. It must contain not more than 16% ash and not more than 4% acid-insoluble ash.

poultry feathers, hydrolized. The product resulting from the treatment under pressure of clean, undecomposed feathers from slaughtered poultry, free of additives and/or accelerators. Not less than 75% of its crude protein content must be digestible by the pepsin digestibility method.

rice bran. The pericarp or bran layer and germ of the rice, with only such quantity of hull fragments, chipped, broken or brewer's rice, and calcium carbonate as is unavoidable in the regular milling of edible rice. It must contain not more than 13% crude fiber. When the calcium carbonate exceeds 3%, the percentage must be declared in the brand name.

rice mill by-product. The total offal obtained in the milling of rice. It consists of rice hulls, rice bran, rice polishings, and broken rice grains. Its crude fiber content must not exceed 32%.

safflower meal. The ground residue obtained after extracting the oil from whole safflower seed, by a mechanical or solvent extraction process.

screenings. Obtained in the cleaning of grains that are included in the U.S. Grain Standard Act and other agricultural seeds. It may include light and broken grains and agricultural seeds, weed seeds, hulls, chaff, joints, straw, elevator or mill dust, sand. and dirt. It must be designated as grain screenings, mixed screenings, and chaff and/or dust.

No grade of screenings must contain any seeds or other material in amount that is either injurious to animals or will impart any objectionable odor or flavor to their milk or flesh. The screenings must contain not more than four whole prohibited noxious weed seeds per pound and must contain not more than 100 whole restricted noxious weed seeds per pound. The prohibited and restricted noxious weed seeds must be those named as such by the seed control law of the state in which the screenings are sold or used.

All grades of screenings must bear minimum guarantees of crude protein and crude fat and maximum guarantees of crude fiber and ash.

soybean meal. The product obtained by grinding the cake, chips, or flakes that remain after removal of most of the oil from soybeans by a mechanical or solvent extraction process. It must contain not more than 7% crude fiber. It may contain a nonnutritive inert, nontoxic conditioning agent to reduce caking and improve flowability in an amount not to exceed that necessary to accomplish its intended effect and in no case exceed 0.5%.

Soybean meal, dehulled, solvent extracted, is obtained by grinding the flakes remaining after removal of most of the oil from dehulled soybeans by a solvent extraction process. It must contain not more than 3.3% crude fiber.

soybean mill feed. Composed of soybean hulls and the offal from the tail of the mill that results from the manufacture of soy grits or flour. It must contain not less than 13% crude protein and not more than 32% crude fiber.

soybeans, heat processed. The product resulting from heating whole soybeans without removing any of the component parts. It may be ground, pelleted, flaked or powdered. It must be sold according to its crude protein content. Similar term: dry roasted soybeans.

sunflower meal. Obtained by grinding the residue remaining after extraction of most of the oil from whole sunflower seed, by mechanical or solvent extraction process. May also be obtained by grinding the residue after extraction of oil from dehulled sunflower seed by either process.

urea. Predominantly urea but may contain other nontoxic nitrogenous compounds that are present as by-products from the commercial synthesis and processing of urea. It must contain not less than 45% nitrogen (equivalent to 281.25% crude protein). If it contains less than 45% N but 41% or more N, it must be designated as urea and conditioner(s). If the name of the conditioner(s) does(do) appear in the product name, the ingredient listing must contain the specific name of the conditioner(s).

If the urea and condition(s) contribute more than 0.5% conditioner(s) to the mixed feed, the conditioner(s) must be named in the mixed ingredient list.

wheat bran. The coarse outer covering of the wheat kernel as separated from cleaned and scoured wheat in the usual process of commercial milling.

wheat mill run. Coarse wheat bran, fine particles of wheat bran, wheat shorts, wheat flour, and the offal from the "tail of the mill." This product must be obtained in the usual process of commercial milling and must contain not more than 9.5% crude fiber.

wheat middlings. Fine particles of wheat bran, wheat shorts, wheat germ, wheat flour, and some of the offal from the "tail of the mill." This product must be obtained in the usual process of commercial milling and must contain not more than 9.5% crude fiber.

wheat shorts. The fine particles of wheat bran, wheat germ, wheat flour, and the offal from the "tail of the mill." This product must be obtained in the usual process of commercial milling and must contain not more than 7% crude fiber.

BREEDING, GENETIC, AND PRODUCTION SYSTEMS TERMS

abortion. The delivery of fetuses or fetal membranes between date of service and up to and including 109th day of pregnancy.

accuracy (of selection). Correlation between an animal's unknown actual breeding value and a calculated estimated breeding value.

adjusted 21-day litter weight (W). Total weight of all pigs from one litter adjusted to 21 days of age, to a 3-6 parity sow and 10 pigs/litter after transfer.

ad lib feeding. No limit placed on amount of feed intake. Self-feeding or allowing pigs to consume feed on a free-choice basis.

age of weaning. The number of days between farrowing and weaning (farrowing = day 0), i.e., farrowing-to-weaning interval.

alleles. Alternate forms of genes. Because genes occur in pairs in body cells, one gene of a pair may have one effect and another gene of that same pair (allele) may have a different effect on the same trait.

anestrous female. A female that is not detected in estrus.

artificial insemination (AI). The technique of placing semen from the male in the reproductive tract of the female by means other than natural service.

average carcass backfat thickness. The average of three backfat measurements taken along the midline at the first rib, last rib, and last lumbar vertebra.

average daily gain (ADG). Amount of weight gained per day per pig during a defined period of time.

average mated female inventory. The average of mated female inventory in the herd as recorded on 12 or more equally spaced occasions during the year.

average number of farrowing crates. The average number of farrowing crates available for use as recorded on 12 equally spaced occasions during the year. (Multiple recordings of number of crates available during the year is not necessary if the number of crates does not change.)

average service boar inventory. The average number of service boars in the herd as recorded on 12 or more equally spaced occasions during the year.

average total boar inventory. The average number of service boars and replacement boars in the herd as recorded on 12 or more equally spaced occasions during the year.

average total female inventory. The average total female inventory in the herd as recorded on 12 or more equally spaced occasions during the year.

backcross. The mating of a two-breed crossbred offspring back to one of its parental breeds. Example: A Yorkshire-Landrace cross sow bred back to a Yorkshire boar.

birth weight (BW). The weight of a pig taken within 24 hours after birth.

boar. A replacement boar is any intact (uncastrated) male pig intended for use in the breeding herd which has not yet been used for breeding. A service boar is any boar that is being, or has been, used for breeding purposes.

breeder. In most swine breed associations, the owner of the sow at the time she was mated or bred to produce a litter of pigs.

breeding herd. The inventory of the mated females and replacement gilts, service boars, and replacement boars.

breeding program goals. The objective or "direction" of breeders' selection programs. Goals may vary among breeders due to relative genetic merit of their swine, their resources, and their markets.

breeding value. Value of an animal as a parent. The working definition is twice the difference between a very large number of progeny and the population average when individuals are mated at random within the population and all progeny are managed alike. The difference is doubled because only a sample half (one gene of each pair) is transmitted from a parent to each progeny. Breeding value exists for each trait and is dependent on the population in which the animal is evaluated. For a given trait, an individual can be an above average producer in one herd and a below average producer in another herd.

carcass evaluation. Techniques of measuring components of quality and quantity in carcasses.

carcass merit. Desirability of a carcass relative to quantity of muscle, fat, bone, and quality of lean tissue.

carcass quality. An estimate of palatability based primarily on marbling, color, firmness, and texture of lean.

carrier. A heterozygous individual having one recessive gene and one dominant for a given pair of genes (alleles).

central test. A location where animals are assembled from several herds to evaluate differences in certain performance traits under uniform management conditions.

chromosome. Chromosomes are long DNA molecules on which genes (the basic genetic codes) are located. Swine have 19 pairs of chromosomes.

closed herd. A herd in which no outside breeding stock are introduced.

collateral relatives. Relatives of an individual that are not its ancestors or descendants. Brothers and sisters are an example of collateral relatives.

conception. The fertilization of the ovum (egg). The act of conceiving or becoming pregnant.

congenital. Acquired during prenatal life. Condition exists at or dates from birth. Often used in the context of congenital (birth) defects.

contemporary group. A group of swine that are of the same breed and sex and have been raised in the same management group (in the same location and on the same feed). Contemporary groups should include as many pigs as can be accurately compared.

correlation. A measure of how two traits vary together. A correlation of +1.00 means that as one trait increases the other also increases—a perfect positive relationship. A correlation of −1.00 means that as one trait increases the other decreases—a perfect negative, or inverse, relationship. A correlation of 0.00 means that as one trait increases, the other may increase or decrease—no consistent relationship. Correlation coefficients may vary between +1.00 and −1.00.

crossbreeding. The mating of animals of different breeds within a species. Crossbreeding usually results in heterosis (hybrid vigor).

culling. The process of eliminating less productive or less desirable animals from a herd.

dam. The female parent.

date of service. The first mating date (or for group breeding, the first potential mating date) within a service period. For group breeding programs, the potential mating date equals the date a boar was first placed in the group.

deviation. A difference between an individual record and the average for that trait for that contemporary group. These differences sum to zero when the correct average is used.

dominance. Dominant genes affect the phenotype when present in either homozygous or heterozygous condition. A dominant gene need only be obtained from one parent to achieve expression.

economic value. The net return within a herd for making a unit change in the trait in question.

embryo. A fertilized ovum (egg) in the earlier stages of prenatal development usually prior to development of body parts.

embryonic mortality. The loss in number of live, fertilized ova during pregnancy.

environment. All external (nongenetic) conditions that influence the reproduction, production, and carcass merit of swine.

estimate. The process of calculating a particular value from data (verb). The value itself obtained from data (noun). The idea is that the true value is being obtained from the calculated value within limits of sampling variation.

estimated breeding value (EBV). An estimated of an individual's true breeding worth for a trait based on the performance of the individual and close relatives for the trait. EBV is a systematic way of combining available performance information on the individual, brothers, and sisters of the individual, and the progeny of the individual. EBV's for growth traits are generally expressed in ratio units; however, EBV's can be expressed in actual units (pounds, cm, percent). *See* definition of "breeding value" above.

estrus (heat). The recurrent, restricted period of sexual receptivity in sows and gilts. Nonpregnant sows and gilts usually come into heat 18 to 24 days following their previous estrus.

expected progeny difference (EPD). The difference in performance to be expected from future progeny of a sire, compared with that expected from future progeny of the average boar in the same test. EPD is equal to one-half the estimate of breeding value obtainable from records. EPD's for growth traits are generally expressed in pounds, either as a plus difference or minus difference from the population average. EPD's are generally reported in the units of measure of the trait (e.g., pounds, inches, square inches, percent, etc.).

farrowing. Production of a litter of one or more live or dead pigs on or after the 110th day of pregnancy, i.e., parturition.

farrowing difficulty. Abnormal or difficult labor, causing difficulty in delivering the fetus and/or placenta.

farrowing frequency. A calendarized system for initiating group farrowings expressed in days or weeks.

farrowing interval. The number of days between two consecutive farrowings for an individual sow.

feeder pigs. Young swine weighing 30 to 100 pounds.

F_1 offspring. Offspring resulting from the mating of a purebred (straighbred) boar to purebred (straightbred) females of another breed.

feed conversion (feed efficiency). Units of feed consumed per unit of weight gained. Also, the production (meat, milk) per unit of feed consumed.

generation interval. Average age of the parents when the offspring destined to replace them are born. A generation represents the average rate of turnover of a herd.

genes. The basic units of heredity that occur in pairs and have their effect in pairs in the individual, but which are transmitted singly (one or the other gene at random of each pair) from each parent to offspring.

genetic correlations. Correlations between two traits that arise because some of the same genes affect both traits. When two traits (i.e., number pigs born alive and 21-day litter weight) are positively and highly correlated to one another, successful selection for one trait will result in an increase in the other trait. When two traits are negatively and highly correlated to one another, successful selection for one trait will result in a decrease in the other trait.

genotype. Actual genetic makeup (constitution) of an individual determined by its genes or germplasm.

genotype-environment interaction. Variation in the relative performance of different genotypes from one environment to another. For example, the "best" hogs (genotypes) for one environment may not be the "best" for another environment.

gestation. Period of time between conception and farrowing.

gilt. A female of swine specie less than one year of age that has not produced a litter of pigs.

gilt pool. A group of replacement gilts available for breeding purposes. Should be counted as part of the total female inventory as soon as they have been selected for breeding purposes *and* their unselected contemporaries have been sent to slaughter. (Gilts must be counted in the gilt pool by the time they have reached seven months of age.)

gonad. The organ that produces the reproductive cells, the testicle in the male and the ovaries in the female.

half-sibs. Individuals having the same sire or dam. Half-brothers and/or half-sisters.

heredity. The transmission of genetic or physical traits of parents to their offspring.

heritability. The proportion of the differences among individuals, measured or observed, that is transmitted to the offspring. Heritability varies from zero to one. The higher the heritability of a trait, the more accurately does the individual performance predict breeding value and the more rapid should be the response due to selection for that trait.

heritability estimate. An estimate of the proportion of the total phenotypic variation between individuals for a certain trait that is due to heredity. More specifically, hereditary variation due to additive gene action.

heterosis (hybrid vigor). Amount by which measured traits of the crossbreds exceed the average of the two or more purebreds that are mated to produce the crossbreds.

heterozygous. Genes of a specific pair (alleles) are different in an individual.

homozygous. Genes of a specific pair (alleles) are alike in an individual.

hybrid. Progeny of genetically diverse parents of purebred breeding or crossbred parents.

inbreeding. Production of offspring from parents more closely related than the average of a population. Inbreeding increases the proportion of homozygous gene pairs and decreases the proportion of heterozygous gene pairs. Also, inbreeding increases prepotency and facilitates expression of undesirable recessive genes.

independent culling levels. Selection based on animals meeting specific levels of performance for each trait included in the breeder's selection program. For example, a breeder could cull all boars with scrotal hernias.

linebreeding. A form of inbreeding in which an attempt is made to concentrate the inheritance of some one ancestor, or line of ancestors, in a herd. The average relationship of the individuals in the herd to this ancestor (outstanding individual or individuals) is increased by linebreeding.

linecross. Offspring produced by crossing two or more inbred lines within a breed.

litter. Pigs born to a sow during one farrowing.

litter birth weight. Total birth weight of all live pigs weighed in a litter.

live pigs born. Total pigs born, minus the stillborn and mummified pigs.

loin muscle area. The cross sectional area of the loin muscle cut between the 10th and 11th ribs.

marbling. The specks of fat (intramuscular fat) distributed in muscular tissue. Marbling is usually evaluated in the loin muscle between the 10th and 11th rib.

market hogs. Finished hogs sold for slaughter.

mated females. All females in the breeding herd that have had at least one opportunity to be mated.

mating. The act of insemination.

metabolism. The transformation by which energy is made available for body uses.

mummified pigs. The pigs that are born degenerate (discolored and shriveled or decomposed) that died sometime during gestation.

nonadditive gene effects. Favorable effects or actions produced by specific gene pairs or combinations. Nonadditive gene action is the primary cause of heterosis. Nonadditive gene action occurs when the heterozygous genotype is not intermediate in phenotypic value to the two homozygous genotypes.

number of contemporaries. The number of animals of similar breed, sex, and age, against which an animal was compared in performance tests. The greater the number of contemporaries, the greater the accuracy of comparisons.

number of sows weaned. The number of females that farrowed a live litter and had their pigs removed from them, regardless of how long they nursed. Should approximately equal the number of litters farrowed over a year's time.

number of sows nursed. The number of females that completed a normal lactation period.

nursing pig. Any pig not weaned.

open. A term commonly used to indicate a nonpregnant female.

outcrossing. Mating of individuals that are less closely related than the average of the breed. Commercial breeders and some purebred breeders should be outcrossing by periodically adding new sires that are unrelated to their herd. This outcrossing should reduce the possibility of loss of vigor due to inbreeding.

parity. The number of times a female has farrowed, i.e., a 4th parity female has farrowed four times.

parturition. The act of giving birth; farrowing.

pedigree. A tabulation of names of ancestors, usually only those of three closest generations.

performance data. The record of the individual animal for reproduction, production, and possibly carcass merit. Traits included would be number of pigs born alive, 21-day litter weight, days to 230 pounds, backfat, etc.

performance pedigree. A pedigree that includes performance records of ancestors, half and full sibs, and progeny in addition to the usual pedigree information.

performance testing. The systematic collection of comparative production information for use in decision making to improve efficiency and profitability of swine production.

phenotype. The visible or measurable expression of a character; for example, 21-day litter weight, postweaning gain, reproduction, etc. Phenotype is influenced by genotype and environment.

phenotypic correlations. Correlations between two traits caused by both genetic and environmental factors influencing both traits.

post-weaning death loss. Pigs that die between weaning and market. May be further subdivided into nursery, growing/finishing, etc., using the appropriate method of calculating death loss percentage.

pounds of lean pork retail cuts per day of age. A measure of cutability and growth combined, it is calculated from predicted carcass lean divided by age in days.

prepotent. The ability of a parent to transmit its characteristics to its offspring so they resemble that parent, or each other, more than usual. Homozygous dominant individuals are prepotent. Inbred swine tend to be more prepotent than outbred swine.

pre-weaning death losses. Pigs born alive that die before weaning.

profitability. A criterion for measuring the relationship between income and expenses for different investments, operations, or firms.

progeny records. The average, comparative performance of the progeny of sires and dams.

progeny testing. Evaluating the genotype of an individual by a study of its progeny records.

puberty. The age at which the reproductive organs become functionally operative and secondary sex characteristics begin to develop.

purebred. An animal of known ancestry within a recognized breed that is eligible for registry in the official herdbook of that breed.

qualitative traits. Those traits in which there is a sharp distinction between phenotypes, such as hair coat color. Only one or few pairs of genes are involved in the expression of qualitative traits.

quantitative traits. Those traits in which there is no sharp distinction between phenotypes, with a gradual variation from one phenotype to another, such as backfat. Usually, many gene pairs are involved, as well as environmental influences.

random mating. A system of mating where every female has an equal or random chance of being assigned to any male used for breeding in a particular breeding season. Random mating is required for accurate progeny tests.

rate of genetic improvement. Rate of improvement per unit of time (year). The rate of improvement is dependent on: (1) heritability of traits considered; (2) selection differentials; (3) genetic correlations among traits considered; (4) generation interval in the herd; and (5) the number of traits for which selections are made.

reach. *See* Selection differential.

recessive gene. Recessive genes affect the phenotype only when present in a homozygous condition. Recessive genes must be received from both parents before the phenotype caused by the recessive genes can be observed.

reference sire. A boar designated to be used as a benchmark in progeny testing other boars (young sires). Progeny by reference sires in several herds enables comparisons to be made between boars not producing progeny in the same herd(s).

regression (regressed). A measure of the relationship between two variables. The value of one trait can be predicted by knowing the value of the other variable. For example, easily obtained carcass traits (hot carcass weight, fat thickness, and loin muscle area) are used to predict percent lean. Likewise, breeding value estimates based on limited data are regressed back toward the population average to account for the imperfection of this relationship.

repeat breeder. A female that returns to estrus before the anticipated farrowing date.

replacement gilt. A female that has been transferred to the gilt pool and has not yet had the opportunity to be mated.

return to service. Occurrence of another estrus after a previous service period (at least 10 days previously) within a single parity. Regular return to service is return to estrus 18 to 24 days after previous service. Early return to service is return to estrus less than 18 days after previous service period. Delayed return to service is return to estrus more than 24 days after previous service period.

rotational crossbreeding. Systems of crossing two or more breeds where the crossbred females are bred to boars of the breed contributing the least genes to that female's genotype. Rotation systems maintain relatively high levels of heterosis and produce replacement females from within the system.

seedstock breeders. Producers of breeding stock for purebred and commercial breeders. Progressive seedstock breeders have comprehensive programs designed to produce an optimum or desirable combination of economical traits (genetic package) that will ultimately increase the profitability of commercial swine production.

selection. Causing or allowing certain individuals in a population to produce offspring in the next generation.

selection differential (reach). The difference between the average for a trait in selected animals and the average of the group from which they came. The expected response from selection for a trait is equal to selection differential times the heritability of the trait.

selection index. A formula that combines performance records from several traits or different measurements of the same trait into a single value for each animal. Selection indexes weigh the traits for their relative net economic importance and their heritabilities plus the genetic associations among the traits.

service period. A period of time during which one or more matings or breedings can take place.

services per boar per time period. Includes the total number of times the boar ejaculates in that time period.

sibs. Brothers and sisters of an individual.

sow. Any breeding female that has farrowed at least one litter or has reached 12 months of age.

stillborn pigs. Fully developed pigs found dead behind the sow, or in the afterbirth, after farrowing. Lungs of stillborn pigs do not float in water.

strain. A pig's reaction against stress.

stress. Any force causing or tending to cause a change in a pig's function, structure, or behavior.

systems approach. An approach to evaluating alternative individuals, breeding programs, and selection schemes that involves assessment of these alternatives in terms of their net impact on all inputs and output in the production system. This approach specifically recognizes that intermediate optimum levels of performance in several traits may be more economically advantageous than maximum performance for any single trait.

tenth rib backfat. Backfat taken over the loin muscle at the 10th rib. Divide the longest axis of the loin muscle into quarters. Measure the fat depth opposite a point three-fourths the distance along the long axis toward the belly. The measurement is from the edge of the muscle to the outer edge of, and perpendicular to, the skin.

terminal sires. Sires used in a crossbreeding system where all their progeny, both male and female, are marketed. For example, F_1 crossbred dams could be bred to sires of a third breed and all pigs marketed. Although this system allows maximum heterosis and complementary effects of breeds, replacement females must come from other herds.

total female inventory. Sum of replacement gilts and mated females.

total pigs born. All pigs born in a litter, which includes those born live, stillborn, and mummified.

trait ratio. An expression of an animal's performance for a particular trait relative to the herd or contemporary group average.

ultrasonics. Technique for estimating certain aspects of body composition and for pregnancy detection.

variance. Variance is a statistic that describes the variation we see in a trait. Without variation, no genetic progress is possible, since genetically superior animals would not be distinguishable from genetically inferior ones.

weaning. The act of separating the pigs and the sow.

weaning-to-service interval. The time interval from date of weaning to next date of service.

weight per day of age (WDA). Weight of an individual divided by days of age.

APPENDIX 5a

Pork Industry Handbook: Table of Contents, January, 1991

Production Systems		*Publication Date**
Pork production systems with business analyses—the two litter pasture system (farrow-to-finish)	PIH- 13 (REV)	6/86
Pork production systems with business analyses—the low investment, low intensity system (2 groups of sows farrow-to-finish)	PIH- 14 (REV)	6/86
Pork production systems with business analyses—the high investment, high intensity confinement system (3 groups of sows farrow-to-finish)	PIH- 15	12/76
Pork production systems with business analyses—producing feeder pigs (low investment)	PIH- 16	12/76
Pork production systems with business analyses—feeding purchased pigs	PIH- 17	12/76
Pork production systems with business analyses—selecting the "right" system	PIH- 48	2/78
Pork production systems with business analyses—determining capital requirements	PIH- 49	2/78
Pork production systems with business analyses—the high investment, high intensity confinement system (5 groups of sows farrow-to-finish)	PIH- 75	1/81
Pork production systems with business analyses—the high investment, high intensity system (continuous farrowing with 21 groups of sows, farrow-to-finish)	PIH- 83 (REV)	1/91
Employer-employee relationships on hog farms	PIH- 88	6/83
Financial leasing of capital assets in pork production	PIH-115	6/88

(*continued*)

Breeding and Genetics

Boar selection guidelines for commercial producers	PIH- 9 (REV)	12/87
Guidelines for choosing replacement females	PIH- 27 (REV)	10/85
Crossbreeding systems for commercial pork production.........	PIH- 39 (REV)	12/87
Selection guidelines for the seedstock producer	PIH- 58 (REV)	12/88
Swine genetic abnormalities	PIH- 97	10/84
Selection for feet and leg soundness	PIH-101	6/85
Genetic principles and their application	PIH-106	6/86

Reproduction

Management of the boar	PIH- 1 (REV)	11/81
Managing sows and gilts for efficient reproduction..............	PIH- 8 (REV)	11/89
Reproductive efficiency in managing the breeding herd	PIH- 45 (REV)	12/89
Artificial insemination in swine..............................	PIH- 64 (REV)	10/84
Management of developing gilts and boars for efficient reproduction..	PIH- 74	1/81
Managing the gilt pool	PIH- 89	10/83
Troubleshooting swine reproductive failure....................	PIH- 96	11/84

Nutrition

Vitamins for swine	PIH- 2 (REV)	1/84
Energy for swine..	PIH- 3 (REV)	5/84
Protein and amino acids for swine	PIH- 5 (REV)	1/87
Principles of balancing a ration..............................	PIH- 7 (REV)	11/87
Swine diets	PIH- 23 (REV)	1/91
Feed additives for swine	PIH- 31 (REV)	4/89
Minerals for swine..	PIH- 52 (REV)	1/86
Physical forms of feed—feed processing for swine	PIH- 71	5/80
High-moisture grains for swine..............................	PIH- 73	9/80
Management to prevent drug residue problems in pork..........	PIH- 86 (REV)	6/90
Calibrating meter-type feed mills	PIH- 94	3/84
By-products in swine diets	PIH-108	6/87
Determining the relative value of feeds........................	PIH-112	11/87
Forages for swine	PIH-126	6/90

Management

On farm feed processing	PIH- 4	12/76
Baby pig management—birth to weaning......................	PIH- 18	12/76
Starting purchased feeder pigs	PIH- 20 (REV)	1/83
Care of the sow during farrowing and lactation.................	PIH- 46 (REV)	9/86
Performance guidelines for the swine operation	PIH-100 (REV)	1/91
Controlling rats and mice in swine facilities....................	PIH-107	1/87
Management and nutrition of the newly weaned pig	PIH-111	11/87
Calculating swine schedules	PIH-113	11/87
Processing baby pigs..	PIH-114	3/88
Handling hogs	PIH-116	11/88
A guide for fly control around swine facilities..................	PIH-121	6/89

Housing		*Publication Date**
Swine farrowing units	PIH- 10 (REV)	4/84
Swine confinement growing-finishing units	PIH- 11	12/76
Combining swine housing units into a system of buildings	PIH- 22	5/77
Confinement sow gestation and boar housing	PIH- 28	6/77
Building materials and equipment for swine facilities	PIH- 32 (REV)	12/88
Maintenance and operation of ventilation fans for hog barns	PIH- 41	2/78
Flooring for swine	PIH- 53 (REV)	12/84
The environment in swine housing	PIH- 54	7/78
Space requirements for swine	PIH- 55 (REV)	11/86
Supplemental heat for swine	PIH- 57	9/78
Mechanical ventilation of swine buildings	PIH- 60 (REV)	1/91
Insulation for swine housing	PIH- 65 (REV)	5/87
Floor heat for swine	PIH- 66	11/79
Swine confinement breeding units	PIH- 69	11/79
Swine nursery units	PIH- 70	1/80
Decking for nursery pigs	PIH- 77	3/81
Remodeling ideas for farrowing facilities	PIH- 78	3/81
Troubleshooting mechanical ventilation systems	PIH- 84	12/82
Cooling swine	PIH- 87	6/83
Solar heating for swine buildings	PIH- 90	11/83
Energy conservation in ventilating and heating swine buildings	PIH- 92	10/83
Earth tempering of ventilation air	PIH-102	5/85
Safety in swine production systems	PIH-104	1/86
Electrical wiring for swine buildings	PIH-110	6/87
Planning considerations for the construction of a new swine unit	PIH-117	12/88
Non-mechanical ventilation of MOF swine buildings	PIH-120	6/89
Heat exchangers in swine facilities	PIH-124	12/89
Waste Management		
Systems of runoff control	PIH- 21	7/77
Fertilizer value of swine manure	PIH- 25 (REV)	11/83
Controlling odors from swine buildings	PIH- 33 (REV)	6/88
Legal guidelines for swine waste management	PIH- 35	7/77
Lagoon systems for swine waste treatment	PIH- 62	3/79
Flushing systems for swine buildings	PIH- 63 (REV)	9/82
Swine waste management alternatives	PIH- 67	9/79
Methane gas from swine manure	PIH- 76	1/81
Pumping liquid manure from swine lagoons and holding ponds	PIH- 91	11/83
Gravity drain gutters for swine manure systems	PIH- 95 (REV)	1/91
Scraper systems for removing manure from swine facilities	PIH-105	4/86
Herd Health		
Mycoplasmal diseases of swine	PIH- 29 (REV)	12/85
Enteric colibacillosis of newborn pigs	PIH- 30 (REV)	6/84
Baby pig anemia	PIH- 34	7/77

(continued)

Herd Health

Swine arthritis	PIH- 36	9/77
Lactation failure in the sow	PIH- 37 (REV)	11/86
Pseudorabies (Aujeszky's disease, mad itch)	PIH- 38 (REV)	9/85
External parasite control on swine	PIH- 40 (REV)	5/87
Edema disease (gut edema; E.*coli* enterotoxemia)	PIH- 43	11/77
Internal parasites	PIH- 44 (REV)	6/84
Transmissible gastroenteritis (TGE)	PIH- 47	12/77
Atrophic rhinitis	PIH- 50 (REV)	3/86
Swine dysentery (bloody scours, vibrionic dysentery, black scours)	PIH- 56 (REV)	1/87
Infectious swine reproductive diseases	PIH- 59	11/78
Rotaviral diarrhea in pigs	PIH- 61 (REV)	5/87
Guidelines for the development of a swine herd health calendar	PIH- 68 (REV)	11/89
Environmental sanitation and management in disease prevention	PIH- 79	9/81
Selection and use of disinfectants in disease prevention	PIH- 80 (REV)	6/90
Swine coccidiosis	PIH- 81	1/82
Actinobacillus (Haemophilus) pleuropneumoniae	PIH- 82 (REV)	11/88
Slaughter checks—an aid to better herd health	PIH- 93 (REV)	6/90
Porcine proliferative enteritis	PIH- 98	11/84
Mycobacteriosis (tuberculosis, TB)	PIH- 99	12/84
Trichinosis	PIH-103	6/85
Streptococcus suis disease in pigs	PIH-118	5/89
Basics of immunology	PIH-122	10/89

Marketing

Producing and marketing hogs under contract	PIH- 6 (REV)	6/88
Choosing a slaughter hog market	PIH- 12	12/76
Using futures markets for hedging	PIH- 19 (REV)	5/89
Optimal weight to market slaughter hogs	PIH- 24	5/77
How and where is price established	PIH- 51 (REV)	6/90
Feeder pig marketing techniques	PIH- 72	6/80
The structure of the U.S. pork industry	PIH- 85 (REV)	6/90
Commodity options as price insurance for pork producers	PIH-109	3/87
Understanding hog production and price cycles	PIH-119	6/89
Marketing cull sows	PIH-123	12/89

Pork & Pork Quality

Porcine stress syndrome	PIH- 26 (REV)	5/85
Carcass evaluation	PIH- 42 (REV)	11/87
Composition and nutritive value of pork	PIH-125	6/90
Pork quality	PIH-127	1/91
Pork by-products	PIH-128	1/91

* Fact sheets published prior to 6/78 do not have the publication date listed on the fact sheet.

APPENDIX 5b

National Swine Improvement Federation Swine Genetics Fact Sheets: Table of Contents

Series No.	*Title*	*Publication Date*
NSIF-FS1	The economic impact of genetic improvement	(REV) 3/90
NSIF-FS2	Basic concepts of genetic improvement	3/88
NSIF-FS3	Genetic parameters and their use in swine breeding	12/87
NSIF-FS4	Inbreeding in swine	4/88
NSIF-FS5	Performance records for selection programs	8/88
NSIF-FS6	Environmental effects and adjustment factors	3/88
NSIF-FS7	Performance records on relatives	8/88
NSIF-FS8	Estimating genetic merit	(REV) 3/90
NSIF-FS9	Application of selection concepts for genetic improvement	9/90
NSIF-FS10	Multiple trait selection for pork improvement	10/90
NSIF-FS11	Utilizing central boar test stations to enhance genetic progress	8/89
NSIF-FS12	Genetic evaluation programs	7/90
NSIF-FS14	Genetic improvement of sire and dam lines for enhanced performance of terminal crossbreeding systems	3/90

Index

See pages 385–431 for glossaries of terms used in swine production, feed processing, and in breeding, genetics, and production systems, and for directories of the swine industry and USDA standards for feeder and slaughter pigs.

Kolibri 360 GmbH
Ferdinand-Jühlke-Straße 7
99095 Erfurt, Deutschland
produktsicherheit@kolibri360.de